# Irrigation Show & Education Conference 2017



Orlando, Florida, USA
8 – 10 November 2017

ISBN: 978-1-5108-5538-0

**Printed from e-media with permission by:**

Curran Associates, Inc.
57 Morehouse Lane
Red Hook, NY 12571

**Some format issues inherent in the e-media version may also appear in this print version.**



Printed by Curran Associates, Inc. (2018)



Irrigation Association (IA)
8280 Willow Oaks Corporate Drive
Suite 400
Fairfax, Virginia 22031
USA

Phone: 703.536.7080
Fax: 703.536.7019

info@irrigationshow.org

**Additional copies of this publication are available from:**

Curran Associates, Inc.
57 Morehouse Lane
Red Hook, NY 12571 USA
Phone: 845-758-0400
Fax: 845-758-2633
Email: curran@proceedings.com
Web: www.proceedings.com

# Agriculture Track - Table of Contents

# Landscape Track - Table of Contents

# Ogallala Aquifer Program
# Center Pivot Irrigation Technology Transfer Effort

**Dr. Freddie R. Lamm**
Professor and Research Irrigation Engineer
Kansas State University
Northwest Research-Extension Center
Colby, Kansas
Voice: 785-462-6281
flamm@ksu.edu

**Dr. Dana O. Porter**
Professor, Extension Program Leader and
Associate Department Head
Texas A&M University
AgriLife Research and Extension Center
Lubbock, Texas
Voice: 806-746-6101
d-porter@tamu.edu

## ABSTRACT

The year 2018 will mark the 40th anniversary of research and development with Low Energy Precision Application (LEPA) for use with center pivot sprinkler irrigation systems. Since that time, researchers and extension specialists in the Ogallala region have continued development of multiple types of technologies that are suitable for mobile lateral irrigation platforms. A two-year technology transfer effort with funding from the USDA-ARS Ogallala Aquifer Program (OAP) was initiated in January 2017 to promote adoption of advanced and efficient irrigation technologies and to highlight recommended practices for these mobile irrigation platforms [center pivots (CP) and lateral move systems (LMS)]. This paper will report on pertinent mobile irrigation history and the progress and future plans of the project with a particular focus on the current status of the technology and research and educational needs.

## INTRODUCTION

The Kansas and Texas High Plains / Southern Ogallala Aquifer Region are noted for limited and declining groundwater resources (Konikow, 2013) and relatively high rate of adoption of efficient advanced irrigation technologies (Wagner, 2012; Colaizzi, et al., 2009). One of the earliest advanced mobile sprinkler irrigation technologies, Low Energy Precision Application (LEPA), was first researched in the OAP region near Halfway, Texas by William Lyle and James Bordovsky beginning in 1978 (Lyle and Bordovsky, 1981, 1983). Low pressure center pivot irrigation, including (LEPA), Low Elevation Spray Application (LESA), Mid-Elevation Spray Application (MESA), and other variations have become the most widely practiced irrigation methods in the region (Colaizzi, et al., 2009). This is due in large part to the suitability of the technologies to the crop production systems in the region; relevant applied research programs; collaborations among research and extension programs and with industry; effectiveness of cost-share programs, and the willingness of agricultural producers in the region to adopt technologies and BMPs to adapt to limited water conditions (Wagner, 2012). From the early work on development on LEPA that began in 1978 and later Low Elevation Spray Application (LESA) irrigation and Mid-Elevation Spray Application (MESA) to the newer integrated sensor/control systems mounted on CP and LMS systems, OAP affiliated programs have made important contributions to the advancement of irrigation using mobile platforms.

While low pressure center pivot irrigation is widely practiced in the region, applied research continues to refine recommendations, so this technology transfer effort is providing opportunities for end-users to hear up-to-date recommendations to aid in their irrigation decisions. There is much less understanding by "non-practitioner audiences (absentee landlords, ag lenders, crop insurance agents, policy makers) about the most appropriate uses of these technologies, so this effort will help to improve their understanding of the state of the art, considerations for irrigation management, and appreciation for the advances in agricultural irrigation technology, management and efficiency. The technology transfer effort will also provide a good opportunity for the engineers and scientists to collaborate and synthesize "what we know" into more accessible publications and media as well as to provide a venue to brainstorm additional improvements to systems and technologies.

## A BIT OF HISTORY OF LOW PRESSURE CENTER PIVOT IRRIGATION

Although by no means do the OAP project participants plan to limit their technology transfer effort to low pressure center pivot irrigation, some historical discussion is warranted to illustrate how the science and conceptualization of LEPA and its prodigies (i.e., LESA and MESA) can lead and has led to improved irrigation management in the OAP region and beyond.

Original development of the LEPA system coincided with a period of relatively high energy costs and concerns about energy availability in the late 1970s, thus low energy usage was a key objective in its development. In Texas where LEPA was originally developed under semi-arid conditions, air and canopy evaporative losses from sprinkler irrigation can be appreciable, reducing crop yields in water-limited operations with low capacity irrigation systems, so reduction or elimination in these losses were assets to the LEPA system. Original design issues were development of an application system adaptable to flowrates from 100-1000 gpm with operating pressures between 5 and 20 psi (Lyle and Bordovsky, 1981). The system was to be adaptable to all soil types, and since there are great differences in water infiltration rates across soil types, runoff was to be controlled by using micro-basin tillage techniques (Lyle and Bordovsky, 1981). Early development of LEPA was on land slopes of less than 1 percent and physical geometry limitations of the micro-basins imposes some limitation on their effectiveness on greater slopes. For example, runoff from LEPA sprinklers was negligible on 1% sloping silt loam soils in eastern Colorado but exceeded 30% when slopes increased to 3% (Buchleiter, 1991). Scientifically, LEPA has always been considered to be a system of technologies with both center pivot hardware and adoption of specific farming practices (Lyle, 1992). Application efficiencies in Texas for LEPA and conventional sprinkler irrigation were measured at 99 and 84%, respectively, when micro-basin tillage was practiced as compared to 88 and 81% when conventional tillage was used (Lyle and Bordovsky, 1983). The worldwide annual benefit of LEPA has been estimated to be $US 1.1 billion with a $US 0.477 billion benefit to consumers in the United States (Lacewell, 1998).

Failure to adopt the underlying LEPA system principles will usually result in unsuccessful application of the technology. Producers' reluctance to adopt some of the guiding principles or land considerations have led to alternate in-canopy or near-canopy application systems such as LESA and MESA, which are spray applications at low and mid elevations, respectively. These systems with a larger wetting pattern reduce the chance of excessive runoff, particularly when used in conjunction with conservation tillage (Lamm et al., 2017). The adoption of LESA and MESA systems as compared to LEPA is more prevalent moving northward in the southern and central Great Plains, particularly on tighter soils, greater land slopes and with greater capacity groundwater wells.

Briefly summarizing the history, the science and conceptualization of low pressure center pivot irrigation technologies led to multiple adaptations of the overall technology that have been adopted on a relatively wide scale. When the implementation knowledge was ignored or discarded much of the potential water and energy saving benefits were not realized.

## CENTER PIVOT BRAINSTORMING AND BRAIN STRETCHING RETREAT

In the spring of 2017, an invitation was sent out to a broad range of irrigation engineers, scientists, USDA NRCS specialists, and industry representatives associated with center pivot technologies to participate in a brainstorming retreat sponsored by the OAP CP Technology Transfer Project to be held in Amarillo, Texas on March 28-29. A total of 39 individuals from 16 U.S. states (Alabama, Arizona, California, Colorado, Georgia, Idaho, Kansas, Louisiana, Mississippi, Missouri, Nebraska, Oklahoma, Texas, South Carolina, South Dakota, and Virginia) were able to participate in the retreat. There were several goals of the retreat including networking opportunities for both more experienced and less experienced individuals, electronic distribution of large bodies of CP-related publications from the Central Plains Irrigation Conference and the USDA-ARS Conservation and Production Research Laboratory, discussion of past and current research, identification of research, extension and educational needs, and discussion of industry status and information gaps.

Although it is impossible to fully capture the richness and value of this two day event in this brief report, an attempt to tabulate the key topics, their status and the important knowledge gaps was concluded by these two authors. No attempt to prioritize any of the key topics was intended through this tabulation (Table 1), nor should it be considered inclusive of all topics discussed during this two-day event.

## OTHER PLANS FOR THE TECHNOLOGY TRANSFER EFFORT

### Technical Sessions at Conferences

The Irrigation Association (IA) technical session for which this paper is a part was developed and coordinated through the USDA-ARS OAP Center Pivot Technology Transfer Effort. Through coordinating of this session, the project brings together engineers, scientists, agency staff, and industry and the general public for networking and further technology transfer about CP technologies. Further technical sessions are being planned and coordinated for regional conferences such as the High Plains Conference in Amarillo, Texas on February 7, 2018 and at the Central Plains Irrigation Conference in Colby, Kansas on February 20-21, 2018. These sessions are geared toward producers, consultants, irrigation professionals and agency staff and they leverage annual educational events and ongoing programs. Additional technical sessions at national professional conferences are being proposed for IA in Long Beach, California in December 2018, the American Society of Agricultural and Biological Engineers (ASABE) in Detroit, Michigan in July 2018 and the Agronomy, Crop Science (ASA-CSSA) meeting in Baltimore, Maryland in November 2018. These meetings are geared more toward scientist to scientist/industry interchanges.

### Review or Summary Papers

Participants in the technology transfer effort have agreed to prepare literature reviews or summary papers during the coming year. Topics that have been agreed upon thus far are a summary paper on history and development of LEPA, a conceptual discussion of all in-canopy and near-canopy sprinkler irrigation and a summary paper on irrigation decision support systems. Other possibilities include a state of the art discussion on remote sensing, UAVs and their role in CP management, a review or summary paper on sprinkler chemigation, a summary paper on VRI and a summary paper on future needs for CP. There are opportunities for non-project participants to lead or collaborate on some of these efforts.

### Tours and Field Days

Specific CP technology transfer field days are being planned for the summer of 2018 in both Texas and Kansas. Dates and locations have not been finalized as of this time. Additionally portions of other tours and regular university field days will likely encompass some of our presentations. It is anticipated that the center pivot technology industry will be approached for support of these activities. If you are interested in supporting this project, feel free to contact either of the authors who are the project's principal investigators.

### Website and Activity Listing

The project can be followed at this link http://www.ksre.k-state.edu/irrigate/cptt/index.html

The project has been very active to date with 1 book chapter, 3 refereed journal articles, 11 national or international conference papers, 14 regional conferences papers, and 51 additional miscellaneous technology transfer activities documented at http://www.ksre.k-state.edu/irrigate/cptt/TechTranCPTTT.pdf

## ACKNOWLEDGEMENTS

This technology transfer effort is supported by the Ogallala Aquifer Program, a consortium between USDA Agricultural Research Service, Kansas State University, Texas A&M AgriLife Research, Texas A&M AgriLife Extension Service, Texas Tech University, and West Texas A&M University.

Watch for our project logo.

## REFERENCES

Buchleiter, G. W. 1991. Irrigation with LEPA. In: Proc. Central Plains Irrigation Shortcourse, North Platte, NE., Feb. 5-6, 1991. pp. 64-68.

Lacewell, R. D. 1998. Precision agriculture and value of water. In: Proc. The Great Plains Symposium – Determining the value of water, March 10-12, 1998, Lubbock, TX. Great Plains Foundation, Overland Park, Kansas. pp. 46-50.

Lyle, W. M. 1992. LEPA, concept and system. In: Proc. Central Plains Irrigation Short Course, Goodland, KS, Feb. 4-5, 1992. Available from KSU Extension Biological and Agricultural Engineering, Manhattan, KS. pp. 14-16.

Lyle, W. M. and J. P. Bordovsky. 1981. Low energy precision application (LEPA) irrigation system. Trans ASAE 24(5):1241-1245.

Lyle, W. M. and J. P. Bordovsky. 1983. LEPA irrigation system evaluation. Trans ASAE 26(3):776-781.

Lamm, F. R., T. A. Howell, and J. P. Bordovsky. 2017. Erraticity of sprinkler-irrigated corn under drought. In: Proc. 2017 Irrigation Association Technical Conference, Orlando, Florida, November 6-10. Available from the Irrigation Association, Fairfax, Virginia. 13 pp.

Colaizzi, P. D., P. Gowda, T. H. Marek, and D. O. Porter. 2009. Irrigation in the Texas High Plains: A brief history and potential reductions in demand. J. Irrigation and Drainage. 58(3):257-274.

Konikow, L. F. 2013. Groundwater depletion in the United States (1900–2008): U.S. Geological Survey Scientific Investigations Report 2013–5079, 63 p., http://pubs.usgs.gov/sir/2013/5079.

Wagner, K. 2013. Status and Trends of Irrigated Agriculture in Texas. Special report EM-115 by the Texas Water Resources Institute. Texas A&M University System, College Station.

**Table 1. Key topics, comments and information status and key knowledge and/or implementation gaps identified at a center pivot irrigation brainstorming retreat in Amarillo, Texas, March 28-29, 2017.** *The order or extent of the listing does not indicate any priority, nor should it be considered inclusive of all ideas discussed during this two day event. This tabular listing is meant only to portray the wide range of topics and some key gaps that were identified.*

| Key Topic | Comments and Information Status | Key Knowledge and/or Implementation Gaps |
|---|---|---|
| **Variable Rate Irrigation (VRI ) or Site Specific Irrigation (SSI)** | Emerging technology, still uncertainty about extent of future needs and adoption.<br>Three types identified (Sector Control, Speed Control, VRI Zone or Individual Sprinkler Control).<br>Many current (commercially available) CP systems have more capabilities than recognized by system owner.<br>VRI not needed by all and some producers will not recoup costs of implementation. | Hardware development has outpaced development of management information.<br>Although many teams working on dynamic prescriptions, continued work is needed to remove this impediment.<br>Uncertainty about producer expectations.<br>Abandonment can be high in absence of appropriate support to producers from industry, universities, consultants, and/or USDA-NRCS.<br>Continued need for research and education. |
| **Sprinkler Packages** | Maturing technology, many different types of packages are provided by industry to meet needs of producers.<br>Selection should consider crop, soil, water source/quality and energy.<br>LEPA, LESA and MESA have specific requirements that need consideration.<br>Greater interest and adoption of in-canopy and near-canopy application when evaporative losses are higher, irrigation capacity is lower and land slope is lower. | Although maturing technology, still many implementation mistakes.<br>"One size fits all" mentality ignores the knowledge we have.<br>Runoff must be controlled first for any realistic success with in-canopy and near-canopy sprinkler application.<br>Educational needs of producers still remain. |
| **Sprinkler Uniformity** | Hydraulics can be modeled, but catch can results are still instructive and can point out hardware and implementation problems.<br>Catch can tests are still time and labor intensive.<br>Mismatch of nozzle package and operating pressure is commonplace.<br>Need to remember that crop can integrate some minor uniformity problems. | Uncertainty of continued status of some modeling efforts.<br>CPED is now available from USDA-NRCS in a MS-Excel format.<br>Producers still need to monitor and respond to the basic information of system flowrate and pressure. |
| **Mobile Drip Irrigation** | Emerging technology with just a few research studies to date.<br>Can reduce wheel track problems (rutting). | Scope of appropriate applicability of the technology (e.g. soil type, slope, crops) is still unknown.<br>Rodents can be a problem.<br>Forces applied on CP systems may be concern.<br>Maintenance issues, filtration needs and other concerns. |
| **Wheel tracks, rutting and getting stuck.** | Primarily anecdotal or industry-held information.<br>Actually may negatively affect irrigation management, such as early end-of season irrigation termination. | Need for generic (non-brand specific) publication or guidance on span selection and wheel/flotation system selection. |

| | | |
|---|---|---|
| **Chemigation** | Maturing technology, but perhaps not as much recent research efforts by Universities.<br>Uncertainty of audience (i.e., end-users, regulators or chemical industry) may result in inertia | Sprinkler packages and sprinkler spacings.<br>VRI interactions with chemigation.<br>Safety and standards needs; associated educational needs. |
| **Microbursts /Tornadoes and CPs** | No known resources identified.<br>Student project or modeling effort?? | What direction to park CP?<br>Loaded with water for downforce or not? |
| **Center Pivot Safety** | Maturing knowledge base<br>USDA-NRCS has some materials and trains their own staff about approaching CP systems. | Producers and installers still need education.<br>Need for lay-oriented publications.<br>Who has expertise/presentations? |
| **Remote Sensing** | Emerging area with large amount of interest<br>Can interface with VRI research needs but standalone research area as well.<br>UAVs are of considerable interest to producers now.<br>Remote sensing could encompass weather, soil, or plant information and combinations of the three types. | Lots of approaches are necessary for research but make selection of approach difficult for producer.<br>Hardware offerings may presently outpace development of management information.<br>Continued need for research and education. |
| **Variable Frequency Drives (VFDs)** | Technology is maturing and interest is growing due to more usage of electricity as sole energy source for CPs.<br>Still not economical for many cases.<br>Economic feasibility will depend on field slopes and other changes in pressures, time of operation, and price of energy. | Some evaluations have been done in region but more are needed.<br>More modeling is needed.<br>VRI will further complicate the need for VFDs |
| **Publications and Information Needs** | Mature, yet continuing evolving topic area.<br>Fewer attendees at traditional university-led workshops, tours, and field days.<br>Not just agricultural problem with attendance, landscape having similar issues.<br>Grower panels can be useful when remaining sufficiently unbiased and scientifically sound.<br>Younger generation audiences are definitely more open to electronic media.<br>Fewer, but better, regional conferences may be an option for "sounding" the knowledge but may still have attendance issues. | How well are we targeting audiences?<br>Do we adjust to the audience (i.e., professional, producers, regulators, industry, legislators, urban audiences, genders and age).<br>Could public/private partnerships be used to greater advantage?<br>Individual companies may have material that could be packaged better for broader industry-wide educational material.<br>Technology farms or large plots research may be better at information delivery. |
| **University Degree Programs and Certificate Programs** | Small and decreasing number of agricultural irrigation programs in USA and attracting fewer US-born students.<br>Importance of agriculture is not always reflected at universities.<br>Community colleges may be able to fill some staffing needs.<br>USDA-NIFA may need to provide irrigation fellowships to help build capacity. | Industry needs well-educated staff that are willing to live in agricultural regions.<br>Universities need well-trained faculty and funding to retain good faculty.<br>Universities need to develop students to find food and fiber solutions for 9.6 billion people by 2050. |

# Discussion of Why VRI May Not Being Used to Its Full Potential

**Jacob LaRue, PE**
**Valmont Irrigation, Valley, NE**
**jlarue@valmont.**

**Abstract.** *VRI, variable rate irrigation, sometimes call site specific irrigation was introduced to farmers in 2001 by the University of Georgia and early versions became available in 2004 commercially. The major OEM manufacturers of center pivots began to introduce center pivots equipped with VRI in 2010. Many different groups have touted the concept of water savings and improved water use efficiency when VRI is used. However after discussions with a number of farmers in 2016 and 2017 using center pivots equipped with VRI zone or individual sprinkler control it was found some had never used the VRI function or had tried to use and stopped for a variety of reasons. This paper will present the findings on why VRI is not being used to its full benefit exploring both hardware and software. In addition some suggested recommendations to improve the use of this technology will be presented.*

**Keywords.** VRI, variable rate irrigation, site specific irrigation, center pivot, linear, water use efficiency, water savings

## Background

Since the introduction of the center pivot in the mid-1950s, the mechanical move industry has continued to improve and develop products to better meet the needs of production agriculture. The overall goal has been to provide cost-effective, uniform irrigation across the field with a specific application depth. With the introduction and acceptance of precision agriculture, more information has become available for a particular field and areas in the field, including yield, EC maps, soil and grid sampled fertility maps. Farmers now have data indicating the variability across the field, which was already suspected but not generally considered at a sub-field level. The challenge became how to use this data to change the depth of irrigation application for different areas of the field. The goal is to apply the 'right' amount of irrigation to each management zones within a field.

Early research with commercial fields by the University of Georgia indicated substantial water savings as many fields have areas that require less water or no water at all (Perry and Milton, 2007). However a number of researchers have identified significant barriers to adoption. The foremost need is for the development of guidelines and tools to assist consultants and grower in predefining standards for economically defining sizes and numbers of management areas and writing basic prescriptions, (Evans 2013). And more recently Dr. Troy Peters, Washington State University said "The data collection, analysis, and creation of optimal VRI prescriptions for a specific field's needs can be complex, time consuming and expensive, especially since many field situations require these prescriptions to vary both in time, and in space" (Peters 2017).

Growers continue be as efficient as possible and do not have time to spend on using highly detailed or difficult to use software.

## Discussion

The following discussion focuses on VRI zone and individual sprinkler control. Speed control is not considered.

Beginning in 2010 with the introduction of VRI zone control Valmont's team worked closely with each grower adopting the product. Over time as the hardware matured less emphasis was placed on supporting the customer with creating prescriptions after the VRI hardware was installed and operational. A very simplified prescription software was provided with each VRI package. It was presumed the grower and/or their consultant would handle the prescriptions. Growers and/or consultants collected data from soil maps, yield maps and electro-conductivity maps generated from Veris or Dual EM scans of the field to generate prescriptions. Valmont early in the commercial roll-out of VRI provided consultants access to the format required for uploading a prescription via telemetry products

During this same time in general researchers utilized VRI to simplify their irrigation projects by providing easy control of water delivery by plot. Limited emphasis on field trials using the hardware to manage irrigation to the spatial variability of the field has been done. .

An early observation was rarely were prescriptions changed during the growing season. Also few changed their overall irrigation scheduling program.

While the following does not specifically mention VRI, it seems to be a very apt description of the current situation with VRI - The overall driving force for adopting irrigation scheduling is economics – scheduling is used by the farmer because it makes or saves money. Nonetheless, even irrigators who find scheduling profitable will discontinue its use if it becomes too burdensome (Hennegler, 2013).

During 2016 to 2017 an unscientific survey was done where a number of owners of center pivots with VRI zone control or individual sprinkler control purchased between 2010 and 2014 were contacted either by phone or in a personal interview. The rough total ended up being about 41% of the fields where VRI was installed. This was to understand why and how they were using the VRI package and any challenges they were facing.

The focus of this work was on those already owning VRI. It is worth mentioning common reasons cited by growers why they are not interested in VRI are:

- Long perceived payback
- Limited perceived value

The planned use of VRI by growers can be broken broadly into two different categories

- To shutoff water delivery completely in different areas of the field for ponds, water ways and other non-cropped areas.
    - Definitely reduces water use
    - Stop application of crop production products in non-crop areas
    - Generally one prescription meets the needs and no adjustments are made unless the non-cropped area changes
    - See figure 1

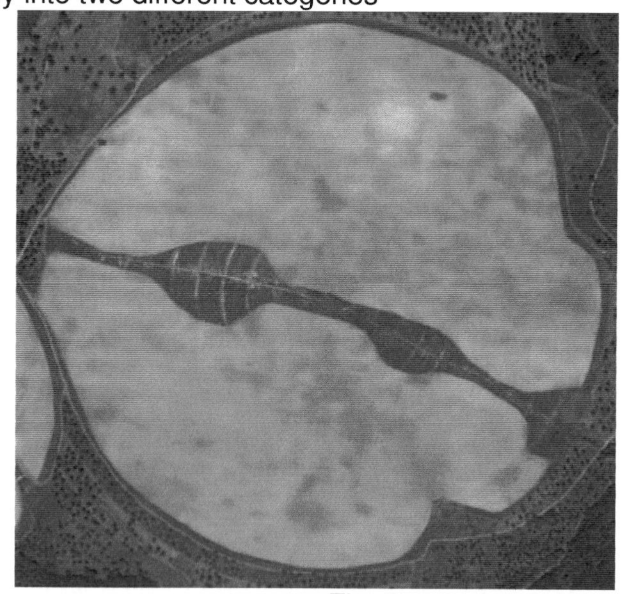

Fig. 1

- To vary the water depth applied to different parts of the field
    - May reduce water use depending on goals
    - Based on specific characteristics of the field base
        - Soils
        - Topography
        - Yield
        - Other
    - Unlikely one prescription is suitable for the entire season
    - See figure 2

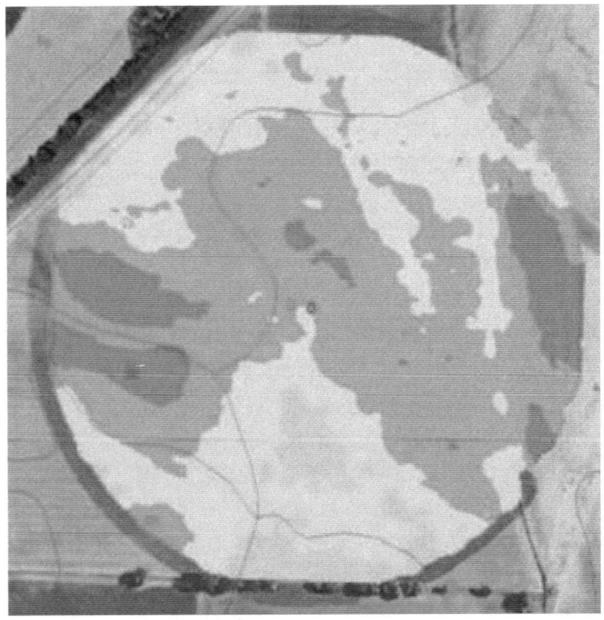

Fig. 2

From discussions with growers it was found their use of VRI can be broken into three broad categories

- Had never used for a full year – 3% of those contacted
- Had used for a few years and quit – 27% of those contacted
- Still using – 70%
    - Rarely if ever change the prescription – 47% of those contacted
    - Change three or more times during the growing season – 23% of those contacted

Reasons given for not using after installation or stopping use include

- Hardware challenges – 17%
    - Loading prescriptions
    - Valves
- Challenges with prescriptions – 72%
    - Knowing what to base the prescription on
    - Creating the initial prescription
    - When and what to base changing the prescriptions on
- Consultant quit offering or became too expensive – 11%
    - The change in commodity prices was cited as a reason

This was information was very eye opening and may help explain the why adoption of VRI has not been as rapid as expected.

**Conclusions**

The adoption and use of variable rate irrigation zone and/or individual sprinkler control has not grown at the rate the irrigation industry expected.

Information from an unscientific poll conducted in 2016 and 2017 indicated growers who had purchased VRI zone or individual sprinkler control current status of use to be:

- Did not use a full year – 3%
- Had used some and quit – 27%
- Still using - 70%

Of those who indicated they had not used a full year and/or had used some and quit they cited the reasons to be:

- Hardware issues 17%
- Challenges with prescriptions 72%
- Consultants quitting or cost 11%

Based on the information collected the following are suggested for future work by both for the public and private sectors:

Hardware

- Dependable and reliable sprinkler controls
- Easy to troubleshoot problems
- Provides as applied information

Creation of initial prescriptions

- Easy to prepare
- Automated based on a variety of inputs

Creation of future prescriptions

- Easy to use requiring minimal grower intervention
- Easy to incorporate data collected during the growing season
- Automated based on a choice of inputs

Information on the economics of using VRI

While the survey did not collect data on VRI speed control it should be mentioned most if not all of the center pivots computerized control panels being manufactured in the United States today come with the software for speed control.

Speed control has not received attention from the public sector but does offer easy access to a form of variable rate irrigation and has the potential to provide improved field irrigation for growers. The preparation of prescriptions and economics certainly deserve attention.

**Acknowledgements**

- Personal communication with a number of owners of center pivots equipped with variable rate irrigation
- Personal communication with a number of Irrigation and County Extension Agents

**References**

Evans, Robert G, LaRue, Jake, Stone, Kenneth C, King Bradley, A, 2013 Adoption of Site-Specific Variable Rate Sprinkler Irrigation Systems, USDA-ARS / UNL Faculty Publication #1245

Henggeler, J., 2013, Irrigation Scheduling, University of Missouri Extension, #13

Perry, Calvin D, Milton, Andrea W, 2007, Variable-Rate Irrigation: Concept to Commercialization: Presented at the 29th Southern Conservation Agricultural Systems Conference

Peters, Troy, Flury, Markus, Variable Rate Irrigation: The Potential for Profitability, 2017, Presented at the International Meeting of ASABE, paper 1701354, St. Joseph, MI

# Evaluation of Variable Rate Irrigation in Humid Region

**Ruixiu Sui, Research Agricultural Engineer**
USDA-ARS Crop Production Systems Research Unit, Stoneville, Mississippi, USA

**Abstract.** Variable rate irrigation (VRI) is a new irrigation method in irrigation industry. VRI technologies allow producers to site-specifically apply irrigation water at variable rates within a field to adjust the temporal and spatial variability in soil and plant characteristics. Adoption of VRI has the potential to improve water use efficiency. VRI method was evaluated in soybean and corn in Mississippi Delta. Soil apparent electrical conductivity (EC) was used to delineate VRI management zones and create VRI prescription maps. Irrigation was scheduled using soil moisture content measured by soil moisture sensors. Crop yields and irrigation water productivity in VRI treatment was compared to that in the uniform rate irrigation (URI) treatment. Results demonstrated that the VRI saved 25% irrigation water in soybean and 21% in corn. Irrigation water productivity (WP) of VRI in soybean was 31% higher than the URI. WP of the VRI in corn was 27% higher than the URI. VRI management was superior to the URI in terms of irrigation water use efficiency. Soil EC coupled with soil physical properties could be used to establish irrigation management zones for VRI practice.

**Keywords.** Irrigation, soil electrical conductivity, variable rate irrigation, water management

## Background

Irrigation plays a critical role in crop production. Irrigated crops produced more and stable yields than dryland crops. Irrigated agriculture in US is a major consumer of freshwater, accounting for 80% of the nation's consumptive water use (Schaible and Aillery, 2015). Limited water resources are becoming an increasing constrain in agriculture. To meet global demands in food and fiber while maintaining agricultural production sustainable, crop water use efficiency has to be increased.

In recent years, acreage of irrigated land in US has increased rapidly in the humid regions including the Mississippi Delta (MD). MD is one of the major crop production regions in the United States. Main row crops in this region are corn, soybean, and cotton. Uncertainty in the amount and timing of precipitation has become one of the most serious risks to crop production in MD. Studies demonstrated that supplemental irrigation in this humid region could increase crop yield and reduce production risk (Cassel et al., 1985, Boquet, 1989, Sui et al., 2014). The producers have become increasingly reliant on supplemental irrigation to ensure adequate yields. In this region, approximately 90 percent of irrigated cropland relies on the groundwater supply from the Mississippi River Valley Alluvial Aquifer. Excessive withdrawal of the groundwater resulted in a decline in aquifer levels across the region. Ongoing depletion and stagnant recharging of the aquifer jeopardize the long-term availability of the aquifer and place irrigated agriculture in the region on an unsustainable path.

Variable rate irrigation (VRI) is a new irrigation method. VRI technologies allow the producers to site-specifically apply irrigation water at variable rates within the field to adjust the temporal and spatial variability in soil and plant characteristics. Adoption of VRI has the potential to improve water use efficiency. VRI technologies are normally implemented on self-propelled center-pivot and linear-move sprinkler irrigation systems. Similar to other variable rate application systems in precision agriculture, VRI practices require specialized hardware and software. VRI hardware requirements include a GPS receiver to determine the spatial position of the irrigation system and an intelligent electronic device to control individual sprinklers or groups of sprinklers to deliver the desired amount irrigation water on each specific location within the field according to the VRI prescription. The software required includes the algorithms to calculate the water application rates and the computer programs to create VRI prescription maps. Two control methods can be used for VRI, the speed control and the duty-cycle control (LaRue and Evans, 2012). The speed control method changes the travel speed of the sprinkler irrigation system to vary the water application depth. The speed control is able to vary the application rate only in the travel direction of the irrigation system, not along the lateral pipeline, resulting in difficulty to develop VRI for randomly-shaped management zones to address the variability of soil and plant characteristics across the field. The duty-cycle control method changes the duty cycle of individual sprinklers or groups of sprinklers installed along the lateral pipeline. The duty-cycle control method is capable of varying the irrigation rate in the system's travel direction and along the lateral pipeline, which offers more flexibility in development of the management zones. VRI practice requires a prescription map. A prescription map provides the information to the controller of a VRI system for how much water to deliver at each specific management zone within the field. The VRI prescription map should include spatial coordinates of each management zone and the irrigation water depth associated with each management zone within the field. Normally the prescription map can be created using the software associated with the VRI system.

One or multiple inputs including soil properties, plant water stress, crop yield potential, field topography, and other relevant parameters could be used with geographical information system (GIS) software to delineate each management zone and determine the irrigation water application rate. Currently, VRI systems are commercially available. However, development of algorithms and models using various inputs for calculating the appropriate amount of water to site-specifically apply is a bottleneck of VRI technologies and one of the great challenges faced by VRI researchers.

The objective of this study was to develop and evaluate VRI method to improve water use efficiency in crop production.

## Procedures

The study was conducted for two years in 2014 and 2015 in two adjacent fields (Field A and Field B) in Stoneville, Mississippi, USA (latitude: 33°26'30.86", longitude: -90°53'26.60"). Each field is 6.7 ha with a 1% slope from West to East. Soil samples were taken from Fields A and B in a 0.3-ha grid and 15-cm depth, and analysed for soil physical properties in 2013. Though silt loam was the predominant soil type, variability in clay and sand content existed across the fields. Fields A and B were under the coverage of a VRI centre pivot irrigation system, and occupied half of the pivot's full circle between 0 to 180 degree (clockwise from north). Field A was in the circular angle 0° to 90° while Field B was in 90° to 180°.

The experiment layouts in the fields were showed in Figure 1. In 2014 and 2015 season, each field was equally divided into two sectors. One sector was assigned to VRI treatment, another one to URI treatment, and the remaining area not covered by the pivot in each field was assigned to the rainfed treatment.

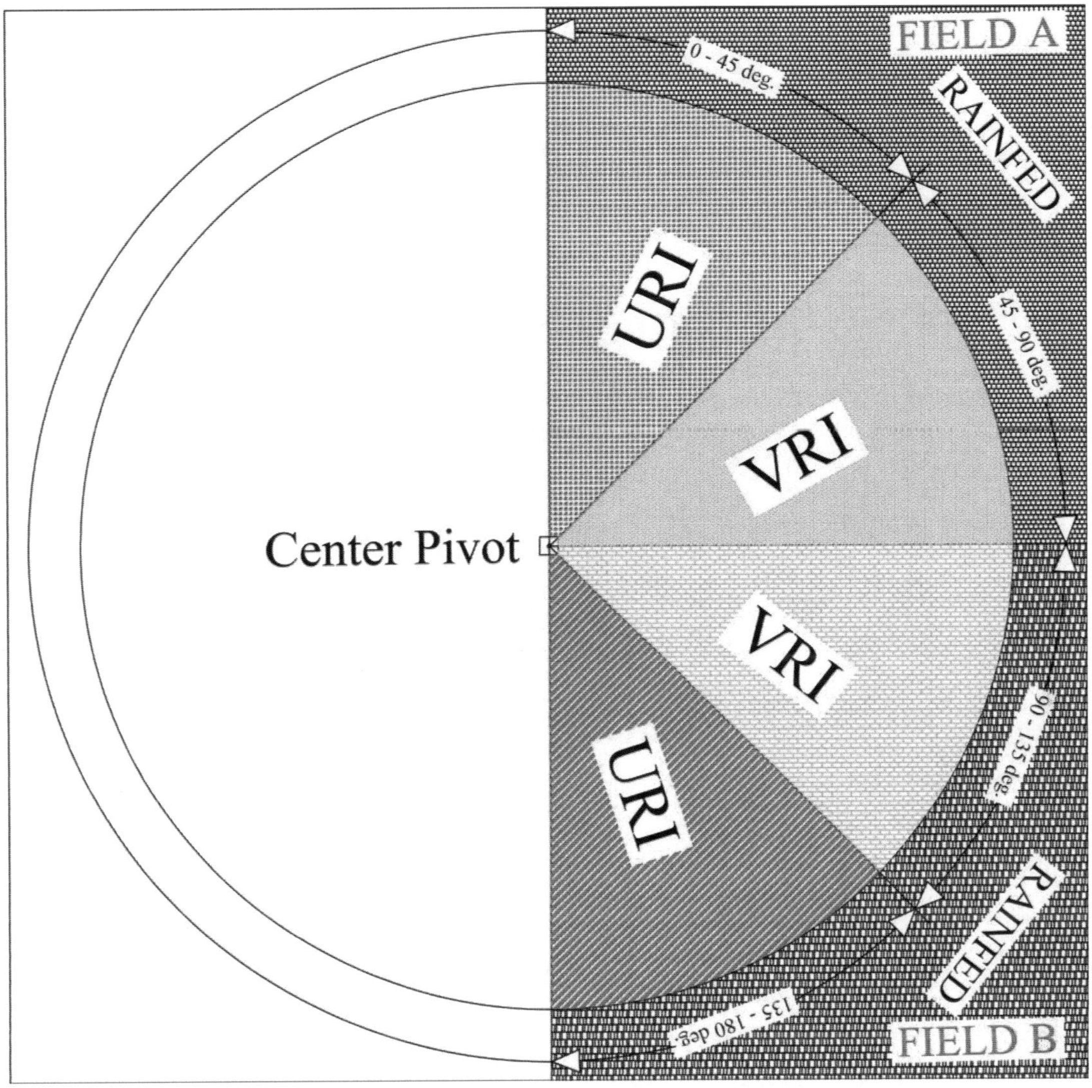


Fig. 1. Experimental layout in 2014 and 2015

The irrigation system used in this study consisted of a Valley 8000 Standard Pivot coupled with the Valley VRI zone control package (Valmont Irrigation, Valley, NE, USA). Field tests showed that this centre pivot VRI system had a coefficient of uniformity of 86.5% with constant rate application and 84.3% with variable rate application (Sui and Fisher, 2015). The system was configured in 4 spans with a total length of 233 m. Sprinklers along the length of the centre pivot were divided into 10 control zones, with each zone covering the same surface area of 1.7 ha. The Valley VRI controller included the zone control units, solenoid valves, a GPS receiver, and software. The zone control unit controlled the duty cycle of the sprinklers by turning electric solenoid valves on and off to achieve desired application depths in individual control zones. The GPS receiver determined the pivot's position in the field for identification of control zones in real time. VRI prescriptions were created using the software provided with the VRI system.

Management zones for VRI management were created based on soil electrical conductivity (EC). Soil EC of Field A and Field B was measured using the Veris 3100 soil EC mapping system. An $EC_{dp}$ map of Field A and B was created using software ArcMap (version 10.2.1, Esri, CA) (Fig. 2)

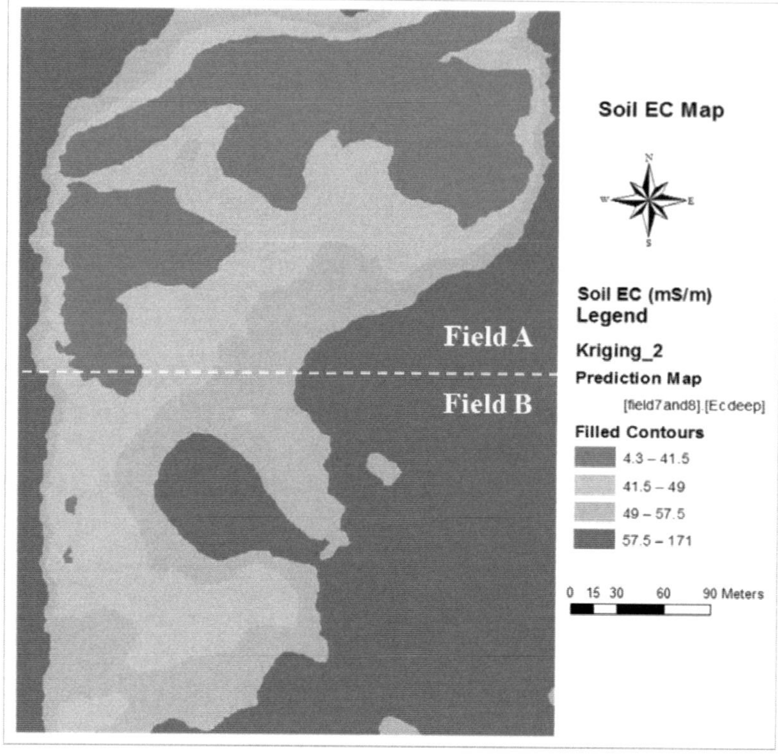


Fig. 2. Soil electrical conductivity map of Field A and Field B. The filled contours correspond to soil $EC_{dp}$ categories 1-4 (in blue to red on legend).

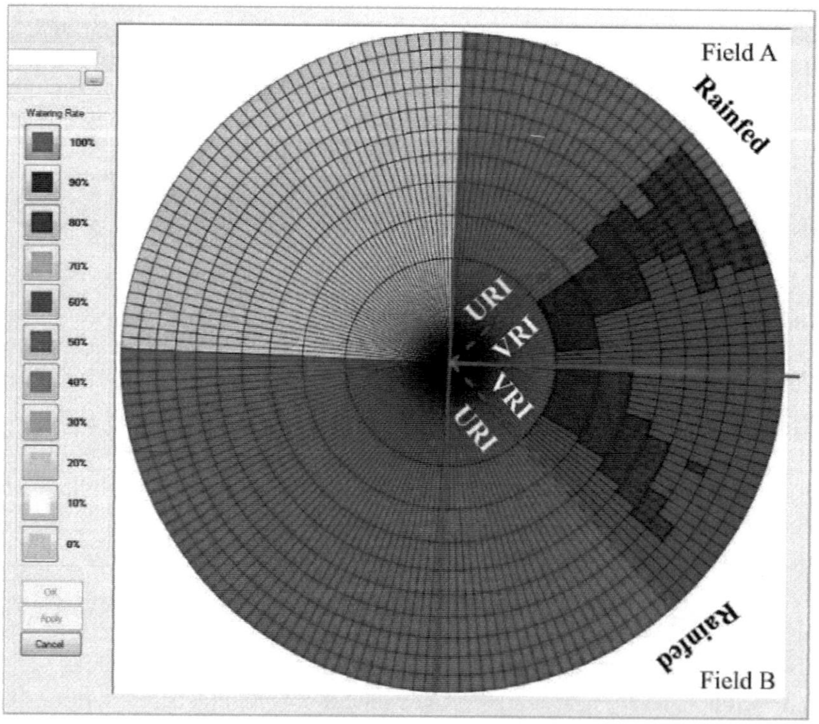


Fig. 3. Prescription map for variable rate irrigation in 2014 and 2015. Irrigation water application rates were indicated by different colours on the map.

Three management zones were created based on the soil $EC_{dp}$. In Field A, areas in $EC_{dp}$ category 1 and 2 were assigned as management zones A (MZ-A) and B (MZ-B), respectively. Areas under $EC_{dp}$ category 3 and 4 were combined together to be assigned as management zone C (MZ-C). In Field B, areas in $EC_{dp}$ category 1 and 2 were merged and assigned as MZ-A, and the areas in the category 3 and 4 were assigned as MZ-B, and MZ-C, respectively.

On account of their soil properties under the $EC_{dp}$ categories and previously observed yield potential, irrigation rates of 100% (R100), 80% (R80), and 60% (R60) were respectively applied to MZ-A, MZ-B, and MZ-C in the VRI treatment. Irrigation rate R100 was applied to the entire URI treatment. No irrigation was applied to the rainfed treatment. Irrigation rate R100 represented the irrigation rate that was determined using soil water content measured by soil moisture sensors and the application rates of the other management zones were scaled based on their percentages. With soil $EC_{dp}$ map as the background image, a VRI prescription was generated using software provided by the VRI system manufacturer (Valmont Irrigation, Valley, NE, USA). In the VRI prescription map, various depths of irrigation water were applied to different management zones according to the irrigation rate assignments (Fig. 3).

Soil water content sensors were installed at depths of 15 cm, 30 cm and 61 cm in the predominant soil of the field to measure soil water content (SWC). The sensors were calibrated with the soil from the field. The weighted average of the soil water contents in the three depths was used for irrigation scheduling. Percent plant available water (PPAW) is calculated using equation 1 to trigger irrigation events.

$$PPAW = \frac{(Sensor - measured\ SWC) - (SWC\ at\ wilt\ point)}{(Field\ Capacity - SWC\ at\ wilt\ point)} \qquad \text{(Eq. 1)}$$

Irrigation was triggered when PPAW dropped approximately to 50%.

The amount of irrigation water used in the VRI and URI treatments was measured using a water flow meter installed at the inlet of lateral pipeline of the centre pivot. Crop yield data from 18 sampling locations in each crop-year of 2014 and 2015 were collected and analysed to compare the effect of the irrigation treatment on yield and irrigation water productivity (WP). WP was defined as follows.

$$WP\ (\frac{kg}{m^3}) = \frac{Amount\ of\ grain\ produced\ with\ Irrigation\ Water\ (kg)}{Amount\ of\ Irrigation\ Water\ used\ (m^3)} \qquad \text{(Eq. 2)}$$

## Results and Discussion

In soybean, VRI treatment used 25% less irrigation water than the URI. There was no significant differences between the yields in VRI and URI. The yield of the rainfed treatment significantly differed from that of the VRI and URI. Compared with the URI and rainfed treatment, VRI management increased soybean yield by 2.8% and 37.2%, respectively.

In corn, there was no significant yield difference among the irrigation treatments, VRI used 21% less irrigation water than the URI. Yield comparison across management zones indicated no difference between VRI and URI treatments. However, yield in both the VRI and URI treatments significantly differed from the yield of the rainfed. Irrigation increased the corn yield by 18%.

The WP in soybean was 0.84 kg/m$^3$ in the VRI management and 0.64 kg/m$^3$ in the URI, which indicated that the WP in the VRI was 31.2% higher than that in the URI. In 2014 corn, the VRI treatment had the highest WP of 2.49 kg/m$^3$ because only 2.54 cm irrigation water applied made 3.2% yield increase. In 2015 corn, the WP in the VRI treatment was 1.69 kg/m$^3$, which was 27.1% greater than the WP in the URI. This result was consistent with the result in soybean, showing the VRI management was able to use irrigation water more efficiently.

# Conclusion

There was no significant difference between the yields in the VRI and URI treatment. However, the amount of irrigation water applied to the VRI treatment was 25% and 21% less than the URI treatment in soybean and corn, respectively. It was obvious that the VRI management resulted in significant water savings. The yield of the rainfed treatment significantly differed from that of the VRI and URI treatment in a dry year. In soybean, WP in the VRI was 31.2% higher than that in the URI. In corn, the WP in the VRI was 27.1% greater than the URI. Results indicated the VRI management was able to use irrigation water more efficiently in Mississippi Delta region.

With a large spatial variability of soil EC in a field and understanding the relationships among the soil EC, soil properties, and yield potential of the field, the method reported in this article has the potential to be used in other climates and fields to improve irrigation management.

Even though the use of soil EC to generate irrigation management zones could be an easy-to-use method in VRI management, researches on the algorithms with multiple input variables for delineating VRI management zones and determining VRI application rates are needed because there are many factors affecting crop water requirements for irrigation.

**Acknowledgments**

Author thanks Mr. Jonnie Baggard for his assistance in this study.

**Disclaimer**

Mention of a trade name, proprietary product, or specific equipment does not constitute a guarantee or warranty by the U. S. Department of Agriculture and does not imply approval of the product to the exclusion of others that may be available.

# References

Boquet, D.J. 1989. Sprinkler irrigation effects on determinate soybean yield and lodging on a clay soil. *Agron. J.* 81, 793-797.

Cassel, D.K., C.K. Martin, and J.R. Lambert. 1985. Corn irrigation scheduling in humid regions on sandy soils with tillage pans. *Agron. J.* 77, 851-855.

LaRue, J., and R. Evans. 2012. Considerations for variable rate irrigation. In: Proc. the 24th Annual Central Plains Irrigation Conference, 111-116, Colby, Kansas, February 21-22, 2012.

Schaible, G., and M. Aillery. 2015. *Irrigation & water use.* Retrieved from http://www.ers.usda.gov/topics/farm-practices-management/irrigation-water-use.aspx.

Sui, R., R.K. Byler, D.K. Fisher, E.M. Barnes, and C.D. Delhom. 2014. Effect of supplemental irrigation and graded levels of nitrogen on cotton yield and quality. *J. Agric. Sci.* 6(2), 119-131. doi:10.5539/jas.v6n2p119.

Sui, R., and D.K. Fisher. 2015. Field test of a center pivot irrigation system. *Appl. Eng. in Agric.* 31(1), 83-88. doi 10.13031/aea.31.10539. .

# Site specific irrigation management of a center pivot irrigation system using a sensor based decision support system

**Manuel A. Andrade**
Oak Ridge Institute for Science and Education, Sponsored by USDA-Agricultural Research Service, P.O. Drawer 10, 2300 Experimental Station Road, Bushland, TX 79102, USA, Alejandro.Andrade@ars.usda.gov

**Susan A. O'Shaughnessy**
USDA-Agricultural Research Service, P.O. Drawer 10, 2300 Experimental Station Road, Bushland, TX 79102, USA, Susan.O'Shaughnessy@ars.usda.gov

**Steven R. Evett**
USDA-Agricultural Research Service, P.O. Drawer 10, 2300 Experimental Station Road, Bushland, TX 79102, USA, Steve.Evett@ars.usda.gov

**Abstract.** For farmers to take full advantage of center pivot Variable Rate Irrigation (VRI) systems, a decision support system (DSS) incorporating site specific irrigation scheduling methods must be integrated into their operation. This can be done with a Supervisory Control and Data Acquisition (SCADA) system implementing irrigation scheduling methods based on plant stress. This Irrigation Scheduling SCADA System (ISSCADAS) has been effectively used for site specific irrigation management of center pivot VRI systems. The ISSCADAS automates the collection of data from plant and microclimate sensing systems, as well as the application of algorithms to process those data. A software, named ARSPivot (ARSP), has been developed to facilitate the irrigation management of VRI center pivots using an ISSCADAS. This paper describes how a sensor based DSS consisting of ARSP and the ISSCADAS was used in the summer of 2016 to operate a six-span VRI center pivot near Bushland, TX. Corn was planted on one half of the field (the other half was left fallow) with three deficit irrigation treatments: 100, 50, and 30% of full replenishment of soil water depletion to 1.5 m. Irrigations were scheduled using either manual weekly neutron probe readings or the irrigation scheduling method implemented by the ISSCADAS, which incorporated a set of thermal stress thresholds and varying irrigation amounts for each irrigation treatment. No significant differences were found between overall means of dry grain yield and crop water use efficiency at the higher irrigation levels. At the lowest irrigation level dry grain yield and water use efficiency of plots using the plant stress based method were greater and significantly different than those obtained with the neutron probe measurements.

**Keywords:** center pivot irrigation, sensors, software, site specific irrigation scheduling.

status, parental status, religion, sexual orientation, genetic information, political beliefs, reprisal, or because all or part of an individual's income is derived from any public assistance program. (Not all prohibited bases apply to all programs.) Persons with disabilities who require alternative means for communication of program information (Braille, large print, audiotape, etc.) should contact USDA's TARGET Center at (202) 720-2600 (voice and TDD). To file a complaint of discrimination, write to USDA, Director, Office of Civil Rights, 1400 Independence Avenue, S.W., Washington, D.C. 20250-9410, or call (800) 795-3272 (voice) or (202) 720-6382 (TDD). USDA is an equal opportunity provider and employer.

The mention of trade names of commercial products in this article is solely for the purpose of providing specific information and does not imply recommendation or endorsement by the U.S. Department of Agriculture.

## Introduction

According to Evans et al. (2013), adoption of site specific VRI technology has been slow. Among other needs required for the sustained adoption of site specific VRI technology, they mentioned a need for "the development and testing of easy-to-use basic decision support systems for simple site specific irrigation scheduling scenarios." This paper describes the use of a sensor based Decision Support System (DSS) for the site specific irrigation management of a center pivot VRI system. At the core of the DSS lies an Irrigation Scheduling Supervisory Control and Data Acquisition System (ISSCADAS) developed and patented by scientists with the USDA-ARS Conservation & Production Research Laboratory, Bushland, TX (Evett et al. 2014). The ISSCADAS incorporates irrigation scheduling methods based on plant stress and functions to automate the collection and analysis of data required by these methods. The ISSCADAS has been used in previous studies near Bushland for irrigation management of center pivot VRI systems irrigating soybean (*Glycine max* (L.) Merr.) (Peters and Evett, 2008), cotton (*Gossypium hirsutum* L.) (O'Shaughnessy and Evett, 2010), grain sorghum (*Sorghum bicolor* (L.) Moench) (O'Shaughnessy et al., 2013), and corn (*Zea mays* L.) (O'Shaughnessy et al., 2017). Results from these studies demonstrated that the ISSCADAS can be used to intensely manage center pivot VRI systems with little labor, resulting in yield and Water Use Efficiency (WUE) values comparable to those obtained using labor-intensive neutron probe (NP) measurements, and greater than county-wide averages.

The DSS used in this study is accessible to users without in-depth knowledge of sensing systems or irrigation scheduling methods thanks to a user friendly software developed as an embodiment of the ISSCADAS. The software, named ARSPivot (ARSP), automatically collects and processes data from plant, soil, and microclimate sensing systems to generate site specific prescription maps that users can visualize and modify using ARSP's Graphical User Interface (GUI). ARSP incorporates multiple tools commonly found in Geographic Information System (GIS) software to assist in the spatial and temporal analysis of data collected from the sensing systems (Andrade et al., 2016), and is also a flexible tool that can be used under a wide range of conditions (Andrade et al., 2017). This paper presents the results of using a sensor based DSS for the integrated irrigation management of a VRI center pivot system located in a field planted with corn near Bushland, TX. Some of the key benefits of the DSS are illustrated by analyzing the performance of experimental plots for which irrigation was scheduled using a plant stress method implemented in ARSP and plots for which irrigation was triggered by NP measurements.

## Methodology

A DSS consisting of ISSCADAS/ARSP was used during the summer of 2016 for the site specific irrigation management of a six-span center pivot (260 m) equipped with a Pro2 control panel and a commercial VRI zone control system (Valmont Industries, Inc., Valley NE). Drought tolerant corn hybrid Pioneer® P0157AM was planted on the North-Northeast side of the field on June 16, 2016, day of year 168. The field was divided into six sectors of 28° each and six concentric plots spaced 18.3 m (30 ft) apart, for a total of 36 plots (Fig. 1). Irrigations were applied using low elevation spray application (LESA) and furrows were diked to reduce runoff. Plots were assigned one of three irrigation levels (100, 50, and 30%) and their irrigation was either scheduled by a plant stress method or by weekly NP measurements. The three irrigation levels were used in order to establish data for production functions of yield versus water use. For plots assigned the former method these levels represented percentages of a specified water depth of 38.1 mm (1.5 in), corresponding to 3.5 times the daily peak water use for corn in the region (assuming a 3.5-day return period for the center pivot). For plots assigned the latter method, irrigation levels represented percentages of replenishment of soil water depletion to field capacity in the top 1.5 m of soil. Irrigation scheduling of plots using this method was determined based on NP (model 503DR1.5, Instrotek, Campbell Pacific Nuclear, Martinez, CA) measurements in 0.2 m increments from 0.1 to 2.3 m (center of measurement) in treatment plots with the highest irrigation level. The NP was calibrated in the field using methods described by Evett (2008). Any precipitation occurring prior to irrigation was subtracted from the total amount required for the week.

The plant stress method implemented in ARSP is based on the estimation of an integrated Crop Water Stress Index (iCWSI) obtained as the sum of theoretical Crop Water Stress Indices (CWSIs) calculated at discrete intervals during daylight hours. ARSP estimates a CWSI for any given location in the field at time interval $t$ using the normalized difference between the crop canopy temperature in the location and the air temperature at time $t$. Hence, values of CWSI so calculated range from 0 for well-watered crops to 1 for severely-stressed crops, and the iCWSI estimated for any location in the field can be described in an intuitive way as the total number of time intervals during daylight hours that the crop in that position experiences the most severe stress. Additional details of the iCWSI method and the formulas used for its calculation can be found in O'Shaughnessy et al. (2013) and O'Shaughnessy et al. (2017).

A network of 12 wireless infrared thermometers (IRTs) (model SapIP-IRT, Dynamax, Inc., Houston, TX) was mounted on the center pivot to measure crop canopy temperatures required by the iCWSI method. IRTs were located forward of the drop hoses, at an oblique angle from nadir and distributed along the center pivot in such a way that data from two IRTs with opposing views of a concentric plot were averaged to estimate the canopy temperature inside of the concentric plot. Six wireless IRTs were placed in the field to provide a reference canopy temperature for a well-watered crop. These IRTs were scattered in wetlands located in the inner and outer areas of the North-Northeast side of the field. Time intervals of one minute were used for the estimation of canopy temperatures using data collected from the IRTs. Scans of the field were performed periodically by running the pivot dry. Canopy temperatures collected within daylight hours during the scans (9 am to 7:30 pm – avoiding hours near sunrise and sunset per Peters and Evett, 2004) were used to schedule the irrigation of plots assigned the iCWSI method.

Experimental plots were organized using a Latin square design shown in Figure 1, as they are displayed in ARSP's GUI. Plots irrigated with the iCWSI method are labeled as C100, C50, or C30, where 'C' stands for iCWSI-based control and numbers correspond to irrigation levels. Similarly, plots irrigated with the NP method are labeled as U100, U50, and U30, where 'U" indicates that irrigation scheduling is controlled by the user. Irrigation scheduling of 'C' plots was determined by comparing all the iCWSIs calculated inside of a plot with a set of pre-established thresholds (Table 1). These thresholds were determined from mean seasonal iCWSI values obtained by O'Shaughnessy et al. (2017) for two corn hybrids during the 2013 and 2014 growing seasons at Bushland. Irrigation was triggered for plots where most iCWSIs were greater than 100 and no irrigation was triggered otherwise. The irrigation amounts prescribed were then assigned based on the irrigation level of each plot and by the values of iCWSIs obtained inside the plot. A plot received a 'low' irrigation depth if most of its iCWSIs were higher than 100 but less than or equal to 150 (column 3 in Table 1), a 'medium' depth if most of its iCWSIs were larger than 150 but less than or equal to 250 (column 4 in Table 1), or a 'high' depth if most of its iCWSIs were larger than 250 (column 5 in Table 1).

ARSP consists of two programs running simultaneously on an embedded computer located at the pivot point. Both programs operate using a client-server architecture, where one program (named as the 'client') contains functions to collect data from sensing systems and to communicate with the Pro2 control panel; the other program (named as the 'server') incorporates ARSP's GUI and algorithms to process the data collected by the client program, such as the iCWSI irrigation scheduling method. Figure 2 illustrates the interaction of the DSS with its user and the VRI center pivot system for the generation and application of a site specific prescription map. The process occurs through the following stages (Fig. 2): once at the beginning of the season the user provides information of the field, crop, characteristics of the VRI center pivot system, etc., to the DSS through ARSP's GUI; the user then initiates ARSP to (1) collect data from a weather station, the wireless network of IRTs mounted on the center pivot, and the IRTs in the field is stored in the embedded computer; (2) those data are processed by the client program and then (3) sent to the server program that uses the iCWSI method to generate a site specific prescription map that is (4) presented in a friendly format to the user through the GUI implemented in the server program; (5) the user looks at the prescription map and, if needed, modifies it before it is (6) sent to the client program, which in turn (7) submits it to the Pro2 panel for its (8) application by the VRI center pivot; (9) the process starts again after the irrigation with the collection (1) of a new set of canopy temperatures and weather data. Additional details of how site specific prescription maps can be visualized and modified using ARSP's GUI, and how color scaled maps of iCWSIs can be generated to analyze such prescription maps can be found in Andrade et al. (2016).

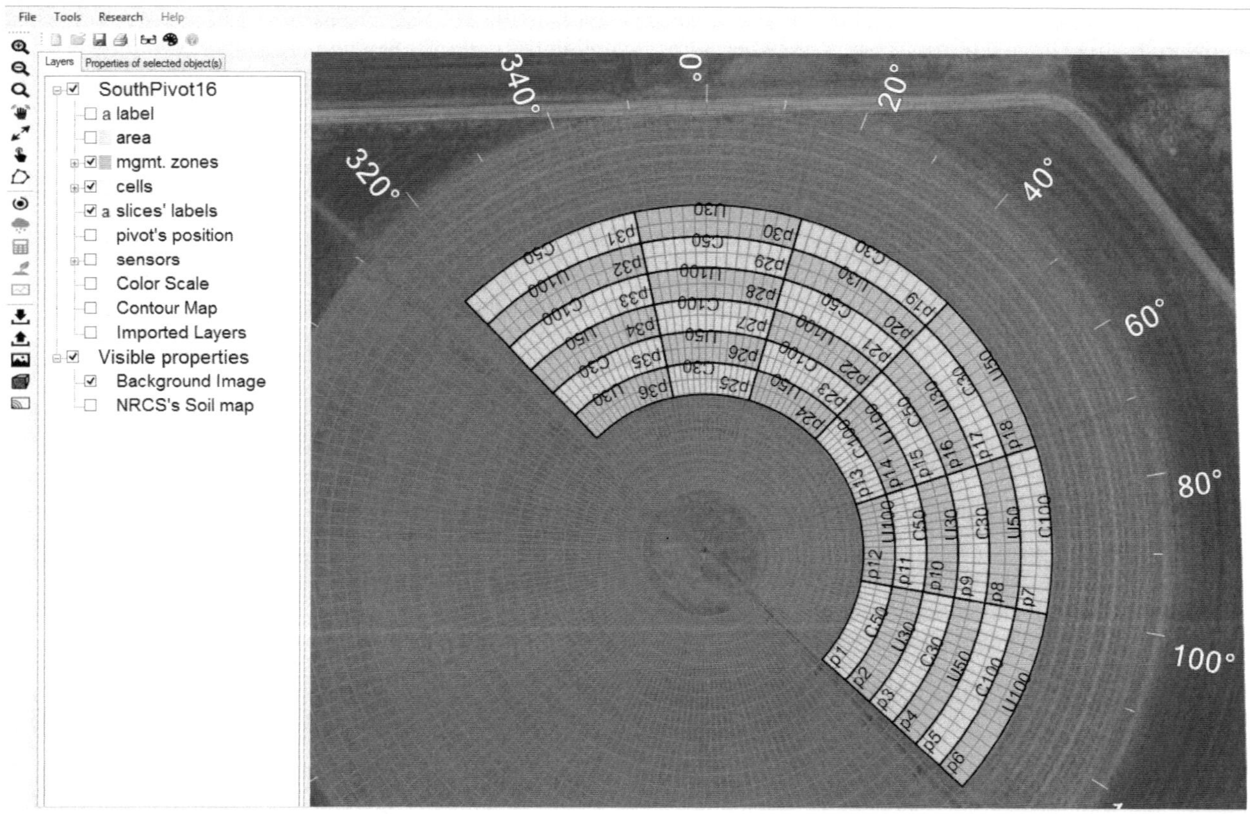


**Fig. 1. Experimental setup of 36 plots in the six-span center pivot irrigation system, as displayed in ARSP's GUI. Letters C and U inside a plot indicate, respectively, that irrigation scheduling of the plot is CWSI-based (C) or controlled by the user (U) through weekly NP measurements. Similarly, numbers 100, 50, and 30 (located next to letters C or U) indicate the irrigation level assigned to the plot (100, 50, or 30%). Numbers inside plots preceded by the letter 'p' indicate the number of plot.**

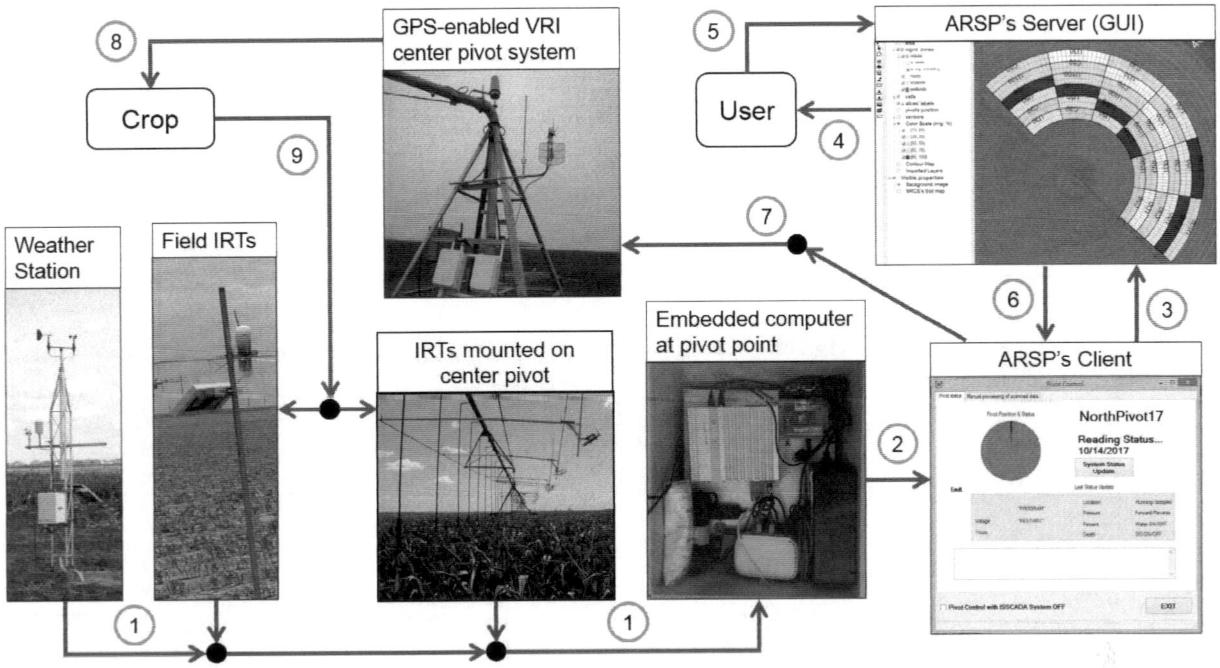


**Fig. 2. Interaction of a sensor based DSS with its user and a VRI center pivot system for the generation and application of a site specific prescription map. The process occurs through nine stages noted by numbers.**

**Table 1. Irrigation depths and iCWSI thresholds used for the irrigation scheduling of experimental plots assigned the iCWSI method**

| Irrigation Level (%) | No irrigation iCWSI<=100 | Low depth 100<iCWSI<= 150 | Medium depth 150<iCWSI<=250 | High depth iCWSI>250 |
|---|---|---|---|---|
| 100 | 0 | 19.1 mm (0.75 in) | 25.4 mm (1 in) | 38.1 mm (1.5 in) |
| 50 | 0 | 12.7 mm (0.5 in) | 19.1 mm (0.75 in) | 25.4 mm (1 in) |
| 30 | 0 | 6.4 mm (0.25 in) | 12.7 mm (0.5 in) | 19.1 mm (0.75 in) |

## Results

Tables 2, 3, and 4 summarize experimental results in terms of mean dry grain yield (Mg/ha), crop water use (mm), and WUE (kg/m$^3$). Crop water use displayed in these tables includes a total precipitation of 236 mm measured during the season. Crop response to irrigation scheduling methods is presented in Table 2. Larger mean yield and WUE were obtained from plots irrigated using the iCWSI method ('C' plots) and their values were significantly different than those obtained from plots irrigated with the NP method ('U' plots). Although the mean seasonal crop water use of the 'U" plots was smaller than the mean crop water use of 'C' plots, their difference was not significant. Crop response to irrigation levels is presented in Table 3. Mean yields and crop water use values increased with irrigation levels and their differences were significant. Mean WUE values, on the other hand, were not significantly different for any irrigation level, a result which may be explained by the use of a drought tolerant hybrid for the experiment.

**Table 2. Mean dry grain yield (Mg/ha), seasonal crop water use (mm), and WUE (kg/m$^3$), grouped by irrigation scheduling method[(a)]**

| Irrigation Method | Dry grain yield (Mg/ha) | Crop water use (mm) | WUE (kg/m$^3$) |
|---|---|---|---|
| Neutron Probe (U) | 10.95a | 562a | 1.95a |
| iCWSI (C) | 12.11b | 592a | 2.05b |

(a) Mean values in each column followed by a different letter are significantly different at $p<0.05$

**Table 3. Mean dry grain yield (Mg/ha), seasonal crop water use (mm), and WUE (kg/m$^3$), grouped by irrigation level[(a)]**

| Irrigation Level (%) | Dry grain yield (Mg/ha) | Crop water use (mm) | WUE (kg/m$^3$) |
|---|---|---|---|
| 100 | 12.86a | 644a | 2.0a |
| 50 | 11.51b | 569b | 2.03a |
| 30 | 10.22c | 518c | 1.97a |

(a) Mean values in each column followed by a different letter are significantly different at $p<0.05$

Crop response to the six irrigation treatments resulting from the interaction of two irrigation scheduling methods and three irrigation levels is presented in Table 4. Mean yields of 'C' plots were consistently larger than mean yields of 'U' plots with the same irrigation levels. However, mean yields were only significantly different at the smallest irrigation level. Although mean crop water use was not significantly different at the largest irrigation level, at the smaller irrigation levels the mean crop water use of 'C' plots was larger and significantly different than the mean crop water use of 'U' plots. Mean WUE was only significantly different at the smallest irrigation level, where 'C30' plots obtained a larger mean WUE than 'U30' plots.

**Table 4. Mean dry grain yield (Mg/ha), seasonal crop water use (mm), and WUE (kg/m$^3$), grouped by irrigation treatment[a]**

| Irrigation treatment | Dry grain yield (Mg/ha) | Crop water use (mm) | WUE (kg/m$^3$) |
|---|---|---|---|
| U100 | 12.43ab | 646a | 1.92ab |
| C100 | 13.29a | 642a | 2.07a |
| U50 | 11.02c | 535c | 2.06a |
| C50 | 12.0bc | 603b | 1.99ab |
| U30 | 9.4d | 505d | 1.86b |
| C30 | 11.04c | 531c | 2.08a |

(a) Mean values in each column followed by a different letter are significantly different at $p<0.05$

ARSP's GUI incorporates multiple tools to assist the irrigation management of VRI center pivot systems. Among these tools is the generation of color scale maps of different numerical properties of management zones (experimental plots in this study). These maps are displayed in the GUI as layers drawn in top of (or below) other layers to facilitate the visual analysis of spatial information collected by the DSS. Layers of information can be overlaid because the GUI generates a scaled representation of center pivots and their management zones that is geo-referenced using the Universal Transverse Mercator projection (UTM). ARSP's GUI can also download and display other useful layers, such as a background satellite image of the terrain that is obtained using Google Maps application programming interface (API) (see Fig. 1 and Fig. 3), and a layer containing soil map units in the field obtained using the USDA-National Resources Conservation Service's Web Soil Survey website (USDA-NRCS, 2017).

Figure 3 shows color scaled maps of dry grain yield (Mg/ha) and seasonal crop water use (mm) obtained for all 36 experimental plots used in this study. These maps help to identify if particular conditions in portions of the field had an effect on crop performance. However, no spatial pattern seems to be involved in either yield or crop water use, since their values seem to be mainly influenced by irrigation level and (to a lesser degree) by the irrigation method assigned to each plot. Spatial distribution of dry grain yield is further explored in Figure 4, where a color scaled map of yield is displayed on top of an elevation contour map generated by ARSP using the kriging interpolation method. Elevations were obtained using Google Maps API through an external script written in the R programming language and the 'googleway' package for the same language (Cooley, 2017). Elevation doesn't seem to play a significant role in crop performance in terms of yield. If sectors are numbered in a counter-clockwise direction, sectors 4 and 6 are the sectors with the lowest and highest elevations in the experimental area, respectively. Nevertheless, a visual inspection of yields in these sectors shows that similar values were obtained from plots with the same irrigation treatments. Users of the DSS can perform similar analyses to assist the decision making process before the start of a growing season. For example, a soil contour map can be displayed in ARSP's GUI on top of yield and elevation contour maps to delineate the management zones that will be used during the season.

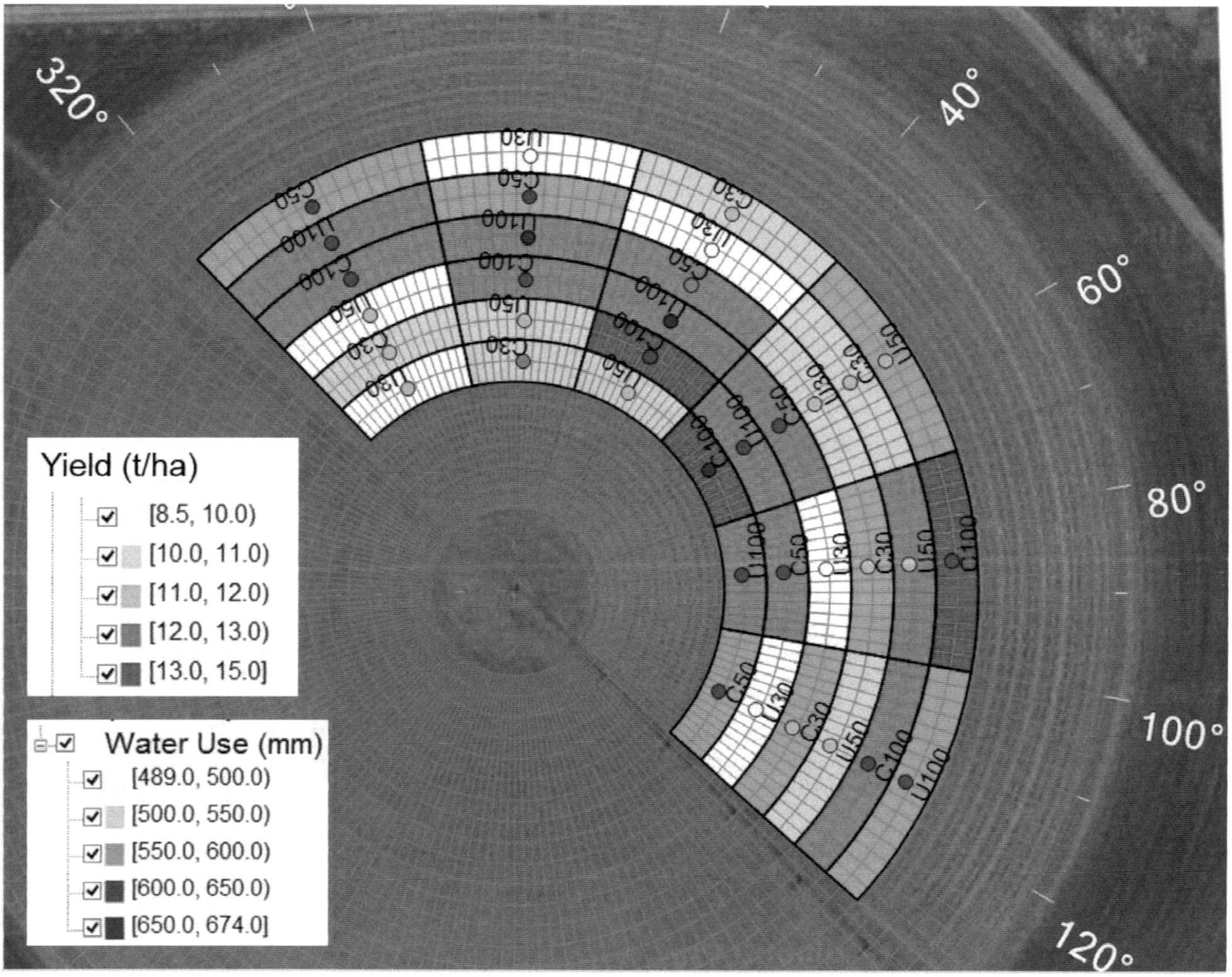


**Fig. 3. Color scaled maps of dry grain yield (Mg/ha) and crop water use (mm) generated in ARSP's GUI. Yields are displayed using a color scale that progresses from white (least yield) to green (largest yield). Crop water use values are represented by circles inside plots; the color scale used for these circles progresses from white (least water use) to blue (largest water use).**

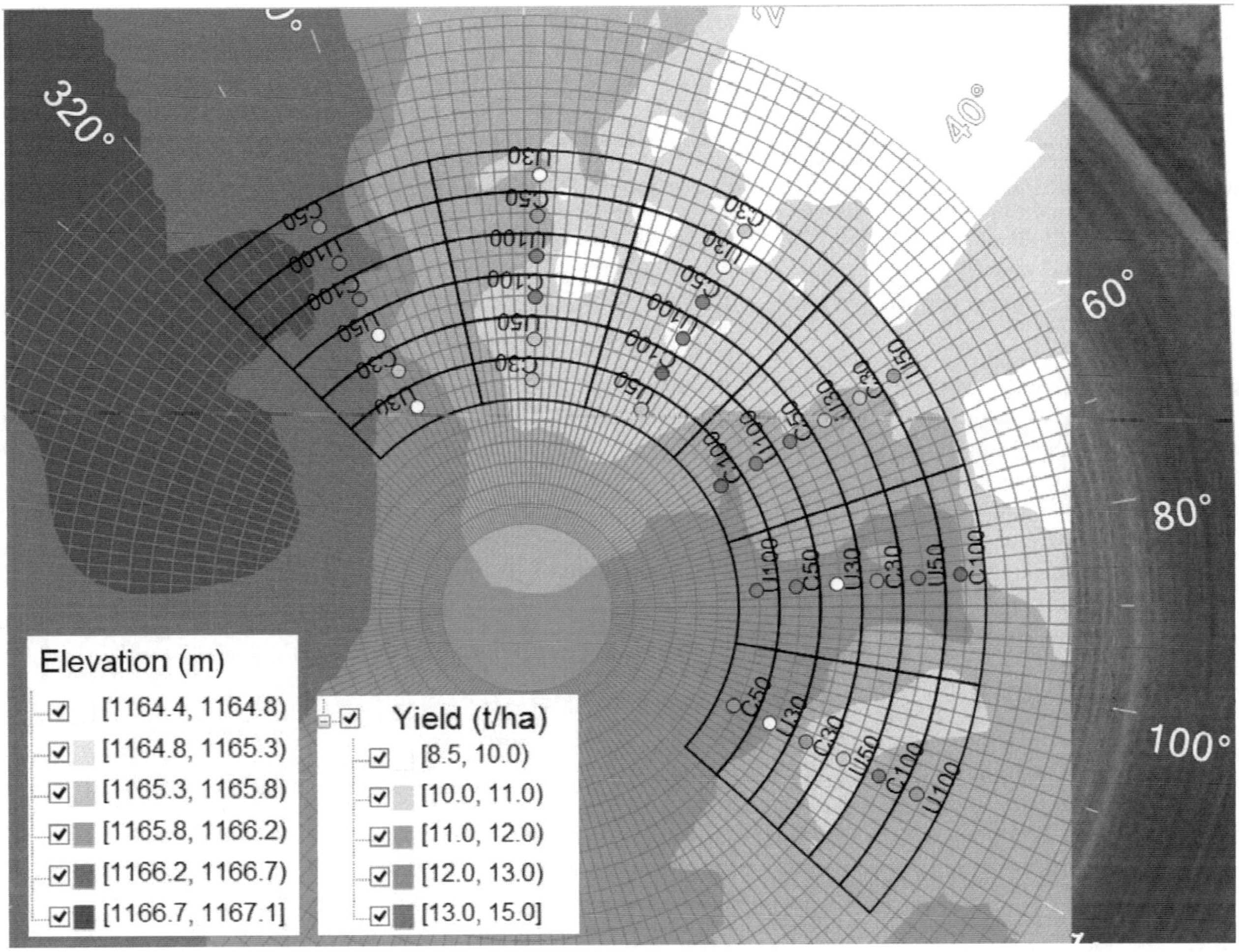


**Fig. 4. Color scaled map of dry grain yield (Mg/ha) and elevation (m) contour map generated by ARSP using the kriging interpolation method.**

## Conclusions

A user-friendly DSS was applied to operate a complex system comprised of a VRI center pivot system and a network of plant and microclimate sensing systems supporting a site specific irrigation scheduling method based on plant stress. A software developed as part of the DSS allows the seamless operation of such a complex system. A post-harvest analysis of an experiment carried out in Bushland, TX, during the summer of 2016 demonstrated how the DSS can be used to assist the irrigation management of a VRI center pivot system. Results from this experiment showed that, at the higher irrigation levels, mean yield and WUE values obtained from plots using the iCWSI method were not significantly different than those obtained using the time consuming neutron probe method.

## Acknowledgements

This material is based upon work that is supported by the National Institute of Food and Agriculture, U.S. Department of Agriculture, under award number 2016-67021-24420. Work reported in this paper was also accomplished as part of a Cooperative Research and Development agreement between USDA-ARS and Valmont Industries, Inc., Valley, NE (Agreement No.: 58-3K95-0-1455-M).

## References

Andrade, M., O'Shaughnessy, S., and Evett, S. (2016). A GIS-based Decision Support Tool for Center Pivot Irrigation Systems. *In Proc. 2016 ASABE Annual International Meeting*, St. Joseph, MI.

Andrade, M., O'Shaughnessy, S., and Evett, S. (2017). ARSPivot, a Sensor Based Decision Support Tool for the Integrated Irrigation Management of VRI Center Pivot Systems. *In Proc. 28th Annual Central Plains Irrigation Conference*, Colby, KS.

Cooley, D. (2017). Googleway: Accesses Google Maps APIs to Retrieve Data and Plot Maps. R package version 2.2.0. https://CRAN.R-project.org/package=googleway

Evans, R., LaRue, J., Stone, K., and King, B. (2013). Adoption of site-specific variable rate sprinkler irrigation systems. *Irrigation Science*, *31*, 871–887. http://doi.org/10.1007/s00271-012-0365-x

Evett, S. (2008). Neutron moisture meters. In Field Estimation of Soil Water Content: A practical Guide to Methods, Instrumentation, and Sensor Technology, 39-54. S. R. Evett, L. K. Heng, P. Moutonnet, and M. L. Nguyen, eds. IAEA-TCS-30. Vienna, Austria: International Atomic Energy Agency.

Evett, S., O'Shaughnessy, S., and Peters, T. (2014). Patent application serial No. 13/403,091 Entitled "Irrigation Scheduling and Supervisory Control and Data Acquisition System for Moving and Static Irrigation Systems". Patent allowed August 29, 2014.

O'Shaughnessy, S., Andrade, M., and Evett, S. (2017). Using an integrated crop water stress index for irrigation scheduling of two corn hybrids in a semi-arid region. *Irrigation Science*. 31(5): 451-167. Available at http://doi.org/10.1007/s00271-017-0552-x

O'Shaughnessy, S., and Evett, S. (2010). Canopy temperature based system effectively schedules and controls center pivot irrigation of cotton. *Agricultural Water Management.* 97(9): 1310–1316. Available at http://linkinghub.elsevier.com/retrieve/pii/S0378377410001095.

O'Shaughnessy, S., Evett, S., Colaizzi, P., and Howell, T. (2013). Wireless Sensor Network Effectively Controls Center Pivot Irrigation of Sorghum. *Applied Engineering in Agriculture.* 29(6): 853–864. Available at http://elibrary.asabe.org/abstract.asp?aid=44178&t=3&dabs=Y&redir=&redirType=.

Peters, R., and Evett, S. (2004). Modeling diurnal canopy temperature dynamics using one-time-of-day measurements and a reference temperature curve. *Agronomy Journal*. 96(6):1553–1561.

Peters, R., and Evett, S. (2008). Automation of a Center Pivot Using the Temperature-Time-Threshold Method of Irrigation Scheduling. *Journal of Irrigation and Drainage Engineering*. 134(3): 286-291.

USDA National Resources Conservation Service (NRCS). (2017). Natural Resources Conservation Service, United States Department of Agriculture. Web Soil Survey. Available at http://websoilsurvey.nrcs.usda.gov/. Accessed 01/25/2017.

# Remote Sensing for Variable Rate Irrigation Management

Kenneth Stone and Philip Bauer

USDA-ARS, Coastal Plains Soil, Water, and Plant Research Center,
US Department of Agriculture – Agricultural Research Service, Florence, SC, USA
* Corresponding author. Email: ken.stone@ars.usda.gov

**Abstract**: *Irrigation using overhead sprinkler systems (primarily center pivots) is commonly used for corn and cotton production in the southeast USA. Technology for variable rate water application is available; however, it is not widely used due to increased management requirements. Methods to develop prescriptions in-season in response to changing crop conditions are needed to move this technology forward. The objective of this research was to evaluate the potential of using normalized difference vegetative index (NDVI) to estimate crop coefficients for development of spatial irrigation prescriptions. Field studies were conducted near Florence, SC, USA under center pivot irrigation systems equipped with variable rate technology. Studies in maize were conducted over several years comparing NDVI-based irrigation management to management with soil water sensors. The two methods did not significantly differ for yield or water volume applied. A cotton irrigation study was initiated in 2016 that compared the checkbook method (applying irrigation amount based on age of the crop and weekly precipitation totals) to NDVI-based irrigation prescriptions. Soil water sensors were used to initiate irrigation events. Irrigation amounts during the season for the NDVI-based method often differed from rates prescribed by the checkbook method up until about 70 days after planting when differences in NDVI among plant density treatments and field areas no longer existed. The results, along with the emerging use of air- and land-based remote sensing techniques, suggest continued research into the use of NDVI-based irrigation prescription technology is warranted.*

## Introduction

Agricultural water management in the humid coastal plain region of the southeastern United States is problematic. Although this region generally receives adequate rainfall, the amount and distribution of rainfall is highly unpredictable. Additionally, croplands in the region have varying soil types with differing soil water holding capacities resulting in variable crop growth and yields (Sadler, et al., 2000). Variable rate irrigation systems (VRI) may be a tool to address these production problems. VRI systems are irrigation systems capable of applying different water depths both in the direction of travel and along the length of the irrigation system (Evans, et al., 2010). Thus, VRI systems can be tools for conserving water and spatially allocating water resources while potentially increasing profits (Evans and King, 2012). These VRI systems could also be used as a tool for improving crop water management and efficiency by delivering water to plants where needed, when the crop demands it, and in the appropriate amounts (O'Shaughnessy, et al., 2015). Although spatially variable water application technology is available and has high interest among growers, there has been limited adoption of VRI systems (Evans, et al., 2013). One potential reason for this limited adoption of VRI systems is the lack of science-based information on how to precision-apply water with these systems (Sadler, et al., 2005).

Both Sadler et al. (2005) and Evans, et al. (2013) identified critical research needs that included the development of decision support systems and integrated management systems to sense within-field variability in real time and dynamically define irrigation management zones. Dynamic management zones for VRI system management can be estimated using remote sensing methods including canopy reflectance and crop canopy temperatures. A popular method of estimating vegetation from growing plants is to use remotely sensed spectral vegetative indices. One of the most commonly used vegetative indexes is the normalized difference vegetation index (NDVI). It is defined as the difference between visible and near infra-red (NIR) measurements divided by their sum. The NDVI measurements can be used to assess the overall general health of the plants (Berger, et al., 2010).

NDVI can be used in irrigation management. For example Hunsaker et al. (2005a, 2005b) used NDVI measurements to calculate within-season real-time crop coefficients for irrigating wheat and cotton in Arizona. In Spain, Gonzalez-Piqueras et al. (2004) used NDVI measurements to calculate within-season crop coefficients for corn. Likewise, in South Carolina, Stone et al. (2016) investigated the use of NDVI to spatially irrigate corn and found that NDVI based irrigation has the potential to be effectively used in delineating dynamic management zones in fields irrigated with VRI systems. In this research, our objective was to evaluate the potential of using NDVI to estimate crop coefficients for developing spatial irrigation prescriptions for both corn and cotton crops.

## Methods

Irrigation experiments were conducted at the USDA-ARS Coastal Plain Soil, Water, and Plant Research Center and at Clemson University's Pee Dee Research and Education Center using variable-rate irrigation (VRI) systems.

Corn Experiment: From 2012 to 2014, corn was grown on a 6 ha site under a VRI system at the USDA-ARS site Florence, South Carolina. The soils under the center-pivot irrigation system are highly variable. Two irrigation treatments were evaluated and compared for their potential for spatial irrigation management. A treatment based on measured soil water potentials (SWP) was compared to a treatment based on remotely sensing the crop normalized difference vegetative index (NDVI treatment). The SWP treatment used SWP sensors to maintain SWP values above -30 kPa (approx. 50% depletion of available water) in the top 30 cm of soils. The NDVI treatment used the measured NDVI values to calculate spatial crop coefficients to similar to methods used by Bausch (1993), Hunsaker et al. (2003), and Glenn et al. (2011). These estimated crop coefficients were used in the FAO 56 dual crop coefficient method for estimating crop evapotranspiration ($ET_c$) and irrigation requirements. The VRI system was divided into 4 quadrants with each treatment having three replicates per quadrant. Irrigation management for the SWP treatments was a 12.5 mm irrigation application when the SWP decreases below -30 kPa in the rooting zone. For the NDVI treatment, irrigation depths were applied to 4 sub-plots zones within each quadrant. Crop coefficients of plants in each sub-plot were calculated using the NDVI-Kc relationships (Kc = 1.5*NDVI -0.1) developed by Hunsaker et al. (2005b) and Gonzalez-Piqueras et al. (2004). Irrigation depths were determined by using these Kc values in a 7-day water balance using reference ET calculated from an on-site weather station.

NDVI was measured using a crop circle NDVI sensor (Crop Circle ACS-430 Active Canopy Sensor, Holland Scientific, Lincoln, Nebraska) mounted on the tractor spray boom and collected at 1-2 week intervals starting after planting through full canopy.

Cotton Experiment: At Clemson University's Pee Dee Research and Education Center, an experiment on irrigated cotton was conducted in a field that contains a commercial 305 m long site-specific center pivot irrigation system that has five 60 m long spans with each span configured to provide three 20-m irrigation zones. This center pivot is on a field that contains soils that differ in texture of the A horizon (ranging from sand to sandy loam). This experiment was conducted only under the outer span of the pivot. Three irrigation application amount treatments (spatially dependent application using NDVI, uniform, and rain-fed) were evaluated at low and normal plant densities. Two plant densities were utilized to allow for a greater range of NDVI within each replicate. Low plant density treatment had 5 seeds per m of row while normal plant density had 11.5 seeds per m of row (1.5 and 3.5 plants per foot of row). Plot size was 6° of arc long (ranging from 26 m to 32 m, depending on distance from pivot) and one irrigation management zone (20-m) wide. Seven treatment replicates were used in 84° of pivot travel.

In four of the replicates, SWP was measured (30 cm deep) in the uniform - normal plant density treatment combination. These tensiometers were used to trigger all irrigation events and irrigations were applied when they average -30 kPa. All irrigations were applied in amounts to provide the recommended three-day water amounts. Irrigation amounts for the uniform method were based on recommendations of the University of Georgia - Georgia Cotton Production Guide (http://www.ugacotton.com/vault/file/2017-Georgia-Cotton-Production-Guide.pdf )

Table 1. Cotton Irrigation Schedule (UGA - Georgia Cotton Production Guide).

| **Crop Stage** | **Weekly** | | **Daily** | |
|---|---|---|---|---|
| | **Inches** | **mm** | **inches** | **mm** |
| Week beginning at 1st bloom | 1.0 | 25 | 0.15 | 4 |
| 2nd week after 1st bloom | 1.5 | 38 | 0.22 | 6 |
| 3rd week after 1st bloom | 2.0 | 51 | 0.30 | 8 |
| 4th week after 1st bloom | 2.0 | 51 | 0.30 | 8 |
| 5th week after 1st bloom | 1.5 | 38 | 0.22 | 6 |
| 6th week after 1st bloom | 1.5 | 38 | 0.22 | 6 |
| 7th week and beyond | 1.0 | 25 | 0.15 | 4 |

When SWP readings approached -30 kPa, NDVI in two interior rows of each NDVI irrigation method plot were measured using a handheld Greenseeker NDVI sensor (Trimble Agriculture, Westminster, CO USA). Crop coefficients of the plants in each plot were estimated using the NDVI-Kc relationship (Kc = 1.5*NDVI -0.1) developed by Hunsaker et al. (2005b) and Gonzalez-Piqueras et al. (2004).

## Results

Annual rainfall for the three-year corn irrigation study varied widely from 620 mm in 2013 to 414 mm in 2014. The annual corn yields were significantly different with overall mean annual yields from 2012 to 2014 were 15.6, 10.5, and 13.5 Mg/ha, respectively. Since the annual yields were significantly different, we analyzed them individually and found that for the three-year study the treatment yields were not significantly different.

The NDVI measurements were then used to calculate crop coefficients. NDVI measurements were taken periodically throughout the growing season until tasseling. Figure 1 shows a progression of NDVI measurements as the crop grew from May 14, 2013 to June 21, 2013. The plots show the variability in NDVI throughout the field at various growth stages.

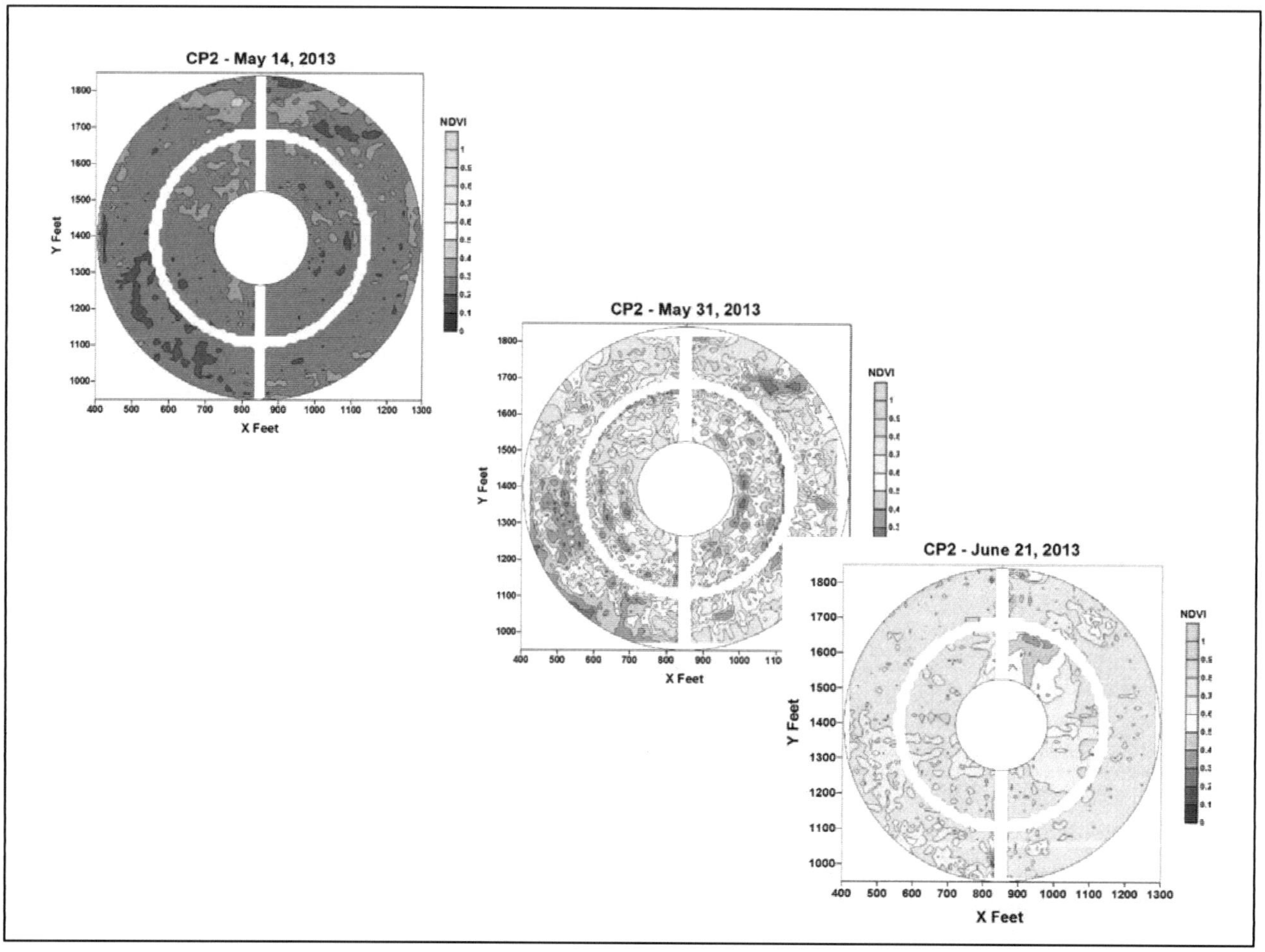


Figure 1. Example field maps of NDVI measurements over the growing season.

Since the treatment corn yields were not significantly different for any year of the study, we combined the NDVI reading across each treatment for analysis. Figure 2 shows a comparison of the yearly NDVI based calculated crop coefficients compared to the typical irrigation requirements based on the FAO-56 recommendations (Allen, et al. 1998). The 2012 growing season had near normal rainfall and the calculated crop coefficients were very similar to the FAO-56 coefficients. However, 2013 and 2014 had early season low temperatures that delayed crop growth and impacted the calculated crop coefficients. The 2013 and 2014 crop coefficients were approximately 1-2 weeks delayed from the recommended FAO-56 coefficients. During these years, if irrigation was based on the standard FAO-56 schedule, it would have applied irrigation in excess of crop demands during the early growing.

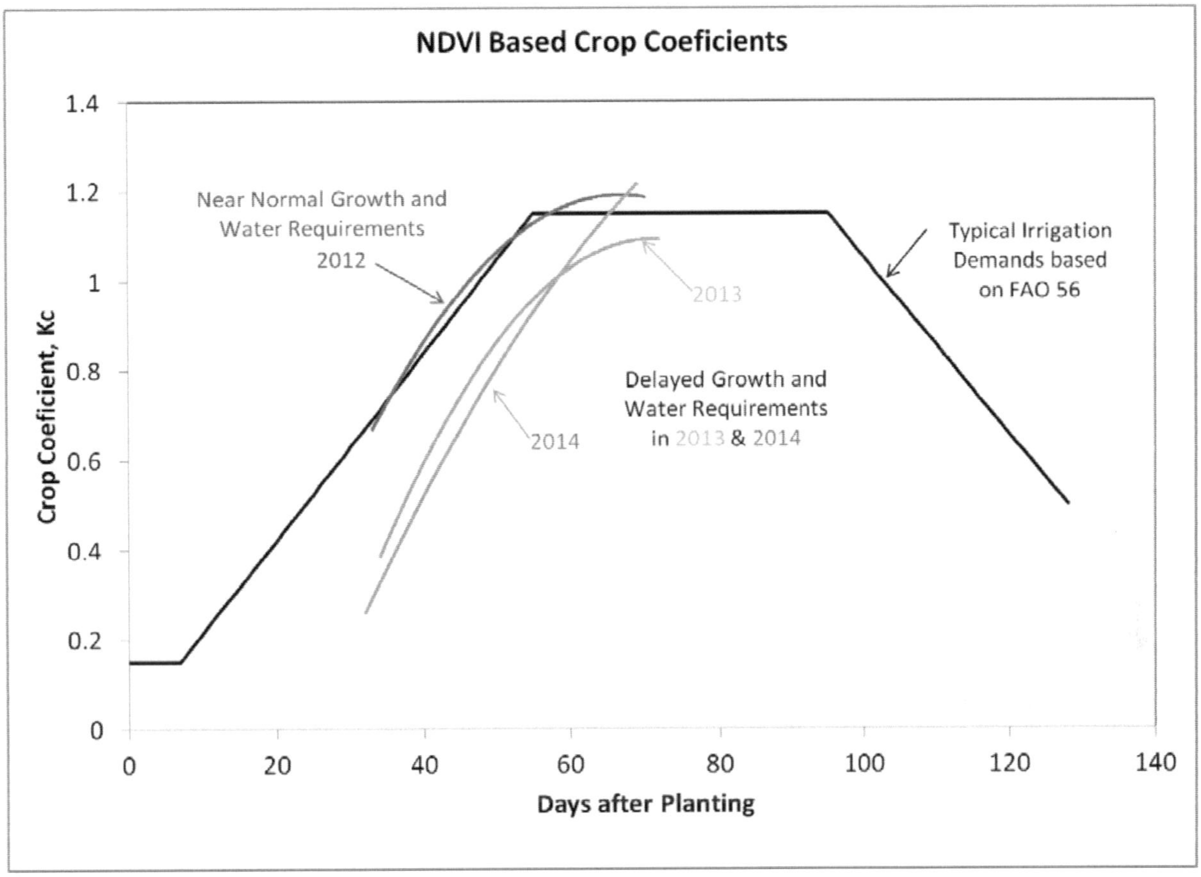


Figure 2. NDVI based crop coefficients for the 2012-2014 corn crops.

In 2016, we initiated an experiment to evaluate spatial irrigation of cotton based on NDVI readings. The calculated NDVI based crop coefficients are show in figure 3 and were compared to the FAO-56 recommended values. The NDVI based crop coefficients were similar to those of the FAO-56 recommendations. The different seeding rates show different NDVI measurements and associated crop coefficients indicating less total crop biomass and potentially reduced water requirements until they both reached full canopy closure.

An example irrigation event on June 21, 2016 for the cotton experiment is as follows. The checkbook treatment had an irrigation of 0.5 inches (13 mm). The VRI high plant population treatment had irrigation depths ranging from 0.2 to 0.6 inches (5-15 mm) with an average application of 0.39 inches (10 mm). The VRI low population treatment had irrigation depths ranging from 0.15 to 0.3 inches (4-8mm) with an average application of 0.24 inches (6 mm). Unfortunately, due to extreme weather at harvest, we were unable to obtain yield data. We plan on repeating the study in subsequent years.

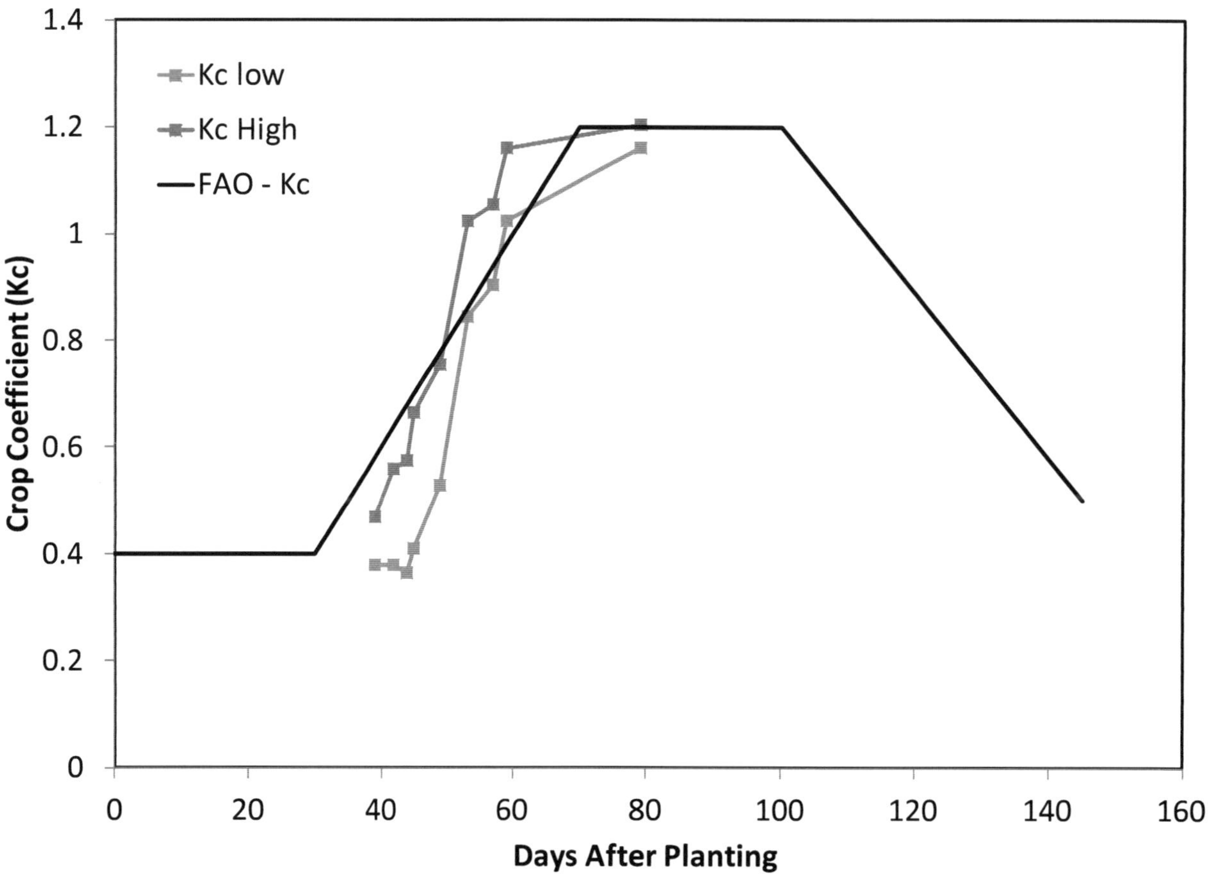


Figure 3. NDVI based crop coefficients for cotton.

## Conclusions

Two experiments were carried out to evaluate the potential of using normalized difference vegetative index (NDVI) to estimate crop coefficients for development of spatial irrigation prescriptions. Field studies were conducted near Florence, SC under variable-rate irrigation systems. Studies in corn were conducted over several years comparing NDVI-based irrigation management to management with soil water sensors did not significantly differ for yield or water volume applied. However, there were annual differences indicating delayed crop water demand in two of three years due to delayed growth. A cotton irrigation study NDVI-based irrigation prescriptions had differing irrigation amounts during the season as compared to rates prescribed by the checkbook method up until about 70 days after planting when NDVI differences among the treatments no longer existed. The results of these two studies along with the emerging use of air- and land-based remote sensing techniques, suggest continued research into the use of NDVI-based irrigation prescription technology is warranted.

## References


Allen, R., Pereira, L., Rais, D., & Smith, M. (1998). Crop evapotranspiration: Guidelines for computing crop water requirements. FAO Irrigation and Drainage Paper 56. Rome, Italy: United Nations FAO.

Bausch, W. C. (1993). Soil background effects on reflectance-based crop coefficients for corn. *Remote Sens. Environ., 46*(2), 213-222.

Berger, B., B. Parent and M. Tester. 2010. High-Throughput Shoot Imaging to Study Drought Responses. Journal of Experimental Botany 61:3519-3528.

Evans, R.G. and B.A. King. 2012. Site-Specific Sprinkler Irrigation in a Water-Limited Future. Transactions of the ASABE 55:493-504.

Evans, R.G., J. LaRue, K.C. Stone and B.A. King. 2013. Adoption of Site-Specific Variable Rate Sprinkler Irrigation Systems. Irrigation Science 31:871-887.

Evans, R.G., W.M. Iversen, W.B. Stevens and J.D. Jabro. 2010. Development of Combined Site-Specific Mesa and Lepa Methods on a Linear Move Sprinkler Irrigation System. Applied Engineering in Agriculture 26:883-895.

Glenn, E. P., Neale, C. M., Hunsaker, D. J., & Nagler, P. L. (2011). Vegetation index-based crop coefficients to estimate evapotranspiration by remote sensing in agricultural and natural ecosystems. *Hydrol. Proc., 25*(26), 4050-4062.

González-Piqueras, J., A. Calera, M.A. Gilabert, A. Cuesta and F. De La Cruz Tercero. 2004. Estimation of Crop Coefficients by Means of Optimized Vegetation Indices for Corn. Proceedings of SPIE - The International Society for Optical Engineering.

Hunsaker, D. J., Pinter, J. P., Barnes, E. M., & Kimball, B. A. (2003). Estimating cotton evapotranspiration crop coefficients with a multispectral vegetation index. *Irrig. Sci., 22*(2), 95-104.

Hunsaker, D.J., E.M. Barnes, T.R. Clarke, G.J. Fitzgerald and P.J. Pinter Jr. 2005a. Cotton Irrigation Scheduling Using Remotely Sensed and FAO-56 Basal Crop Coefficients. Transactions of the ASAE 48:1395-1407.

Hunsaker, D.J., P.J. Pinter and B.A. Kimball. 2005b. Wheat Basal Crop Coefficients Determined by Normalized Difference Vegetation Index. Irrigation Science 24:1-14.

O'Shaughnessy, S.A., S.R. Evett and P.D. Colaizzi. 2015. Dynamic Prescription Maps for Site-Specific Variable Rate Irrigation of Cotton. Agricultural Water Management 159:123-138.

Sadler, E.J., R.G. Evans, K.C. Stone and C.R. Camp. 2005. Opportunities for Conservation with Precision Irrigation. Journal of Soil and Water Conservation 60:371-379.

Stone, K.C., P.J. Bauer and G.C. Sigua. 2016. Irrigation Management Using an Expert System, Soil Water Potentials, and Vegetative Indices for Spatial Applications. Transactions of the ASABE 59(3):941-948.

# Precision Agriculture and Irrigation – Current U.S. perspectives

**Steven R. Evett, Acting Deputy Administrator**
Natural Resources and Sustainable Agricultural Systems, USDA Agricultural Research Service, steve.evett@ars.usda.gov

**Susan A. O'Shaughnessy, Research Agricultural Engineer**
USDA-Agricultural Research Service, P.O. Drawer 10, 2300 Experimental Station Road, Bushland, TX 79102, USA, Susan.O'Shaughnessy@ars.usda.gov

**Manuel A. Andrade, Research Agricultural Engineer**
Oak Ridge Institute for Science and Education, Sponsored by USDA-Agricultural Research Service, P.O. Drawer 10, 2300 Experimental Station Road, Bushland, TX 79102, USA, Alejandro.Andrade@ars.usda.gov

**Abstract.** *Precision Agriculture (PA) as a conceptual framework for farming operations responds to the need to manage inter-field and intra-field variability on farms, within watersheds, regionally and internationally. How PA is used, the objectives involved, and the technologies that support it have changed substantially since the inception of modern PA in the 1980s when the U.S. Global Positioning System (GPS) became available for public use. Coupled with geographical information system (GIS) computer technologies that were first developed for satellite imagery, PA became a mainstream tool for farmers to plan site-specific agricultural operations, early on including fertilizer application, followed by seeding rate, seed variety, pesticide spraying and now site-specific irrigation. Equipment with GPS steering and position-aware supervisory control systems allowed pre-determined site-specific prescription maps to be downloaded into equipment and used, for example, to turn off a spraying system as it passed over a waterway. GPS-enabled harvesting equipment produced yield maps that were some of the first data to be used for site-specific management, often with confusing results due to a lack of co-varying field data and adequate decision support systems (DSS) based on how soil spatiotemporal properties influence plant development. This kind of passive and indirect PA has evolved, however, to provide more capable solutions that, for example, provide for variable rate application of fertilizers based on georeferenced soil sampling that leads to prescription maps of fertilizer need. Or for another example, spatially variable irrigation management based on 30-m resolution maps of crop water use based on multi-satellite sensor fusion. Many of the more successful PA technologies involve on-board sensor systems that feed data to embedded computing platforms that make on-the-fly adjustments to equipment. Such active and direct PA systems use modern technology that provides the ability, for instance, to turn spray equipment on in the presence of weeds and off otherwise, or to turn on variable rate irrigation nozzles where abiotic stress sensors indicate crop water stress. Such supervisory control and data acquisition (SCADA) systems rely on algorithms based on sophisticated understanding of biophysics and biological systems. Today the confluence of computing power, data acquisition and management infrastructure, new modeling paradigms, and spatial decision support systems ushers in new possibilities for PA. Providers of PA services now include government institutions from national to local levels, private providers (often using publically available data from government ground, aerial and satellite sensing systems), university extension systems and farmer cooperatives. Sources of data range from public domain to private data held by farmers or third parties. Questions around data standards, data sharing, data ownership, and public and private rights add further complexity to modern PA, but are actively being addressed by both public and private institutions.*

**Keywords.** Variable rate irrigation, evapotranspiration, decision support system, SCADA, soil water content, remote sensing, precision agriculture

## Introduction

The basic premise of precision agriculture has been around since the first farmer decided to plant here, not there, to graze this area not that area, to irrigate that field not this one; and later on grew in complexity as farmers selected land races for specific environments. For example, farmers in West Africa have a wide variety of land races of sorghum and millet, some of which thrive in the wet lowlands while failing in the dry uplands, and vice versa. Similarly, farmers in the rice lands of Mali in the inland delta of the Niger River have a variety of rice landraces, some adapted to deeper flooding and planting in lower elevations and some adapted to less or intermittent flooding and planting at the upper edges of planting areas. Judicious selection and planting of these varieties helps farmers there grow rice successfully without terraforming to create level rice paddies, and allows considerable rice production despite inter-annual variations in flooding depth.

Site-specific water management likewise found its genesis in the selection of areas for drainage to ameliorate waterlogged soils and the sizing of fields and basins for irrigation according to the perceived infiltration rates in specific parts of the landscape. Because they create structures that persist over long periods and because the land areas affected are relatively large, these irrigation and drainage design practices are not recognized as precision agriculture, even though they are site specific and often based on precise topographic and geophysical data. Mapping of irrigation systems dates back at least to ancient Babylonia, almost 4,000 years (Fig. 1). With the advent of GIS, GPS and modern sensing and irrigation application systems, attitudes about the role of irrigation systems in PA are now changing.

Modern PA began in the 1980s when the GPS became available for public use. Coupled with GIS computer technologies, PA became a mainstream tool for farmers to plan site-specific agricultural operations. Equipment with GPS steering and position-aware supervisory control systems allowed application prescription maps to be downloaded into equipment, for example to turn off a moving irrigation system as it passed over a rock outcrop. GPS-enabled harvesting equipment produced yield maps that were used for site-specific management, often with confusing results due to a lack of co-varying data on soil and landscape properties and lack of adequate decision support systems (DSS) based on how soil spatiotemporal properties and landscape influence plant development.

Many of the more successful PA technologies involve on-board sensor systems allowing on the fly adjustments to equipment, for example to turn spray equipment on in the presence of weeds and off otherwise, or to turn on variable rate irrigation nozzles where abiotic stress sensors indicate crop water stress. These supervisory control and data acquisition (SCADA) based systems are multiplying rapidly and include systems that automatically thin fruit tree blossoms according to bloom density as a system moves through an orchard. Key to these PA systems are wireless data transmission, wireless sensor networks and the internet-of-things (IOT) in which every sensor is a georeferenced node in a larger network, and in which subnetworks are integrated into the internet. Although many successful SCADA systems rely on wireless sensor networks and georeferencing, many are not IOT enabled, although the potential exists. As systems are connected to the internet, issues of data ownership, already extant, become even more prevalent.

Today the confluence of computing power, data acquisition and management infrastructure, new modeling paradigms, and spatial decision support systems ushers in new possibilities for PA. For example, satellite data, initially not deemed useful for PA due to poor temporal and spatial resolution, are now used in computational systems that fuse data from satellites with different spatial and temporal resolutions and with different spectral imagers to provide daily evapotranspiration maps with 30-m resolution (Anderson et al., 2017). Providers of PA services now include government institutions, private providers (often using publically available data from state and federal government on-the-ground, aerial and satellite sensing systems), university extension systems and farmer cooperatives. Sources of data range from public domain to private data held by farmers or third parties. Questions around data sharing, data ownership and public and private rights add further complexity to modern PA. The IT sphere now has such importance to PA that some see PA as, "a suite of IT based tools which allow farmers to electronically monitor soil and crop conditions and analyze treatment options" (Aubert et al., 2012).

MS 3196

**Figure 1. Map on clay tablet of canals and irrigation systems west of Euphrates. Named are Euphrates and three canals. Lengths, widths and depths of the canals are given. Source: The Schøyen Collection, MS 3196, http://www.schoyencollection.com/24-smaller-collections/maps/map-irrigation-ms-3196. (visited on 4 Sept 2017).**

## Examples of PA

There are essentially two paradigms for PA: (1) A passive/indirect method in which data are collected/assembled to produce maps of various state variables, which are then used to guide PA; and (2) An active and typically direct method in which sensor subsystems are parts of SCADA systems that process the data using algorithms to guide control of machinery for input and practice applications. These SCADA systems typically embody a DSS, often one that automatically generates a spatiotemporal prescription for action, which can likewise, but not necessarily, be automatically applied. Examples of the first paradigm include the numerous private and public organizations, including large agribusinesses such as Monsanto/Bayer, Cargill and John Deere, as well as a plethora of smaller businesses, that are involved in collecting high resolution spatiotemporal data from farms, evaluating the data, and providing value-added services that promise to increase yield, optimize input use and increase profitability and sustainability through spatially- and temporally-varying application of agricultural inputs and practices. Examples of the second paradigm include sensor feedback systems, such as herbicide sprayers, fertilizer application systems, and plant and soil feedback based irrigation systems, which automatically acquire sensor data, analyze the data to determine actions, and direct machinery to carry out the actions.

**Prescription Fertilization.** Site-specific fertilizer application was the earliest widely adopted example of PA practices in the US, and typically still follows the passive/indirect paradigm. Presently, the 4R concept (Right source, Right rate, Right time, Right place) is used to both promote and explain the importance of precision fertilizer management for increased nutrient use efficiency and decreased environmental impact (Sposari and Flis, 2017). In 2016 the USDA Economic Research Service (ERS) reported that nearly half of U.S. corn and soybean growers used GPS yield monitoring, greater than 20% used yield maps, and 16-19% used GPS soil fertility mapping (Schimmelpfennig, 2016). Of these, 20% used variable rate fertilization; but this practice was applied on 26% of corn and 34% of soybean acres, which indicates that adoption was greater on larger farms. Since 2011, yearly surveys of agricultural retail service providers by Purdue University showed increasing adoption of GPS soil mapping, yield monitoring and soil bulk electrical conductivity (EC) mapping (Erickson and Lowenberg-Deboer, 2017). Soil sampling with GPS mapping is more highly adopted than other practices and is closely tied to adoption of variable rate fertilizer application (Griffin et al., 2016); both farmers and dealers report positive returns on investments in PA fertilizer practices and equipment (Erickson and Lowenberg-Deboer, 2017). PA fertilizer practices are most commonly applied to corn, soybean and wheat in the US (Snyder, 2016).

Despite much research on the use of optical sensors of canopy reflectance for guiding fertilizer applications, this is still considered an advanced and emerging technology that is most often used later in the growing season to guide supplemental fertilizer applications (Snyder, 2016). While N-sensors may improve profitability by preventing over- and under-fertilization, the literature reports mixed results (Ondoua and Walsh, 2017). Like other methods, PA nitrogen fertilization guided by sensors fails when something other than N (most commonly water) is limiting. As with other PA technologies, the availability of precision application equipment outstrips the availability of DSS and the multiple sources of data required to make DSS reliable and the outcomes of following DSS-based application prescriptions successful.

**Prescription Irrigation.** A recent example in site-specific variable-rate irrigation (VRI) is the Irrigation Scheduling Supervisory Control and Data Acquisition (ISSCADA) system of Evett et al. (2014) (Fig. 2). This is an example of the active/direct PA paradigm. Motivated by the rapid increase in pressurized irrigation systems amenable to control in the US, and designed to work with linear move and center pivot irrigation systems that cover 65% of U.S. irrigated lands, this system uses plant sensors mounted on the irrigation system lateral pipe to scan plant water stress in the field and produce maps prescribing variable rate irrigation according to stress level (O'Shaughnessy et al., 2015).

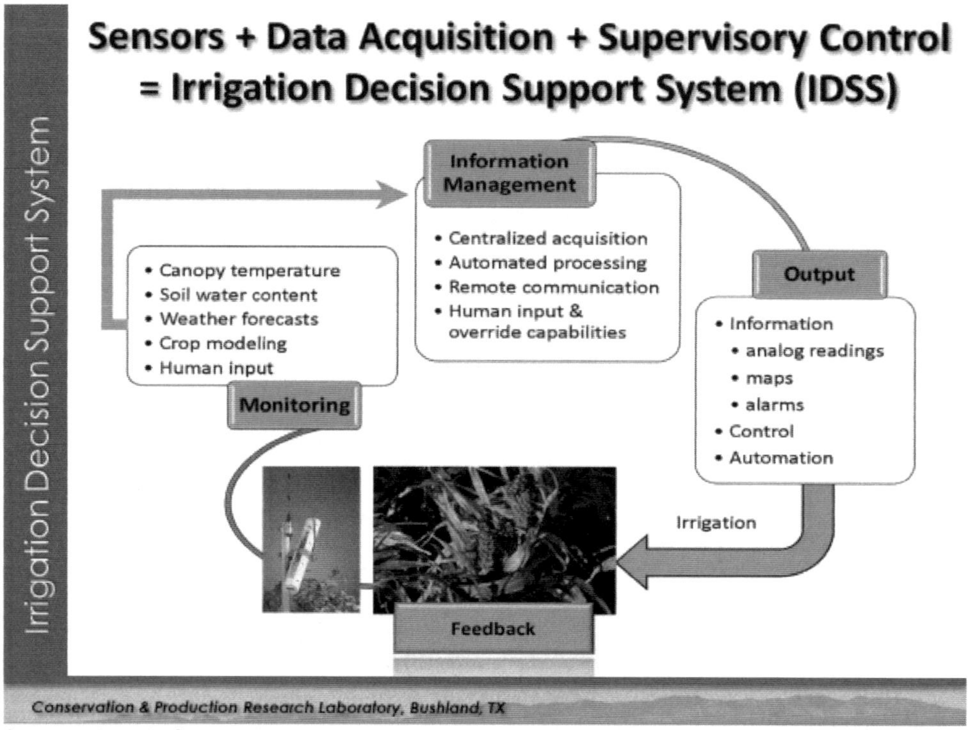


**Figure 2. The sensing, information management, prescription mapping, irrigation control, and plant feedback loop for an Irrigation Scheduling Supervisory Control and Data Acquisition System that directs a variable rate center pivot irrigation system to apply water when, where and in the quantity needed.**

A subset of sensors is fixed in the field for reference stress sensing and soil water sensors are buried to provide feedback on irrigation effects in the soil. Data from all crop and soil sensors and from weather sensors is automatically collected wirelessly by an embedded computer at the irrigation system pivot point. Novel algorithms allow conversion of plant stress measurements taken at one time of day in a specific location in the field to a diurnal curve of plant stress, which is then converted to an integrated crop water stress index for the day. This process is repeated for each control zone, producing a map of crop water stress (Fig. 3, Left). Control zones may be as small as 2 degrees of arc with radial increments defined by adjacent pairs of crop sensors pointing at the control zone from opposite sides (to control for sun angle and sensor zenith angle effects). The crop water stress map is converted into a prescription map defining irrigation amounts for each control zone (Fig. 3, Right), which may automatically guide the irrigation system, or be modified by the irrigation manager before automatic application. The infrared thermometer sensors and sensor network were commercialized from research prototypes (O'Shaughnessy et al., 2013), and the soil water sensors were also developed with a commercial partner (Evett et al., 2015; Schwartz et al., 2016). This amounts to a 3R system for irrigation: Right place, Right amount, and Right time; and it results in improved crop water productivity for several field crops in the U.S. Great Plains (cotton, maize, sorghum, soybean) (O'Shaughnessy et al., 2016).

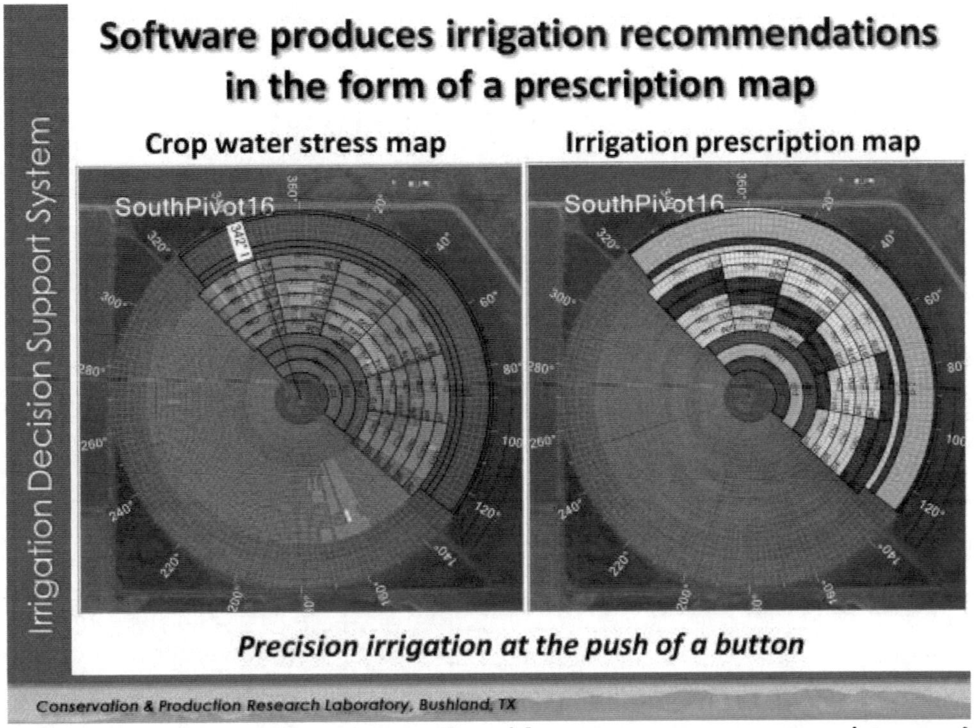


**Figure 3. (Left) a crop water stress map produced from canopy temperature data acquired by a wireless infrared thermometer system deployed on a center pivot irrigation system lateral. The lower left half of the field was fallowed. (Right) An irrigation prescription map automatically produced by the ISSCADA DSS system from those data.**

The ISSCADA system puts the PA sensors, IT system and application equipment in the hands of the producer. Although it can make use of secondary data such as SSURGO (NRCS, 2017) soil mapping units or soil EC maps to fine tune prescriptions, it doesn't require them.

A contrasting PA irrigation system is the newly developed "pixelated" irrigation management system used in vineyards in California (Semmens et al., 2015; Xia et al., 2016). This system is of the passive/indirect type. Data from multiple satellite remote sensing platforms is fused to produce daily, 30-m pixels of surface temperature and reflectance, which are combined with local microclimate data to produce evapotranspiration (ET—crop water use) data (Anderson et al., 2012; Cammalleri et al., 2013, 2014). In trials in California, daily 30-m data of vineyard ET were used to manage vineyard irrigation systems, reducing spatial variation of crop water status and yield, and improving crop quality. Expected operational products of this USDA-ARS-NASA collaboration, called GRAPEX, include datacubes of daily ET at 30-m resolution for selected growing areas (Fig. 4). E.J. Gallo Co. has developed a toolkit for using ET datacubes to determine the start of the irrigation season and weekly irrigation recommendations (Fig. 5). As a result of this work, ARS developed an ET toolkit that has been used in South Dakota, Maryland, Nebraska, North Carolina, and elsewhere to help solve site specific water management problems (Sun et al., 2017; Yang et al., 2017a,b,c)

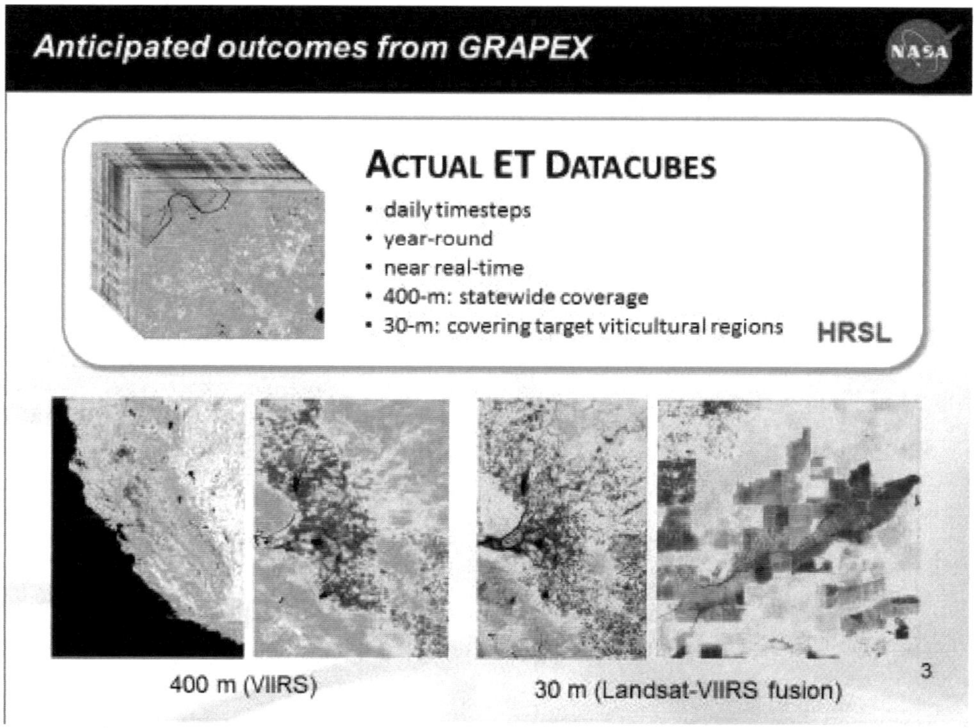


Figure 4. From left to right: California state-wide daily ET image at 400-m resolution, detail of the Napa Valley, same detail of the NAPA Valley but at 30-m resolution, and a closer look at a few vineyards and other fields at 30-m resolution.

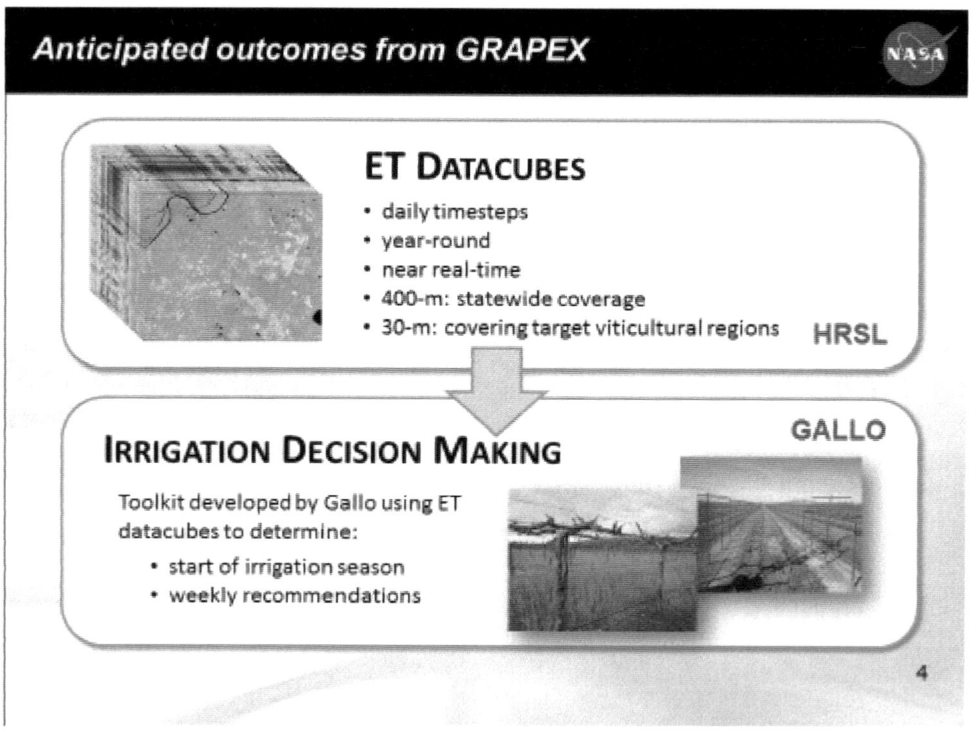


Figure 5. ET datacubes are made available at daily time steps in near real time and with 30-m resolution in targeted growing areas. E.J. Gallo has developed a toolkit for using ET datacubes to determine the start of the irrigation season and weekly irrigation recommendations.

**PA for Other Specialty Crops**. Specialty crop production, including grapes, accounts for greater than 38% of U.S. crop value production (USDA-NASS, 2015), and tree fruit production accounts for 39% of that production (USDA-NASS, 2014). But specialty crop production is labor intensive and thus costly, while labor availability varies inter-annually. Fruit thinning is one of the most costly steps in production, but ensures profitable production through optimal fruit size and quality. The same is true for crops such as lettuce, which is why precision planting and thinning systems are also being increasingly developed and put into practice in the US (Shearer and Pitla, 2014), and new ones are being developed (Lyons et al., 2015).

## A new paradigm for PA – G × E × M

The consideration of plant genetics, environmental factors, and management practices in research on sustainable farming systems has led to the paradigm of G × E × M research (Hatfield and Walthall, 2015), which harkens back to the very beginnings of farmer recognition of environmentally-adapted land races and the management practice of spatial sensitivity in planting them. The G × E × M paradigm also looks forward to new technologies of rapid plant breeding; big data sets that include spatial and temporal landscape, soil, plant and weather data; and modern computing power applied to data analysis, agroecosystems modelling, and development and application of algorithms for decision making in the near term and even real time.

USDA-ARS is using the G × E × M paradigm and is extending it with a post-harvest component that relates strongly to the first three through yield quality, value and the production system's relationship to socioeconomics: G × E × M × S. The paradigm is a key part of USDA-ARS's Long Term Agroecosystem Research (LTAR) network research plan and is taking hold in other ARS national programs due to its power to guide research objectives to outcomes that are productive for stakeholders because it takes into account the entire farming operation and physical and hydrologic landscapes. These outcomes are naturally realized through PA because of its power to manage inputs in relation to the crop and environment in space and time to produce the crop yield and quality desired.

Because G × E × M × S research pulls in large amounts of interrelated environmental, genetic, management and yield and quality data, it produces large data sets that are well suited for use in developing and testing simulation models and sub-models of several kinds – crop growth and yield, canopy and cover development, canopy reflectance-emittance-temperature, soil water balance and plant water uptake, energy and water balance, and so forth. While crop simulation models are notoriously unreliable for real-time, site-specific prediction (e.g., Webber et al., 2017), the use of near real-time data assimilation techniques can render them sufficiently accurate for management purposes, with the added advantage of being able to predict at least short term future outcomes of applied management practices. When used with data assimilation, simulation models become the basis for PA decision support systems.

The accuracy and usefulness of simulation models is also improved through use of large data sets in multi-model comparison studies that explore the reasons for model inaccuracies and lead to model improvements (Liu et al., 2016; Webber et al., 2017). The Agricultural Model Intercomparison and Improvement Project (AgMIP, http://www.agmip.org/) is demonstrating how G × E × M datasets can lead to better understanding of model deficiencies and to model improvements (e.g., Maiorano et al., 2017; Pauli et al., 2017; Wang et al., 2017). This international research effort is strongly supported by UKaid and USDA, plus a variety of in-kind contributions by universities and other organizations internationally. Although not directly intended to support PA, the potential for AgMIP to improve PA DSS is clear, particularly when improved models are combined with wireless sensor networks.

## Role of IT, including "big data"

**Data standards for Information Technology (IT).** A primary problem with data that could be used in PA DSS is that data follow either one of many disparate data standards that exist today, or no accepted standard at all. For example, a datum as simple as soil water content has no meaning for crop management unless one knows at what depth the reading was taken, the support volume for the reading, where the reading was taken, and what the error limits of the datum are. Similarly for crop canopy temperature data, which may be used to guide irrigation, a datum has no utility for that purpose unless one knows whether the view was oblique (and at what angle) or nadir, the zenith angle it was taken at, the time of day and day of year (sun angle effects), the area covered, the crop growth stage (for estimation of soil background interference), and the error limits. Beyond these data characteristics, users need to know units of reported measurements; the metadata should include details of what was actually measured as well as what was reported. For example, many soil water sensors report volumetric water content, but none measure that; they measure either in the frequency or time domains; and knowing which can tell the user a lot about data reliability. We are living in a data Babel – shades of ancient Babylonia.

Application of data standards and data management plans are not keeping up with the Internet cloud and the IOT that encompasses rapidly burgeoning wired and wireless sensor networks. Traditional sources of environmental data—national weather networks (daily, subdaily); state and regional weather networks (daily and often subdaily); and hydrologic networks--are being surpassed by new soil moisture networks; ecosystem and agroecosystem research networks; satellite platforms (biweekly to daily and subdaily); data fusion systems applied to satellite data; genomics and plant breeding programs; ad hoc and commercially proprietary sensing and data manipulation networks; etc.

Nonetheless, bright spots are emerging. The USDA-ARS LTAR network has adopted data standards, including metadata standards, similar to those of EPA and USGS. All U.S. federal government data standards are transitioning to the ISO suite of standards; ISO 19115 and its accompanying standards will replace prior standards as the official metadata standard for U.S. federal agencies. Data management planning is now required for all USDA-ARS research projects, and for those funded by USDA National Institute of Food and Agriculture (NIFA), as well as most other federal agencies. In the commercial sector, AgGateway (http://www.aggateway.org/), a non-profit with greater than 230 member companies, is a leader in PA data standards, including the SPADE (Standardized Precision Ag Data Exchange) project creating standards for data exchange between farm management systems and field equipment. Its Precision Ag Irrigation Language (PAIL) project sets standards for field data used to develop irrigation management plans, operate irrigation equipment according to plans, and record the results (Ferreyra et al., 2017). AgGateway works with standards groups, including GS1, ASABE, AEF, OAGi and USDA. The AgGateway Global Network is a non-profit recently formed to expand the successful AgGateway collaborative framework outside the US.

Because of the rich, rapidly expanding and changing commercial sector of data providers, interpreters and users, data standards are key for interoperability of the sensors, IT systems, DSS and SCADA systems that provide value in the agricultural market through mechanized PA. Many potential partners inhabit this space: USDA, universities and cooperative extension, NOAA, NASA, USGS, DOE, NEON, FLUX-NET and many others. Turning potential to actual partners is the business of an ad-hoc consortium of many players, including commodity groups, Farmers Business Network, Field to Market, Ag Data Coalition, Ag Gateway, Open Ag Data Alliance, and private data integrators. Because a substantial part of these data are collected using systems that farmers own, or are observations made using government resources that relate directly to privately owned land, issues of data ownership and privacy arise. Some U.S. farm groups have suggested that USDA become the repository for such data, with appropriate privacy safeguards in place.

# Future directions in research and technology transfer

## Research

Almost every aspect of agricultural research has some application in precision agriculture. Geostatistical investigations of soil and plant attributes have long been established, but inclusion of temporal variations involves the application of ever more sophisticated models of plant growth and yield in response to the environment, management and dynamics of water and nutrients. As noted previously, supercomputing holds promise for not only the more deterministic simulation modeling approaches, but also for investigation of overall system behavior and identification of key variables through hypercube data analyses by means of network analysis methods.

AgMIP is one example of ongoing research needs in agricultural modeling (Rosenzweig et al., 2013); and it has yielded new insights into the need for not only better simulation models but also better data to support the development and testing of those models. For example, of 46 models tested, none was consistent in accurately simulating crop ET using high quality data sets. Since ET is a key covariate with yield, and one that is sensitive to climate forcing, it is of great interest to get this right. And because data are increasingly available at appropriate spatial and temporal resolutions for in-field management, the potential application to PA is clear.

There is much yet to be done in the development of sensors that can help identify plant biotic and abiotic stresses more accurately and quickly, at low cost and with low power consumption. The assembling of these sensors into wireless networks that are themselves low cost and low power yet reliable over long distances is a continuing challenge, but greatly aided by technology coming out of the smart phone industry. New wireless data transmission protocols and commercial systems are announced almost weekly. Increasingly, agricultural research requires true interdisciplinary teams that include crop physiologists, soil scientists, computational scientists, proximal and remote sensing scientists, agricultural and biological engineers, and electrical engineers. Such teams will be needed to develop the next generation of more capable GPS-guided SCADA systems for PA.

Unmanned aircraft (UA), also known as unmanned aerial vehicles (UAV), are increasingly used for crop, pest and irrigation systems management in the US. There were more than 1.1 million UAs in the US in 2016, and the FAA estimates that number will at least triple by 2021. Commercial UAs numbered approximately 42,000 in 2016 and are expected to number at least 442,000 by 2021—and may number as much as 1.6 million. Rules and waivers are in place to allow UA use in agriculture. Low cost UAs are the result of a confluence of miniaturized electromechanical technologies similar to those that allow low-cost wireless sensor networks: microelectromechanical systems (MEMs) sensors (gyros, accelerometers, etc.), GPS modules, low power-long range (LoRa) radios, and multi-band cameras. Thanks to a competitive smart phone market, these technologies have become very inexpensive and small, yet powerful. Research progress is rapid in both university and private venues, aided by an open-source community sharing computer code such as DIY Drones (http://diydrones.com/), and by 3-D printers for rapid prototyping and production, also with an open-source user community. Code for image stitching, orthogonal correction and image processing is readily available. While imaging fields is increasingly easy and inexpensive, even on a daily basis, there are continuing impediments to progress in delivering useful PA DSS. These include sensor calibration, image correction, image analysis and reliable decision support generation software. However, in many cases images are directly useful, for example in showing problems with an irrigation system, or a pest incursion.

While most wireless sensor networks operate with sensors above ground or embedded in the soil surface, there is increasing interest in sensing networks beneath the soil surface that can characterize the state and dynamics of chemical, physical and biological aspects of the rhizosphere. An upcoming National Science Foundation-sponsored workshop on the "Subterranean MacroScope" will focus on the many problems involved in developing the needed sensors and communications networks (https://ime.uchicago.edu/subterranean_macroscope/). Disciplines involved include microbiology, genomics, biochemistry, plant and microbial physiology, physical chemistry, biophysics, soil physics, MEMS, microfluidics and electrical engineering.

## Technology transfer

The commercial sector is increasingly involved in PA technical transfer because most PA technologies are too complex for on-farm development. Manufacturers are involved in every phase to produce and market sensors and sensor network systems, build SCADA systems into agricultural equipment and produce and sell the equipment, often through dealers who to varying degrees take on the role of system support. This is also true in the development and marketing of new crop varieties that fit environmental and management scenarios and may be useful in PA planting systems. Commercial entities are increasingly active in the provision of actionable PA data and prescriptions to farmers. Companies such as Climate Corporation assemble data from multiple public sources (Landsat and other satellites, NOAA weather data, NRCS SURGGO soils data, etc.) and use large computing systems and statistical analysis to deliver recommendations. Many companies are providing aerial imagery at high resolution and in multiple visible and infrared light bands to guide PA farming, although data interpretation and decision support still are a work in progress.

NGOs are increasingly providing assistance for PA. Trade Industry groups such as the Irrigation Association provide certified training at their annual meetings and via webinar. Scientific societies are involved through their meetings and outreach to the commercial sector. For example, the American Society of Agronomy is currently developing a Precision Agriculture specialization within its Certified Crop Advisor (CCA) training program that reaches several thousand crop consultants in North America.

Cooperative extension also plays a role. In the US, cooperative extension was the predominant paradigm for transferring technology to farmers in the 1900s. Today, extension is hampered by budget cuts and to some degree cut out of the picture due to the expanding role of commerce. There are still valid roles for extension however, in conducting public trials of new PA DSS technologies, running publicly accessible demonstration farms, developing cost-benefit analyses of technology adoption, and publishing guides on new technologies.

### Keys to Future Success

A 2016 Roundtable hosted by USDA-ARS's Office of International Research Programs and the International Society of Precision Agriculture identified 10 keys to ensuring successful DSS for PA (Yost et al., Submitted):

- Increase research documentation of PA outcomes
- Enhance funding for PA research
- Facilitate public-private partnerships
- Develop more IP-neutral relationships between public and private research
- Improve involvement of NGOs and others in PA research and application efforts
- Generate more PA projects that encompass the four goals of sustainable agriculture
- Achieve better balance between basic and applied research, short and long term funding, and small versus large grants
- Include more stakeholder involvement and retrospective assessments
- Enhance research relevance to smallholder farms, especially internationally
- Continue regular roundtable discussions

While these higher level concerns are certainly important considerations, it will be the constant process of developing and testing sensors and sensing networks, IT systems and software, and control systems and application hardware, coupled with a robust and open user community, that develops useful PA DSS and application technologies.

## Conclusion

The fundamentals of precision agriculture were employed thousands of years ago in manual fashion, but it is only since circa 1980 that GPS technology has allowed easy and efficient mapping of soil, landscape and crop properties that can be used to guide precision application of practices and inputs. Originally conceived as a passive/indirect process of first measuring and mapping the state variables of interest, then using the map to make decisions about PA practices, PA today involves practices using a mix of the older paradigm and a newer one of active and direct response to variations in crop and environmental properties detected using sensor networks, often wireless, and often on the go. The definition of what PA is has been greatly widened by the availability of inexpensive, wireless and often mobile sensors, coupled with modern IT, sophisticated algorithms for data processing and decision support, and computer control systems guiding machinery to apply practices and inputs. Renewed focus on sensors of soil chemical, physical, biological and microbiological properties in the rhizosphere, and ways to wirelessly transmit data out of the soil, promise to engender the next generation of sensing systems for guiding PA. The great increase in data from private and public sources is opening up new avenues for both PA research and application. Simulation model improvements are proceeding and future models promise to be competent enough to be the internal engines of PA decision support systems, particularly if they are made self-correcting through assimilation of data from the plethora of internet-of-things sensors. Because most PA technology will be manufactured and made available to farmers through retailers, technology transfer is steadily moving from the public to the private sector, but there remains a place for public sector technology transfer, both from research to commercial production, and by extension services field testing, demonstrating and analyzing the economics of the new technologies.

## Acknowledgements

This material is based upon work that is supported by the National Institute of Food and Agriculture, U.S. Department of Agriculture, under award number 2016-67021-24420. Work reported in this paper was also accomplished as part of a Cooperative Research and Development agreement between USDA-ARS and Valmont Industries, Inc., Valley, NE (Agreement No.: 58-3K95-0-1455-M).



## References

Alston, J. M. and P.G. Pardey. 2008. Public funding for research into specialty crops. HortSci. 43(5):1461-1470.

Anderson, M.C., W.P. Kustas, J.G. Alfieri, C.R. Hain, J.H. Prueger, S.R. Evett, P.D. Colaizzi, T.A. Howell and J.L. Chavez. 2012. Mapping daily evapotranspiration at Landsat spatial scales during the BEAREX'08 field campaign. Adv. Water Resour. 50:162-177

Aubert, B.A., A. Schroeder, and J. Grimaudo. 2012. IT as enabler of sustainable farming: An empirical analysis of farmers' adoption decision of precision agriculture technology. Decision Support Systems 54(1):510-520. https://doi.org/10.1016/j.dss.2012.07.002

Bronson, K.F., White, J.W., Conley, M.M., Hunsaker, D.J., Thorp, K.R., French, A.N., Mackey, B.E., Holland, K.H. 2017. Active optical sensors in irrigated durum wheat: Nitrogen and water effects. Agronomy Journal. 109:1060-1071.

Cammalleri, C., M.C. Anderson, F. Gao, C.R. Hain and W.P. Kustas. 2013. A data fusion approach for mapping daily evapotranspiration at field scale. Water Resources Res. 49:1-15, doi:10.1002/wrcr.20349

Cammalleri, C., M.C. Anderson, F.H. Gao, C.R. Hain and W.P. Kustas. 2014. Mapping daily evapotranspiration at field scales over rainfed and irrigated agricultural areas using remote sensing data fusion. Agric. For. Meteorol. 186:1-11

Erickson, B., and J. Lowenberg-Deboer. 2017. Making the turn toward decision agriculture: 2017 dealership survey. CropLife 180(6):8-14.

Evett, S.R., S.A. O'Shaughnessy and R.T. Peters. 2014. "Irrigation Scheduling and Supervisory Control and Data Acquisition System for Moving and Static Irrigation Systems", U.S. Patent No. 8,924,031.

Evett, S.R., S.K. Anderson, J.J. Casanova and R.C. Schwartz. 2015. "Soil Water and Conductivity Sensing System". U.S. Patent No. US 8,947,102 B1.

Ferreya, R.A., D. Berne, and C. Hillyer. 2017. PAIL supports precision irrigation. Irrigation TODAY (July 2017):24-25.

Griffin, T., N. Miller, and C. Torrez. 2016. Precision agriculture technology and obsolescence. KSU-AgEcon-TG-2016.1. Kansas State University Department of Agricultural Economics Extension Publication.

Hatfield, J.L. and C.L. Walthall. 2015. Meeting global food needs: Realizing the potential via genetics × environment × management interactions. Agron. J. 107:1215–1226. doi:10.2134/agronj15.0076.

Liu, B., Asseng, S., Muller, C., Ewert, F., Elliott, J., Lobell, D.B., Marte, P., Ruane, A.C., Wallach, D., Jones, J.W., Rosenzweig, C., Aggarwal, P.K., Alderman, P.D., Anothai, J., Basso, B., Biernath, C., Cammarano, D., Challinor, A., Deryng, D., De Sanctis, G., Doltra, J., Fereres, E., Folberth, C., Garcia-Vila, M., Gayler, S., Hoogenboom, G., Hunt, L., Izaurralde, R.C., Jabloun, M., Jones, C.D., Kersebaum, K.C., Kimball, B.A., Koehler, A., Kumar, S., Nendel, C., O'Leary, G., Olesen, J.E., Ottman, M.J., Palosuo, T., Prasad, P., Priesack, E., Pugh, T.A., Reynolds, M., Rezaei, E., Rotter, R.P., Schmid, E., Semenov, M.A., Shcherbak, I., Stehfest, E., Stockle, C.O., Stratonovitch, P., Streck, T., Supit, I., Tao, F., Thorburn, P., Waha, K., Wall, G.W., Wang, E., White, J.W., Wolf, J., Zhao, Z., Zhu, Y. 2016. Similar negative impacts of temperature on global wheat yield estimated by three independent methods. Nature Climate Change. 6:1130-1138.

Lyons, D.J., P.H. Heinemann, J.R. Schupp, T.A. Baugher and J. Liu. 2015. Development of a selective automated blossom thinning system for peaches. Trans. ASABE. 58(6):1447-1457. DOI 10.13031/trans.58.11138

Maiorano, A., Martre, P., Asseng, S., Ewart, F., Muller, C., Rotter, R.P., Ruane, A., Semenov, M.A., Wallach, D., Wang, E., Aldeman, P.D., Kassie, B.T., Biernath, C., Basso, B., Cammarano, D., Challinor, A.J., Doltra, J., Dumont, B., Gayler, S., Kersebaum, C.K., Kimball, B.A., Koehler, A., Liu, B., O'Leary, G.J., Olesen, J.E., Ottman, M., Priesack, E., Reynolds, M.P., Rezaei, E.E., Stratonovitch, P., Streck, T., Thornburn, P.J., Waha, K., Wall, G.W., White, J.W., Zhao, Z., Zhu, Y. 2017. Crop model improvement reduces the uncertainty of the response to temperature of multi-model ensembles. Field Crops Research. 202:5-20.

NRCS. 2017. Soil Survey Geographic Database (SSURGO). USDA Natural Resources Conservation Service, https://catalog.data.gov/dataset/soil-survey-geographic-ssurgo-database-for-various-soil-survey-areas-in-the-united-states-.

Ondoua, R. N., and O. Walsh. 2017. Precision agriculture advances and limitations: Lessons to the stakeholders. Crops Soils 50:40-47. doi:10.2134/cs2017.50.0408

O'Shaughnessy, S.A.,S.R. Evett, P.D. Colaizzi, and T.A. Howell. 2013. Wireless sensor network effectively controls center pivot irrigation of sorghum. Appl. Engr. Agric. 29(6):853-864.

O'Shaughnessy, S.A., S.R. Evett and P.D. Colaizzi. 2015. Dynamic prescription maps for site-specific variable rate irrigation of cotton. Agric. Water Manage. 159:123–138.

O'Shaughnessy, S.A., S.R. Evett, A. Andrade, F. Workneh, J.A. Price and C.M. Rush. 2016. Site-specific variable rate irrigation as a means to enhance water use efficiency. Trans. ASABE 59(1):239-249. DOI 10.13031/trans.59.11165.

Pauli, D., White, J.W., Andrade-Snachez, P., Conley, M.M., Huen, J., Thorp, K.R., French, A.N., Hunsaker, D.J., Carmo-Silva, E.A., Wang, G., Gore, M.A. 2017. Investigation of the influence of leaf thickness on canopy reflectance and physiological traits in upland and Pima cotton populations. Frontiers in Crops Science and Horticulture. 8:1405. doi: 10.3389/fpls.2017.01405.

Rosenzweig, C., J.W. Jones, J.L. Hatfield, A.C. Ruane, K.J. Boote, P. Thorburn, J.M. Antle, G.C. Nelson, C. Porter, S. Janssen, S. Asseng, B. Basso, F. Ewert, D. Wallach, G. Baigorria, and J.M. Winter. 2013. The Agricultural Model Intercomparison and Improvement Project (AgMIP): Protocols and pilot studies. Agric. For. Meteorol. 170:166–172. doi:10.1016/j.agrformet.2012.09.011

Schimmelpfennig, D. 2016. Farm profits and adoption of precision agriculture. Economic Research Report no. 2017. USDA Economic Research Service.

Schwartz, R.C., S.R. Evett, S. Anderson and D. Anderson. 2016. Evaluation of a direct-coupled TDR for determination of soil water content and bulk electrical conductivity. Vadose Zone J. 15(1)2016. doi: 10.2136/vzj2015.08.0115. 2016.

Semmens, K.A., M.C. Anderson, W.P. Kustas, F. Gao, J.G. Alfieri, L. McKee, J.H. Prueger, C.R. Hain, C. Cammalleri, Y. Yang, T. Xia, M. Vélez, L. Sanchez and M. Alsina. 2015. Monitoring daily evapotranspiration over two California vineyards using Landsat 8 in a multi-sensor data fusion approach. Remote Sens. Environ., doi:10.1016/j.rse.2015.1010.1025

Shearer S.A. and S.K. Pitla. 2014. Precision Planting and Crop Thinning. Pp 99-124 In: Young S., Pierce F. (eds.) Automation: The Future of Weed Control in Cropping Systems. Springer, Dordrecht. ISBN: 978-94-007-7511-4. https://doi.org/10.1007/978-94-007-7512-1_6.

Snyder, C. 2016. Suites of 4R nitrogen management practices of sustainable crop production and environmental protection. Issue Review Ref. no. 16057. International Plant Nutrition Institute.

Sposari, M., and S. Flis. 2017. 4R framework implementation: precision ag adoption by farmers and dealers. Crops Soils 50:24-26. doi:10.2134/cs2017.50.0507

Sun, L., M.C. Anderson, F. Gao, C.R. Hain, J.G. Alfieri, A. Sharifi, G. McCarty, Y. Yang and Y. Yang. 2017. Investigating water use over the Choptank River Watershed using a multi-satellite data fusion approach. *Water Resources Res., in press*

USDA-NASS. (2014). 2012 Census of Agriculture. United States Summary and State Data, Vol 1 – Geographic Area Series – Part 51, AC-12-A-51. Washington, D.C.: USDA National Agricultural Statistics Service.

USDA-NASS. (2015). 2012 Census of Agriculture. Specialty Crops, Vol 2 – Subject Series – Part 8, AC-12-S-8. Washington, D.C.: USDA National Agricultural Statistics Service.

Wang, E., P. Martre, S. Assenge, F. Ewert, Z. Zhao, A. Maiorano, R.P. Rotter, B.A. Kimball, M.J. Ottman, G.W. Wall, J.W. White, P.K. Aggarwal, P.D. Alderman, A. Jakarat, B. Basso, C. Biernath, D. Cammarano, A.J. Challinor, G. De Sanctis, J. Doltra, E. Fereres, M. Garcia-Vila, G. Sebastian, G. Hoogenboom, L.A. Hunt, R.C. Izaurralde, M. Jabloun, C.D. Jones, K.C. Kersebaum, A. Koehler, C. Muller, L. Liu, S.N. Kumar, C. Nendel, G. O'Leary, J.E. Olesen, T. Palosuo, E. Priesack, M.P. Reynolds, E.E. Rezaei, D. Ripoche, A.C. Ruane, M.A. Semenov, I. Shcherbak, C. Stockle, P. Stratonovitch, T. Streck, I. Supit, F. Tao, P. Thorburn, K. Waha, D. Wallach, J. Wolf and Y. Zhu. 2017. The uncertainty of crop yield projections is reduced by improved temperature response functions. Nature Plants. 3:17102. doi: 10.1038/nplants.2017.102.

Webber, H., Martre, P., Asseng, S., Kimball, B.A., White, J.W., Wall, G.W., Ottman, M., De Sanctis, G., Doltra, J., Grant, R., Kassie, B. 2017. Canopy temperature for simulation of heat stress in irrigated wheat in a semi-arid environment: a multi-model comparison. Field Crops Research. 202:21-35.

Xia, T., W.P. Kustas, M.C. Anderson, J.G. Alfieri, F. Gao, L. McKee, J.H. Prueger, H.M.E. Geli, C.M.U. Neale, L. Sanchez, M.M. Alsina and Z. Wang. 2016. Mapping evapotranspiration with high-resolution aircraft imagery over vineyards using one-and two-source modeling schemes. Hydrology and Earth System Sciences, 20, 1523-1545.

Yang, Y., M.C. Anderson, F. Gao, C. Hain, W.P. Kustas, T. Meyers, W. Crow, R.G. Finocchiaro, J.A. Otkin, L. Sun and Y. Yang. 2017a. Impact of tile drainage on evapotranspiration (ET) in South Dakota, USA based on high spatiotemporal resolution ET timeseries from a multi-satellite data fusion system. *J. Selected Topics in Applied Earth Obs. and Remote Sensing, 10*, 2550 - 2564

Yang, Y., M.C. Anderson, F. Gao, C.R. Hain, K.A. Semmens, W.P. Kustas, A. Normeets, R.H. Wynne, V.A. Thomas and G. Sun. 2017b. Daily Landsat-scale evapotranspiration estimation over a managed pine plantation in North Carolina, USA using multi-satellite data fusion. *Hydrol. Earth Syst. Sci., 21*, 1017-1037

Yang, Y., M.C. Anderson, F. Gao, B. Wardlow, C.R. Hain, J.A. Otkin, Y. Yang, L. Sun and W. Dulaney. 2017c. Field-scale mapping of evaporative stress indicators of crop yield: an application over Mead, NE. *Remote Sens. Environ., in revision*

Yost, M.A., K.A. Sudduth, C.L. Walthall and N.R. Kitchen. International review of collaborative public-private precision agriculture research, education, and development opportunities, Submitted to Precision Agric.

# Dashboard for Irrigation Efficiency Management (DIEM)

**James Bordovsky, Senior Research Scientist and Agricultural Engineer**
Texas A&M AgriLife Research, Lubbock/Halfway, TX

**Blake Stoker, Software Developer**
Texas Center for Applied Technology, College Station, TX

**Paul Bilnoski, Software Developer**
Texas Center for Applied Technology, College Station, TX

**Cesar Garcia, Software Developer**
Texas Center for Applied Technology, College Station, TX

**Dana Porter, Professor**
Texas A&M AgriLife Extension and Research, Lubbock, TX

**Srinivasulu Ale, Associate Professor**
Texas A&M AgriLife Research, Vernon, TX

**Keith Biggers, Director of Computing and Information Technology**
Texas Center for Applied Technology, College Station, TX

**James Wall, Executive Director**
Texas Center for Applied Technology, College Station, TX

**Abstract.** *In areas with limited irrigation capacity and irregular seasonal rainfall, methods to efficiently distribute available irrigations are challenging due to the inability to control soil profile water levels at times of critical crop water needs, the environmental effects on irrigation distributions, and changes in irrigation capacity within a growing season. The Dashboard for Irrigation Efficiency Management (DIEM) was developed to help make irrigation decisions in the short-season, semi-arid environment of the Texas High Plains where irrigation capacities from the Ogallala Aquifer are rapidly decreasing and cotton is the predominant row crop. DIEM is a web-based software solution for forecasting optimized field-specific irrigation schedules based on water availability. This is accomplished by configuring basic historic, near real-time, and future agronomic information including weather, soil and crop characteristics, and irrigation parameters into an interactive data visualization for portraying key aspects supporting irrigation decisions. The public beta-version of DIEM is available at* https://diem.tamu.edu.

**Keywords**. Irrigation, Cotton, Evapotranspiration, Texas High Plains, Irrigation Scheduling, Dashboard

## Introduction

Evapotranspiration (ET) based methods have traditionally been used to efficiently schedule irrigation applications. With ample irrigation capacity, irrigation scheduling can be rather straightforward, i.e., determine crop ET ($ET_c$) since the last irrigation, subtract effective precipitation, and apply this net amount at regular intervals or at set soil moisture thresholds. With the availability of cellular automation and regional weather networks, growers have immediate access to daily estimates of irrigation requirements for specific crops at specific locations (TAWC, 2017; Texas A&M AgriLife Extension, 2017; Kansas State Research and Extension, 2017; CSU Extension, 2017; University of Missouri Extension Service, 2017; Vellidis, et al., 2014). In areas with limited irrigation capacity and irregular seasonal rainfall, methods to efficiently distribute available irrigation applications can be more challenging due to reduced control of soil water levels related to crop water needs, environmental effects on irrigation distribution, and the changes in irrigation capacity within a growing season.

The continuing decline in well capacities within the Ogallala Region of the central US has resulted in available irrigation being truly supplemental to seasonal rainfall on much of the available cropland. For example, peak cotton production is hampered on approximately two million acres in the Texas High Plains (THP) by falling irrigation capacities, irregular rain events, and irrigation volume limits. Ongoing field experiments conducted at Texas A&M AgriLife Research Center, Halfway, have focused on irrigation productivity within specific cotton growth periods using center pivot (LEPA) irrigation (Bordovsky, et al., 2015). One of the most notable findings has been that early season irrigation applications intended to increase water available in the root zone for later seasonal use was only marginally effective in increasing final cotton yields. For example, at the 0.25 inch/d irrigation capacity, over the 4-year test period, the traditional "try filling the root zone early" treatment used 14.1 inches of irrigation per year to produce 1255 lb. lint/acre compared to the "just keep the plant going during the first period" treatment which resulted in 1246 lb. lint/acre with 11.3 inches/year, a 20% reduction in irrigation water use. Similar results were seen at lower irrigation capacities. The results were attributed to the typically harsh early season environmental conditions, contributing primarily to high evaporation losses, and not having sufficient irrigation capacity in the latter periods to support “large” plants resulting from irrigation in excess of plant needs during the vegetative period. Preseason rainfall evaporation losses from large weighing lysimeters at the nearby USDA-ARS Conservation and Production Research Laboratory in Bushland, TX support this conclusion (Marek, et al., 2015).

The Dashboard for Irrigation Efficiency Management (DIEM) was developed to capture these and other research results in a single integrated, web-based software solution for forecasting field-specific irrigation schedules that will help THP cotton growers optimize rainfall and irrigation productivity in a short-season, limited irrigation environment. This was accomplished by configuring basic historic, near real-time, and future agronomic information including weather, soil and crop characteristics, and irrigation parameters into an interactive data visualization for portraying key aspects supporting irrigation decisions. The remainder of this paper describes elements of the DIEM irrigation scheduling tool.

## DIEM Development

DIEM originated from spreadsheets using the daily soil water balance method (Allen, et al., 1998) to schedule cotton irrigation for experiments at Texas A&M AgriLife Research at Halfway. These spreadsheets evolved over time resulting from changes in deficit irrigation treatments and experiments. Accommodations were made in scheduling methods to include the way irrigation applications were limited (fraction of full $ET_c$, irrigation capacity, irrigation volume limits, or by combinations of these) and by the method of irrigation delivery (low elevation spray application, LESA; low energy precision application, LEPA; or subsurface drip irrigation, SDI). Differences in soil surface evaporation among delivery methods, particularly early in the growing season, and reductions in crop transpiration rates as root zone water depletion occurred with low capacity wells were considered. This scheduling method was used in a 4-year field experiment having 27 replicated irrigation regimes composed of combinations of irrigation capacities and growing season irrigation periods (Bordovsky, et al., 2015). Among other outcomes, this experiment resulted in relationships among season-long soil profile (root zone) water contents and cotton yields that could be used to systematically adjust irrigations so that yield or water productivity are optimized within the water constraints of a specific field.

DIEM is a visual web-based tool that provides a season-long "prescription" for irrigation applications within specific fields. DIEM uses a customizable information dashboard that leverages a highly robust technology called the Information Dashboard Framework (IDF). IDF was developed by researchers at the Texas A&M Engineering Experiment Station's Texas Center for Applied Technology (TCAT) and has been used in the development of AgConnect® for the Institute for Infectious Animal Diseases (IIAD), a Department of Homeland Security Science and Technology Center of Excellence. This technology framework which also incorporates visual analytic tools, was used to create an integrated irrigation information dashboard. The public beta-version of DIEM is available at https://diem.tamu.edu .

Within DIEM, once a field and scenario are identified, a soil water chart is updated showing effective rain, irrigation applications, target soil water content, transpiration limiting soil water content, and estimated soil water content over a user defined growing season. Each time DIEM is accessed during a current growing season, projected weather data (from historical weather records) are replaced by updated, near term data from specified weather stations and future irrigation applications reconfigured to optimize the crop yield score. Rainfall, irrigation, and soil water editors allow user override / in-season corrections. At any time, users can change field specific parameters and immediately observe their effects on soil water distribution, yield, and water productivity scores.

## The Dashboard

When DIEM is opened, the dashboard will appear and load the previously active scenario. A scenario is a collection of factors that define an irrigated field through model inputs and ultimately leads to an irrigation schedule. Scenarios are grouped by the field from which they get their soil and geographic parameters. From this screen, the user can view the current outputs of an activated scenario; edit

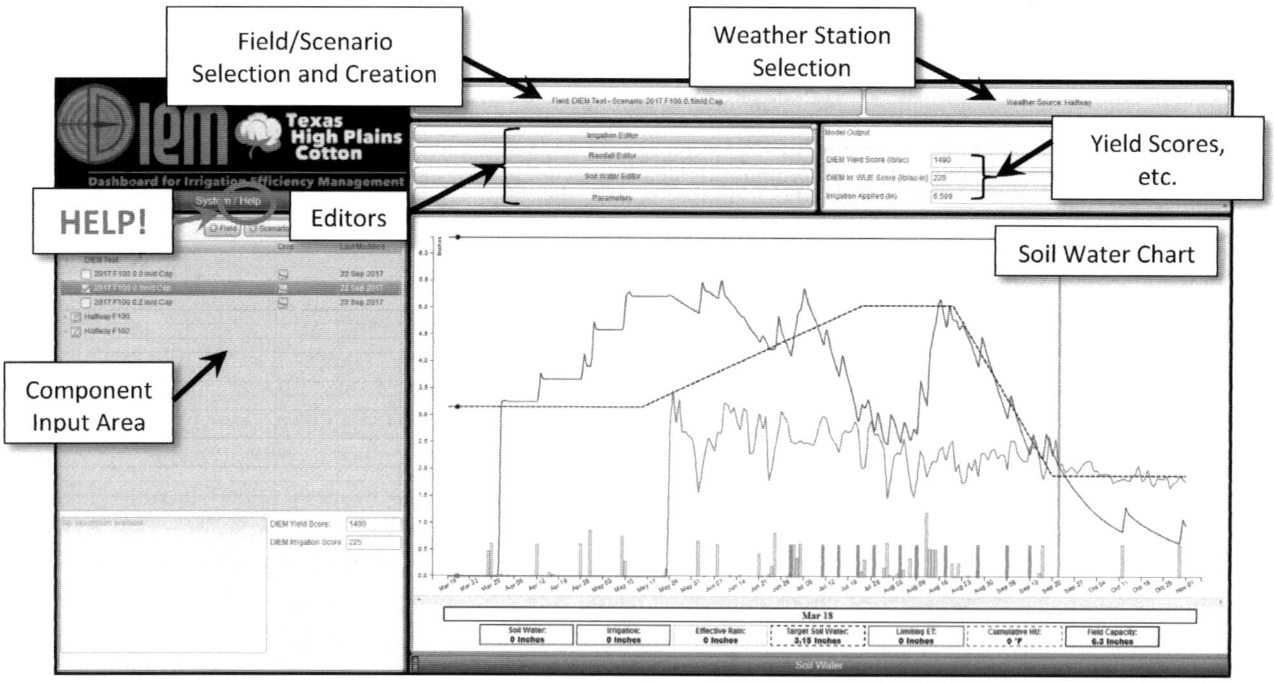


Figure 1. DIEM dashboard containing access to fields and scenarios, weather station selection, parameter editing, as well as dynamic outputs that include year-long profile water estimates and yield scores.

individual parameters of a field or the active scenario; create new scenarios for an existing field; or create new fields and associated scenarios. The soil water chart displays a season-long estimate of soil profile water status and a combination of actual and recommended irrigation applications for the active scenario. Other outputs include irrigation quantity and crop yield scores.

## Field and Scenario

At the top of the DIEM dashboard, the name of the currently active field and scenario are displayed. This panel will open the Field and Scenario Manager (Figure 2) allowing the activation of previously defined scenarios and the creation, editing, or deletion of scenarios and fields. An unlimited number of fields and scenarios can be described for single DIEM accounts; however, users may have difficulties managing large numbers (i.e., hundreds) of scenarios and dashboard response times may be reduced with increased scenario numbers. In newly created scenarios, initial dashboard outputs in the Model Output and Soil Water chart result from default parameters and weather station values that must be modified to describe the new field and scenario.

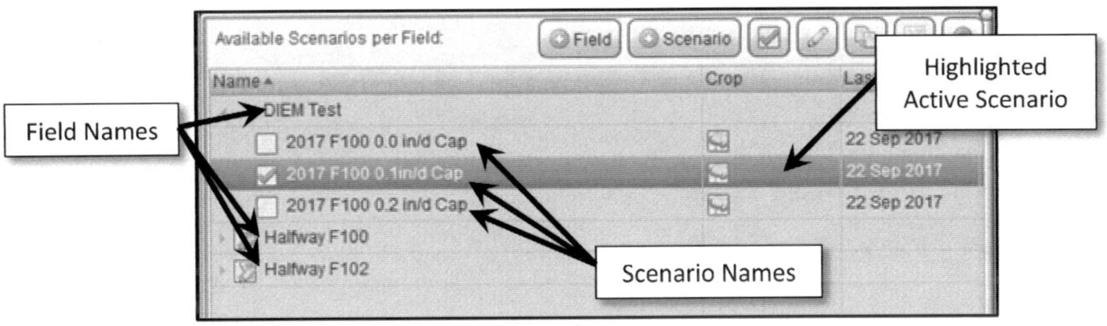


Figure 2. Field and Scenario Manager component displayed in the DIEM dashboard.

## Weather Station

The name of the primary weather station used for the current scenario is displayed at the top right of the DIEM dashboard (Figure 1). This panel opens the Weather Station Manager (Figure 3) where weather data sources are selected. Custom weather sources may also be managed by uploading data sets for use in scenarios. At least two weather data sets are required for DIEM to correctly function in an ongoing growing season. DIEM accesses the current year's weather data up to the current date from the first (upper most listed) weather data set; daily records from the second or subsequent weather data sets are employed only if information is not available from the first weather station. Therefore, for a current year, the first data set would typically be the current year data from the closest weather station. The second station could be a "backup" current year station located nearby providing daily measurements if data availability becomes an issue from the first station. The second or third (and subsequent) weather station data sets would be historic season-long data, such as a file of "average" data from the closest weather station with this data used to project the weather (and indirectly the soil water content) for the remainder of the growing season. As an alternative to "average" data, the user may choose from years characterized as being "wet" or "dry" or weather data from a specific year in the past to initialize the projected weather scenario. To analyze a previous year's water use and compare the actual crop yield to the DIEM's estimated yield score, the weather data set from the previous year at the appropriate location would be positioned as the first data set.

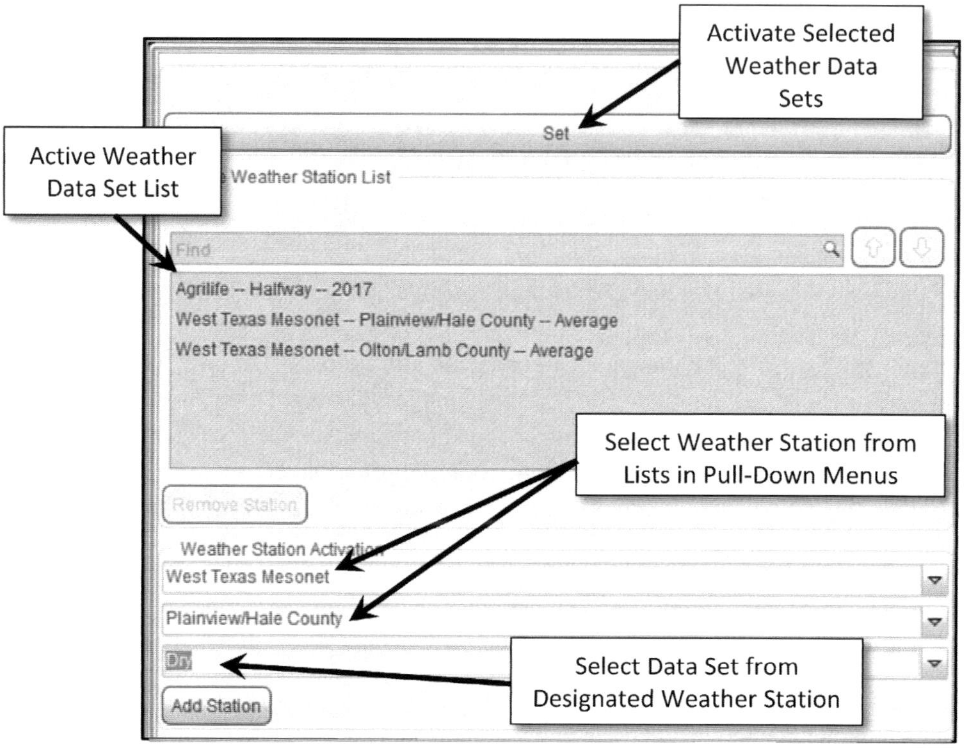


Figure 3. Weather Station Manager component displayed in the DIEM dashboard.

## Field and Crop Parameters

Specific field and basic crop information for an activated scenario can be modified by accessing the Parameters screen (Figure 4). This screen is divided into four sections entitled Model Control Dates, Soil Profile, Planting, and Irrigation Inputs. Each group of parameters initially contains default or dated values that must be replaced with relevant data. Parameter values are generally entered during initial scenario development or at the beginning of a growing season; however, certain irrigation values might be changed as the season progresses. All parameters can be changed prior to planting to investigate various "what-if" scenarios or can be modified, along with the appropriate weather data set, so that DIEM simulates previous crop years. Graphic and numerical outputs are revised following date changes or with updated numerical inputs followed by "Model Execution." Irrigation schedules are printed by activating the "Export Report" button.

Model Control Dates. To provide boundary conditions for model execution, DIEM requires *Model Start* and *End* dates. This period can cover 365 days starting on January 1, however computational speed increases and uncorrected soil water estimates will be more precise by simulating the spring and summer crop growth period (rather than the full year). The *Model Focus* value is a date, between the model start and end dates, that sets a graphic marker in the Soil Water output chart; for user convenience, its default value is the current calendar date.

Soil Profile. The *Root Zone Depth* sets the depth of the soil profile as it relates to plant water availability for the crop considered in the active scenario. *Water Holding Capacity* is the plant available water holding capacity of the soil profile with default values based on soil texture. *Initial Soil Water and Date* provide an estimate of early season soil profile water content and is used to initiate soil water balance calculations. Initial estimates might be a best guess using the soil feel and appearance method (USDA-

NRCS, 1996) or measurements using soil moisture sensors or gravimetric methods. After model initialization, corrections to DIEM soil water can be made in the *Soil Water Editor* (discussed below).

<u>Planting</u>. Currently, cotton is the only *Crop* available for irrigation scheduling within DIEM. Irrigation scheduling for grain sorghum will be added soon with other crops added as appropriate crop functions are developed from local field evaluation and other relevant documented sources. *Planting* and *Emergence Dates* are used in the plant development functions that directly impact soil water calculations.

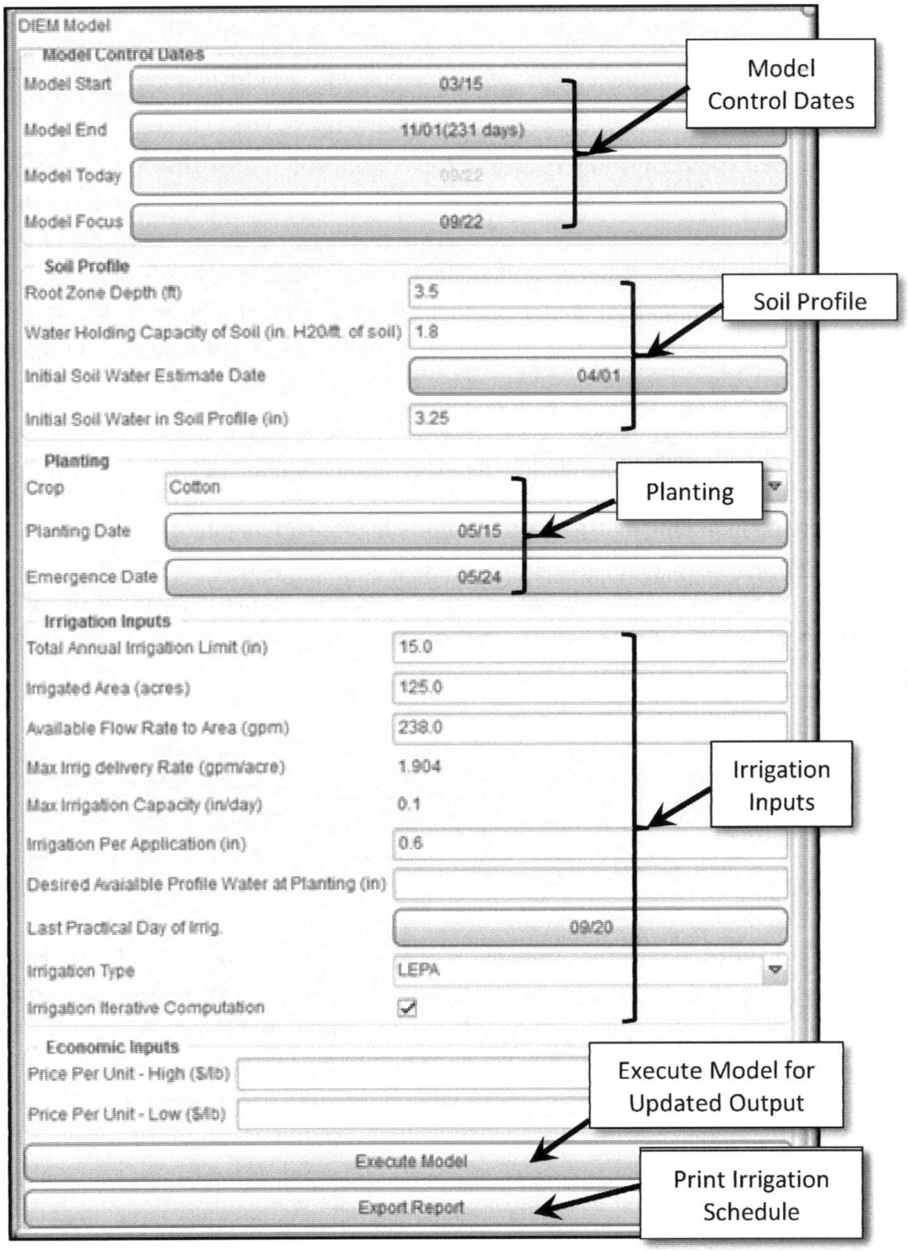


Figure 4. Field and Crop Parameter editor component displayed in the DIEM dashboard.

Irrigation Inputs. The *Total Annual Irrigation Limit* is an irrigation volume restriction that can be set due to policy or user imposed limits. *Irrigated Area* and *Available Flow Rate* determine maximum irrigation delivery rate or irrigation capacity. Following confirmation of actual irrigation applications (*Irrigation Editor*, Figure 5), the *Available Flow Rate* can be adjusted during the growing season in response to seasonal flow rate reductions such as water table declines or flow rate additions as when irrigation capacity is added from a new water well. *Irrigation Depth per Application* is based on irrigation system type, applicator characteristics, and user experience. For center pivot irrigation, this value is generally the maximum irrigation depth that does not cause irrigation runoff issues; for SDI, soil texture and irrigation system management may affect this value. This parameter can be adjusted during the growing season. The *Desired Available Profile Water at Planting* value might be entered when simulating early season conditions where pre-plant irrigation is required to elevate soil profile water to reasonable levels for crop establishment. The *Last Practical Day of Irrigation* sets the calendar date for last DIEM simulated irrigation regardless of developmental state of the crop. This date not only ends scheduled irrigations, but can also affect unconfirmed irrigation distributions, the rate of target soil water decline in the profile at the end of the crop season, and total irrigation volume. The *Irrigation System Type* directly impacts soil surface evaporation during and immediately following an irrigation event, and therefore affects daily changes in estimated soil water content and ultimately simulated crop productivity. DIEM currently considers the soil surface to be wetted at 0, 40, and 100 percent when irrigations are with SDI, LEPA and spray systems, respectively. The *Irrigation Iterative Computation* option allows alternative irrigation distributions to be automatically considered, which may result in higher yield score and slightly increases computation time.

## Irrigation, Rainfall and Soil Water Editors

Irrigation Editor. Growing season irrigation applications are scheduled to achieve high yield and water productivity based on imposed water constraints. Irrigation amounts on specific days are displayed in calendar format in the irrigation editor screen (Figure 5). When actual irrigations differ from DIEM recommendations, corrections on appropriate dates can be entered and confirmed. As the growing season progresses, confirmation of actual irrigations up to the current date allows DIEM to reconfigure future irrigations for the remaining growing season, based on current data.

Rainfall Editor. Rainfall amounts from the selected weather station data sets are automatically downloaded and displayed in the rainfall editor. If necessary, this editor allows the correction of rainfall data with more accurate field measurements up to the current date. When simulating the current year, each time a scenario is activated, the rainfall editor is updated.

Soil Water Editor. The daily estimated available soil profile water content within the designated model period is displayed in the soil water editor. Daily values can be changed on any date to better reflect the actual soil water content in the field being simulated with values obtained through field observations or measurements.

## Technical Help

Directions to the DIEM User Guide and e-mail help desk are accessed through the "System / Help" button (Figure 1).

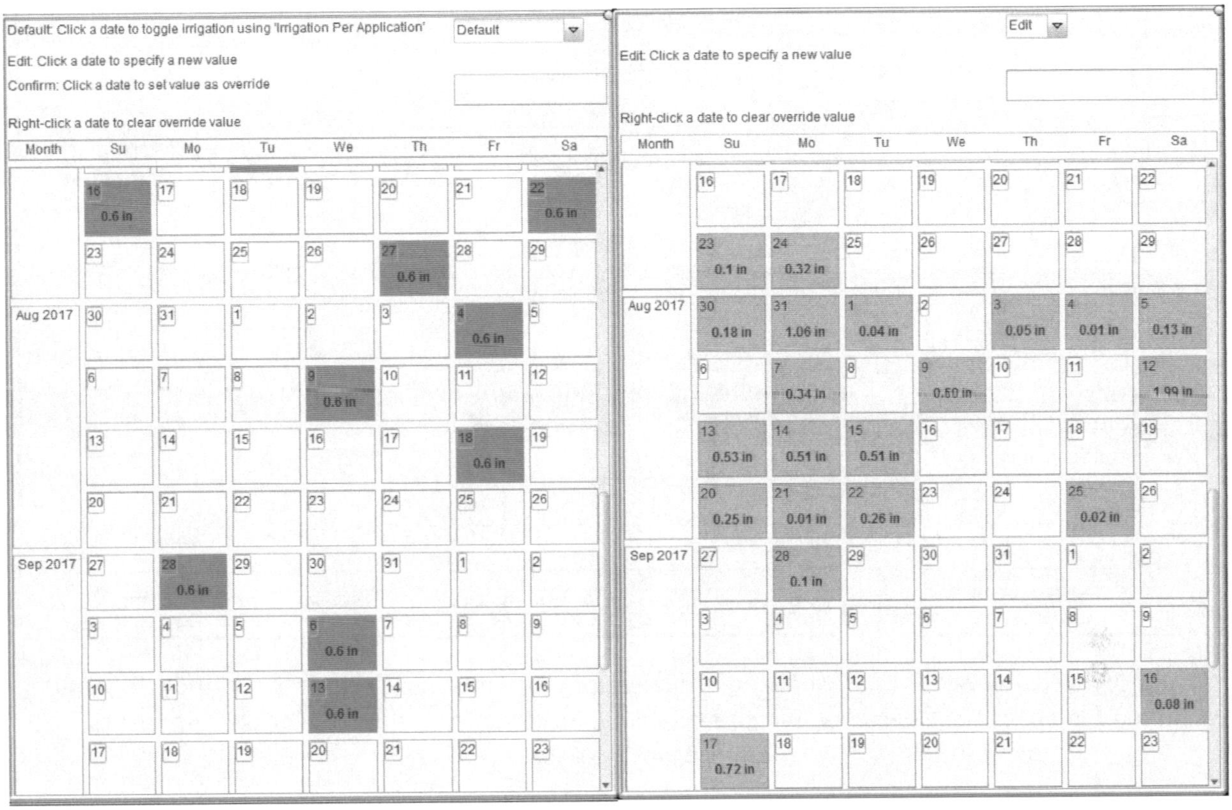


Figure 5. Irrigation (left) and Rainfall (right) Editors component displayed in the DIEM dashboard.

## Outputs

Among other outputs, the main DIEM dashboard provides an integrated soil water chart, a numeric score block, and an option to print a one-page irrigation summary. The soil water chart (Figure 6) displays daily data over the date range specified in the model start and end dates (x-axis). The y-axis ranges from zero to the maximum plant available soil water in the root zone for the active scenario. The *Target Soil Water* is a dynamic, crop-specific, soil profile (root zone) water content above which irrigations are not recommended. This target is a function of the crop emergence date, heat unit accumulation, soil water holding capacity, and local empirical data. For cotton, a planned decline of soil water after peak bloom reduces the possibility of late-season excess vegetative growth, promotes maturity of existing cotton bolls, and allows plants to consume remaining stored soil water (Bordovsky and Lyle, 1996). Daily *Estimated Soil Water* is determined from weather station or field rainfall measurements, irrigation quantities, and crop ET ($ET_c$). $ET_c$ is derived from weather station data and experimentally derived dual crop coefficients. As estimated soil water declines below crop stress levels, transpiration is reduced. This stress level is determined daily and is shown in the soil water chart, *Limiting Transpiration*. The adjusted $ET_c$ ($ET_{cadj}$) is determined by multiplying $ET_c$ by a dimensionless transpiration reduction factor ($K_s$) that is a function of a crop depletion factor (p) according to Allen, et al. (1998). *Irrigation* and effective *Rainfall* are also charted. Any change in a parameter value, an amount in any of the three editors, or weather station input results in a revised soil water chart. *Yield* and *Water Use Efficiency Scores* for the active scenario are updated. Yield scores are derived from functions that relate to the availability of soil water through the growing season and to crop water stress, and the irrigation productivity score is the result of dividing the yield score by the estimated seasonal irrigation.

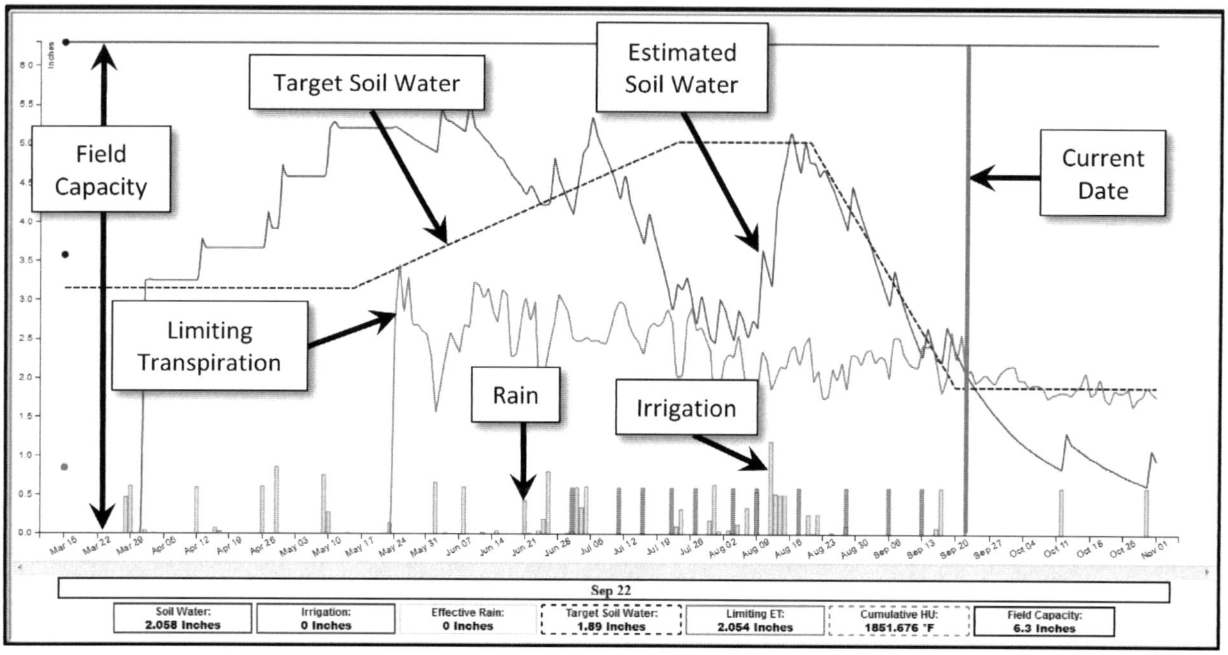


Figure 6. DIEM Soil Water Chart showing estimates of past and future soil water relationships and recommended irrigations based on water availability, irrigation system type, soil characteristics, and crop water needs.

At any point during a growing season, a printed irrigation schedule (Figure 7) can be obtained by executing the "Export Report" button in the Parameters panel (Figure 4). The report provides basic information concerning the field and scenario as well as weekly rainfall, crop ET and irrigation amounts up to the current week and recommendations or estimates thereafter.

## Conclusions

The Dashboard for Irrigation Efficiency Management (DIEM) was developed to support irrigation decisions in the short season, semi-arid environment of the Texas High Plains where the irrigation capacity of the Ogallala Aquifer is rapidly decreasing and cotton is the predominant row crop. DIEM is a web-based software solution for forecasting optimized field-specific irrigation schedules base on water availability. This is accomplished using minimal inputs by configuring basic historic, near real-time, and future agronomic information including weather, soil and crop characteristics, and irrigation parameters into an interactive data visualization for portraying key aspects supporting irrigation decisions. Upcoming plans include accommodating other crops or crop types, adding complementary soil and/or plant based data streams to the DIEM output, and the deployment of a mobile friendly DIEM application.

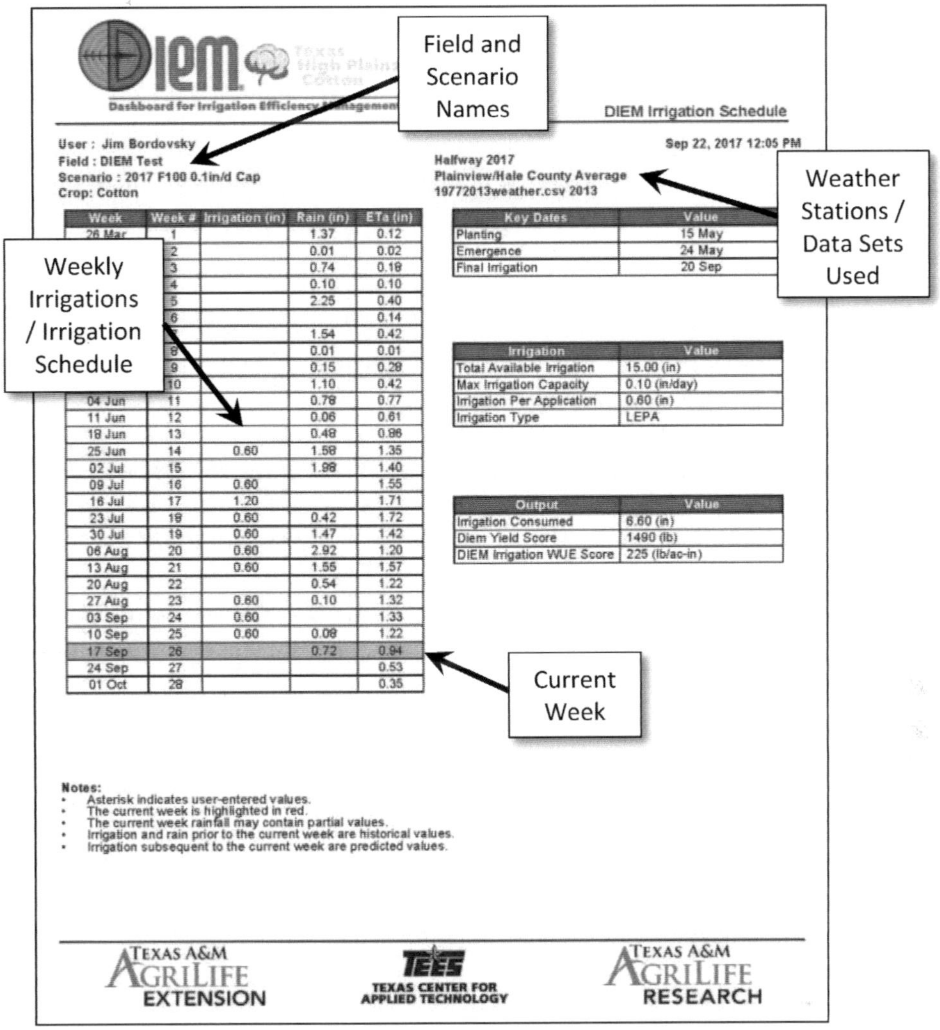


Figure 7. Exported DIEM report providing historic, recommended, or predicted values based on the most recent available data.

**Acknowledgements**

The authors would like to gratefully acknowledge Scott Jordan, Research Engineering Associate for his support. Initial funding for this project was through the Water Seed Grant Initiative, Texas A&M AgriLife Research, Texas A&M AgriLife Extension, and the Texas A&M Engineering Experiment Station, College Station, TX. In addition, funding is being provided by USDA Project No. 2016-68007-25066, through the National Institute for Food and Agriculture's Agriculture and Food Research Initiative, Water for Agriculture Challenge Area. Project website: ogallalawater.org. Also, this work is supported by the USDA National Institute of Food and Agriculture, Hatch project 1007927.

## References

Allen, R.G., Pereira, L.S, Raes, D., and Smith, M., 1998. Crop evapotranspiration – guidelines for computing crop water requirements – FAO Irrigation and Drainage Paper 56. Food and Agricultural Organization of the United Nations. Rome, Italy.

Bordovsky, J., and W. Lyle. 1996. Protocol for planned soil water depletion of irrigated cotton. Intl. Conf. Evapotranspiration and Irrig. Scheduling. St. Joseph, Mich.:ASABE. Pp 201-206.

Bordovsky, J., Mustian, J., Ritchie, G., and Lewis, K. 2015. Cotton irrigation timing with variable seasonal irrigation capacities in the Texas South Plains: *Applied Engineering in Agriculture,* 31(6), 883-897.

Colorado State University Extension. 2017. Water irrigation scheduler for efficient application (WISE) Irrigation Scheduler. Retrieved from: http://wise.colostate.edu/

Kanas State Research and Extension. 2017. Kansched3. Retrieved from: http://www.bae.ksu.edu/mobileirrigationlab/kansched3

Marek, G., P. Gowda, T. Marek, B. Auvermann, S. Evett, P. Colaizzi, and D. Brauer. 2016. Estimating preseason irrigation losses by characterizing evaporation of effective precipitation under bare soil conditions using large weighing lysimeters. Agricultural Water Management 169, 115-128.

Texas Alliance for Water Conservation. 2017. ET/Resource Tools. Retrieved from: http://www.tawcsolutions.com/ .

Texas A&M AgriLIfe Extension. 2017. TexasET Network. Retrieved from: http://texaset.tamu.edu/ .

University of Missouri Extension Service. 2017. Crop water use application. Retrieved from: www.cropwater.org.

USDA-NRCS. 1996. Estimating soil noisture by feel and appearance. Retrieved from: http://irrigationtoolbox.com/IrrigationToolBox/Section6_SoilMoistureMeasurement/soilmoist.pdf

Vellidis, G., V. Liakos, C. Perry, M. Tucker, G. Collins, J. Snider, J. Andreis, K. Migliaccio, C. Fraisse, K. Morgan, D. Rowland, E. Barnes. 2014. A smartphone app for scheduling irrigation on cotton. In S. Boyd, M. Huffman and B. Roberson (eds) Proceedings of the 2014 Beltwide Cotton Conference, New Orleans, LA, National Cotton Council, Memphis, TN (paper 15551), p. 175-186.

# Evaluating Water Use and Management of Center Pivot Drag-line Drip Irrigation Systems

Charles Swanson – Extension Program Specialist
Texas A&M AgriLife Extension Service
College Station, Texas

Dr. Guy Fipps, PE – Extension Agricultural Engineer
Texas A&M AgriLife Extension Service
College Station, Texas

Dr. Charles Hillyer – Extension Specialist
Texas A&M AgriLife Extension Service
Amarillo, Texas

**Abstract.** *Center pivot irrigation is the most popular irrigation method used in Texas. Over the last 15 years, more farmers have gradually switched from furrow to center pivot irrigation. Center pivot drag line drip irrigation systems or Precision Mobile Drip Irrigation (PMDI) have gained attention in recent years because of their ability to combine the application efficiency of drip irrigation with the operational efficiency of center pivot systems. However many farmers are hesitant to implement this technology, unsure if the costs of installation and management will result in significant water and costs savings. This paper will review farmer management of such systems and attempt to quantify irrigation water use and perceived water savings from implementing this technology concept.*

**Keywords.** Center pivot, drip irrigation, drag-line systems

## Background

The use of drip irrigation technology with center pivot irrigation machines is not a new concept. However, recently manufacturers and local irrigation dealers in Texas have started marketing drip irrigation packages for center pivots, often referred to as "drag-line drip irrigation systems" or "precision mobile drip irrigation". These systems drag varying lengths of drip irrigation tubing behind the machine applying water slowly on the surface. The combination of these two technologies has potential to combine the higher application efficiency related to drip with the operational and maintenance benefits of center pivot machines.

A survey of growers with installed center pivot drag line drip irrigation system was conducted to evaluate the field level operation and management of these systems. Growers were identified through extension outreach and collaborating with dealers and installers of these systems. The survey addressed as installed system design, operational advantages and disadvantages and grower measured/perceived water use and savings. To Date, the survey has been completed by five growers.

## System Design

One grower converted all 8 spans of his pivot to drip irrigation. The other two growers installed one and two spans on their existing center pivots, converting from LESA and LEPA water application technologies. Two growers installed the system on new pivots. Design of the systems varied. Factors such as crop type, planting layout (ie straight or circle rows), and designer/installer influenced the design. One type uses a secondary pvc manifold positioned below the pivot main which is held in place by guidewires attached to each pivot tower span. A series of pvc pipes or flexible pivot drop hose were used to connect drip lines to the manifolds. Systems either have a 30 to 80 inch drop/drip line spacing. Drip line length varies based upon the flow rate needed and is matched to the pivot printout. These three systems used Netafim DripNet PC Dripline. Drip line flow rates varied from 1 gallon per hour per foot to 2 gallons per hour per foot.

Figure 1. Example Design of Manifold Assembly.

Filtration and pressure regulation are typically standard practice when operating conventional drip irrigation systems. However, the use of filtration and pressure regulation varied across all the installed systems. Four of the systems evaluated had filters installed. Filter location varied from one at each drip line to only 1 from each pivot drop to the manifold. Pressure regulation also varied and 4 of the systems had pressure regulators on each drop/drop line. All three installations used pressure compensating drip tubing.

## System Operation & Maintenance

None of the growers reported any major maintenance problems with their systems. One grower reported that the plugs on the end of the drip lines would pop off during operation. The original compression plugs were later switched out and replaced with "twist-locking" caps and no further problems were reported. Three growers reported some minor rodent damage to the drip tubing that required repair.

The growers expressed two operational concerns about the system. Adjusting the drip lines when changing pivot direction was required to avoid damaging the crop or having the dripline become entangled in the crop canopy. One grower noted that he had to move the move the drip lines by hand at the end of the field so he could perform tillage operations. One grower expressed the interest/need to roll up the drip lines in the winter to prevent damage and make moving the pivot much easier.

Figure 2. Drag-line system parked on end of field.

Figure 3. Drag Line System Design from the pivot mainline

Figure 4. Drag Line System Design as with 2 drip lines per pivot drop.

## Advantages and Disadvantages

The biggest advantage all three growers found with the use of the drag-line systems was the decreased depth of wheel tracks compared to sprinkler irrigation. One grower noted the ability to irrigate during colder weather and avoid freezing concerns. Growers also noted that use of drip irrigation reduced runoff from the field and evaporative and/or wind losses from sprinklers.

(a) (b)

Figures 5a & 5b. Pivot Wheel Tracks.

One grower noted his advantage to use of the system is being able to maintain use of his pivot and maintain irrigated operation as pumping capacity decreased to flowrates that were difficult to maintain with a sprinkler package system.

Figure 6. Drip line being dragged along crop row, Field Planted in Circle.

Figure 7. Drip line being dragged across crop rows, Field Planted Square.

The biggest problem observed by the grower who did not plant his crop in a circle was that the drip lines would pull across the top of the crop canopy (cotton plants) when traveling perpendicular to the rows. The grower did note some leaf damage but could not verify if it impacted crop yield.

Figure 8. Drip line being dragged over crop canopy.

## Water Use and Yield

Not all growers in the survey were able to provide details of their water use and yield. Some challenges to survey water use were experienced due to the above normal rainfall received during the 2017 growing season. Figures 9 & 10 summarize the frequency and total depth of rainfall received during the growing season.

For the grower in the Brazos Valley, he applied 3 irrigation events of an estimated 1 inch of water per irrigation to his cotton crop. Average yield of the field was 1.5 bales per acre, respectively. Grower noted that rainfall had a severe impact on crop quality and yield. Another grower in the high plains used the system to operate for a total of 498 hours to apply about 2.2 inches of preseason water for his field, however the crop did not come to harvest due to damage from a hail storm. One grower was able to produce water use and yield records from 2016. He ran the system for 96 days total and the system had a flow rate of 360 gallons per minute. This resulted in a total of 15.27 inches of water being applied and a corn crop yield of 236 bushels per acre.

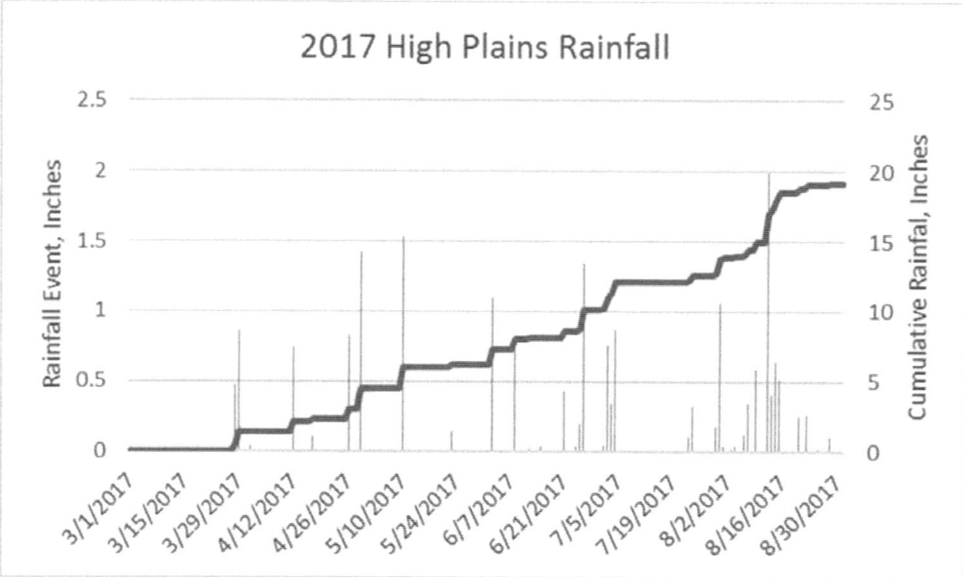


Figure 9. 2017 High Plains Rainfall Summary

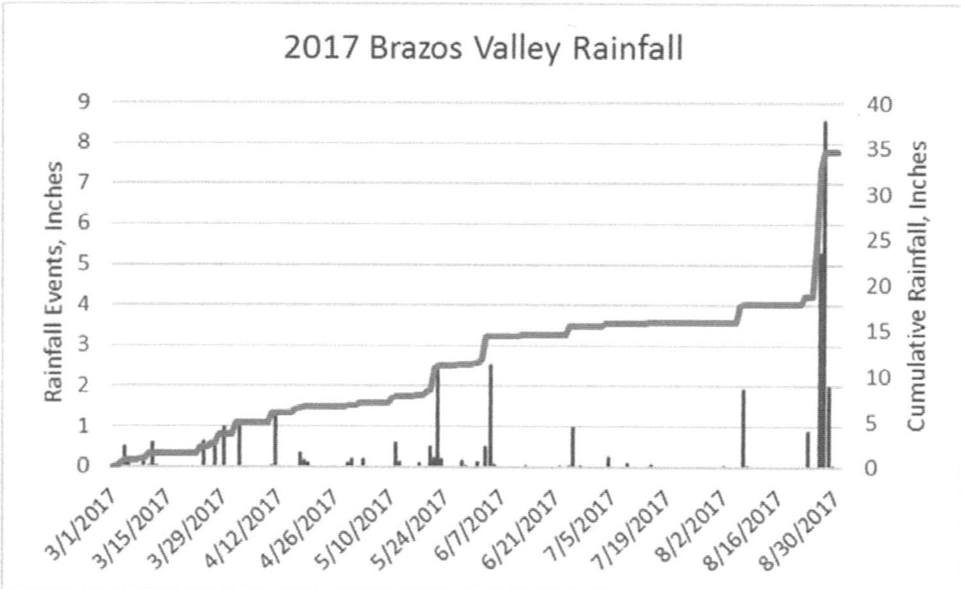


Figure 10. 2017 Brazos Valley Rainfall Summary

## Conclusion

Many growers in Texas are interested in adopting the drag-line drip irrigation concept on their farms but have reservations regarding the performance, operation, management and costs. The information collected from the three growers is helpful in addressing potential grower concerns. Further evaluation and documented successful systems are needed before many growers will implement this system. As all systems had different designs, further evaluation of each systems design using different types of manifolds and drops is needed to determine which design is the most practical and cost beneficial for converting existing systems.

# ERRATICITY OF SPRINKLER-IRRIGATED CORN UNDER DROUGHT

**Freddie R. Lamm**
Research Irrigation Engineer
Kansas State University
Colby, Kansas
Voice: 785-462-6281
Fax: 785-462-2315
flamm@ksu.edu

**Terry A. Howell**
Agricultural Engineer
USDA Agricultural Research Service
Bushland, Texas
Voice: 806-356-5746
Fax: 806-356-5750
terry.howell@ars.usda.gov

**James P. Bordovsky**
Agricultural Engineer
Texas AgriLife Research, Texas A&M
Lubbock / Halfway, Texas
Voice: 806-889-3315
Fax: 806-889-3416
j-bordovsky@tamu.edu

## INTRODUCTION

A definition:

> erraticity (ĭr·ə·tĭs′·ĭ·tē) *n.* The quality or state of being erratic, characterized by the lack of consistency, regularity or uniformity.

That's correct, there is no such word, but you sure know it when you see it. Unfortunately, we saw a lot of it during some extreme droughts in sprinkler irrigated corn.

Figure 1. Nonuniformity of sprinkler irrigated corn under extreme drought conditions in southwest Kansas in 2011.

These instances of erraticity resulted in low quality, low- or non-yielding corn production. Crop water stress caused by the extreme drought in portions of the central and southern Great Plains is ultimately responsible for the erraticity. However, there may be ways to reduce erraticity and its harmful effects by improvements in design and management of center pivot sprinklers for corn production that can minimize water losses.

# SPRINKLER PACKAGE EFFECTS ON WATER LOSSES

Center pivot sprinkler management techniques to avoid water losses begin at the design and installation stages with selection of an appropriate sprinkler package. Typical sprinkler packages in use to today are medium and high pressure impacts which are located on top of the sprinkler span (approximately 12 to 15 ft height above soil surface), low pressure rotating spray nozzles which are typically located on the span or at least above the crop canopy, low pressure fixed spray applicators that are located above and within the crop canopy and LEPA (low energy precision application) that are located near the ground surface (usually 1 to 2 ft maximum height above soil surface). Commercial LEPA applicators often can apply water in multiple modes (e.g., bubble mode with little or no wetting of the canopy, fixed spray mode, and chemigation mode that sprays the undersides of the leaves). The popular low pressure fixed spray applicators have also been categorized by their location with respect to the canopy with the terms LESA (low elevation spray application, 1 to 2 ft maximum height) and MESA (mid elevation spray application, 5 to 10 ft maximum height) (Howell, 1997). Application with MESA is typically above the crop canopy for all or most of the crop season depending on the crop (e.g., MESA application occurs within top portions of corn canopy in last 30 to 40 days of irrigation season). There are numerous water loss pathways using center pivot sprinklers and each type of sprinkler package has advantages and disadvantages as outlined by Howell (2006) that must be balanced against the water loss hazards (Table 1).

Table 1. Water loss components associated with various sprinkler packages. Adapted from Howell (2006).

<table>
<tr><th rowspan="2">Water Loss Component</th><th colspan="4">Sprinkler Package</th></tr>
<tr><th>Overhead<br>(Impact sprinklers, rotating or fixed spray applicators)</th><th>MESA</th><th>LESA</th><th>LEPA</th></tr>
<tr><td>Droplet evaporation</td><td rowspan="3">Yes</td><td rowspan="5">Yes</td><td>Yes</td><td rowspan="2">No</td></tr>
<tr><td>Droplet drift</td><td>No</td></tr>
<tr><td>Canopy evaporation</td><td>Yes<br>(not major)</td><td>No<br>(chemigation mode only)</td></tr>
<tr><td>Impounded water evaporation</td><td>No</td><td>Yes</td><td>Yes<br>(major)</td></tr>
<tr><td>Wetted soil evaporation</td><td>Yes</td><td>Yes</td><td>Yes<br>(limited)</td></tr>
<tr><td>Surface water redistribution</td><td rowspan="2">No,<br>(but possible)</td><td>Yes,<br>(not major)</td><td>Yes</td><td rowspan="2">Yes<br>(not major unless surface storage is not used)</td></tr>
<tr><td>Runoff</td><td>Yes</td><td>Yes</td></tr>
<tr><td>Percolation</td><td>No</td><td>No</td><td>No</td><td>No</td></tr>
</table>

Windy and hot conditions during the growing season affect center pivot sprinkler irrigation uniformity and evaporative losses. As a result many producers in the southern and central Great Plains have adopted sprinkler packages and methods that apply the water at a lower height within or near the crop canopy height, thus avoiding some application nonuniformity caused by wind and also droplet evaporative losses.

In-canopy and near-canopy sprinkler application can reduce evaporative losses by nearly 15% (Table 2), but introduce a much greater potential for irrigation nonuniformity. These sprinkler package systems are often adopted without appropriate understanding of the requirements for proper water management, and thus, other problems such as runoff and poor soil water redistribution occur.

Table 2. Partitioning of sprinkler irrigation evaporation losses with a typical 1 inch application for various sprinkler packages. (Adapted from Howell et al., 1991; Schneider and Howell, 1993).

| Sprinkler package | Air loss, % | Canopy loss, % | Ground loss, % | Total loss,% | Application efficiency, %* |
|---|---|---|---|---|---|
| Impact sprinkler ≈ 14 ft height | 3 | 12 | -- | 15 | 85 |
| MESA ≈ 5 ft height | 1 | 7 | -- | 8 | 92 |
| LEPA ≈ 1 ft height | -- | -- | 2 | 2 | 98 |

* Ground runoff and deep percolation are considered negligible in these data.

Traditionally, center pivot sprinkler irrigation systems have been designed to uniformly apply water to the soil at a rate less than the soil intake rate to prevent runoff from occurring (Heermann and Kohl, 1983). These design guidelines need to be either followed or intentionally circumvented with appropriate design criteria when designing and managing an irrigation system that applies water within the canopy or near the canopy height where the full sprinkler wetted radius is not developed. Peak application rates for in-canopy sprinklers such as LESA (low elevation spray application) and LEPA (low energy precision application) might easily be 5 to 30 times greater than above-canopy sprinklers (Figure 2).

Runoff from LEPA sprinklers was negligible on 1% sloping silt loam soils in eastern Colorado but exceeded 30% when slopes increased to 3% (Buchleiter, 1991). Runoff from LEPA with basin tillage was approximately 22% of the total applied water and twice as great as MESA (mid elevation spray application at 5 foot applicator height) for grain sorghum production on a clay loam in Texas (Schneider and Howell, 2000). Basin tillage created by periodic diking of crop furrow (2 to 4 m spacing), rather than reservoir tillage created by pitting or digging small depressions (0.5 to 1 m), is often more effective at time averaging of LEPA application rates, and thus, preventing runoff (Schneider, 2000).

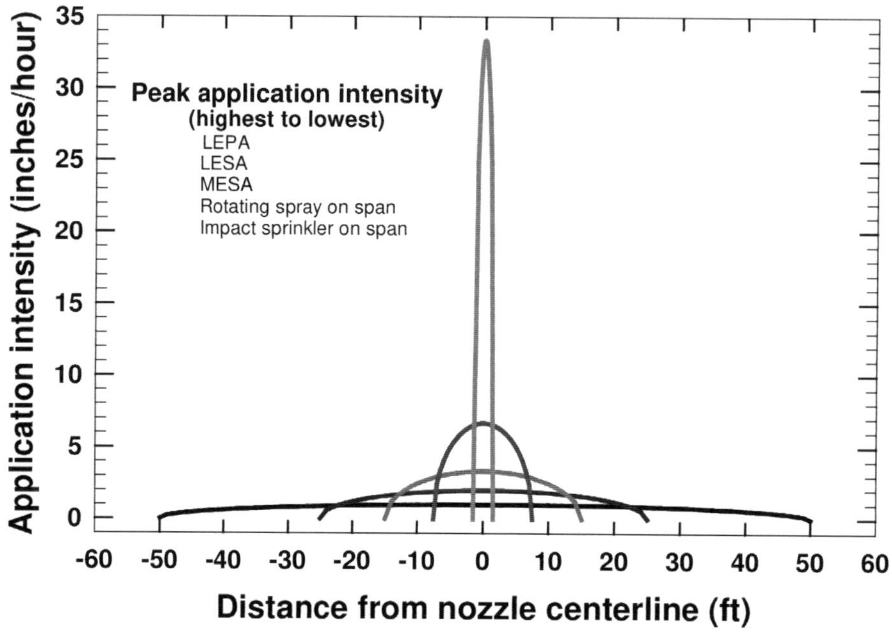


Figure 2. Application intensities for LEPA, LESA, MESA, rotating sprays on span and impact sprinklers on the span as related to the typical size of their wetting pattern.

Decreasing the application intensity is the most effective way to prevent irrigation field runoff losses and surface redistribution within the field (Figure 3.) When runoff and surface redistribution occurs using in-canopy sprinklers because of a reduced wetting pattern, one solution would be to raise the sprinkler height.

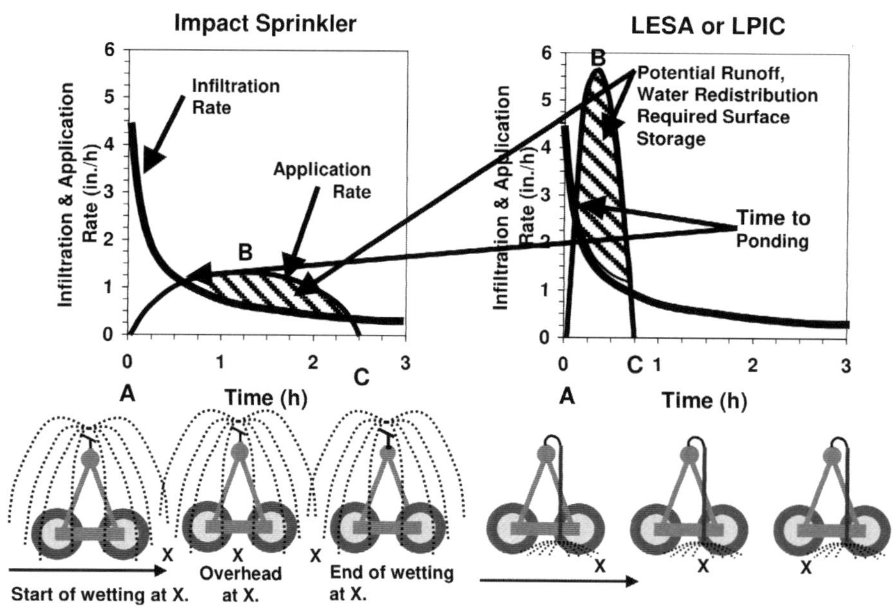


Figure. 3. Illustration of runoff or surface water redistribution potential for impact and LESA sprinkler application packages for an example soil. After Howell (2006).

One might assume that the erraticity observed under drought in sprinkler irrigated fields was primarily associated with the evaporative water loss components shown in Table 1, but that is probably not the case. When using fixed plate applicators near or within the canopy (MESA, LESA and LEPA), the magnitude of field runoff and particularly surface redistribution within the field may overwhelm the evaporative loss reductions possible with these packages. Surveys conducted by Kansas State University have indicated that approximately 90% of the center pivot sprinkler systems in western Kansas use fixed plate applicators and nearly 60% have sprinkler nozzle height less than 4 ft above the soil surface (Rogers et al. 2009). The erraticity can be caused by failure to follow appropriate guidelines for irrigation with near- and in-canopy sprinklers.

## SOME GUIDING PRINCIPLES FOR IN-CANOPY APPLICATION

A prototype of the LEPA system was developed as early as 1976 by Bill Lyle with Texas A&M University. Jim Bordovsky joined the development effort in 1978 (McAlavy and Dillard, 2003) and the first scientific publication of their work was in 1981 (Lyle and Bordovsky, 1981). Although, originally LEPA was used in every furrow, subsequent research (Lyle and Bordovsky, 1983) demonstrated the superiority for alternate furrow LEPA. The reasons are not always evident, but they may result from the deeper irrigation penetration (twice the volume of water per unit wetted area compared with every furrow LEPA), possible improved crop rooting and deeper nutrient uptake, and less surface water evaporation (~30-40% of the soil is wetted). The seven guiding principles of LEPA were given by Lyle (1992) as:

1) *Use of a moving overhead tower supported pipe system (linear or center pivotal travel)*
2) *Capable of conveying and discharging water into a single crop furrow*
3) *Water discharge very near the soil surface to negate evaporation in the air*
4) *Operation with lateral end pressure no greater than 10 psi when the end tower is at the highest field elevation*
5) *Applicator devices are located so that each plant has equal opportunity to the water with the only acceptable deviation being where nonuniformity is caused by nozzle sizing and topographic changes*
6) *Zero runoff from the water application point*
7) *Rainfall retention which is demonstratively greater than conventionally tilled and managed systems.*

The other types of in-canopy and near-canopy sprinkler irrigation do not necessarily require adherence to all of these seven guidelines. However, it is unfortunate that there has been a lack of knowledge or lack of understanding of the importance of these principles because many of the problems associated with in-canopy and near-canopy sprinkler irrigation can be traced back to a failure to follow or effectively "work around" one of these principles. In-canopy and near-canopy application systems can definitely reduce evaporative losses (Table 2), but these water savings must be balanced against runoff and within field water redistribution, deep percolation and other soil water nonuniformity problems that can occur when the systems are improperly designed and managed.

## PROVIDING PLANTS EQUAL OPPORTUNITY TO SOIL WATER

The No. 5 LEPA guiding principle listed earlier emphasizes the importance of plants having equal opportunity to root-zone soil water. Ensuring this equal opportunity requires sufficient uniformity of water application and/or soil water infiltration. Key issues that must be addressed are irrigation application symmetry, crop row orientation with respect to center pivot sprinkler direction of travel, and the seasonal longevity of the sprinkler pattern distortion caused by crop canopy interference.

## SYMMETRY OF SPRINKLER APPLICATION

Increased sprinkler application uniformity will often result in increased yields, decreased runoff, and decreased percolation (Seginer,1979). Improved sprinkler uniformity can be desirable from both economic and environmental standpoints (Duke et al., 1991). Their study indicated irrigation nonuniformity can result in nutrient leaching from over-irrigation and water stress from under-irrigation. Both problems can cause significant economic reductions.

Sprinkler irrigation does not necessarily have to be a uniform broadcast application to result in each plant having equal opportunity to the irrigation water. Equal opportunity can still be ensured using a LEPA nozzle in the furrow between adjacent pairs of crop rows provided runoff is controlled (Figure 4).

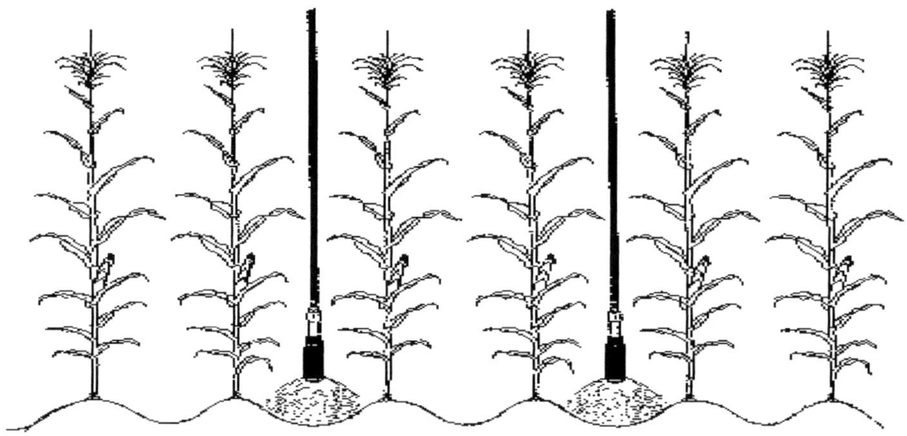

Figure 4. LEPA concept of equal opportunity of plants to applied water. LEPA heads are centered between adjacent pairs of corn rows. Using a 5–ft nozzle spacing with 30-inch spaced crop rows planted circularly results in plants being approximately 15 inches from the nearest sprinkler. After Lamm (1998).

Some sprinkler application nonuniformity can also be tolerated when the crop has an intensive root system (Seginer, 1979). When the crop has an extensive root system, the effective uniformity experienced by the crop can be high even though the actual resulting irrigation system uniformity within the soil may be quite low. Additionally, when irrigation is deficit or limited, a lower value of application uniformity can be acceptable in some cases (von Bernuth, 1983) as long as the crop economic yield threshold is met.

Many irrigators in the U.S. Great Plains are using wider in-canopy sprinkler spacings (e.g., 7.5, 10, 12.5, and even 15 ft) in an attempt to reduce investment costs (Yonts et al., 2005). Surveys from western Kansas in 2005 and 2006 indicated only 34% of all sprinkler systems with nozzle height of less than 4 ft had consistent nozzle spacing less than 8 ft (Rogers et al. 2009). Sprinkler nozzles operating within a fully developed corn canopy experience considerable pattern distortion and the uniformity is severely reduced as nozzle spacing increases (Figure 5).

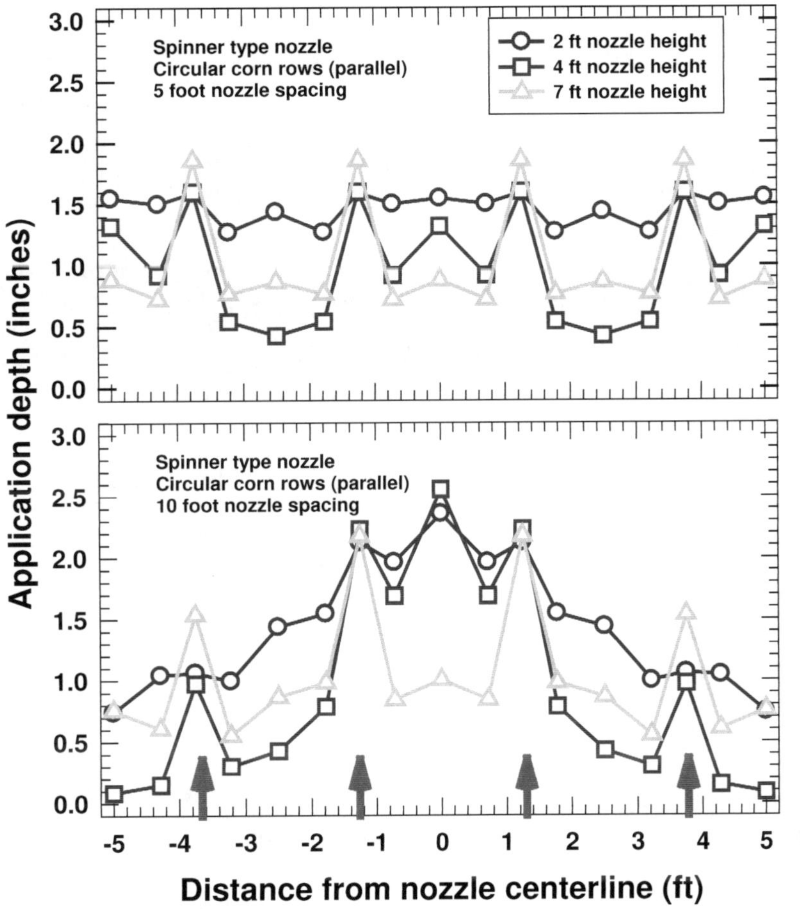


Figure 5. Differences in application amounts and application patterns as affected by sprinkler nozzle height and spacing. Center pivot sprinkler lateral is traversing parallel to the circular corn rows. Data are from a fully developed corn canopy, July 1996, KSU Northwest Research-Extension Center, Colby, Kansas. Data are mirrored about the nozzle centerline for display purposes. Arrows on X-axis represent location of corn rows and thus the location for higher stemflow amounts.

Although Figure 5 indicates large application nonuniformity, these differences may or may not always result in crop yield differences. Hart (1972) concluded from computer simulations that differences in irrigation water distribution occurring over a distance of approximately 3 ft were probably of little overall consequence and would be evened out through soil water redistribution.

Some irrigators in the Central Great Plains contend that their low capacity systems on nearly level fields restrict runoff to the general area of application. However, nearly every field has small changes in land slope and field depressions which do cause field runoff, in-field redistribution or deep percolation in ponded areas when the irrigation application rate exceeds the soil infiltration rate. In the extreme drought years of 2000 to 2003 that occurred in the U. S.

Central Great Plains, even small amounts of surface water movement affected sprinkler-irrigated corn production (Figure 6). Similarly some of the worst erraticity in sprinkler-irrigated corn observed in the summer of 2011 was for sprinklers with 10 ft spaced in-canopy sprinkler packages (Figure 7).

Figure 6. Large differences in corn plant height and ear size for in-canopy sprinkler application over a short 10-ft. distance (4 crop rows) as caused by small field microrelief differences and the resulting surface water movement during an extreme drought year, Colby, Kansas, 2002. The upper stalk and leaves have been removed to emphasize the ear height and size differences.

Figure 7. Erraticity of sprinkler irrigated corn in southwest Kansas in 2011 under extreme drought conditions thought to be related to a nozzle spacing too wide (10 ft) for in-canopy application (2 ft nozzle height).

## CROP ROW ORIENTATION WITH RESPECT TO DIRECTION OF SPRINKLER TRAVEL

When using in-canopy sprinkler application, it has been recommended that crop rows be planted circularly so that the crop rows are always perpendicular to the center pivot sprinkler lateral. Matching the direction of sprinkler travel to the row orientation satisfies the important LEPA Principles 2 and 5 noted by Lyle (1992) concerning water delivery to one individual crop furrow and equal opportunity to water by for all plants. Producers are often reluctant to plant row crops in circular rows because of the cultivation and harvesting difficulties of narrow or wide "guess" rows. However, using in-canopy application for center pivot sprinkler systems in non-circular crop rows can pose two additional problems (Figure 8). In cases where the CP lateral is perpendicular to the crop rows and the sprinkler spacing exceeds twice the crop row spacing, there will be nonuniform water distribution because of pattern distortion. When the CP lateral is parallel to the crop rows there may be excessive runoff due to the great amount of water being applied in just one or a few crop furrows. There can be great differences in in-canopy application amounts and patterns between the two crop row orientations (Figure 9).

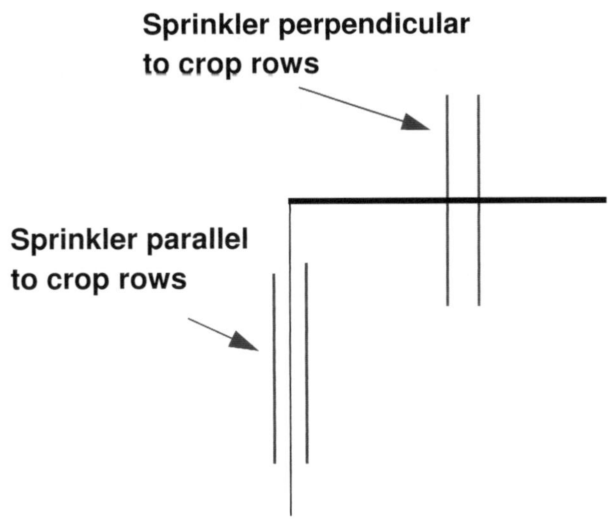


Figure 8. Two problematic orientations for in-canopy sprinklers when crops are not planted in circular rows.

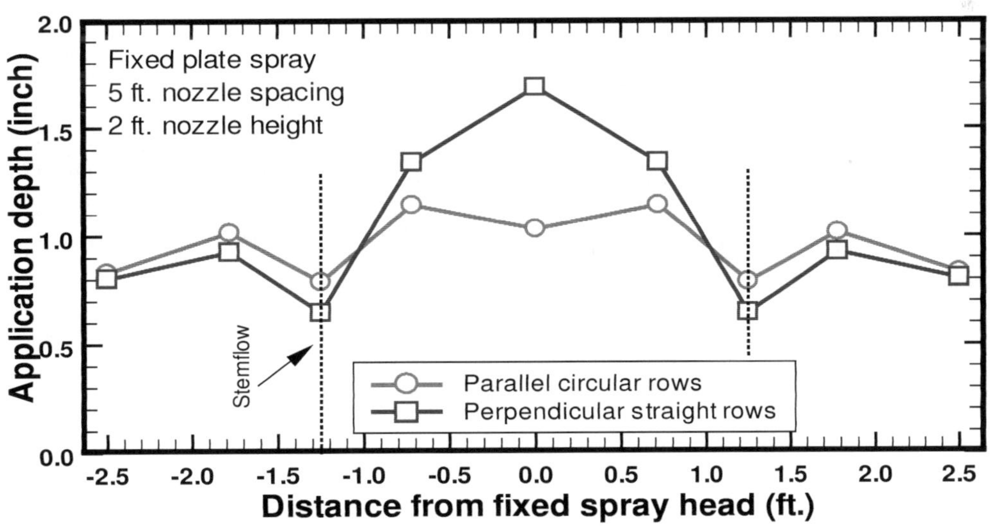


Figure 9. Differences in application amounts and application patterns as affected by corn row orientation with respect to the center pivot sprinkler lateral travel direction. Dotted lines indicate location of corn rows and stemflow measurements. Data are from a fully developed corn canopy, July 23-24, 1998, KSU Northwest Research-Extension Center, Colby, KS. Data are mirrored about the centerline of the nozzle.

## PATTERN DISTORTION AND TIME OF SEASON

Drop spray nozzles just below the center pivot sprinkler lateral truss rods (approximately 7-10 ft height above the ground) have been used for over 30 years in northwest Kansas. This configuration rarely has had negative effects on corn yields although the irrigation pattern is distorted after corn tasseling. The reasons are that there is only a small amount of pattern distortion by the smaller upper leaves and tassels and this distortion only occurs during the last 30 to 40 days of the irrigation season. In essence, the irrigation season ends before a severe soil water deficit occurs. Compare this situation with spray heads at a height of 1 to 2 ft that may experience pattern distortion for more than 60 days of the irrigation season. Under dry and elevated evapotranspiration conditions in 1996, row-to-row corn height differences developed rapidly for 10-ft spaced sprinkler nozzles at a 4 ft nozzle height following a single one-inch irrigation event at the KSU Northwest Research-Extension Center, Colby Kansas (Figure 10). A long term study (1996-2001) at the same location on a deep silt loam soil found that lowering an acceptably spaced (10 ft) spinner head from 7 ft further into the crop canopy (e.g., 4 or 2 ft) caused significant row-to-row differences in corn yields (Figure 11).

Figure 10. Crop height difference that developed rapidly under a widely spaced (10 ft) in-canopy sprinkler (4 ft height) following a single 1 inch irrigation event at the KSU Northwest Research-Extension Center, Colby, Kansas. Photo taken on July 6, 1996.

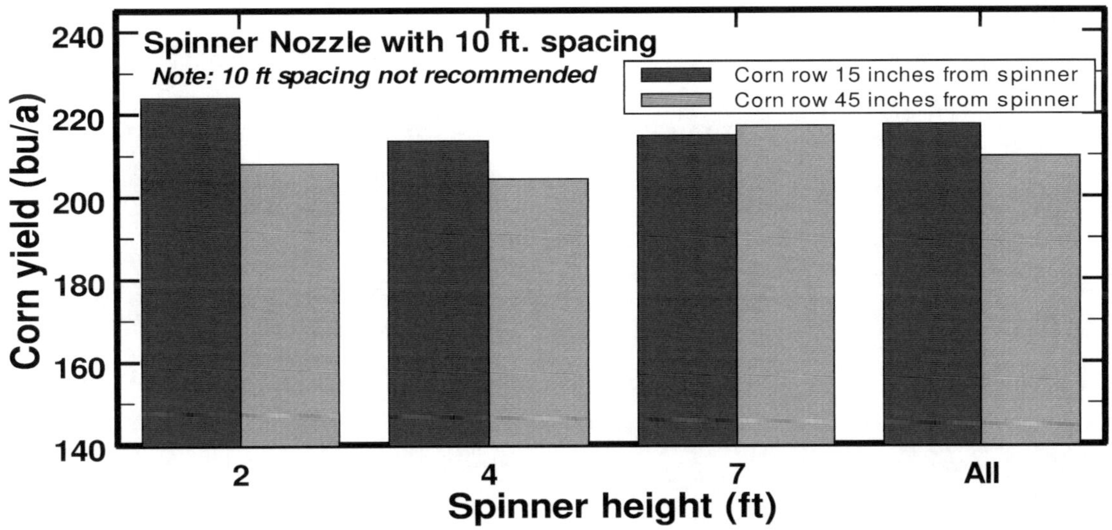


Figure 11. Row-to-row variations in corn yields as affected by sprinkler height for 10 ft. spaced in-canopy sprinklers. Sprinkler lateral travel direction was parallel to crop rows. Data was averaged from four irrigation levels for 1996 to 2001, KSU Northwest Research-Extension Center, Colby, Kansas.

## COMBINATION OF EFFECTS CAN CAUSE ERRATICITY

Sometimes poor design, installation or maintenance problems can exist for years before they are visually observed as sprinkler irrigation erraticity. It may take severe drought conditions for some of these subtle effects to combine to such an extent to be noticeable erraticity. In addition, smaller row-to-row differences in crop yield cannot be measured with yield monitors on commercial-sized harvesters. An example of a combination several of these subtle effects was observed during the severe drought of 2002 in northwest Kansas (Figure 12). The small nozzle height difference on this sprinkler allowed at least three small effects to combine negatively to cause the sprinkler erraticity:

1. *Since there are no pressure regulators, the small height difference results in unequal flow rates for these low pressure spray nozzles.*
2. *There is an incorrect overlap of the sprinkler pattern due to the height difference with one sprinkler within the canopy while the other two nozzles are above the canopy.*
3. *Evaporative losses would be greater for the nozzles above the crop canopy.*

Figure 12. Erraticity of sprinkler-irrigated corn near Colby, Kansas during an extreme drought.

## CONCLUSIONS

The severe droughts of the early 2000s and 2011-2012 in Kansas was devastating to production on many sprinkler irrigated corn fields, but the erraticity did highlight some design and management issues that producer might need to address before the next irrigation season:

1. *Does the selected sprinkler package strike the correct balance in reducing evaporative losses without increasing irrigation runoff or in-field water redistribution?*
2. *Does the sprinkler package and its installation characteristics provide the crop with equal opportunity to applied or infiltrated water?*
3. *Are the sprinkler nozzle heights and spacings appropriate for the intended cropping?*
4. *Should planting of taller row crops such as corn be in circular patterns if in-canopy sprinklers are used?*
5. *Are there subtle irrigation system characteristics (design, installation, or maintenance) that might combine negatively to reduce crop yields?*

These design and management improvements won't change the weather conditions, but they might change how the crop weathers future droughts.

## ACKNOWLEDGEMENTS

This paper is part of a center pivot irrigation technology transfer effort is supported by the Ogallala Aquifer Program, a consortium between USDA Agricultural Research Service, Kansas State University, Texas A&M AgriLife Research, Texas A&M AgriLife Extension Service, Texas Tech University, and West Texas A&M University.

# REFERENCES


Buchleiter, G. W. 1991. Irrigation with LEPA. In: Proc. Central Plains Irrigation Shortcourse, North Platte, NE., Feb. 5-6, 1991. pp. 64-68.

Duke, H. R., D. F. Heermann, and L. J. Dawson. 1991. Selection of appropriate applied depth from center pivots. Presented at the Int'l. Summer Meeting of the ASAE. ASAE paper no. SW91-2052, ASAE, St. Joseph, MI. 18 pp.

Heermann, D. F. and R. A. Kohl. 1983. Fluid dynamics of sprinkler systems. Chapter 14 in Design and Operation of Farm Irrigation Systems, ASAE Monograph No. 3, pp. 583-618, ASAE, St. Joseph, MI.

Howell, T.A. 1997. What's in a name. Wetting Front 1(2);1-3.

Howell, T. A. 2006. Water losses associated with center pivot nozzle packages. In: Proc. Central Plains Irrigation Conf., Colby, KS., Feb. 21-22, 2006. Available from CPIA, 760 N.Thompson, Colby, KS. pp. 12-24.

Howell, T. A., A. D. Schneider, and J. A. Tolk. 1991. Sprinkler evaporation losses and efficiency. In: Proc. Central Plains Irrigation Shortcourse, February, 5-6, 1991, North Platte, NE. pp. 69-89.

Lamm, F.R. 1998. Uniformity of in-canopy center pivot sprinkler irrigation. Presented at the Int'l. Meeting of the ASAE, July 12-16, 1998, Orlando FL. ASAE Paper No. 982069. ASAE, St. Joseph, MI. 7 pp. Also at http://www.ksre.ksu.edu/irrigate/UICCP98.html

Lyle, W. M. 1992. LEPA, concept and system. In: Proc. Central Plains Irrigation Short Course, Goodland, KS, Feb. 4-5, 1992. Available from KSU Extension Biological and Agricultural Engineering, Manhattan, KS. pp. 14-16.

Lyle, W. M., and J. P. Bordovsky. 1981. Low energy precision application (LEPA) irrigation system. Trans. ASAE 24(5):1241-1245

Lyle, W. M., and J. P. Bordovsky. 1983. LEPA irrigation system evaluation. Trans. ASAE 26(3):776-781.

McAlavy, T. and P. Dillard. 2003. LEPA leaps forward – A crop irrigation revolution born on the Texas Plains. Lifescapes Spring 2003. Texas A& M University, College Station, TX. 3 pp. Also at http://agcomwww.tamu.edu/lifescapes/spring03/lepa.htm

Rogers, D. H., M. Alam, and L. K. Shaw. 2009. Kansas center pivot survey. Kansas State Research and Extension. Irrigation Management Series MF-2870. April 2009. 8 pp. Also at http://www.ksre.ksu.edu/library/ageng2/mf2870.pdf

Seginer, I. 1979. Irrigation uniformity related to horizontal extent of root zone. Irrig. Science 1:89-96.

Schneider, A. D. 2000. Efficiency and uniformity of the LEPA and spray sprinkler methods: A review. Trans. ASAE 43(4):937-944. Schneider, A. D. and T. A. Howell. 1993. Reducing sprinkler water losses. In: Proc. Central Plains Irrigation Shortcourse, February, 2-3, 1993, Sterling, CO. pp. 43-46. methods: A review. Trans. ASAE 43(4):937-944.

Schneider, A. D. and T. A. Howell. 2000. Surface runoff due to LEPA and spray irrigation of a slowly permeable soil. Trans. ASAE 43(5):1089-1095.

von Bernuth, R. D. 1983. Uniformity design criteria under limited water. Trans ASAE 26(5):1418-1421.

Yonts, C. D., F. R. Lamm, W. Kranz, J. Payero and D. Martin. 2005. Impact of wide drop spacing and sprinkler height for corn production. In: Proc. Central Plains Irrigation Conf., Sterling, CO, Feb. 16-17, 2005. Available from CPIA, 760 N.Thompson, Colby, KS. pp. 99-106. Also at http://www.ksret.ksu.edu/irrigate/OOW/P05/Yonts2.pdf

## Triggering drip irrigation onset by soil water tension

Clinton C. Shock, Oregon State University Malheur Experiment Station, Ontario, Oregon clinton.shock@oregonstate.edu

Feng-Xin Wang, China Agricultural University, Beijing, China, fxinwang@cau.edu.cn

Alan D. Campbell, SmartVineyards LLC, Portland, Oregon, alan@smartvineyards.net

Hector Dominguez-Aguirre, SmartVineyards LLC, Portland, Oregon, hector@smartvineyards.net

Sustainable irrigated agricultural production depends on high and stable productivity using limited inputs of water and nutrients. When irrigation events are triggered by soil water tension (SWT) criteria, the criteria should be adjusted to precisely meet plant needs and provide increments of applied water consistent with soil properties. Drip irrigation systems are often used for triggered irrigation due to the potential to precisely control the irrigation system. Here we review the processes of collecting, filtering, and using SWT data for irrigation onset. A compilation of the successful triggering of irrigation onset using SWT criteria for many crops is discussed.

Introduction

Precise irrigation is becoming extremely important due to the shortages of water, competition for water, increase in population, consumer preferences for consumption of meat in the developing countries, and environmental pollution resulting from the excessive application of irrigation water. There are several irrigation scheduling techniques that can be used to obtain excellent irrigation scheduling. These include the use of estimates of crop evapotranspiration, SWT, soil water content, or plant stress. In the following paragraphs we will discuss applications of the use of SWT for precise irrigation scheduling.

Six years ago two of us (Shock and Wang, 2011) summarized the literature on the use of SWT as an irrigation onset criteria. Since that time additional research has fine-tuned irrigation criteria for several other crops. Also we have become aware of parallel research in agricultural engineering where the emphasis has been on triggering irrigation rather than the emphasis on optimal plant responses. Consequently in this brief summary we seek to combine the literature that emphasizes the idea of water tension to trigger irrigation with the literature that emphasizes the idea of water tension criteria for the ideal plant response.

The units of SWT in bars or kilopascals are the force necessary for plant roots to extract water from the soil. The higher the SWT number the dryer the soil. Different plant species have different ranges of ideal SWT irrigation onset criteria. Also the onset criteria to trigger irrigation can vary with the soil type and climate as we have shown previously (Shock and Wang, 2011; see tables 1 and 2 below). Tables 1 through 4 are from Shock et al. (2013). For the detailed references of the research results in tables 1-4 see Shock and Wang (2011).

There are a variety of different instruments that can be used to measure SWT that vary in price and in the range of SWT where they are most accurate. Soil moisture sensors for SWT can be read manually, integrated into an automated reading system, or be integrated into a totally automated system where the SWT feedback information controls the irrigation (Shock et al. 2002).

Soil water tension information can only lead to precise irrigation if plant responses are known and the soil properties are well defined. Critical for precise irrigation is the application of the correct amount of water to refill the root zone without creating undue water percolation. In both manual and automated systems, knowledge of soil properties and characteristics of the irrigation system govern the duration of the irrigation following the crop reaching its irrigation onset criteria. In this way nutrients that are in the soil are retained in the root zone as well as fertilizer inputs supplied by fertigation.

Sensor placement

In any irrigation decision system, the location used to determine whether or not irrigation of a particular block of crops is necessary is a subjective decision. With the use of sensors subjective decisions need to be made as to what part of the field or subpart of the field is appropriate for governing the entire irrigated block within the field. Knowledge needs to be applied as to the number of sensors required to provide an accurate estimate since the soil moisture always varies from spot to spot (Stieber and Shock, 1995). Sensor readings can vary tremendously by sensor position in the soil, positioning of plants in the field, and the relationship between sensor placement and water application. These placements are extremely critical in the case of drip irrigation where the soil more distant from the drip tape or emitters receives less water.

Automated systems

In automated systems not only is there a need for appropriate sensor placement and sensor number, but there is also the need to filter sensor data before each set of sensor readings is averaged and used for automated irrigation. Suppose there are sets of six sensors installed to provide information for the irrigation of a particular zone of a field. The technique that one of us (Shock) developed in 1995 to reduce automated irrigation errors was to sort the data so that errant sensor readings would not cause errors in automated irrigation. Many things can go wrong with wiring or electronics. Reasonable ranges of device readings can be determined and if the readings for a particular sensor are outside of the reasonable range that sensor needs to be flagged and not used in calculations until the operator can check the device for proper operation. The readings from the other devices that are within that zone can be used to provide automated irrigation.

The automated system must make readings of the field sufficiently frequently to catch the moment when the SWT in the field reaches the onset criteria to trigger irrigation. Then the automated program runs the irrigation system in that zone for the time needed to replace the water in the soil according to the irrigation system's properties and the soil water retention characteristics of the soil. As a matter of brief review, the units of SWT (kPa = cb) are the force necessary for plant roots to extract water from the soil and send the water up into the plant. The higher the SWT number the dryer the soil. Different plant species have different ranges of ideal SWT irrigation onset criteria (Tables 3 and 4). For the detailed references of the research results tabulated here see Shock and Wang (2011).

Recent advances

Progress has continued at a rapid pace. One example is the work by Paris et al. (2017) where *Stevia rebaudiana* (stevia) leaf quantity and stevia leaf quality as measured by the natural non-caloric sweet steviol glycosides were closely related to the SWT criteria triggering irrigation. Stevia leaf yield was highest at relatively wet SWT (10 to 20 kPa) and several sweet leaf constituents were also closely related to similar wet irrigation criteria. Contreras et al. (2017) showed that zuchinni was best irrigated at 25 kPa. Seidel et al. (2017) report that cabbage was best irrigated at 25 kPa. Muller et al. 2016 showed that eggplant growth could be divided into two stages, where early plant development would be irrigated at 15 kPa and then fruit development would be best irrigated at 40 kPa. Kumar et al. (2016) examined flooded rice and recommended and irrigation criteria of 30 kPa, considerably drier than flooded conditions. Felix et al. (2015) studied sweet potato irrigation criteria and found that the best irrigation criteria was 25. In the first year of the study of the sweet potatoes the wettest treatment tested was (40) kPa, which was better than the drier treatments. Létourneau et al. (2015) examined, strawberry irrigation and recommended a triggering irrigation onset criteria of 10 kPa. Xi et al. (2014) studied the irrigation of the popular specie *Populus tomentosa*, and recommended a triggering onset criteria of 50 to 75 kPa. Rekika et al. (2014) worked on a series of vegetable species and recommended triggering irrigation onset criteria of 20, 15 to 30, and 10 kPa for onion, celery, and spinach, respectively. Evangelista et al. (2013) studied coffee and found ideal irrigation onset criteria of 32 kPa during flowering and fruit formation and 38 kPa during fruit maturation.

We are currently examining the drip irrigation of vineyards by and are studying the use of SWT criteria for triggering irrigations in widely different environmental circumstances.

Literature cited

Contreras, J.I.., F. Alonso, G. Cánovas, and R. Baeza. 2016. Irrigation management of greenhouse zucchini with different soilmatric potential level. Agronomic and environmental effects. Agricultural Water Management 183:26–34.

Evangelista, A.W.P., L.A. Lima, A.C. Da Silva, C. De P. Martins, and M.S. Ribeiro. 2013. Soil water potential during different phenological phases of coffee irrigated by center pivot. Eng. Agric., Jaboticabal 33(2):269-278.

Felix, J., C.C. Shock, J. Ishida, E.B.G. Feibert, L.D. Saunders. 2015. Irrigation criteria and sweetpotato cultivar performance in the Treasure Valley of Eastern Oregon. HortScience 50(7):1011-1017.

Kumara, A., A.K. Nayak, S. Mohanty, and B.S. Das. 2016. Greenhouse gas emission from direct seeded paddy fields under different soil water potentials in Eastern India. Agriculture, Ecosystems and Environment 228:111–123.

Létourneau, G., J. Caron, L. Anderson, and J. Cormier. 2015. Matric potential-based irrigation management of field-grown strawberry: Effects on yield and water use efficiency. Agricultural Water Management 161:102–113

Müller, T., C. Ranquet Bouleaua, and P. Perona. 2016. Optimizing drip irrigation for eggplant crops in semi-arid zones using evolving thresholds. Agricultural Water Management 177:54–65

Parris, C.A., C.C. Shock, and M. Qian. 2017. Soil water tension irrigation criteria affects *Stevia rebaudiana* leaf yield and leaf steviol glycoside composition. HortSci. 52(1):154–161. doi: 10.21273/HORTSCI11352-16

Rekika, D., J. Caron, G.T Rancourt, J.A. Lafond, S.J. Gumiere, S. Jenni, and A. Gosselin. 2014. Optimal irrigation for onion and celery production and spinach seed germination in histosols. Agronomy Journal 106(2):981-994.

Shock, C.C., E.B.G. Feibert, L.D. Saunders, and E.P. Eldredge. 2002. Automation of Subsurface Drip Irrigation for Crop Research. American Society of Agricultural Engineers. World Congress on Computers in Agriculture and Natural Resources. Iguazu Falls, Brazil. pp. 809-816.

Shock, C.C. and F.X. Wang. 2011. Soil water tension, a powerful measurement for productivity and stewardship. *HortScience* 46:178–185.

Shock, C.C., F.X. Wang, R.J. Flock, E.B.G. Feibert, C.A. Shock, and A.B. Pereira. 2013. Irrigation monitoring using soil water tension. Sustainable Agriculture Techniques, Oregon State University Extension Service. EM 8900 10p. http://ir.library.oregonstate.edu/xmlui/bitstream/handle/1957/37569/em8900.pdf

Shock, C.C., F.X. Wang, R.J. Flock, E.B.G. Feibert, C.A. Shock, and A.B. Pereira. 2013. Irrigation monitoring using soil water tension. Sustainable Agriculture Techniques, Oregon State University Extension Service. EM 8900 10p. http://ir.library.oregonstate.edu/xmlui/bitstream/handle/1957/37569/em8900.pdf

Stieber, T.D., and C.C. Shock. 1995. Placement of soil moisture sensors in sprinkler irrigated potatoes. Am Potato J. 72:533-543.

Xi, B., G. Li, M. Bloomberg and L. Jia. 2014. The effects of subsurface irrigation at different soil water potential thresholds on the growth and transpiration of Populus tomentosa in the North China Plain, Australian Forestry 77:(3-4):159-167. doi: 10.1080/00049158.2014.920552

*Table 1. Soil water tension (SWT) as irrigation criteria for onion bulbs as reviewed by Shock and Wang, 2011.*

| SWT (cb) | Location | Soil type | Irrigation system | Soil moisture sensor depth (inches) |
|---|---|---|---|---|
| 8.5 | Piauí, Brazil | Sandy | Microsprinkler | — |
| 10 | Pernambuco, Brazil | — | Flood | — |
| 15 | São Paulo, Brazil | — | Furrow | — |
| 10–15 | Malheur County, Oregon | Silt loam | Drip | 8 |
| 17–21 | Malheur County, Oregon | Silt loam | Drip | 8 |
| 27 | Malheur County, Oregon | Silt loam | Furrow | 8 |
| 30 | Texas | Sandy clay loam | Drip | 8 |
| 45 | Karnataka, India | Sandy clay loam | — | — |

*Table 2. Soil water tension (SWT) as irrigation criteria for potato as reviewed by Shock and Wang, 2011.*

| SWT (cb) | Location | Soil type | Irrigation system | Soil moisture sensor depth (inches) |
|---|---|---|---|---|
| 20 | Western Australia | Sandy loam | Sprinkler | — |
| 25 | Maine | Silt loam | Sprinkler | — |
| 25 | Luancheng, Hebei Province, China | Silt loam | Drip | 8 |
| 30 | Lethbridge, Alberta, Canada | Sandy loam | Sprinkler | — |
| 30 | Malheur County, Oregon | Silt loam | Drip | 8 |
| 50 | California | Loam | Furrow | — |
| 50–60 | Malheur County, Oregon | Silt loam | Sprinkler | 8 |
| 60 | Malheur County, Oregon | Silt loam | Furrow | 8 |

*Table 3. Soil water tension (SWT) as irrigation criteria for cole crops as reviewed by Shock and Wang, 2011.*

| Common name | SWT (cb) | Soil type | Irrigation system or measurement equipment | Soil moisture sensor depth (inches) | Location, season |
|---|---|---|---|---|---|
| Broccoli (*Brassica oleracea* var. *italica*) | 10–12 | Sandy loam | Subsurface drip | 12 | Maricopa, AZ; fall–winter |
| Broccoli | 50, 20[1] | Silt loam | Lysimeters in rain shelter | 4 | Agassiz, British Columbia, Canada; spring |
| Cabbage (*Brassica oleracea* var. *capitata*) | 25 | Loamy sand and sand | Lysimeters in rain shelter | 4 | Tifton, GA; spring and fall |
| Cauliflower (*Brassica oleracea* var. *botrytis*) | 10–12 | Sandy loam | Subsurface drip | 4 | Maricopa, AZ; fall–winter |
| Cauliflower | 25[2] | Sandy loam | Furrow and flood | 7 | Bangalore, India; winter |
| Cauliflower | 20–40 | Sandy loam | — | — | Skierniewice, Poland; spring–summer |
| Collard | 9 | Sandy loam | Subsurface drip | 12 | Maricopa, AZ; fall–winter |
| Mustard, greens | 6–10 | Sandy loam | Subsurface drip | 12 | Maricopa, AZ; fall–winter |
| Mustard, greens | 25[2] | Loamy sand and sand | Lysimeters in rain shelter | 4 | Tifton, GA; spring and fall |

[1]SWT of 50 cb during plant development, then 20 cb during head development.
[2]Twenty-five cb was the wettest irrigation criterion tested.

*Table 4. Soil water tension (SWT) as irrigation criteria for other field and vegetable crops as reviewed by Shock and Wang, 2011.*

| Common name | SWT (cb) | Soil type | Irrigation system or measurement equipment | Soil moisture sensor depth (inches) | Location, season |
|---|---|---|---|---|---|
| Alfalfa grown for seed | 200–800 | Fine sandy loam, loam, silt loam | Sprinkler and surface flood | 4–72 | Logan, UT; summer season of the perennial crop |
| Beans, snap (*Phaseolus vulgaris*) | 25[z] | Loamy sand | Lysimeters in rain shelter | 4 | Tifton, GA; spring and fall |
| Beans, snap | 45 | Sandy clay loam | — | 6 | Bangalore, India; fall–winter |
| Beans, snap | 50 | Clay loam | Furrow and drip | 12 | Griffin, NSW, Australia; summer |
| Carrot | 30–50 | — | Sprinkler | — | Nova Scotia, Canada; spring–summer |
| Carrot | 40–50 | — | Microsprinkler | 6 | Nova Scotia, Canada; spring–summer |
| Celery | 10 | Sandy loam | Drip | 8 | Santa Ana, CA; fall–winter |
| Corn for sweet corn | 10–40 | Sand | Drip | 6 | — |
| Corn for sweet corn | 30 | Carstic soils | Drip | 12 | Champotón, Campeche, Mexico; spring–summer |
| Corn for sweet corn | 50 | — | — | — | Utah; spring–summer |
| Corn for grain | 30 | Loamy fine sand | Sprinkler | 6 | Quincy, FL; spring–summer |

**Table 4 continues**

**continued**—*Table 4. Soil water tension (SWT) as irrigation criteria for other field and vegetable crops as reviewed by Shock and Wang, 2011.*

| Common name | SWT (cb) | Soil type | Irrigation system or measurement equipment | Soil moisture sensor depth (inches) | Location, season |
|---|---|---|---|---|---|
| Corn for grain | 50 | — | — | — | Utah[5] |
| Cucumber | 15–30 | Fine sand and sandy clay | Drip | 8 | Piikkio, Finland; spring–summer |
| Lettuce, romaine | <6.5 | Sandy loam | Subsurface drip | 12 | Maricopa, AZ; fall–winter |
| Lettuce, leaf | 6–7 | Sandy loam | Subsurface drip | 12 | Maricopa, AZ; fall–winter |
| Lettuce | <10 | Red earth | Drip | 12 | NSW, Australia |
| Lettuce | 20 | Clay loam, sandy loam | Sprinkler, drip | 6 | Las Cruces, NM; summer-fall |
| Lettuce, romaine | 30[1] | Clay loam | Surface | 12 | — |
| Lettuce, crisphead and romaine | 50 | Sandy loam | Sprinkler | 6 | Salinas, CA; spring–summer |
| Radish | 35 | Silt loam | Drip | 8 | Luancheng, Hebei Province, China; summer–fall |
| Radish | 20 | Sandy clay loam | Control basin and furrow | 7 | Bangalore, India; winter |
| Rice | 16 | Sandy loam | Flood | 6–8 | Punjab, India; summer–fall |
| Spinach | 9 | Sandy loam | Drip | — | Maricopa, AZ |
| Squash, summer | 25[1] | Loamy sand and sand | Lysimeter | — | Tifton, GA; spring, summer, and fall |
| Sweet potato | 25, then 100[2] | Loamy sand and sand | Lysimeters in rain shelter | 9 | Tifton, GA; summer |
| Sweet potato | 25–40 | Silt loam | Drip | 8 | Ontario, OR; summer |
| Tomato | 10 | Fine sand | Drip | 6 | Gainesville, FL; spring |
| Tomato | 20 | Sand | Drip | 6 | Coruche, Portugal; spring–summer |
| Tomato | 12–35[3] | Clay | Drip | 4–8[4] | Federal District, Brazil; fall–winter |
| Tomato | 50 | Silt loam | Drip | 8 | Yougledian, Tongzhou, Beijing, China; summer |
| Watermelon | 7–12.6 | Sandy loam | Drip | 12 | Maricopa, AZ; spring–summer |

[1]Twenty-five cb or 30 cb was the wettest irrigation criterion tested.
[2]SWT of 25 cb during plant development, then 100 cb during root enlargement.
[3]Thirty-five, 12, and 15 cb during vegetative, fruit development, and maturation growth stages, respectively.
[4]Tensiometer depth was 4" during the vegetative growth stage, 6" in the beginning of the fruit development stage, and 8" from thereon until the irrigations were stopped.
[5]Taylor, S.A., D.D. Evans, and W.D. Kemper. 1961. *Evaluating Soil Water.* Utah Agricultural Experiment Station Bulletin 426.

# Reducing Nitrous Oxide Emissions with Sub-surface drip Irrigation and Split Fertilizer Applications

Dave Goorahoo, PhD
California State University, Fresno. 2415 E. San Ramon Ave MS AS 72. Fresno, CA. 93740.

Navreet Mahal, MS
Iowa State University, 2017 Agronomy Hall, Ames, IA. 50011.

Florence Cassel Sharma, PhD
California State University, Fresno. 2415 E. San Ramon Ave MS AS 72. Fresno, CA. 93740.

Touyee Thao, MS
California State University, Fresno. 5370 N. Chestnut M/S OF 18, Fresno, CA. 93740.

Govind Seepersad, PhD
The University of the West Indies, St. Augustine, Trinidad & Tobago, West Indies.

**Abstract**. *Of the three biogenic greenhouse gases, (i.e., carbon dioxide ($CO_2$), methane ($CH_4$) and nitrous oxide ($N_2O$), $N_2O$ is considered to be most potent. The overall goal of this study was to determine detailed time series of soil $N_2O$ fluxes at crucial management events for tomatoes subjected to deficit subsurface drip irrigation (SDI) regime and multiple fertilizer application rates. Flux chamber measurements were conducted using an EPA approved methodology to collect air samples that were ultimately analyzed using a Gas Chromatograph. Significant differences in the $N_2O$ fluxes due to the irrigation and/or fertilizer treatments generally peaked within two hours after fertilizer application. Overall, there was a moderate positive correlation between the amount of N2O-N emitted and the fertilizer applied (r= 0.64) and with the volume of water applied (r= 0.74). More importantly, these emission rates were relatively constant in both years at 0.002 kg N2O-N per ha per lb of N fertilizer and would imply that the incremental addition of both fertilizer and water through SDI could be highly efficient management practices to minimize the N2O emissions in tomato cropping systems.*

**Keywords**: Sub-surface drip irrigation; nitrous oxide; greenhouse gases; deficit irrigation.

## Introduction

The effects of the anthropogenic increase in atmospheric greenhouse gas (GHG) concentrations on climate change are beyond dispute (IPCC, 2007), and agriculture does play a key role in this issue, both as a source and a potential sink for GHG (California Energy Commission, CEC, 2005). Of the three biogenic GHGs (i.e., $CO_2$, $CH_4$, and $N_2O$) contributing to radiative forcing in agriculture, $N_2O$ is the most important GHG to be considered, researched, and eventually controlled within intensive and alternative cropping systems. It is estimated that in California, agricultural soils account for 64% of the total $N_2O$ emissions, and $N_2O$ may contribute as much as 50% to the total net agricultural greenhouse gas emissions (CEC, 2005). However, the reliability of these estimates is highly uncertain, which stems, in part, from a lack field measurements in California (CEC, 2005; EPA 2010), and in part, from the inherently high temporal variability of $N_2O$ flux from soils. In a statistical analysis of 1125 $N_2O$ studies from all over the world, the average 95% confidence interval was -51% to +107% (Stehfest and Bouwman, 2006). Among California's statewide greenhouse gas emissions, the magnitude of $N_2O$ emissions is the most uncertain (CEC 2005).

Episodes of high $N_2O$ fluxes are often related to soil management events like N fertilization, irrigation, or incorporation of crop residue, but the magnitude of the responses to such field operations also depends on soil physical and chemical factors, climate and crop system. Meta-analyses based on over 1000 studies found that fertilizer N application rates have significant effects on $N_2O$ emissions, in addition to other factors like fertilizer type, crop type, or soil texture (Bouwman et al., 2002 a and b; Stehfest and Bouwman, 2006). Many of California's high-value crops are intensively managed in terms of N fertilizer use and irrigation, which are factors that have the potential to contribute to substantial $N_2O$ emissions. Furthermore, California's mild winter temperatures and erratic rainfall patterns may be conducive to sporadic high $N_2O$ emissions in the winter. The intensive management of cropland and the dependence on irrigation might also present opportunities to optimize management practices in order to mitigate $N_2O$ emissions (CDFA, 2012).

With the rapid improvement of irrigation technologies and as vegetable cropping systems continue to transition from furrow to sub- surface drip irrigation (SDI), there is a need to evaluate the impact of SDI on $N_2O$ emissions. Furthermore, it is also essential to investigate the combined effect of SDI and other agronomic and cultural management practices. For example, Kallenbach et al. (2010) compared effects of SDI versus flood irrigation and winter cover crop system versus no cover crop system in tomato on $N_2O$ emissions using a flux chamber method, and concluded that SDI showed promise in reducing overall $N_2O$ emissions in crop rotations with legume cover crops. Similar evaluations are needed for management systems that implement SDI with fertilizer management strategies, such as split application of Nitrogen (N) fertilizers throughout the growing season.

## Objective

The overall goal of on-going research is to determine detailed time series of $N_2O$ fluxes and underlying factors at crucial management events (irrigation, fertilization, etc.) in representative vegetable cropping systems in the San Joaquin Valley (SJV) of California. The objective of the current study was to determine the $N_2O$ fluxes from a tomato crop subjected to three SDI irrigation rates (100, 80 and 60 % of total Evapotranspiration (ET)) and three N fertilizer rates of Urea Ammonium Nitrate (UAN-32 at 100, 150 and 200 lbs N/acre).

## Materials and Methods

The tomato study was conducted on Center for Irrigation Technology (CIT) research plots at California State University, Fresno (Fresno State) located at GPS co-ordinates latitude 36° 81' 51.63" N, longitude -119° 73'21.38" W. Two tomato trials were conducted in 2012 and 2013 with a fresh market cultivar, *Quali T-47*, that is well adapted to the hot summer conditions in the SJV. Soils at the experimental site were characterized as Hanford Fine sandy loam soil.

Fresh market tomato cultivar *Quali T-47*, which is a beefsteak, determinate and late maturity type was hand transplanted in late May in 2012 and in mid-June in 2013 on beds that were 5 feet wide and 75 feet long. Plant spacing was 12 inches. The crop was harvested in August 2012 and in September 2013, equivalent to 100 days after transplanting (DAT) by hand picking the fruits. The fruits were separated into green, breaker and red fruits. The total yield, marketable and non-marketable yields were recorded. In addition, the Brix values, a measure of the total soluble sugars (TSS), of red fruits were also recorded.

The experimental layout was a split plot design with SDI rates (I) being the major factor and fertilizer rates (F) being the sub plot factor. The irrigation rates comprised of one standard rate and two deficit irrigation rates where the I1 treatment was equivalent to 100% of the daily evapotranspiration rate (ET), I2 was 80% ET and I3 was 60% ET. The ET was calculated using the California Irrigation Management Information System (CIMIS) station number 80 located on the CSU, Fresno campus as a reference ET, which was then converted to use with a tomato cropping system using published crop coefficients (Amayreh and Al-Abed, 2005). A manifold with three irrigation lines for the three irrigation rates controlled by electronic valves in connection with automated data logger system. An electronic meter was used to calculate the amount of water added to each irrigation treatment. Irrigation was performed using a sub-surface drip irrigation system, with drip lines buried at six inches.

Urea Ammonium Nitrate (UAN 32) was used at three different rates 100 lbs/acre (F1), 150 lbs/acre (F2) and 200 lbs/acre (F3) as fertilizer rate treatments. In 2012, the fertilizer was applied by splitting net application rate into 10, 15, 20, 20, 20 and 15% of at 9, 21, 27, 45, 56 and 65 days after transplanting (DAT). In 2013, a basal rate of 15lbs N/ac was applied to all plots. Then, the remainder of the fertilizer for the three treatment rates were applied at rates equivalent to 10, 10, 20, 25 and 20% of the total N rate at 13, 27, 40, 47 and 54 DAT. Typical nitrogen application rate in California used by growers is 125-250 lbs/acre.

Rectangular stainless steel chamber bases (50 x 30 x 8 cm) were installed in each plot to a depth of approximately 5 cm. These chambers were left in place throughout the growing season. Flux measurements were performed, following the USDA-ARS GRACEnet project protocols (Parkin and Venterea, 2010), by placing stainless steel chamber tops lined with a rubber gasket on the chamber bases and collecting gas samples after 0, 20 and 40 minutes. Air samples were collected from the chamber's headspace with a needle and a 20 ml syringe, and were stored at room temperature (20°C) in 12 ml Labco glass vials until analyzed with a Gas Chromatograph (GC). Chamber and air temperatures were measured during each gas sampling time, and the ppm data derived from the GC was adjusted for the chamber temperature variation and converted to flux data by following the protocol recommended by the California Air Resources Board (CARB).

A total of 10 sampling events occurred over the 2012 season. Of these 10 events, 9 were centered around fertilizer applications with sampling events at DAT 27, 43 and 64 occurring a day prior to fertilizer application, events at DAT 28, 45 and 65 occurring the same day as fertilizer application and events at DAT 29, 46 and 66 occurring one day after fertilizer application. The final sampling event occurred at harvest and corresponded to DAT 100.

In 2013, there was a total of 22 sampling events. Generally, flux measurements were conducted a day before the fertilizer application, and then at 2 hours, 24 hours and 48 hours after the fertilizer application during drip irrigation. Sampling events were centered around fertilizer applications on the following DAT: 12, 26, 40, 47, 54, and 64. The final sampling event occurred prior to harvest and corresponded to DAT 83.

$N_2O$ fluxes were calculated from the rate of change of the concentration of $N_2O$ in the chamber headspace and for this GRACEnet protocol was followed. According to this protocol, if the rate of change of trace gas concentration in the headspace was constant then linear regression was used to calculate the slope of concentration vs time data otherwise curvi-linear concentration data with time was used (Parkin and Venterea, 2010). For calculation of total $N_2O$–N emissions for different treatments throughout the crop season, flux rates over the entire crop season were interpolated linearly and integrated to determine the cumulative N emissions calculated in the units g N/ha. The final flux data were subjected to analysis of variance (ANOVA) at a probability of 0.05 using Microsoft Excel 2010 software. The separation of means was conducted using Tukey's HSD ($\alpha$ =0.05).

## Results

Tomato Yield: In 2012, there was no significant effect of either fertilizer rate or the interaction between irrigation and fertilizer rates on total fruit yield, non- marketable yield, marketable yield, Green tomato weight, red tomato weight, breaker tomato weight and Brix indices of fruits (Table 1). However, irrigation rates affected total weight, marketable, green tomato and breaker tomato yields with the highest values from the irrigation treatment with 100% ET as compared to those from 80 and 60% ET (Table 2). The Brix values of tomato fruits were highest from the treatment with 60%ET compared to plants that received 80 and 100% of daily ET. In 2013, fertilizer and/or irrigation had no significant effects on any of the tomato yields (Table 3).

Table 1: Level of Significance from ANOVA for tomato yields obtained in 2012.

| Treatment | Total Weight | Non-marketable Weight | Marketable Weight | Green Weight | Breaker Weight | Red Weight | Brix |
|---|---|---|---|---|---|---|---|
| Irrigation | 0.003* | 0.126 | 0.004* | 0.021* | 0.015* | 0.117 | 0.025* |
| Fertilizer | 0.627 | 0.797 | 0.713 | 0.784 | 0.737 | 0.825 | 0.366 |
| Irrigation x fertilizer | 0.666 | 0.451 | 0.848 | 0.412 | 0.475 | 0.594 | 0.489 |

Table 2: Mean weights (lbs per subplot) for tomatoes subjected to the various irrigation rates. *Values followed by the same letters are not significantly different at the α= 0.05 level.*

| ET Rate (%) | Total Wt. | Non-marketable Weight | Marketable Weight | Green Weight | Breaker Weight | Red Weight | Brix |
|---|---|---|---|---|---|---|---|
| 100 | 21.93 a | 1.74 a | 20.183 a | 14.86 a | 2.5 a | 2.82 a | 3.68 b |
| 80 | 14.67 b | 1.75 a | 12.92 b | 9.98 b | 1.43 b | 1.5 a | 3.95 b |
| 60 | 11.9 b | 0.98 a | 10.92 b | 7.47 c | 1.07 b | 2.37 a | 4.56 a |

Table 3: Level of Significance from ANOVA for tomato yields obtained in 2013.

| Treatment | Total Weight | Green Weight | Breaker Weight | Red Weight |
|---|---|---|---|---|
| Irrigation | 0.456 | 0.248 | 0.502 | 0.094 |
| Fertilizer | 0.210 | 0.520 | 0.252 | 0.855 |
| Irrigation x fertilizer | 0.733 | 0.826 | 0.834 | 0.565 |

$N_2O$ Emissions from Tomato Crops in 2012 & 2013: The total fluxes and amount of $N_2O$-N emitted on a kg per ha (or lbs/ac) basis were determined by integrating the area under the time series graphs generated for each growing season. Figures 1 and 2 show the nitrous oxide emissions as a function of (a) irrigation (I) and (b) fertilizer (F) rates throughout the 2012 and 2013 tomato seasons, respectively. A summary of **total** $N_2O$ emissions from the (a) irrigation (I) and (b) fertilizer (F) rates throughout the 2012 and 2013 as a function of fertilizer and irrigation rates is provided in Table 4. A sampling protocol that included continuous monitoring, or at least more frequent sampling events, would have provided a better depiction of seasonal $N_2O$ fluxes.

(a)

(b)

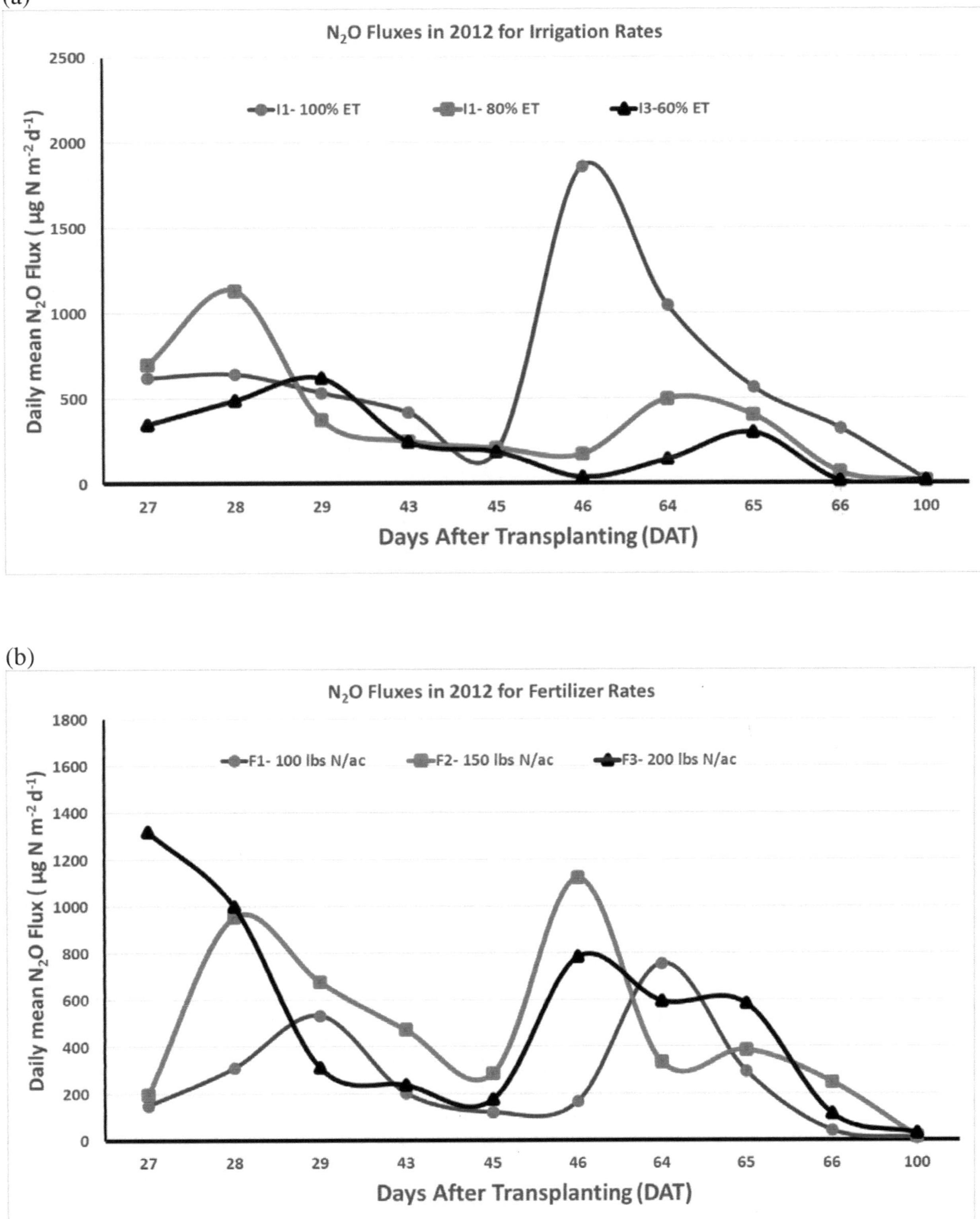


Figure 1. Nitrous oxide emissions as a function of (a) irrigation (I) and (b) fertilizer (F) rates throughout the 2012 tomato season.

(a)

(b)

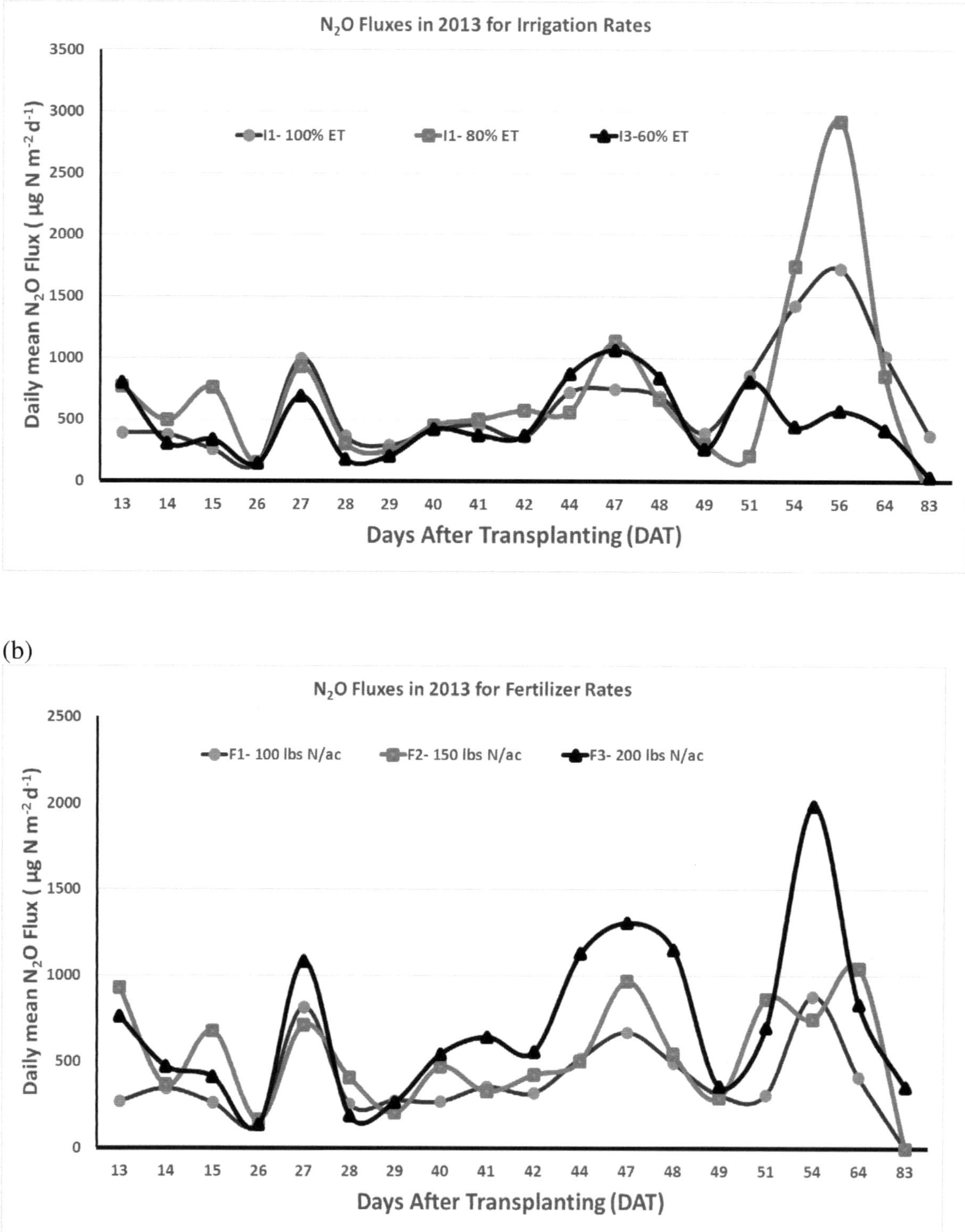


Figure 2. Nitrous oxide emissions as a function of (a) irrigation (I) and (b) fertilizer (F) rates throughout the 2013 tomato season.

Table 4: Summary of total $N_2O$ emissions from the tomato crops in 2012 and 2013 as a function of fertilizer and irrigation rates.

| | 2012 | | | | 2013 | | |
|---|---|---|---|---|---|---|---|
| **Fertilizer** | **F1** | **F2** | **F3** | | **F1** | **F2** | **F3** |
| TOTAL N2O emitted (ug/m2) | 16167.6 | 29134.14 | 22333.74 | | 20306 | 35489.1 | 44350. |
| N2O-N emitted in kg N/ha | 0.162 | 0.291 | 0.223 | | 0.203 | 0.355 | 0.444 |
| N2O-N emitted in lbs N/ac | 0.144 | 0.259 | 0.199 | | 0.181 | 0.316 | 0.395 |
| Total N applied per acre (lbs N/ac) | 100 | 150 | 200 | | 100 | 150 | 200 |
| N2O-N emitted in kgN/ha/ lb | 0.0016 | 0.0019 | 0.0011 | | 0.002 | 0.002 | 0.002 |
| N2O-N emitted in lbs N/ac/lb | 0.0014 | 0.0017 | 0.0010 | | 0.001 | 0.0021 | 0.0020 |
| Relative Change in emissions | NA | 0.06% | -0.15% | | NA | 0.06% | -0.03% |
| | | | | | | | |
| **Irrigation** | **I1-** | **I2-** | **I2-** | | **I1-** | **I2-** | **I2-** |
| TOTAL N2O emitted (ug/m2) | 42753.1 | 14731.38 | 10150.93 | | 45751 | 47634.7 | 26616. |
| N2O-N emitted in kg N/ha | 0.428 | 0.147 | 0.102 | | 0.458 | 0.476 | 0.266 |
| N2O-N emitted in lbs N/ac | 0.381 | 0.131 | 0.090 | | 0.407 | 0.424 | 0.237 |
| Total water applied (mm) | 432 | 346 | 259 | | 444 | 355 | 266 |
| N2O-N emitted in kgN/ha/ mm | 0.0010 | 0.0004 | 0.0004 | | 0.001 | 0.0013 | 0.0010 |
| N2O-N emitted in lbs N/ac/ mm | 0.0009 | 0.0004 | 0.0003 | | 0.000 | 0.0012 | 0.0009 |
| Relative Change in emissions | NA | 0.06% | 0.003% | | NA | -0.03% | 0.03% |

For example, the graphs generated for the 2013 season (Figure 2) which comprised of 22 sampling evens versus that generated for the 2012 season (Figure 1) with 10 sampling events, would allow for a more accurate interpolation of the total fluxes between sampling events.

Based on the summary provided in Table 4, the amount of $N_2O$-N in kg per ha emitted during tomato cropping season ranged from 0.162 to 0.291 in 2012 and from 0.203 to 0.444 in 2013. More importantly, when these emissions were expressed on the basis of the amount of fertilizer applied throughout the season, the emission rates were relatively constant in both years at 0.002 kg $N_2O$-N per ha per lb of N fertilizer. Overall, there was a moderate positive correlation ($r= 0.64$) between the amount of $N_2O$-N emitted and the fertilizer applied, with the correlation being relatively stronger in 2013 ($r = 0.99$) than in 2012 ($r = 0.48$).

With respect to the volume of water applied during the 2012 season, the amount of $N_2O$-N emitted increased from 0.102 kg $N_2O$-N per ha per mm water for plots receiving 60%ET (I3) to 0.428 kg $N_2O$-N per ha per mm water for the 100%ET irrigated plots. In 2013, the amount of $N_2O$-N emitted from the 80%ET (I2) and 100%ET (I1) irrigated plots were approximately 1.7 times greater than the emissions from the plots irrigated at 60%ET (I3). Overall, there was a positive correlation ($r= 0.74$) between the amount of $N_2O$-N emitted and the volume of water applied, with the correlation being relatively stronger in 2013 ($r = 0.92$) than in 2012 ($r = 0.82$).

## Concluding Remarks

For fresh market tomatoes grown on a sandy loam soil, fertilized with UAN-32, and irrigated with subsurface drip irrigation (SDI) the major findings from the current study were:

- Fertilizer and irrigation rates appeared to significantly influence the $N_2O$ emission within 2 hours of fertilizer application;

- The amount of $N_2O$-N in kg per ha emitted during tomato cropping season ranged from 0.162 to 0.291 in 2012 and from 0.203 to 0.444 in 2013. More importantly, when these emissions were expressed on the basis of the amount of fertilizer applied throughout the season, the emission rates were relatively constant in both years at 0.002 kg $N_2O$-N per ha per lb of N fertilizer;
- Overall, there was a moderate positive correlation (r= 0.64) between the amount of $N_2O$- N emitted and the fertilizer applied, with the correlation being relatively stronger in 2013 (r = 0.99) than in 2012 (r = 0.48);
- Overall, there was a positive correlation (r= 0.74) between the amount of $N_2O$-N emitted and the volume of water applied, with the correlation being relatively stronger in 2013 (r = 0.92) than in 2012 (r = 0.82); and,
- The relatively constant emission rates of 0.002 kg $N_2O$-N per ha per lb of N fertilizer determined for the fertilizer and deficit irrigation regimes, would imply that the incremental addition of both fertilizer and water through SDI could be highly efficient management practices to minimize the $N_2O$ emissions in tomato cropping systems.

***Acknowledgements:***
We would like to acknowledge Dr. Martin Burger and the rest of the research group at UC-Davis for their assistance with the analysis of the gas samples. Also a special thank you to the "Grad Lab Team" for the long hours in the field during the collection of data. The California Department of Food and Agriculture (CDFA) and the California State University- Agricultural Research Initiative (CSU-ARI) provided funding for this research.

## Citations & Additional Useful References

Amayreh, J. and Al-Abed, N., 2005: Developing crop coefficients for field-grown tomato under drip irrigation with black plastic mulch. Agricultural Water Mang. 73, 247-254.

ARB CA, 2011; California Environmental Protection Agency Air Resource Board. Available online via http://www.arb.ca.gov/homepage.htm . Last accessed on November 20th 2016.

Bouwman, A. F. 1996. Direct emission of nitrous oxide from agricultural soils. Nutri. Cycl. Agroecosyst., 46(1), 53-70.

Burger, M., L.E. Jackson, E.J. Lundquist, D.T. Louie, R.L. Miller, D.E. Rolston and K.M.Scow. 2005. Microbial responses and nitrous oxide emissions during wetting and drying of organically and conventionally managed soil under tomatoes. Biol. Fertil. Soils. 42(2), 109- 118.

Burger, M., and W.R. Horwath. 2012. Assessment of Baseline Nitrous Oxide Emissions in California Cropping Systems. Sacramento, CA: California Air Resources Board.

Burger, M. and W. R. Horwarth. 2013, Assessment of Baseline Nitrous Oxide Emissions in Response to a Range of Nitrogen Fertilizer Application Rates in Corn Systems. West, L.D. & Lewis-Ortega, E.A., California department of Food and Agriculture, Fertilizer Research & Education Program Conference, Modesto, October 29-30,2013. CDFA-FREP, Sacramento, Pg- 67-70.

Cai, Z., S. Gao, A. Hendratna, Y. Duan, B. Hanson. 2014. Effect of urea application rate and water content on nitrous oxide emission from a sandy loam soil - a laboratory study. California Plant and Soil Conference Proceedings. p.131.

Casciotti, K.L., and B.E.Ward. 2001. Dissimilatory nitrite reductase genes from autotrophic ammonia-oxidizing bacteria. Applied and Environmental Microbiology, 67, 2213–2221.

CDFA 2012. California Department of Food and Agriculture. Available online via. http://www.cdfa.ca.gov/. Last accessed on November 20th 2014.

CEC, 2005. California Energy Commission, Bemis, G.; J. Allen. Inventory of California Greenhouse Gas Emissions and Sinks: 1990 to 2002 Update. California Energy Commission (CEC) Publication, 2005, CEC-600-2005-025.

Chapuis-Lardy, L., N. Wrage, A. Metay, J.L. Chottes, and M. Bernouxs. 2007: Soils, a sink for N2O? A review. Glob. Chang. Biol., 13, 1–17.

DeWall, M. 2014. Effect of fertilizer and irrigation rates on CO2 and N2O emissions from tomatoes grown on a sandy loam soil. Master's Senior Project. Biotechnology Program. California State University Fresno. 51pp.

Epa, 2010: United States Environmental Protection Agency. Overview of Greenhouse Gases. Available online via: http://epa.gov/climatechange/ghgemissions/gases/n2o.html. Last accessed on November 20th 2014.

Garland, G. M., E. Suddick, M. Burger, W.R. Horwath, and J. Six. 2011. Direct N2O emissions following transition from conventional till to no-till in a cover cropped Mediterranean vineyard (Vitis vinifera). Agric., Ecosyst. & Environ., 144(1), 423-428.

IPCC, 2007, Climate Change 2007: Mitigation of Climate Change Contribution of Working Group III to the Fourth Assessment Report of the Intergovernmental Panel on Climate Change, 2007, edited by B. Metz, O.R. Davidson, P.R. Bosch, R. Dave, L.A. Meyer (eds). Cambridge University Press, Cambridge, United Kingdom and New York, NY, USA.

Kallenbach, C. M., D.E. Rolston, and W.R. Horwath. 2010. Cover cropping affects soil N2O and CO2 emissions differently depending on type of irrigation. Agric. Ecosyst. Environ., 137(3), 251-260.

Kennedy, T.L., Suddick, E.C., and Six, J., 2013: Reduced nitrous oxide emissions and increased yields in california tomato cropping systems under drip irrigation and fertigation. Agric. Ecosyst. Environ., 170, 16-27.

Lee, J., J.W. Hopmans, C. van Kessel, A.P. King, K.J. Evatt, D. Louie and J.Six. 2009. Tillage and seasonal emissions of CO2, N2O and NO across a seed bed and at the field scale in a Mediterranean climate. Agric. Ecosyst. Environ., 129(4), 378-390.

Liu, C., X. Zheng, Z. Zhou, S. Han, Y. Wang, K. Wang, W. Liang, M. Li, D. Chen, and Z. Yang. 2010. Nitrous oxide and nitric oxide emissions from an irrigated cotton field in Northern China. Plant and Soil 332, no. 1-2: 123-134.

Mahal, N.K., 2014: Nitrous oxide emissions from cotton fertilized with urea ammonium nitrate. Thesis: California State University, Fresno: Jordan College of Agricultural Sciences and Technology. Available on line at http://cdmweb.lib.csufresno.edu/cdm/ref/collection/thes/id/109463

Mc Taggart, L.P. and K.A. Smith. 1996. Nitrous oxide emissions from arable and grassland soils: Effect of crop type and nitrification inhibitors. Transactions5th Nitrogen Workshop, Braunschweig, Sept. 1996, pp. 523-526.

Mosier, A.R., 1994: Nitrous oxide emissions from agricultural soils. Fertilizer Research. 37, 191- 200.

Parkin, T.T. and R.T.Venterea, November, 2010. USDA-ARS GRACEnet project protocols. Available online htttp://www.ars.usda.gov/SP2UserFiles/person/31831/2011%20Parkin%20and%20Venterea%20Trace%20Gas%20Protocol%20Revision%20Final.pdf Last Accessed on December 3rd, 2016.

Russer, R., H. Flessa and F. Beese. 1996. Plant-type and N-fertilizer dependent variation of nitrous oxide emissions. Transactions 9th Nitrogen Workshop, Braunschweig, Sept. 1996, pp 539-542.

Scheer, C., R. Wassmann, K. Kienzler, N. Ibragimov, and R. Eschanov.2008. Nitrous oxide emissions from fertilized, irrigated cotton (Gossypium hirsutum) in the Aral Sea Basin, Uzbekistan: Influence of nitrogen applications and irrigation practices. Soil Biol. Biochem.40, no. 2: 290-301.

Schmidt, I., R.J.M., van Spanning, M.S.M., Jetten. 2004. Denitrification and ammonia oxidation by Nitrosomonas europaea wild-type, and NirK- and NorB deficient mutants. Microbiology, 150,4107–4114

Stehfest, E., and L. Bouwman. 2006. $N_2O$ and NO emission from agricultural fields and soils under natural vegetation: summarizing available measurement data and modeling of global annual emissions. Nutrient Cycling in Agroecosystems 74:207-228.

# Field Optimization of FlowControl Drip Tapes for Superior Performance and Profitability

For presentation at The Irrigation Association Convention, November 2017, Orlando, FL

Inge Bisconer, Technical Marketing and Sales Manager, MBA, CID, CLIA, Toro Ag, El Cajon, CA inge.bisconer@toro.com

Sky Anderson, Product Manager, MBA, Toro Ag, El Cajon, CA sky.anderson@toro.com

Jim Klauzer, Agronomist, Clearwater Supply, Ontario, Oregon jim@cwsupply.com

**Abstract.**

The Toro Company released Flow Control Emitters to the marketplace in 2015, and is now pleased to report published results detailing how commercial agricultural operations in the Americas have optimized FlowControl™ drip tapes to achieve superior performance and profitability in row crop applications.

As noted in IA/ASABE Paper Number 2144972, Flow Control emitters possess a unique flow exponent of 0.3. Flow Control emitters in FlowControl drip tapes deliver superior drip system uniformity when compared to turbulent flow drip tapes and driplines with flow exponents of approximately 0.5. At the same time, unlike pressure-compensating driplines with a flow exponent of 0, the drip system application rate may be adjusted with FlowControl drip tapes by increasing or decreasing system pressure. This feature is especially critical during extreme weather conditions or periods when water deliveries become challenging or altered – both of which are becoming more common. Finally, FlowControl drip tapes are available in more-affordable 5/8-inch 6 mil and 7/8-inch 8 mil wall thickness, and are priced the same regardless of emitter spacing. Since closely spaced emitters are often preferred to achieve superior wetting patterns, the price point helps optimize results without being cost-prohibitive.

Keywords: drip, irrigation, tapes, flow, pressure, emitters, uniformity.

**Introduction:**

This paper reports how well-known growing operations in the Americas optimized irrigation system performance, efficiency and farm profitability using FlowControl drip tapes. These experiences include using FlowControl drip tapes to achieve superior drip system uniformity in both normal and challenging terrain; to achieve superior drip system uniformity in blocks with hilly terrain and long lengths of run, without the expense and logistical complexity of installing extra submains and/or jumpers (small diameter tubes which reduce pressure and flow variation in severe slopes); to increase yield and quality versus ordinary drip tapes; to optimize initial wetting patterns for near-perfect germination and plant-setting conditions; and to simplify and reduce costs of block designs where water deliveries and/or availability are significantly reduced during critical times of the growing season. In addition, commentary from academia regarding the significant value of FlowControl drip tapes to irrigated agriculture are included, as well as how AquaFlow drip irrigation design software may be used to compare drip system uniformities using FlowControl drip tapes versus ordinary drip tapes.

As noted in IA/ASABE Paper Number 2144972, Flow Control emitters possess a unique flow exponent of 0.3. This results in better irrigation system emission uniformity (EU) when compared to turbulent flow drip tapes and driplines with flow exponents of ~ 0.5 (see Figure 1). When compared to pressure-compensating driplines with a flow exponent of 0, FlowControl drip tapes allow growers to significantly adjust system application rates by increasing or decreasing system pressure (see Figure 2). This feature is especially critical during extreme weather conditions or periods when water deliveries become challenging or altered, both of which are becoming more common.

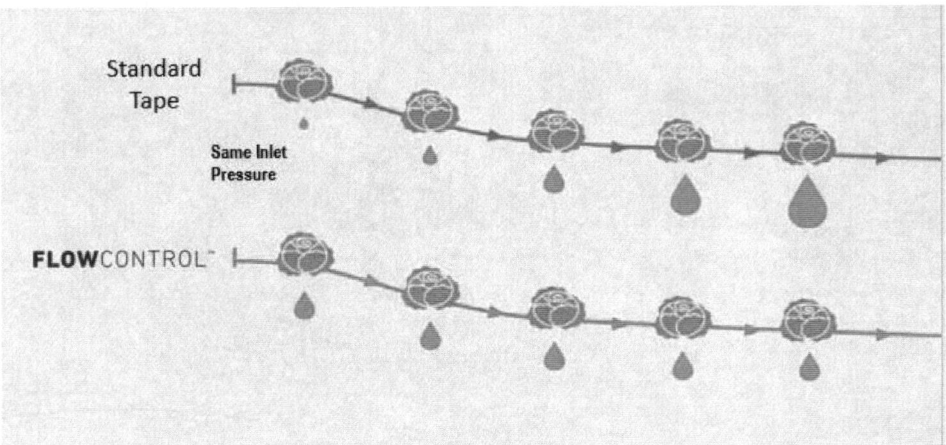


*Figure 1: FlowControl drip tapes, which have a flow exponent of 0.3, provide better uniformity than standard drip tapes, which have a flow exponent of ~0.5.*

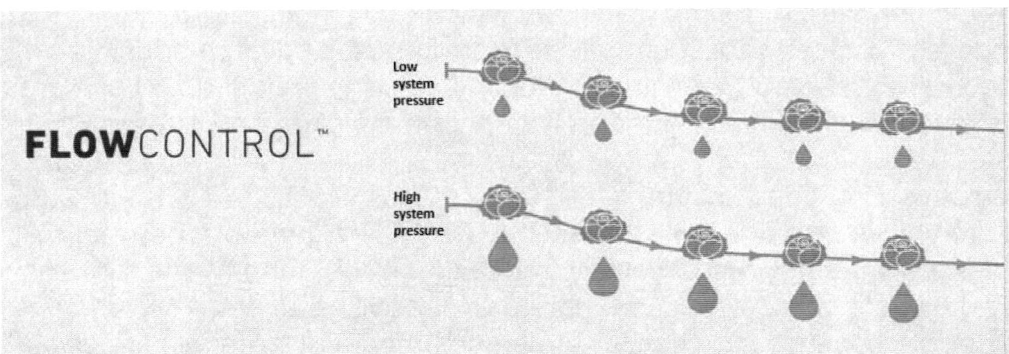


*Figure 2: Unlike pressure-compensating driplines where the application rate is fixed, the application rate of FlowControl drip tapes may be increased or decreased by adjusting pressure. This is important to accommodate changing weather and water availability conditions.*

## Using FlowControl to succeed in challenging terrain

Irrigating uneven, sloped, or hilly terrain is a challenge because low elevations typically are overwatered and high elevations are underwatered. Uneven water distribution often results in disease and poor crop health, and ultimately, poor crop yield, poor crop uniformity and, in vegetables, poor crop quality. Until recently, challenging terrain was often avoided, was addressed with expensive pressure compensating driplines, extra submains or "jumpers", or poor uniformity was simply accepted. Jumpers refers to the practice of inserting smaller diameter tubes in severely sloped tape runs to reduce excessive pressure gain (see Figure 3 and Figure 4).

*Figure 3: The term "jumpers" refers to the practice of inserting small diameter tubes in tape runs to reduce pressure buildup on extremely sloped terrain. Photo courtesy of Jim Klauzer, Clearwater Supply.*

*Figure 4: The use of jumpers on sloped terrain is labor intensive and results in crop damage where the jumpers are installed. Photo courtesy of Jim Klauzer, Clearwater Supply.*

But each of these responses has severe limitations in view of changing market demands, weather patterns, and resource availability:

1. Avoiding challenging terrain is simply a luxury we can no longer afford since there is intense competition for less challenging terrain for purposes such as housing;
2. Accepting poor drip system uniformity is not only financially undesirable but increasingly illegal due to resulting ground and surface water contamination and subsequent regulatory pressures; and
3. Pressure-compensating driplines are
    a. Cost prohibitive because they are only available in expensive, heavier wall thicknesses, and they are priced by the emitter rather than by the foot, making highly desirable closely-spaced emitters expensive, and
    b. Less flexible because they deliver the same amount of water regardless of pressure, thus the system application rate is fixed and can never be adjusted by increasing or lowering pressure to accommodate heat spells or changes in water supply availability. This is a major drawback because this fixed output does not allow for variable conditions that commonly occur for agricultural growers.

4. Using extra submains and/or jumpers increases initial purchase cost, labor and crop damage, and impedes maintenance flushing.

By using FlowControl drip tapes, these challenges may be better met because:

1. FlowControl, with a flow exponent of 0.3, will always provide better uniformity than ordinary drip tapes with a flow exponent of ~0.5.
2. FlowControl application rates may be increased or decreased by adjusting pressure unlike pressure-compensating driplines.
3. FlowControl is available in a wider range of wall thicknesses than pressure-compensating driplines – including more affordable ⅝-inch 6-mil and ⅞-inch 8-mil
4. FlowControl is a cost effective alternative to using extra submains and/or jumpers.

The following case studies provide more detail on these benefits:

- **Trevor Hardy, Brookdale Fruit Farm, Hollis, NH** - "We've seen firsthand how FlowControl helps increase yield and quality. Over the past two years, we have tested it and seen it significantly improve our crops' health, especially in challenging terrain. FlowControl delivers a more uniform crop and higher yields in areas that were once impractical to farm. We've even reduced our pre-plant fertilizer costs, thanks to the improved uniform distribution it offers with fertigation. FlowControl's benefits became especially apparent to us this year – during a major drought we were consistently able to achieve uniformity on long irrigated plasticulture rows."

- **Jim Klauzer, Clearwater Supply, Ontario, OR** - "I have never seen such a fine wetting pattern under such conditions (see Figure 5). In the past we would have had to feed the tape from each side into the swale in an attempt to improve the wetting pattern. This would be expensive and labor-intensive to provide bi-directional water application to a single zone. The benefits of eliminating jumpers and improving uniformity with FlowControl in difficult layout and terrain conditions is better yield and quality, lower fertilizer and water costs, lower system cost, elimination of crop damage from jumper installation, and the ability to properly flush the irrigation system which is impossible with the constriction that jumpers create." (See Figures 6 and 7).

*Figure 5: The wetting pattern on severely sloped terrain is near perfect using FlowControl drip tape. Photo courtesy of Jim Klauzer, Clearwater Supply.*

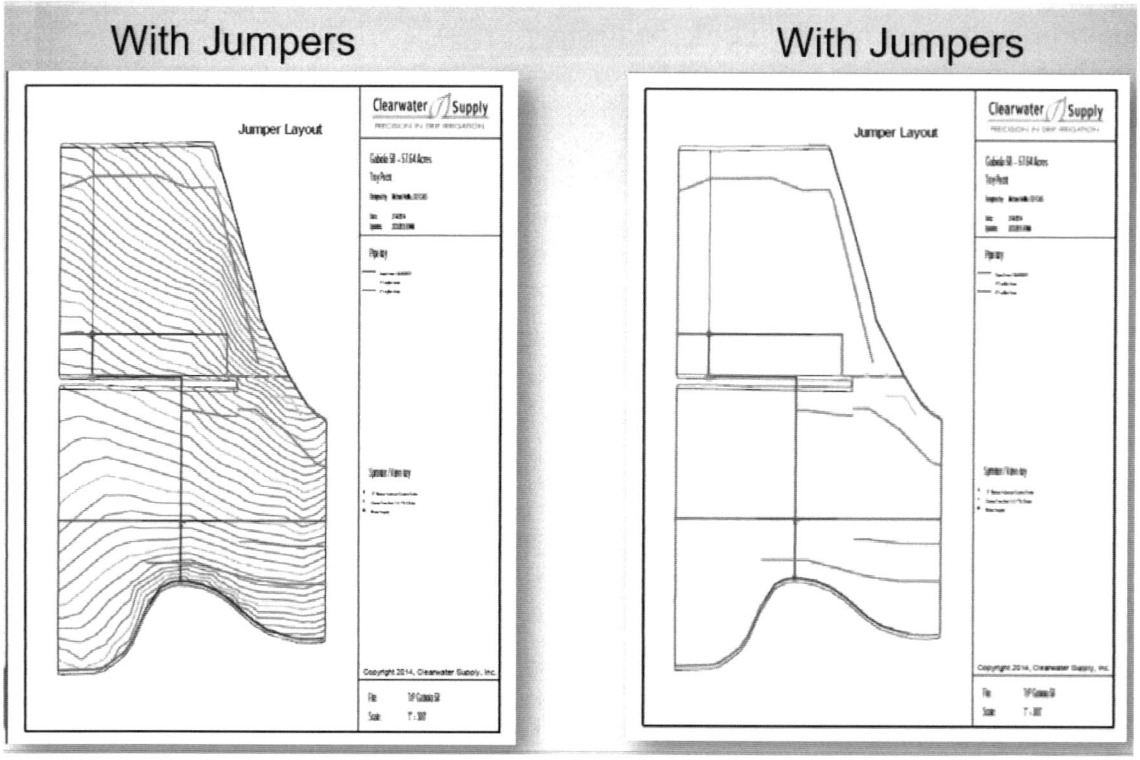


*Figure 6: Topographical map of a field to be irrigated with ordinary drip tape (flow exponent of ~0.5) showing submains in blue and orange, and where jumpers will have to be installed, shown in pink, to reduce pressure buildup on extreme slopes. The use of jumpers significantly increases labor costs, causes field damage, and prevents proper system flushing. Illustration courtesy of Randi Gladwell, Clearwater Supply.*

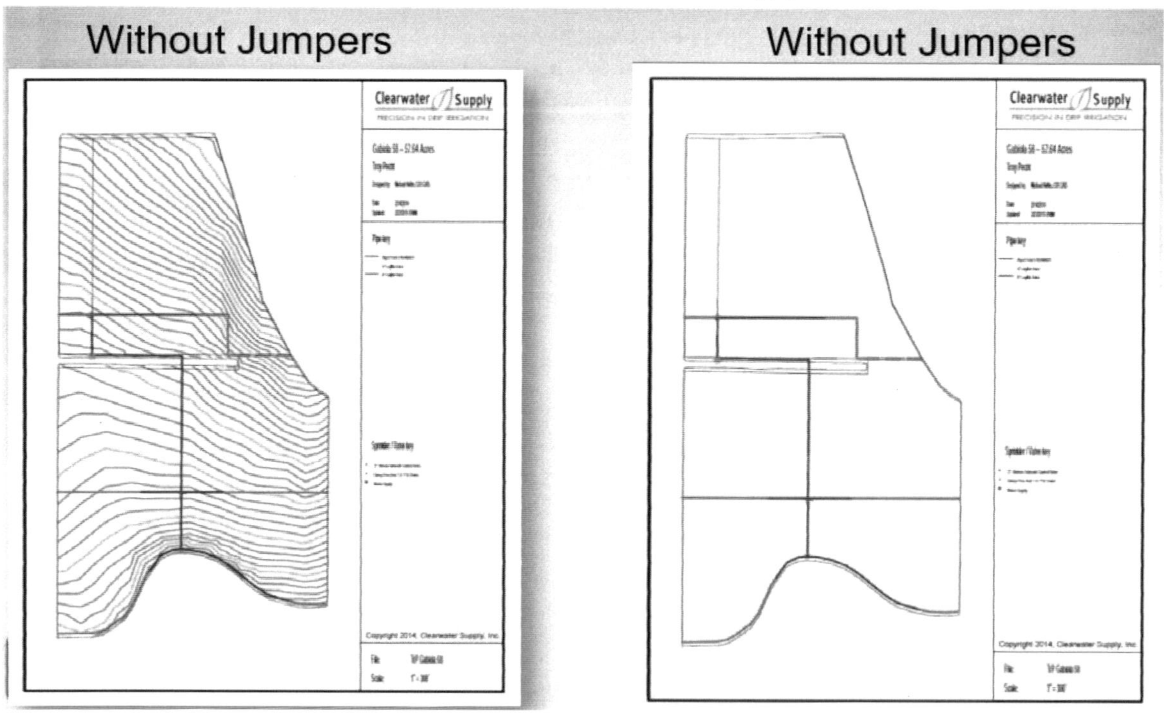


*Figure 7: Topographical map of a field to be irrigated with FlowControl drip tape showing submains in blue and orange. Note that jumpers are not necessary when FlowControl drip tape is used, significantly reducing costs and improving uniformity. Illustration courtesy of Randi Gladwell, Clearwater Supply.*

## Using FlowControl drip tapes to increase uniformity and yields vs. ordinary drip tapes

Ordinary drip tapes stress plants and reduce yield and efficiency by over- or under-watering as pressure changes throughout the field. As a result, water and fertilizer are wasted, and stressed plants reduce yields. Worse yet, challenging terrain isn't farmed at all, or extra expense is incurred installing submains, jumpers or pressure-compensating driplines. FlowControl provides better drip system uniformity than ordinary drip tapes, and is especially beneficial where long lengths of run and/or hilly terrain exist. With better drip system uniformity, improved crop quality and yields are the result.

- **Bianca Pérez-Lizasuain, Peninsula del Rio, Guayanilla, Puerto Rico** "The principal reason we use FlowControl is because our farm is not flat – a 5% to 10% slope. When we tried normal tape, the uniformity was not good. We tried splitting the lots and preparing the land better, but that didn't work. The only thing that improved our uniformity was moving to FlowControl. When we did, we saw the benefits right away – the irrigation and crop growth was uniform, fertilizer-use was better, and we could get longer runs. Before, our maximum run length was 300 feet. With FlowControl we are running 500 feet, which allowed us to eliminate a submain!"

- **Nolan Masser, Red Hill Farms Inc., Pitman, PA** "We wouldn't have moved to drip irrigation if it weren't for FlowControl. Standard drip tape just couldn't provide the results we needed on the slopes of our rolling and uneven terrain. FlowControl not only gives us the uniformity we need, but we've been able to get longer runs as well. Now we get the performance and the water- and energy-efficiency we need – that's what we like about it."

- **Greg Phillips, Oceano Packing Company, Oceano, CA** "With standard drip tape, we had to use up to 2 extra submains to irrigate the whole field. With FlowControl, we're getting longer runs so we can use just one submain. FlowControl is also giving us better uniformity – even in our steep sloping fields. Standard drip tape would under- or over-irrigate in areas."

- **Manuel Paz, Huntington Farms, Soledad, CA** "We are seeing more yield, and the uniformity improvement is visible. The wetting pattern at the beginning of the row is identical to the pattern at the end of the row. When water is equal from the beginning of the field to the end, this means the crop grows evenly and is higher quality."

## Using FlowControl drip tapes to adjust application rates without sacrificing uniformity

Farmers often wish to adjust the application rate of drip systems to quickly accommodate changing weather conditions and/or water availability. For instance, if a heat spell occurs, farmers may want to increase the application rate and apply the necessary water quicker rather than irrigate longer. Since the application rate of pressure-compensating driplines is fixed, farmers typically sacrifice uniformity and opt for the use of ordinary drip tapes whose application rates may be adjusted with pressure. With FlowControl, the application rate may be increased or decreased by adjusting system pressure while at the same time providing better uniformity than ordinary drip tapes. In other words, FlowControl provides more uniform delivery of water and fertilizer while maintaining the flexibility to adjust application rates.

- **Jim Klauzer, Clearwater Supply, Ontario, OR** - "Using FlowControl, we have simplified the block designs and irrigation operation in areas where there are variable output wells and/or where there are temporary water restrictions from the ditch source, such as 25% decreases for a month in the summer. Before FlowControl was available, we would design systems to run in 3 sets when the normal 400 GPM was available from the source (Figure 8), and to run in 4 sets when the restricted 300 GPM was available during peak summer months (Figure 9). This design required extra submains, valves and operational complexity. After FlowControl drip tape became available, the system design was simplified (Figure 10) to run in 4 sets at **14 psi** when the normal 400 gpm was available, and to then run in 4 sets at **8 psi** when delivery flows were reduced. Operational time must increase to deliver a comparable amount of water at 8 psi vs. 14 psi, but this design is much simpler and much less costly than a 3-set and 4-set scenario."

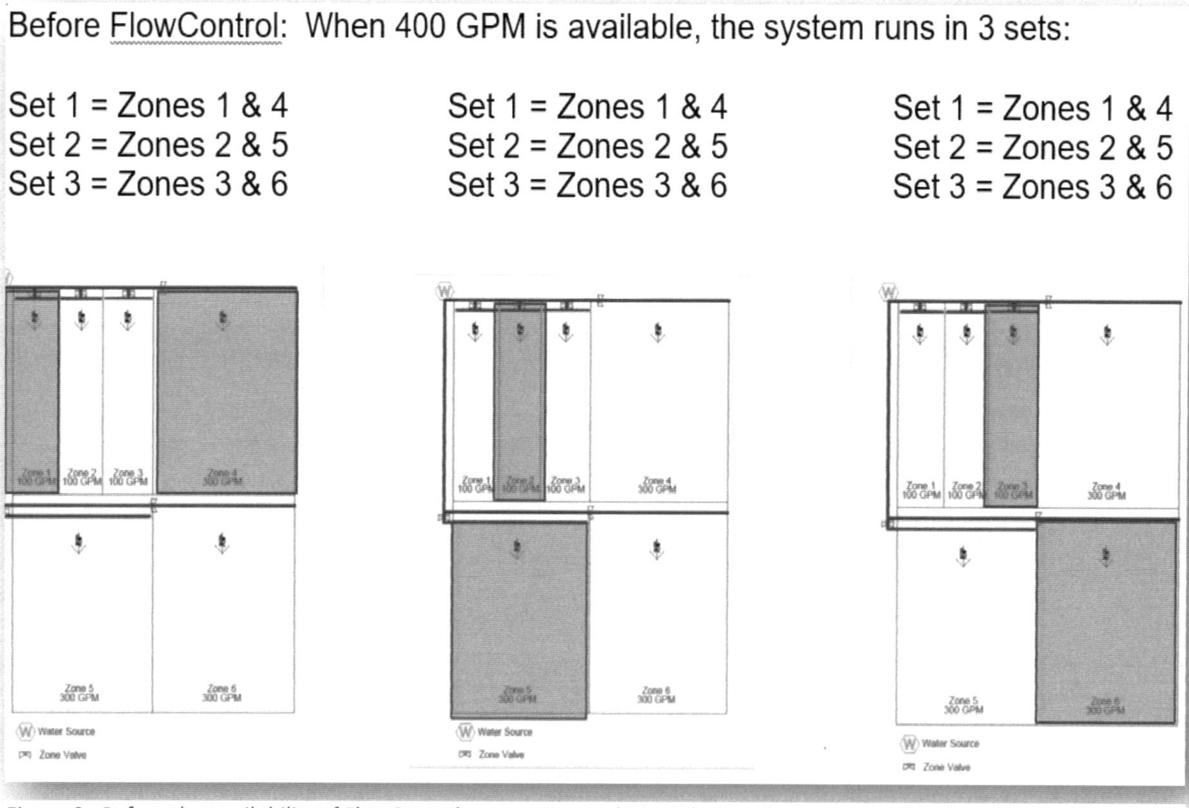


*Figure 8: Before the availability of FlowControl, systems were designed to run in 3 sets when the normal 400 gpm was available from the water source. Illustration courtesy of Randi Gladwell, Clearwater Supply.*

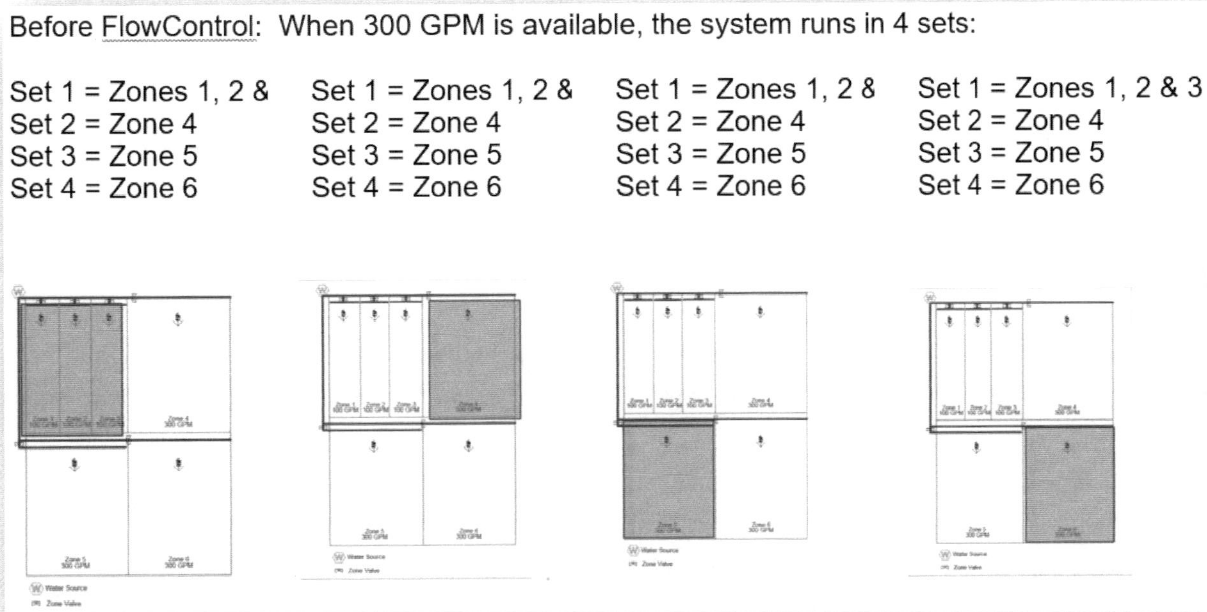


*Figure 9: Before the availability of FlowControl, the system was designed to run in 4 sets when only 300 gpm was available from the water source. Illustration courtesy of Randi Gladwell, Clearwater Supply.*

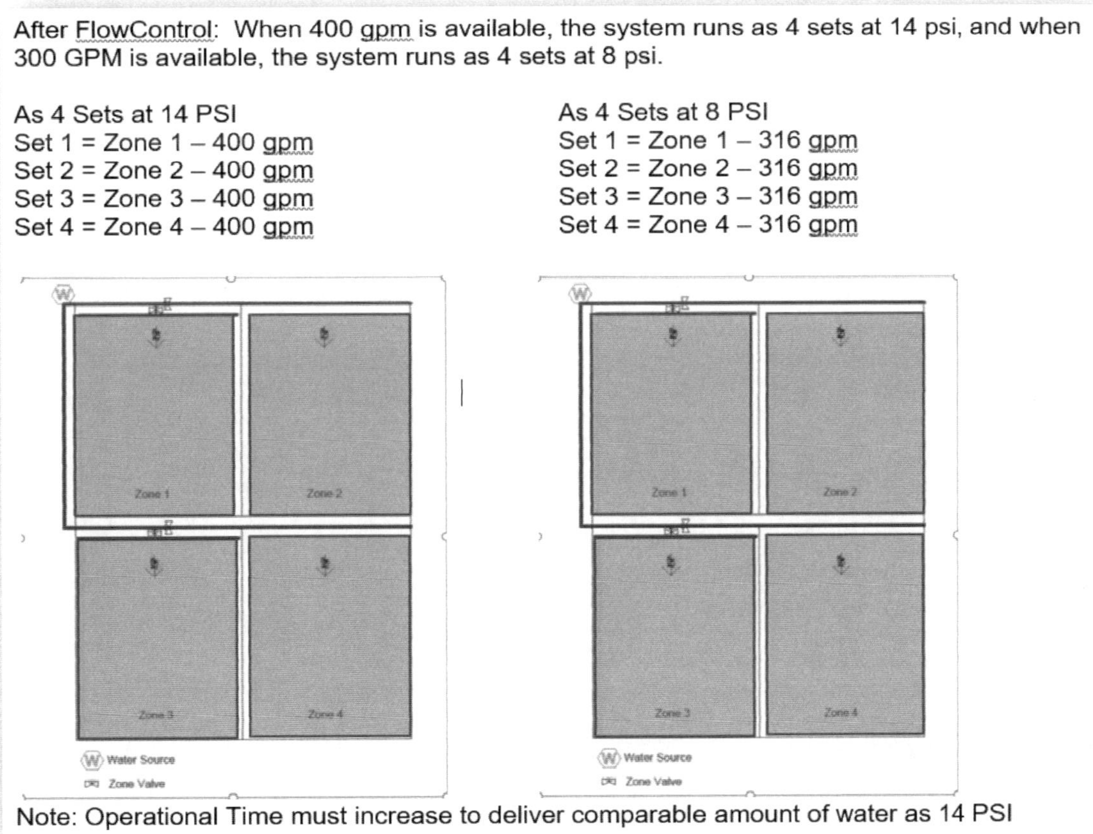


*Figure 10: After FlowControl drip tapes became available, the system design was simplified to run in 4 sets at 14 psi when the normal 400 gpm was available and to run in 4 sets at 8 psi when flows were reduced. Operational time must increase to deliver comparable amount of water as 14 psi, but this is simpler and less costly than a 3-set and 4-set scenario. Illustration courtesy of Randi Gladwell, Clearwater Supply.*

- **Daniele Zaccaria Agricultural Water Management Specialist, University of California, Davis** "FlowControl is an important new technology to provide a uniform application of water. The possibility of adjusting the flow with innovative drip tapes is an important feature that could contribute to increased capacity and flexibility of micro-irrigation systems."

## Using AquaFlow drip irrigation design software to evaluate potential drip system uniformity

Toro's free drip irrigation design software, AquaFlow, allows users to compare different drip laterals against one another while all other design parameters remain constant. AquaFlow's report includes the single lateral and block emission uniformity (EU) and a color-coded Uniformity Map to help illustrate how each lateral choice performs. In this map, the fewer colors displayed indicate higher drip system uniformity. In the example shown below (see Figure 11), FlowControl, with a flow exponent of 0.3, increases drip system uniformity by 5.0% compared to ordinary drip tapes with flow exponents of ~0.5. As a result, the FlowControl drip system will apply water and fertilizer more evenly and require less over-irrigation to mask dis-uniformity.

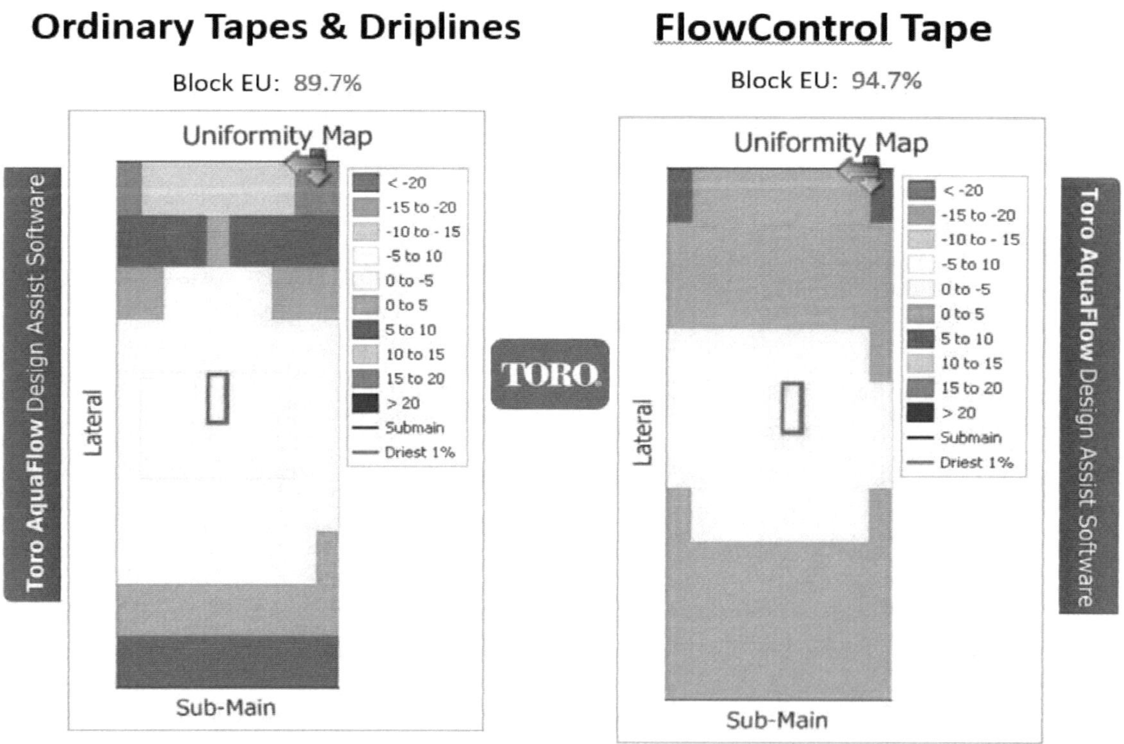


*Figure 11: AquaFlow uniformity map showing superior drip system uniformity using FlowControl drip tape with a flow exponent of 0.3 compared to ordinary drip tapes and driplines with flow exponents of ~0.5.*

## Summary of FlowControl advantages

There are three major advantages to using FlowControl drip tapes. First, FlowControl drip tapes will always provide higher drip system uniformity when compared to ordinary drip tapes because FlowControl has a flow exponent of 0.3. Higher drip system uniformity typically leads to better profitability and resource-use efficiency. FlowControl can help achieve higher drip system uniformity under normal conditions of flat terrain and/or short lengths of run as well as challenging conditions, such as hilly terrain and/or long lengths of run. Specific case studies report that, with FlowControl, costly and cumbersome submains and jumpers are eliminated, wetting patterns are improved, yield and quality is increased, and water, fertilizer, labor and energy input costs are reduced.

Second, FlowControl drip tapes maintain the flexibility to change the drip system application rate by adjusting pressure, unlike pressure-compensating driplines where the drip system application rate is fixed. Case studies report that this feature helps simplify designs, purchase price and operational complexity, especially when weather or water availability conditions change. As a result, flexibility is maintained without sacrificing drip system uniformity.

Third, FlowControl drip tapes are available in a wider range of wall thicknesses than pressure-compensating driplines, including more affordable ⅝-inch 6-mil and ⅞-inch 8-mil. In addition, unlike driplines, FlowControl drip tapes are sold at the same price regardless of emitter spacing, from 6 – 24 inches. Since closely spaced emitters are often preferred to achieve superior wetting patterns, this feature helps optimize results without paying a premium.

## References


The Toro Company: **Grower Reviews from The Field**
https://cdn2.toro.com/en/-/media/Files/Toro/Agriculture/drip-tape-and-dripline/ALT253_FlowControl_Testimonial_ENG_web.ashx

The Toro Company: ***AquaFlow Drip Irrigation Design Software***
http://driptips.toro.com/design-tools/drip-design-software/

# Microirrigation equipment for okra cultivation in the U.S. Virgin Islands

*Rhuanito Soranz Ferrarezi*

University of Florida, Institute of Food and Agricultural Sciences, Indian River Research and Education Center, 2199 South Rock Road, Fort Pierce, FL 34945, E-mail rferrarezi@ufl.edu

*Thomas C. Geiger, Stuart Weiss, Jayar Greenidge and Shamali Dennery*

University of the Virgin Islands, Agricultural Experiment Station, RR#1 Box 10,000, Kingshill, VI 00850, E-mail thomascgeiger@gmail.com

**Abstract.** Drip irrigation presents higher irrigation efficiency when compared to sprinkler irrigation. Proper system design and the use of pressure-compensating emitters plays an important role in irrigation uniformity and efficiency, directly affecting plant growth. This study evaluated the performance of pressure-compensating and non-compensating emitters and the effect of irrigation equipment in different okra varieties in the U.S. Virgin Islands. Experiments were performed in two seasons (Spring and Fall, 2016), and tested four irrigation equipment ("Toro Aqua-Traxx FC", "Eurodrip Thinwall Classic", "Jain Top Drip AS", and "Netafim Dripnet PC") and three varieties of okra ('Clemson Spineless 80', 'Clemson Spineless', and 'Chant'), arranged on a complete randomized block design with three replications. Irrigation was performed based on reference evapotranspiration, measured daily using an automated weather station. Soil moisture, electrical conductivity and soil temperature were monitored using capacitance sensors. The irrigation equipment responses to increasing pressure were evaluated in the lab, on experimental modules using clean water and simulating three different slopes (leveled, uphill and downhill). Yield and leaf physiological parameters were influenced by season ($P<0.05$), while fruit morphological parameters and soluble solids content were variety-dependent ($P<0.01$). The pressure-compensating emitters maintained water flow within the range indicated by the manufacturers. The distribution uniformity decreased overtime in all equipment except "Netafim Dripnet PC" in Fall 2016. Irrigation equipment did not impact plant growth. The equipment should be selected based on price and irrigation efficiency.

**Keywords.** Okra (*Abelmoschus esculentus*), Irrigation efficiency, Drip irrigation, Water-saving technologies, Variety trial, Tropics

## Introduction

Agriculture uses the majority of the potable water available on the planet, with irrigation accounting for 70% of global water withdrawals. The irrigation equipment used plays an important role in water use and irrigation efficiency. For an irrigation system to be considered efficient, water distribution needs to be uniform within the line (≈10%) and pressure variation across the secondary line should be lower than

20% (Burt et al., 1997). After installing the irrigation system, growers must ensure it matches the project design in the field. Pressure, water flow and distribution, and efficiency coefficients are necessary in order to evaluate system performance (Silva and Silva, 2005).

Drip irrigation has become the most common system used in agriculture due to the high irrigation efficiency (>90%) and the application of low water volumes (1 to 150 L/h), resulting in water savings when compared to sprinkler irrigation (Testezlaf, 2011). Drip irrigation applies water directly to the root zone, increasing water and nutrient use efficiency, incrementing yield and crop quality, and maximizing profitability (Borssoi et al., 2012). On the other hand, the initial deployment cost is usually higher than overhead/sprinkler systems. Drip irrigation demands constant maintenance, and requires efficient filtration due to the possibility of emitter clogging (Testezlaf, 2011).

The University of the Virgin Islands (UVI), Agricultural Experiment Station (AES) initiated an irrigation research project in the early 1980s with the objective to increase vegetable production while conserving water resources (Palada et al., 1995). Since then, farmers shifted from sprinkler to drip irrigation (also known as microirrigation). Most of the U.S. Virgin Islands local farmers use drip tapes with non-compensating emitters. Non-pressure compensating emitters' water output vary as the line pressure changes. That may result in inefficient water application (<70%) (Dogan and Kirnak, 2010). Pressure compensating emitters apply the same amount of water at each emitter over a range of different line pressures (i.e. 10-50 psi). These emitters can be used in long lines, irregular or mountainous areas and where precise watering is desired (Dogan and Kirnak, 2010). When water resources start becoming limited, especially in years affected by severe drought, there is a need to improve irrigation management and equipment efficiency in order to save water and pumping energy.

Performance evaluations of several drip irrigation systems are available in the literature (Pereira et al., 2005). However, the application of such tests on commercial operations is still scarce. This is due to the lack of knowledge about the importance of managing irrigation systems properly. Consequences include reduced crop yields and waste of water resources. To improve irrigation performance, it is necessary to promote implementation of irrigation scheduling methods, improve system design and equipment performance, and enhance farmers' skills to manage irrigation systems efficiently (Pereira et al., 2005).

Okra (*Abelmoschus esculentus*) is one of the most important and widely grown crops found throughout the tropical and sub-tropical regions (Eshiet and Brisibe, 2015). It is an annual, erect growing, high yielding crop with numerous cultivars varying in plant height, degree of branching and pigmentation of the various parts, period of maturity, and pod shape and size. Okra is mainly grown for its tender green pods, which are cooked and commonly consumed as boiled vegetables. Despite its enormous economic benefits, okra rarely reaches its maximum yield potential due to several constraints (Eshiet and Brisibe, 2015). Some of the major factors limiting okra production include the use of locally unimproved varieties, high incidence of pests and diseases, a narrow genetic base of existing varieties, and lack of proper irrigation to control plant growth.

Driven by the desire to indicate the best irrigation equipment for local growers to save water and select more adapted genotypes in okra, the objective of the current study is to determine the performance of pressure-compensating and non-compensating emitters and the effect of irrigation equipment on different okra varieties in the U.S. Virgin Islands.

## Material and Methods

*Location*. The studies were evaluated from Mar. 24 to July 11, 2016 (Spring) and from Aug. 12 to Dec 1, 2016 (Fall) at the University of the Virgin Islands (UVI) Agricultural Experiment Station (AES), Kingshill, U.S. Virgin Islands (lat. 17°43'08" N, long. 64°47'46" W, 30 m above sea level).

*Environmental conditions*. Environmental data were recorded throughout the studies using a weather station (ET107; Campbell Scientific, Logan, UT). The equipment was located 50 m from the experiment site, and measured wind speed and direction, rainfall, air temperature, and solar radiation. The vapor pressure deficit (VPD) was calculated from the saturated and actual air vapor pressure using the air temperature and relative humidity data (Fig. 1).

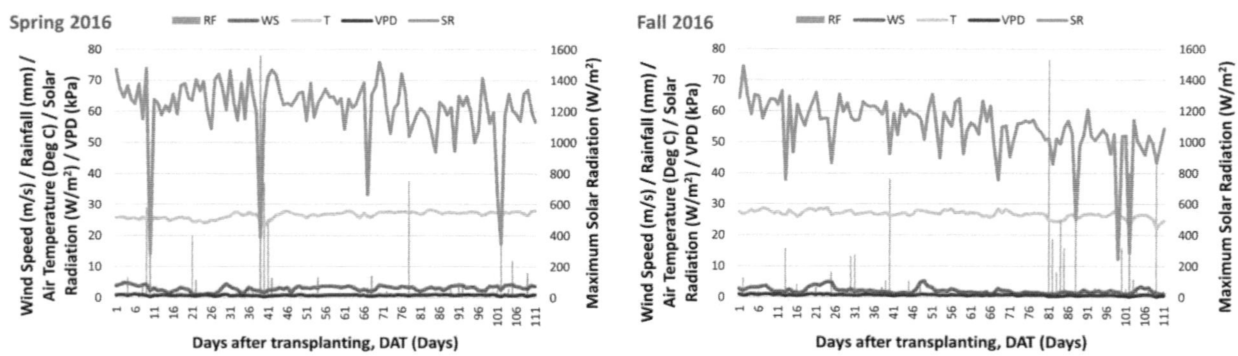


Fig. 1. Wind speed, rainfall, air temperature, solar radiation, and vapor pressure deficit (VPD) during the experiments performed in Spring (left) and Fall (right), 2016. Kingshill, U.S. Virgin Islands.

*Plant material and Cultural practices*. The experimental area had been used for bell peppers preceding the first planting. The crop was terminated by mowing, followed by two passes with a disc harrow and two passes with a rototiller. To improve soil health prior to the okra studies, the field was planted in cover crop by using 'IAC-1' sunn hemp (*Crotalaria juncea*) (800,000 plants/ha) in Fall, 2015. Sunn hemp seeds were inoculated prior to planting with *Bradyrhizobium* sp. inoculant. The cover crop was terminated 90 days after planting by mowing with a rotary mower/shredder.

Okra seeds were sown in 72-cell trays on Feb. 19, 2016 (Spring) and July 25, 2016 (Fall). Transplants were fertigated with a 12N-48P-8K starter fertilizer (Plant Agra; Two-Way Trading Co, Headland, AL). Seedlings were transplanted to the field on Mar. 23, 2016 (Spring) and Aug. 11, 2016 (Fall). Plants were spaced 0.3 m in-row × 1.22 m between-row (representing 26,909 plants/ha).

The fields were scouted for insect pests and plant diseases weekly until first harvest and then at every harvest. Lepidoptera (*Lepidoptera* sp.) was controlled using *Bacillus thuringiensis* (DiPel DF; Valent Biosciences, Walnut Creek, CA) and spinosad (Entrust SC; Dow AgroSciences, Indianapolis, IN). Aphids (Aphidoidea) and leaf miner (*Liriomyza sativae*) were controlled using paraffinic oil (Agri-Dex; Helena Chemical, Collierville, TN) a pyrethrin-based spray (PyGanic Crop Protection EC 1.4II; McLaughlin Gormley King, Minneapolis, MN), and neem oil (Trilogy; Certis USA, Columbia, MD). Powdery mildew was controlled using copper sulphate pentahydrate (Phyton 35; Phyton Corporation, New Hope, MN). Weeds were manually controlled at 55, 66, and 88 (Spring) and 30 days after transplanting (DAT) (Fall).

To reduce the need for manual weeding a 15-cm thick layer of hay was spread on the field at 61 DAT (Spring) and 37 DAT (Fall).

*Soil*. The soil on the experimental site is a Sion clay (SiB) according to the USDA soil survey (USDA, 2015). Samples for soil nutrient concentration were collected approximately 15 d prior transplanting for both seasons. Results are available in Table 1.

Table 1. Nutrient concentration of Sion clay soil in Spring and Fall, 2016. Average of three samples. Kingshill, U.S. Virgin Islands.

| **Nutrient** | **Spring** | **Fall** |
|---|---|---|
| Soil pH | 7.6 | 7.7 |
| Phosphorus (P), lb/acre | 62.0 | 92.7 |
| Potassium (K), lb/acre | 614.7 | 1010.7 |
| Calcium (Ca), lb/acre | 17,172 | 23,058 |
| Magnesium (Mg), lb/acre | 612.0 | 716.7 |
| Sulfur (S), lb/acre | 57.3 | 76.0 |
| Boron (B), lb/acre | 4.5 | 5.9 |
| Copper (Cu), lb/acre | 4.6 | 6.4 |
| Iron (Fe), lb/acre | 66.0 | 38.0 |
| Manganese (Mn), lb/acre | 60.0 | 76.7 |
| Zinc (Zn), lb/acre | 8.8 | 9.9 |
| Sodium (Na), lb/acre | 236.7 | 283.3 |
| Organic Matter, % | 4.0 | 4.3 |
| Nitrate Nitrogen, lb/acre | 40.0 | 52.7 |
| Ammonium Nitrogen, lb/acre | 54.7 | 4.7 |

*Treatments*. We tested four different drip irrigation equipment {"Toro Aqua-Traxx FC" [0.27 gallons per hour (GPH, 1.02 L) at 10 psi], "Jain Top Drip Thin Wall" [0.26 GPH (0.98 L) at 12 psi], "Netafim Dripnet PC" [0.26 GPH (0.98 L) at 12 psi], and "Eurodrip Thinwall Classic" [0.25 GPH (0.94 L) at 12 psi]} and three okra varieties ('Clemson Spineless 80', 'Clemson Spineless', and 'Chant'). We used one drip line and one drip tape with pressure compensating emitters and two drip tapes with non-compensating emitters, all from different manufacturers. Equipment was selected based on water flow to provide the same amount of water to all treatments.

*Irrigation*. The irrigated area had two submain lines (right and left), divided into 16 lateral lines (15.4 m each) per derivation line. The experiments had a dedicated 3,780-L water tank attached to a ½-HP booster pump with 22.7-L pressure tank (94525; Everbilt, Wilmington, DE) to guarantee stable pressure throughout the studies. Each experimental unit had a 12-psi pressure regulator, except for the plots with "Toro Aqua-Traxx FC", which needed a 10-psi pressure regulator to achieve the desired flow rate and an irrigation manifold built using 1-inch (2.54 cm) PVC pipe connected to three lines of drip tape / tubing with 1.02 L per emitter and 30-cm spacing.

Reference evapotranspiration (ETo) was calculated from the environmental data by the American Society of Civil Engineers (ASCE) Environment and Water Resources Institute (EWRI) standardized Penman-Monteith (ASCE-PM) equation (ASCE-EWRI, 2005) Irrigation was performed based on ETo and the water balance method (Fig. 2).

*Fertigation.* The fertilizer solution was based on soil nutrient analysis and okra nutritional requirements. The concentrated stock solution was prepared with a commercial Jack's Professional soluble 20N-20P-20K fertilizer (Peters, Allentown, PA) and applied 112 kg N/ha (final concentration of 100 mg N /L) using a fertilizer injector (D45RE15; Dosatron, Clearwater, FL).

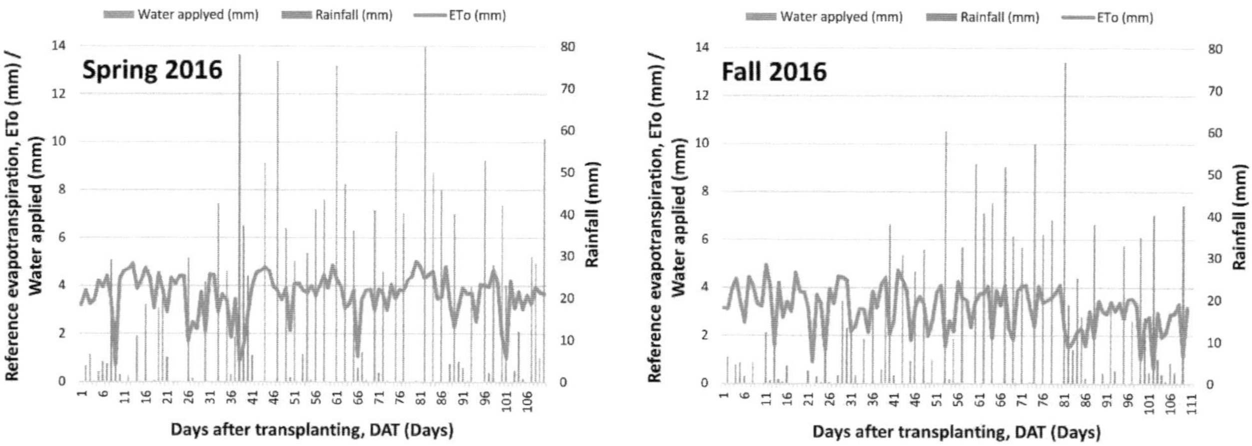


Fig. 2. Reference evapotranspiration (ETo) calculated by the American Society of Civil Engineers (ASCE) Environment and Water Resources Institute (EWRI) standardized Penman-Monteith equation (ASCE-PM), water applied and rainfall in Spring (left) and Fall (right), 2016). Kingshill, U.S. Virgin Islands.

*Lab measurements*. The irrigation equipment water flow responses to increasing pressure using clean water and three different slopes were evaluated in an experimental module assembled in the lab using three 20-ft long roof gutters (Fig. 3A and 3B) attached to a ½-HP booster pump (maximum pressure of 67 psi) with 6-gal pressure tank (94525; Everbilt, Wilmington, DE) (Fig. 3C). We used a 6.1-m long irrigation line with emitters spaced 0.3 m apart. The setup used calibrated pressure gauges to precisely monitor the applied pressure (Fig. 3B).

Incoming pressure was increased in multiple values up to the maximum recommended by the manufacturer ("Toro Aqua-Traxx FC" from 3 to 27 psi in multiples of 3 psi; "Jain Top Drip Thin Wall" from 4 to 40 psi in multiples of 4 psi; "Netafim Dripnet PC" from 6 to 60 psi in multiples of 6 psi; and "Eurodrip Thinwall Classic" from 2 to 22 psi in multiples of 2 psi).

The emitter evaluation was performed at the beginning, 1/4, 1/2, 3/4 and at the end of the line. In each position, the water flow was measured in three sequential emitters for 3 min (Borssoi et al., 2012). Tests were repeated three times and in three different levelling conditions (leveled, uphill and downhill) to simulate field conditions.

Fig. 3. Experimental module designed and installed to evaluate the effect of increasing operational pressures on drip irrigation equipment water flow. Kingshill, U.S. Virgin Islands.

*Field measurements*. The lateral lines were evaluated in the field to determine the irrigation system efficiency. We measured the water flow using catch cans at the initial position, 1/4, 1/2, 3/4 and the end of each line (Keller and Bliesner, 1990). With the emitter water flow information, we determined the Christiansen uniformity coefficient (CUC) [Eq. 1], the statistics (CUE) [Eq. 2] and the distribution uniformity coefficient (CUD) [Eq. 3] (Borssoi et al., 2012). The coefficients were classified according to the ASABE (1994) and ASABE (2001) standards (Table 1).

$$CUC = 100 \cdot \left( 1 - \frac{\sum_{i=1}^{N} \left| Xi - \overline{X} \right|}{N \cdot \overline{X}} \right)$$ [Eq. 1]

where: N = number of samples, Xi = depth of water applied to the n-th point on the soil surface, and X = average depth of water applied.

$$CUE = 100 \cdot \left( 1 - \frac{S}{\overline{X}} \right)$$ [Eq. 2]

where: S = emitter standard deviation, and X = average depth of water applied.

$$CUD = 100 \cdot \frac{x}{X}$$ [Eq. 3]

where: x = average depth of water applied on the 25% lowest volumes of catch cans, and X = average depth of water applied (considering all catch cans).

We also calculated the application efficiency (EA) [Eq. 4] (Bernardo, 1995).

$$AE = 0.9 \times CUD$$ [Eq. 4]

where: AE = application efficiency (%), and CUD = distribution uniformity coefficient (%).

Table 1. Irrigation efficiency parameters classification according to the ASABE (1994) and ASABE (2001) standards. Where CUC: Christiansen uniformity coefficient, AE: application efficiency, and CUD: distribution uniformity coefficient. Kingshill, U.S. Virgin Islands.

| **Classification** | **CUC** | **AE** | **CUD** |
|---|---|---|---|
| | | ------------- % ------------- | |
| Excellent | > 90 | 90 - 100 | > 84 |
| Good | 80 - 90 | 80 - 90 | 68 - 84 |
| Fair | 70 - 80 | 70 - 80 | 52 - 68 |
| Poor | 60 - 70 | 60 - 70 | 36 - 52 |
| Unacceptable | < 60 | < 60 | < 36 |

Soil water content, soil temperature and bulk electrical conductivity were monitored using 36 capacitance sensors (24 10HS and 12 GS3; Decagon Devices, Pullman, WA). The monitoring system was built using a data logger (CR1000; Campbell Scientific, Logan, UT), multiplexer (AM16/32B; Campbell Scientific, Logan, UT) and the capacitance sensors. The controller was powered using a 20-W solar panel (Infinium; ML Solar, Campbell, CA), connected to a 12/24-VDC 10-A Tracer solar charge controller (1210RN; EPSolar, Beijing, China) and two 12-VDC 7.2-Ah rechargeable batteries (Yuasa, Ebbw Vale, United Kingdom). The data collected were transmitted to a computer using a RF401A radio frequency module (Campbell Scientific, Logan, UT). Radios sent the collected information to the computer using Omnidirectional 900 MHz 3 dBd and Yagi 900 MHz 9 dBd antennas (both from Campbell Scientific, Logan, UT).

Total and marketable yield were determined weekly and totalized at the end. Leaf anthocyanin and chlorophyll content indexes (non-destructive analysis) were measured in Spring (day 108) and Fall (day 103). Anthocyanin was measured with a portable anthocyanin content meter (ACM-200 plus; Opti-Sciences, Hudson, NH), and chlorophyll using a chlorophyll concentration meter (MC-100; Apogee Instruments, Logan, UT). Plant growth index {[(height + width 1 + width 2) / 3]}, fruit size (weight, length, and width), fruit hardness using a digital penetrometer (FHP-802; Agriculture Solutions, Strong, ME), and fruit soluble solids content using a refractometer (RF15; Extech Instruments, Nashua, NH) were measured on days 100 (Spring) and 103 (Fall).

*Experimental design and Statistical analysis*. Treatments were arranged on a complete randomized block design with three replications. Each experimental unit had 30 plants / variety for a total of 270 plants per variety and 1,080 per trial. Data were analyzed using a mixed model procedure in SAS (version 9.4; SAS Institute, Cary, NC). Errors were assumed to be normally and independently (NID) distributed. Probability values ≤ 0.05 were considered statistically significant.

**Results and Discussion**

*Lab measurements*. The lab test provided water flow information under different conditions (leveled, uphill and downhill). "Toro Aqua-Traxx FC" and "Eurodrip Thinwall Classic" (non-compensating emitters) increased water flow with increase in pressure. "Jain Top Drip Thin Wall" and "Netafim Dripnet PC" (pressure compensating emitters) provided a steady water flow for pressure > 10 psi (Fig. 4). These results were expected since that is the technology outlined by the equipment manufacturer. Our results clearly indicated that a more uniform water distribution can be achieved by replacing the irrigation equipment. "Toro Aqua-Traxx FC" cost $0.074/ft ($0.24/m), "Eurodrip Thinwall Classic" $0.024/ft ($0.08/m), "Jain Top Drip Thin Wall" $0.046/ft ($0.15/m) and "Netafim Dripnet PC" $0.298/ft ($0.98/m) (cost for the U.S. Virgin Islands with shipping included). "Eurodrip Thinwall Classic" is the cheapest equipment, while "Netafim Dripnet PC" the most expensive. "Jain Top Drip Thin Wall" had proven to be efficient but cost almost the double than the equipment most used in the territory.

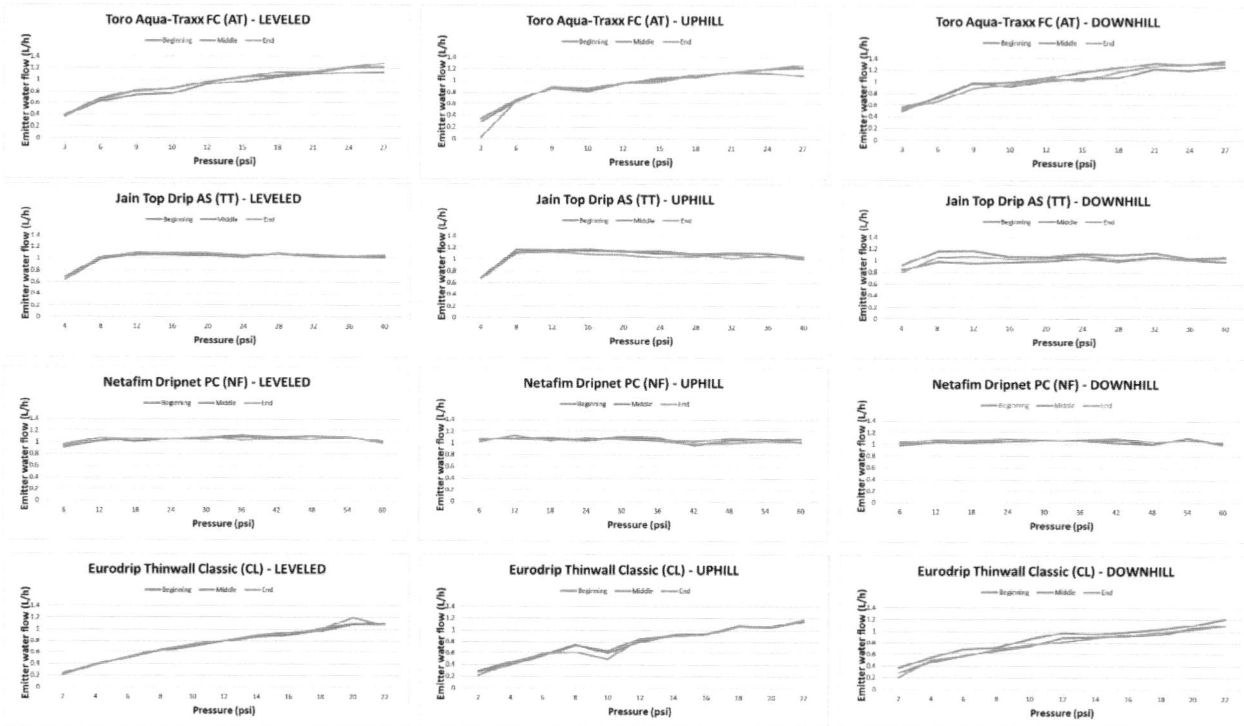


Fig. 4. Emitter water flow of four drip irrigation equipment {[AT] "Toro Aqua-Traxx FC" (1.02 L at 10 psi), [TT] "Jain Top Drip Thin Wall" (0.98 L at 12 psi), [NF] "Netafim Dripnet PC" (0.98 L at 12 psi), and [CL] "Eurodrip Thinwall Classic" (0.94 L at 12 psi)} subjected to increasing operational pressures in three conditions (leveled, uphill and downhill). Kingshill, U.S. Virgin Islands.

*Field measurements*. According to the ASABE standards (Table 1), "Toro Aqua-Traxx FC" and "Eurodrip Thinwall Classic" were classified as fair and good, while "Netafim Dripnet PC" and "Jain Top Drip Thin Wall" as good in Spring 2016. All irrigation tapes were classified as good and excellent in Fall 2016, indicating that our systems met the efficiency requirements for drip irrigation to water the three okra varieties properly (Fig. 5). The efficiency decreased over time for "Toro Aqua-Traxx FC" and "Eurodrip Thinwall Classic" probably due to clogging from suspended solids.

Volumetric water content (Fig. 6A), soil temperature (Fig. 6B) and bulk electrical conductivity (Fig. 6C) presented large variation in the tested treatments. Replication differences are expected, and explained by the use of independent experimental units, variations in moisture caused by soils, sensor position and plants, which was also reported by Ferrarezi et al. (2017). Data collected during Spring 2016 were consistently more stable than Fall 2016. The reason for such variation is unknown. In Fall 2016, sensors malfunctioned in the last 10 days, producing unrealistic measurements.

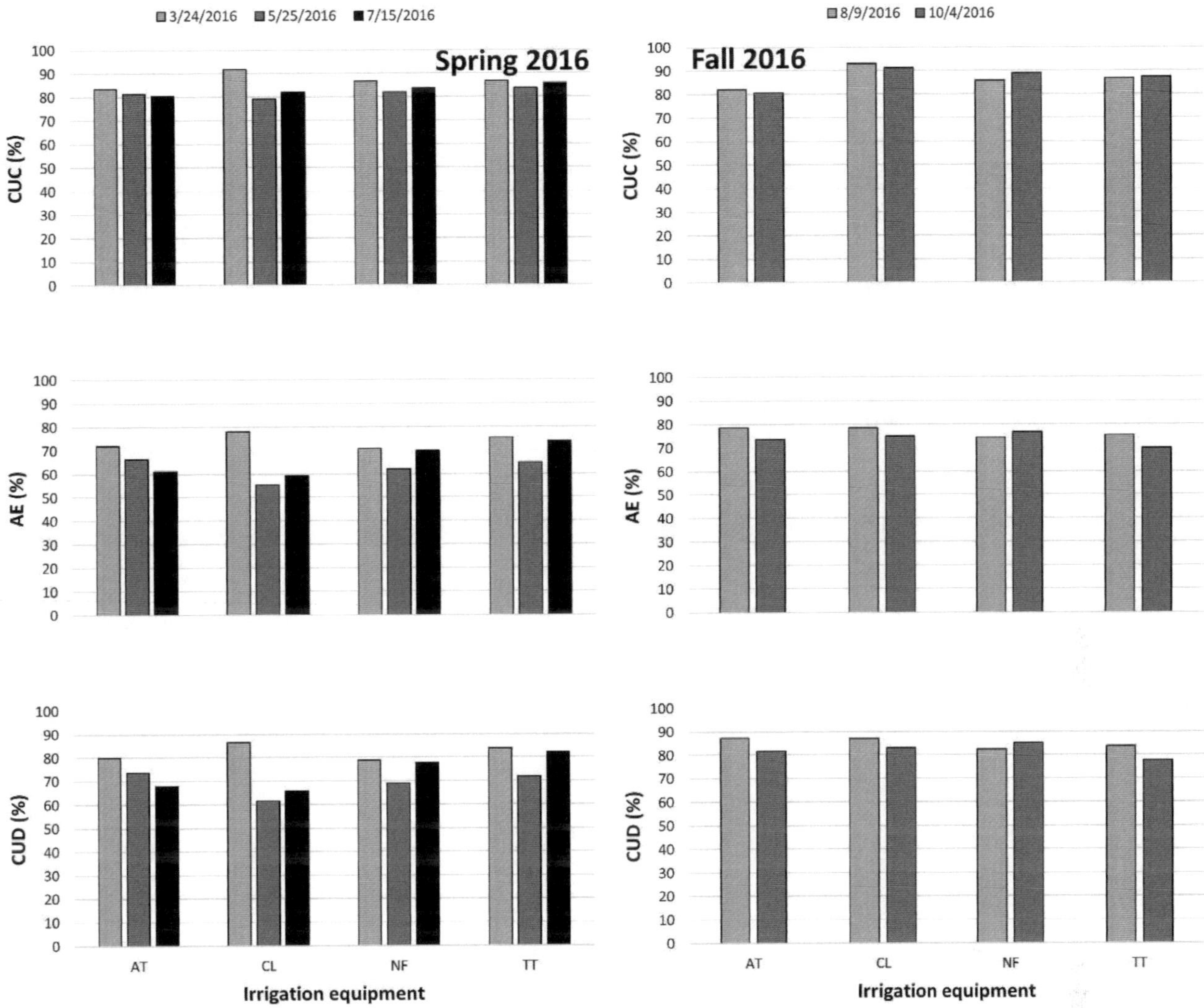


Fig. 5. Drip irrigation equipment efficiency parameters. Where: [AT] "Toro Aqua-Traxx FC" (1.02 L at 10 psi), [TT] "Jain Top Drip Thin Wall" (0.98 L at 12 psi), [NF] "Netafim Dripnet PC" (0.98 L at 12 psi), and [CL] "Eurodrip Thinwall Classic" (0.94 L at 12 psi), CUC: Christiansen uniformity coefficient, AE: application efficiency, and CUD: uniformity distribution coefficient. Kingshill, U.S. Virgin Islands.

Total and marketable yield, leaf anthocyanin, fruit weight, length and width were influenced by seasons ($P<0.05$, Tables 2 and 3). Total and marketable yield, fruit weight, length, width and were higher in Spring 2016 compared to Fall 2016, while leaf anthocyanin and hardness were 13% and 25% lower.

Fruit morphological parameters (length, width and hardness) and soluble solids content were variety-dependent ($P<0.01$, Table 3). 'Chant' presented higher fruit length, while 'Clemson Spineless 80' and 'Clemson Spineless' presented longer and harder fruit, with higher fruit soluble solids content.

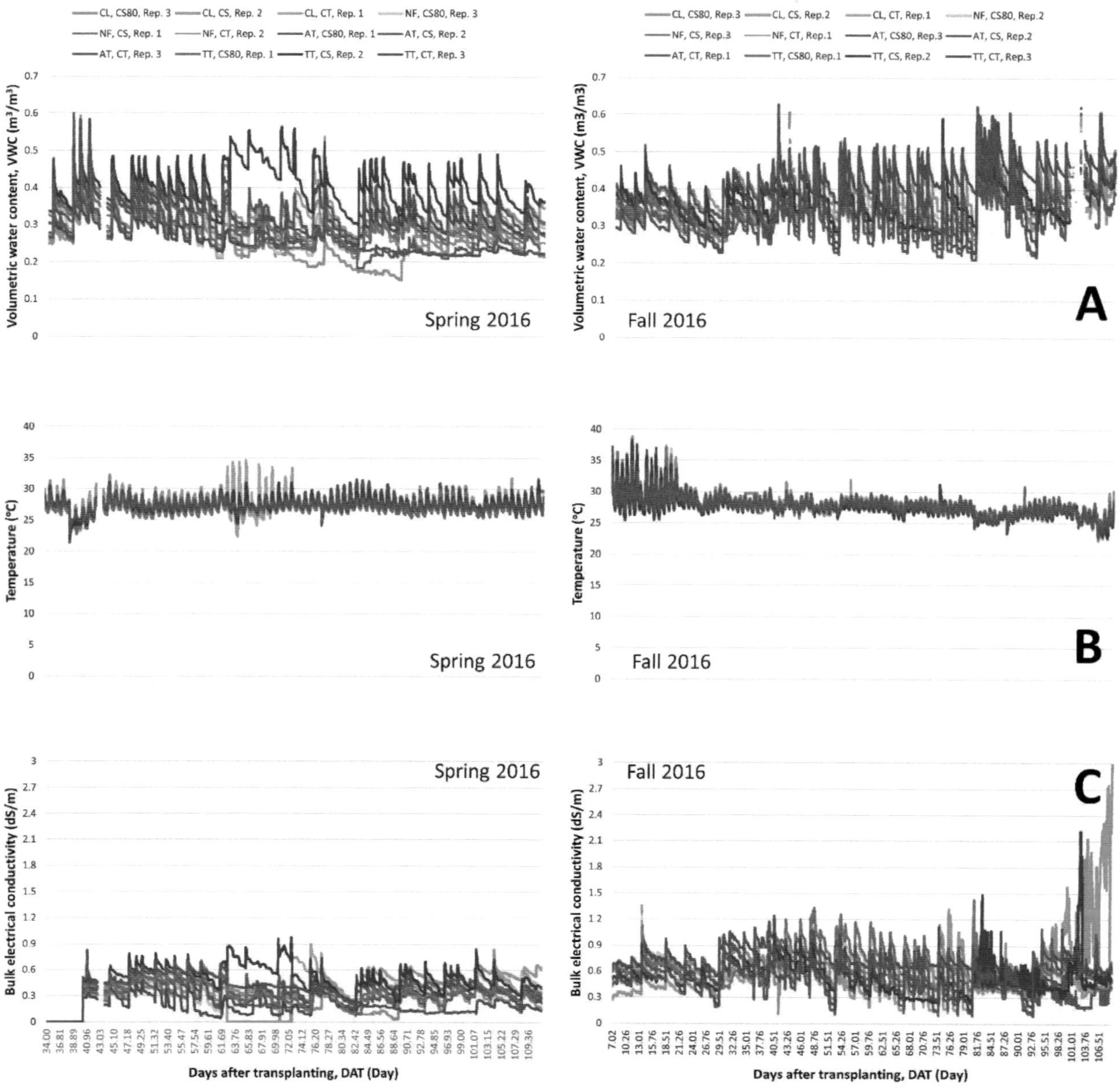


Fig. 6. Volumetric water content (A), soil temperature (B) and bulk electrical conductivity (C) monitored using capacitance sensors (GS3; Decagon Devices, Pullman, WA). Each sensor was positioned in a treatment combination: four drip irrigation equipment {[AT] "Toro Aqua-Traxx FC" (1.02 L at 10 psi), [TT] "Jain Top Drip Thin Wall" (0.98 L at 12 psi), [NF] "Netafim Dripnet PC" (0.98 L at 12 psi), and [CL] "Eurodrip Thinwall Classic" (0.94 L at 12 psi)} and three okra varieties ('Clemson Spineless 80 [CS80]', 'Clemson Spineless [CS]', and 'Chant [CT]'). Kingshill, U.S. Virgin Islands.

Table 2. Total yield, marketable yield, percent marketable yield, leaf anthocyanin, leaf chlorophyll and plant growth index of three varieties of okra ('Clemson Spineless 80 [CS80]', 'Clemson Spineless [CS]', and 'Chant [CT]') cultivated in two seasons (Spring and Fall, 2016) and using four irrigation equipment {[AT] "Toro Aqua-Traxx FC" (1.02 L at 10 psi), [TT] "Jain Top Drip Thin Wall" (0.98 L at 12 psi), [NF] "Netafim Dripnet PC" (0.98 L at 12 psi), and [CL] "Eurodrip Thinwall Classic" (0.94 L at 12 psi)}. Kingshill, U.S. Virgin Islands.

| | **Total yield (kg/ha)** | **Marketa-ble yield (kg/ha)** | **% Marketa-ble yield** | **Leaf anthocya-nin (ACI)** | **Leaf chlorophyll (CCI)** | **Plant growth index (cm)** |
|---|---|---|---|---|---|---|
| **Season** | | | | | | |
| Spring 2016 | 27,315 ± 7,101 a | 20,843 ± 5,368 a | 76.36 ± 4.26 a | 10.46 ± 1.80 b | 17.09 ± 4.05 | 102.51 ± 5.63 |
| Fall 2016 | 15,267 ± 5,305 b | 11,125 ± 4,540 b | 68.40 ± 5.61 b | 11.91 ± 1.34 a | 19.89 ± 3.30 | 104.73 ± 8.54 |
| **Equipment** | | | | | | |
| AT | 22,360 ± 8,090 | 16,567 ± 5,845 | 73.06 ± 5.07 | 11.49 ± 1.43 | 18.70 ± 3.47 | 105.37 ± 7.13 |
| CL | 19,861 ± 6,449 | 14,905 ± 5,153 | 72.82 ± 5.09 | 11.79 ± 2.05 | 16.21 ± 2.76 | 101.91 ± 4.89 |
| NF | 22,856 ± 7,043 | 17,731 ± 6,573 | 72.21 ± 6.71 | 11.32 ± 1.36 | 19.23 ± 3.69 | 104.69 ± 9.00 |
| TT | 20,086 ± 7,339 | 14,732 ± 5,384 | 71.42 ± 5.22 | 10.15 ± 1.59 | 19.83 ± 4.79 | 102.51 ± 7.64 |
| **Variety** | | | | | | |
| CS80 | 20,788 ± 7,893 | 15,309 ± 6,080 | 70.41 ± 6.04 | 10.98 ± 1.24 | 17.39 ± 2.56 | 103.78 ± 5.79 |
| CS | 23,121 ± 7,531 | 16,741 ± 5,748 | 69.96 ± 4.41 | 11.24 ± 1.96 | 17.18 ± 2.65 | 100.82 ± 6.30 |
| CT | 19,963 ± 6,047 | 15,901 ± 5,429 | 76.76 ± 5.11 | 11.34 ± 1.68 | 20.91 ± 5.18 | 106.26 ± 9.07 |
| | | | *p-value* | | | |
| Season (S) | 0.0002* | 0.0001* | 0.0003* | 0.0312* | 0.0549 | 0.4837 |
| Equipment (E) | 0.8436 | 0.7663 | 0.9426 | 0.3198 | 0.3038 | 0.8384 |
| S*E | 0.9609 | 0.8507 | 0.3952 | 0.5584 | 0.1810 | 0.7128 |
| Variety (V) | 0.6626 | 0.8790 | 0.0143* | 0.8940 | 0.0655 | 0.3772 |
| S*V | 0.9813 | 0.9158 | 0.5568 | 0.5573 | 0.6547 | 0.6218 |
| E*V | 0.9248 | 0.9366 | 0.9170 | 0.1375 | 0.2082 | 0.3730 |
| S*E*V | 0.9686 | 0.9311 | 0.4383 | 0.7618 | 0.4116 | 0.9370 |

* Significant at P<0.05.

Table 3. Fruit weight, fruit length, fruit width, fruit hardness, and fruit soluble solids content of three varieties of okra ('Clemson Spineless 80 [CS80]', 'Clemson Spineless [CS]', and 'Chant [CT]') cultivated in two seasons (Spring and Fall, 2016) and using four irrigation equipment {[AT] "Toro Aqua-Traxx FC" (1.02 L at 10 psi), [TT] "Jain Top Drip Thin Wall" (0.98 L at 12 psi), [NF] "Netafim Dripnet PC" (0.98 L at 12 psi), and [CL] "Eurodrip Thinwall Classic" (0.94 L at 12 psi)}. Kingshill, U.S. Virgin Islands.

| | Fruit weight (g) | Fruit length (cm) | Fruit width (cm) | Fruit hardness (kgf) | Fruit soluble solids content (%) |
|---|---|---|---|---|---|
| **Season** | | | | | |
| Spring 2016 | 24.19 ± 2.47 a | 13.73 ± 1.85 a | 2.00 ± 0.09 a | 5.86 ± 0.01 b | 4.35 ± 0.15 |
| Fall 2016 | 15.89 ± 1.48 b | 11.29 ± 1.26 b | 1.74 ± 0.07 b | 7.86 ± 0.37 a | 4.33 ± 0.13 |
| **Equipment** | | | | | |
| AT | 19.85 ± 3.13 | 12.58 ± 1.83 | 1.85 ± 0.11 | 6.89 ± 0.65 | 4.26 ± 0.16 |
| CL | 20.34 ± 3.16 | 12.50 ± 1.57 | 1.89 ± 0.11 | 6.86 ± 0.63 | 4.36 ± 0.14 |
| NF | 20.61 ± 3.41 | 12.84 ± 1.90 | 1.87 ± 0.12 | 6.80 ± 0.61 | 4.37 ± 0.17 |
| TT | 19.37 ± 3.13 | 12.11 ± 1.71 | 1.87 ± 0.11 | 6.90 ± 0.71 | 4.37 ± 0.07 |
| **Variety** | | | | | |
| CS80 | 19.25 ± 2.85 | 10.91 ± 0.93 b | 1.93 ± 0.10 a | 7.11 ± 0.77 a | 4.39 ± 0.15 a |
| CS | 19.88 ± 3.01 | 11.04 ± 0.96 b | 1.95 ± 0.10 a | 6.92 ± 0.66 a | 4.45 ± 0.10 a |
| CT | 20.99 ± 3.60 | 15.57 ± 1.59 a | 1.74 ± 0.08 b | 6.56 ± 0.42 b | 4.18 ± 0.12 b |
| | | | *p-value* | | |
| Season (S) | <0.0001* | <0.0001* | <0.0001* | <0.0001* | 0.6413 |
| Equipment (E) | 0.7740 | 0.6447 | 0.7027 | 0.7725 | 0.3502 |
| S*E | 0.8418 | 0.9293 | 0.4450 | 0.7962 | 0.2094 |
| Variety (V) | 0.2868 | <0.0001* | <0.0001* | <0.0001* | 0.0003* |
| S*V | 0.3479 | 0.0541 | 0.4465 | 0.0701 | 0.8383 |
| E*V | 0.9879 | 0.8395 | 0.9824 | 0.0854 | 0.6660 |
| S*E*V | 0.6410 | 0.5134 | 0.3741 | 0.0695 | 0.5333 |

* Significant at P<0.01.

## Conclusions

The pressure-compensating emitters maintained water flow within the range indicated by the manufacturers. Distribution uniformity decreased overtime in all equipment except "Netafim Dripnet PC" in Fall 2016. Irrigation equipment did not impact plant growth. The equipment should be selected based on price and irrigation efficiency.

## Acknowledgements

We thank Victor Almodovar and Naima Jenkins-El (Horticulture and Aquaculture program), Jose Herrera, K. Paul Beamer, Ephraim Rodriguez and Nelson Benitez (Agronomy program), Jordan Atemazem, and the High School students Nhaquifa Russell, Vernon Sackey and Danah Suarez for their technical assistance.

Funding for this research was provided by USDA-NIFA-Hatch Funds (Accessions no. 227355 and 1010280) and USDA-NIFA-Multi State grant W3128 (Accession no. 1006963).

## References

American Society of Agricultural Engineers (ASABE). 1994. Design and installation of microirrigation systems (EP405.1). St. Joseph: ASABE Standard engineering practices data. 4p.

American Society of Agricultural Engineers (ASABE). 2001. Field evaluation of micro irrigation Systems (EP 458). St. Joseph: ASABE Standard engineering practices data. p.792-797.

American Society of Civil Engineers (ASCE) Environment and Water Resources Institute (EWRI). 2005. The ASCE standardized reference evapotranspiration equation. In: Allen R.G., I.A. Walter, R. Elliott, T. Howell, D. Itenfisu, and M. Jensen (eds.) ASCE-EWRI Task committee report. American Society of Civil Engineers, Reston, VA, 171 pp.

Bernardo, S. 1995. Irrigation manual. Viçosa: UFV, 657p.

Borssoi, A.L., M.A. Vilas Boas, M. Reisdörfer, R.H. Hernández, and F.A. Follador. 2012. Water application uniformity and fertigation in a dripping irrigation set. Engenharia Agrícola 32: 718-726.

Burt, C., A. Clemmens, T. Strelkoff, K. Solomon, R. Bliesner, L. Hardy, T. Howell, and D. Eisenhauer. 1997. Irrigation performance measures: Efficiency and uniformity. Journal of Irrigation and Drainage Engineering 123(6): 423-442.

Dogan, E. and H. Kirnak. 2010. Water temperature and system pressure effect on drip lateral properties. Irrigation Science 28(5): 407-419.

Eshiet A.J. and E.A. Brisibe. 2015. Morphological characterization and yield traits analysis in some selected varieties of okra (*Abelmoschus Esculentus* L. Moench). Advances in Crop Science and Technology 3(5): 1-5.

Ferrarezi, R.S. and R. Testezlaf. 2017. Automated ebb-and-flow subirrigation for citrus liners production. I. Plant growth. Agricultural Water Management 192: 45-57.

Keller, J. and R.D. Bliesner. 1990. Sprinkle and trickle irrigation. New York: van Nostrand Reinhold, 652p.

Palada, M.C., S.M.A. Crossman, and C.D. Collingwood. 1995. Improving vegetable production using microirrigation in the Virgin Islands. Proceedings of 5th International Microirrigation Congress, April 2-6, 1995, Orlando, FL. p.502-509.

Pereira, L.S., I. Cordery, and I. Iacovides. 2013. Improved indicators of water use performance and productivity for sustainable water conservation and saving. Agricultural Water Management 108: 39-51.

Silva, C.A. and C.J. Silva. 2005. Uniformity performance assessment in localized irrigation system. Revista Cientifica Eletrônica de Agronomia 1(8): 1-17.

Testezlaf, R. 2011. Irrigação: Métodos, sistemas e aplicações. Campinas: UNICAMP, 204p.

U.S. Department of Agriculture. 2015. Web Soil Survey. 18 Sept. 2015. http://websoilsurvey.nrcs.usda.gov.

# Greenhouse cucumber production using sensor-based irrigation

*Rhuanito Soranz Ferrarezi*

University of Florida, Institute of Food and Agricultural Sciences, Indian River Research and Education Center, 2199 South Rock Road, Fort Pierce, FL 34945, E-mail rferrarezi@ufl.edu

*Thomas C. Geiger*

University of the Virgin Islands, Agricultural Experiment Station, RR#1 Box 10,000, Kingshill, VI 00850, E-mail thomascgeiger@gmail.com

**Abstract.** Cucumber is the most important vegetable produced in the U.S. Virgin Islands. Crop production is performed outdoors, with no information available regarding cultivation in a closed environment using precision irrigation to increase yield and save water. This study evaluated greenhouse production of different slicing cucumbers under increasing substrate volumetric water contents (VWC) to trigger irrigation automatically based on plant demand. Experiments were conducted in Fall 2016 and Spring 2017. We tested four cucumber varieties ('Boa', 'Corinto', 'Marketmore 76' and 'Verdon') and three substrate VWC to trigger irrigation (0.3, 0.4 and 0.5 $m^3/m^3$), on a split-plot design with three replications. Plants were transplanted into 9.45-L pots spaced 0.46 × 1.22 m, trained on a vertical plastic line, and fertigated with 5N-11P-26K hydroponic fertilizer (Peter's Professional; Everris, Geldermalsen, The Netherlands) and calcium nitrate. Irrigation was performed on-demand using one substrate moisture sensor per experimental unit formed by four pots. The irrigation system applied water automatically when the VWC dropped below the set thresholds. Sensor-based irrigation was effective to water the plants. The number of irrigation events and leaf anthocyanin content differed on both seasons. Total yield and total number of fruit/plant responded to increases in VWC and varieties ($P<0.05$). 'Corinto' cultivated with VWC 0.5 $m^3/m^3$ resulted in the highest yield at 57,445 kg/ha ($P=0.00269$). Marketable yield, fruit width and fruit hardness were variety-dependent. Fruit weight and length, leaf chlorophyll, plant growth index and fruit soluble solids content were influenced by the different seasons and varieties ($P<0.05$). 'Corinto' consistently showed higher yield in all VWCs used to trigger irrigation, producing longer and more fruit than the other varieties, being a promising cultivar for greenhouse cucumber production using sensor-based irrigation in the U.S. Virgin Islands.

**Keywords.** Cucumber (*Cucumis sativus*), Sensor-based irrigation, Water-saving technologies, Drip irrigation, Variety trial, Tropics

## Introduction

Cucumber (*Cucumis sativus*) is one of the most important vegetables in greenhouse production in the U. S., particularly in California, Nevada and Florida. In 2014, cucumbers were cultivated in approximately 102 ha under protected environment, producing 32,940 t of fruit with a sales value of $77.6M (USDA, 2015). The crop is the leading vegetable grown in the U.S. Virgin Islands. According the latest census of agriculture, cucumber was ranked number one in area planted 18.6 ha and amount harvested 51 t (USDA, 2009). However, cucumber production in the U.S. Virgin Islands is mainly outdoors in soil. No information is available regarding the cultivation of cucumbers in protected environment using potting mix on hydroponics. Greenhouse cucumber production can be one alternative for local growers to generate income, expand agricultural production and increase food security in the territory.

One of the main challenges for growers interested in farm crop diversity is which variety to choose in new plantings. Vegetable performance trials are essential for vegetable growers to maximize revenue and reduce the risk of testing new crops (Ferrarezi et al., 2016). Cucumber varieties are classified in slicing and pickling types. Slicing cucumbers produce long, straight fruit with thick skin, with high commercial value and extended shelf life. There are also specialty selections known as burpless, heirloom and greenhouse cucumbers. All types have self-pollinating varieties for indoor cultivation.

There are soil-, weather-, and plant-based methods to determine the volume, frequency, and rate of water for efficient irrigation. Growers generally make irrigation management decisions based on soil and plant visual observations, or use a rigid irrigation schedule set by timers (Nemali et al., 2007). Predefined daily cycles do not apply water and nutrients appropriately, causing water deficit or excess and reducing the yield potential due to the negative effect of stresses on plant physiology, with possibility of environmental contamination (e.g. nitrate percolation). Since water is scarce in the U.S. Virgin Islands, cropping systems that optimize water use are key to guarantee sustainable food production while achieving high yield. Soil-based monitoring systems are easy to implement, efficient and relatively inexpensive. Growers can use water potential (Shock and Wang, 2011) or volumetric water content (VWC) (Blonquist et al., 2005) to determine the amount of water available in the root zone. Automated irrigation based on soil tension has been used for decades (Shock and Wang, 2011), while using the VWC for system automation has become feasible in the recent years with the advent of capacitance sensors (Jones, 2007; Nemali and van Iersel, 2006). Volumetric water content measurements are useful to monitor soil moisture and control irrigation in real-time, allowing precise irrigation management. Sensor-based systems apply water automatically when the VWC drops below set thresholds, precisely irrigating based on plant demand.

This study evaluated greenhouse production of different slicing cucumbers under increasing substrate VWC to trigger irrigation automatically based on plant demand.

## Material and Methods

*Location.* The studies were evaluated from Sept. 21 to Dec. 09, 2016 (Fall) and from Jan. 13 to April 7, 2017 (Spring) at the University of the Virgin Islands (UVI) Agricultural Experiment Station (AES), Kingshill, U.S. Virgin Islands (lat. 17°43'08" N, long. 64°47'46" W, 30 m above sea level).

*Environmental conditions.* Environmental data were recorded inside the greenhouse using a temperature and relative humidity sensor (HMP60; Vaisala, Vantaa, Finland) and quantum sensor (LI190R; Licor, Lincoln, NE). The vapor pressure deficit (VPD) was calculated from the saturated and actual air vapor pressure using the air temperature and relative humidity data (Fig. 1).

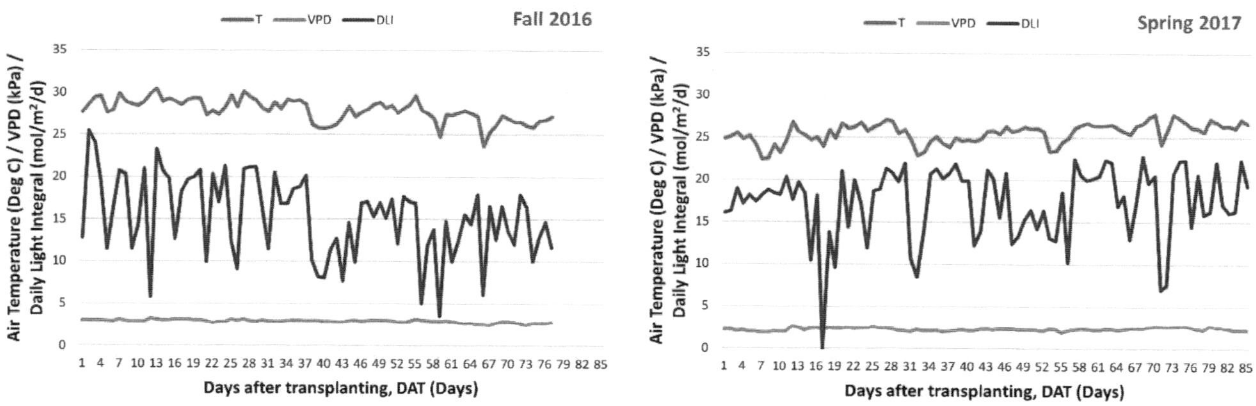


Fig. 1. Air temperature, vapor pressure deficit (VPD), and solar radiation over the two experiments performed in Fall 2016 (left) and Spring 2017 (right). Kingshill, U.S. Virgin Islands.

*Plant material*. Cucumber seeds were sown in 72-cell trays on Sept. 6, 2016 (Fall) and Jan. 3, 2017 (Spring), and transplanted into 9.45-L pots on Sept. 20, 2016 (Fall) and Jan. 12, 2017 (Spring). Pots were spaced at 0.46 × 1.22 m (representing 17,818 plants/ha). The potting mix used was Pro-Mix BX Mycorrhizae: perlite (70%: 30%). Potting mix nutrient concentrations were evaluated prior transplant in three samples. Average pH = 5.95, soluble salts = 0.43 dS/m, and nutrient concentrations (mg/L): nitrate-nitrogen = 0.35, ammoniacal-nitrogen = 5.52, phosphorus = 21.4, potassium = 65.5, calcium = 127.5, magnesium = 18, sulfur = 16.5, boron = 0.2, copper = 0.88, iron = 12.85, manganese = 6.43, zinc = 5.53, sodium = 59 and chloride = 55.53.

*Treatments*. We tested four cucumber varieties ('Boa', 'Corinto', 'Marketmore 76' and 'Verdon') and three substrate VWCs to trigger irrigation automatically (0.3, 0.4 and 0.5 $m^3/m^3$).

*Automated irrigation*. The automated irrigation controller was built using a data logger (CR1000; Campbell Scientific, Logan, UT), multiplexer (AM16/32B; Campbell Scientific), 16-channel AC/DC relay driver (SDM-CD16AC; Campbell Scientific), 36 10HS soil moisture sensors (Decagon Devices, Pullman, WA), and 36 24-VAC 1-inch (2.54 cm) solenoid valves (100DVF; RainBird, Azuza, CA) powered by a 12/24-VDC 500-VA transformer (31EJ02; Dayton, OH). The transformer power line was protected with a surge protector (3400-J 51110-SRG; Leviton, Melville, NY). The controller was powered using a 20-W solar panel (Infinium; ML Solar, Campbell, CA), connected to a 12/24-VDC 10-A Tracer solar charge controller (1210RN; EPSolar, Beijing, China) and two 12-VDC 7.2-Ah rechargeable batteries (Yuasa, Ebbw Vale, United Kingdom).

The irrigation system had 36 independent manifolds – one for each experimental unit. Manifolds were assembled using 1-inch (2.54 cm) PVC pipes. From each solenoid valve, a 2.44-m long × ¾-inch (1.9 cm) diameter polyethylene tubing was laid to receive 1-gallon-per-hour (GPH) (3.78-L) drip emitters. A dribble ring with four holes per plant was connected to the emitter to allow for even water distribution across the substrate surface. Water pressure was maintained at 25 psi using a pressure regulator. Due to the reduced number of drip emitters per irrigation line, additional 2-GPH (7.58-L) emitters were installed on each line to ensure proper closing of the valves after an irrigation event.

Irrigation was performed on-demand and controlled by one substrate moisture sensor per experimental unit. When the substrate VWC dropped below the set thresholds, the irrigation was turned on automatically for 90 seconds.

*Cultural practices*. Plants were trained on a vertical plastic line trellis, and fertigated with 5N-11P-26K hydroponic fertilizer and calcium nitrate (total of 150 mg N/L and 140 mg Ca/L).

*Measurements*. We measured VWC (over time), number of irrigation events (over time), total and marketable yield (determined weekly and totalized at the end), fruit number, fruit size (weight, length, and width), fruit soluble solids content using a refractometer (RF15; Extech Instruments, Nashua, NH), plant growth index [(height + width 1 + width 2) / 3] on days 50 (Fall) and 54 (Spring), and fruit hardness. Leaf anthocyanin and chlorophyll content indexes (non-destructive analysis) were measured in Fall (day 49) and Spring (day 75). Anthocyanin was measured with a portable anthocyanin content meter (ACM-200 plus; Opti-Sciences, Hudson, NH), and chlorophyll using a chlorophyll concentration meter (MC-100; Apogee Instruments, Logan, UT).

*Experimental design and Statistical analysis*. Treatments were arranged on a split-plot design, with three replications. Each experimental unit had four plants / variety for a total of 36 plants per variety and 144 per trial. Data were analyzed using a mixed model procedure in SAS (version 9.4; SAS Institute, Cary, NC). Errors were assumed to be normally and independently (NID) distributed. Probability values ≤ 0.05 were considered statistically significant.

## Results and Discussion

Sensor-based irrigation was effective to water the plants automatically only when the VWC dropped below set thresholds (Fig. 2). Graphs indicate the irrigation was effectively turned on when substrate VWC reached the treatment thresholds (left = VWC of 0.3 $m^3/m^3$; center = VWC 0.4 $m^3/m^3$; and right = VWC 0.5 $m^3/m^3$; Fig. 2). Results are consistent with several studies using sensor-based automated irrigation (Ferrarezi et al., 2014; 2015; Nemali and van Iersel, 2006). Replication differences are expected, and explained by the use of independent experimental units, variations in moisture caused by container positioning in the greenhouse, sensor installation, and the natural variability between plants, which was also reported by Ferrarezi et al. (2017). Data collected during Fall 2016 (Fig. 2A) were more stable than Spring 2017 (Fig. 2B). The reason for such variation is unknown.

The number of irrigation events was 62.5% higher in Spring 2017 compared to Fall 2016 ($P<0.0001$, Table 1). There was no effect of increasing VWC values to trigger irrigation on this variable. Treatments with high VWC tend to have higher substrate moisture and number of irrigations (Ferrarezi et al., 2017).

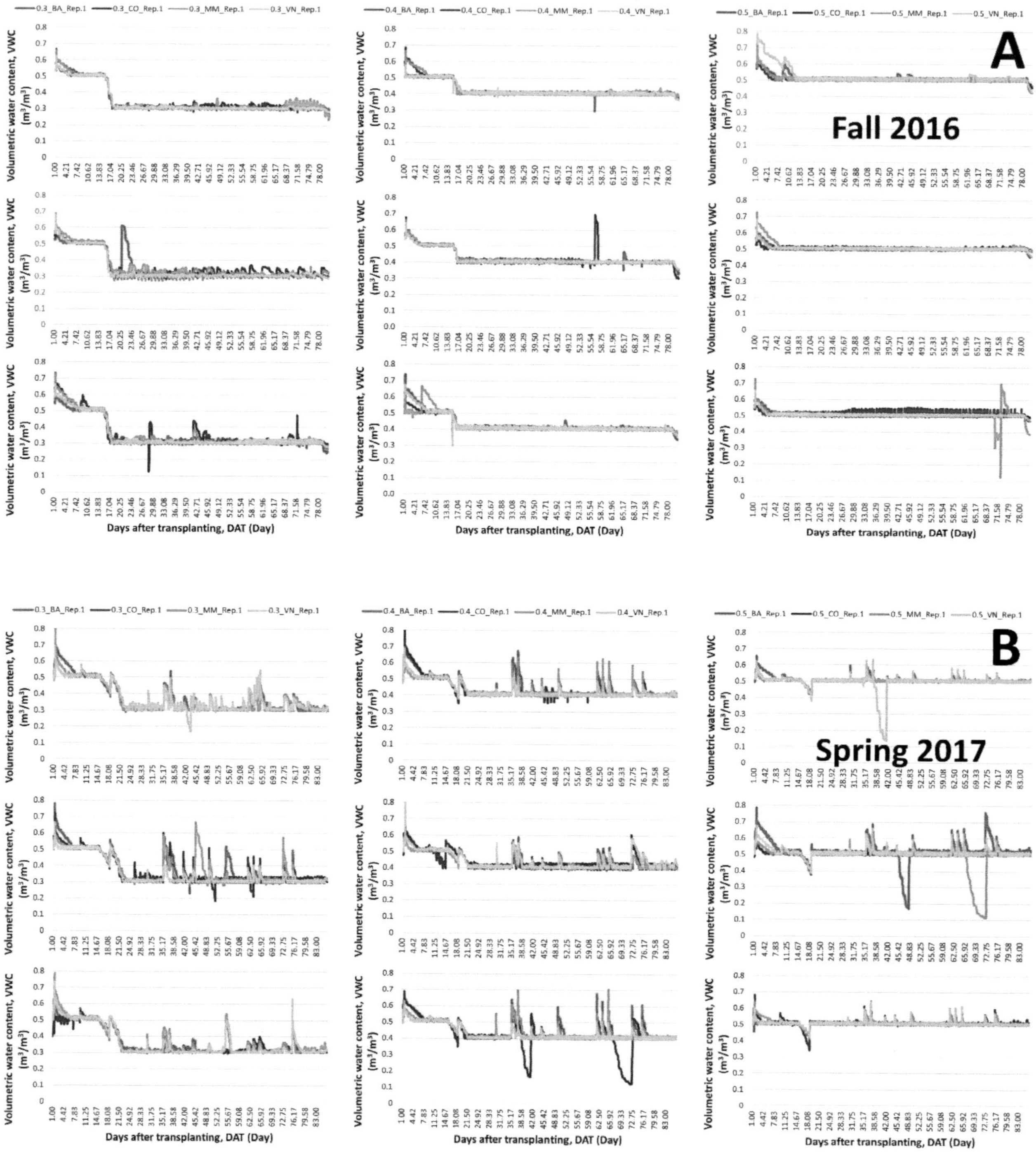


Fig. 2. Volumetric water content (VWC) in Fall 2016 (A) and Spring 2017 (B). VWC treatment thresholds: graphs in the left = VWC of 0.3 $m^3/m^3$ to trigger irrigation; center = VWC 0.4 $m^3/m^3$; and graphs in the right = VWC 0.5 $m^3/m^3$. Kingshill, U.S. Virgin Islands.

Table 1. Number of irrigation events, marketable yield, fruit width and hardness, and leaf anthocyanin of four cucumber varieties ('Boa', 'Corinto', 'Marketmore 76' and 'Verdon') cultivated in two seasons (Fall 2016 and Spring 2017) and under three volumetric water contents (VWC) to trigger irrigation (0.3, 0.4 and 0.5 $m^3/m^3$). * Significant at P<0.05. Kingshill, U.S. Virgin Islands.

| | Irrigation count (nr.) | Marketable yield (kg/ha) | Fruit width (cm) | Fruit hardness (kgf) | Leaf anthocyanin (ACI) |
|---|---|---|---|---|---|
| ***Season*** | | | | | |
| Fall 2016 | 456 ± 67 b | 25,796 ± 7,837 | 4.92 ± 0.26 | 8.04 ± 0.54 | 6.70 ± 0.80 a |
| Spring 2017 | 741 ± 211 a | 23,552 ± 10,224 | 4.84 ± 0.18 | 7.74 ± 0.53 | 5.14 ± 0.37 b |
| ***VWC*** | | | | | |
| 0.3 $m^3/m^3$ | 641 ± 252 | 23,176 ± 6,024 | 4.91 ± 0.25 | 8.04 ± 0.47 | 5.84 ± 0.71 |
| 0.4 $m^3/m^3$ | 535 ± 120 | 21,723 ± 5,991 | 4.86 ± 0.20 | 7.72 ± 0.65 | 6.08 ± 0.84 |
| 0.5 $m^3/m^3$ | 619 ± 125 | 29,122 ± 11,621 | 4.88 ± 0.23 | 7.91 ± 0.49 | 5.83 ± 0.77 |
| ***Variety*** | | | | | |
| 'Boa' | 513 ± 71 | 27,237 ± 7,437 b | 5.14 ± 0.07 a | 8.44 ± 0.44 a | 6.27 ± 0.67 |
| 'Corinto' | 631 ± 151 | 38,243 ± 9,044 a | 4.96 ± 0.09 a | 8.22 ± 0.46 ab | 5.64 ± 0.61 |
| 'Marketmore 76' | 661 ± 270 | 14,356 ± 4,210 c | 4.98 ± 0.14 a | 7.60 ± 0.60 bc | 6.04 ± 0.75 |
| 'Verdon' | 589 ± 157 | 18,858 ± 4,814 bc | 4.45 ± 0.28 b | 7.31 ± 0.42 c | 5.71 ± 0.98 |
| | | | *p-value* | | |
| Season | <0.0001* | 0.3899 | 0.2856 | 0.1153 | <0.0001* |
| VWC | 0.4018 | 0.0560 | 0.8218 | 0.3679 | 0.6612 |
| Season*VWC | 0.5455 | 0.5887 | 0.5547 | 0.4470 | 0.8657 |
| Variety | 0.4371 | <0.0001* | <0.0001* | 0.0002* | 0.2523 |
| Season*Variety | 0.6013 | 0.4143 | 0.4034 | 0.4810 | 0.2804 |
| VWC*Variety | 0.5731 | 0.05040 | 0.6434 | 0.1594 | 0.4542 |
| Season*VWC*Variety | 0.7486 | 0.8300 | 0.9213 | 0.1778 | 0.1733 |

Total yield and total number of fruit per plant responded to increase in VWC and were different among varieties (P<0.05, Fig. 3). 'Corinto' cultivated with VWC 0.5 $m^3/m^3$ resulted in the highest yield at 57,445 kg/ha (P=0.00269, Fig. 3). The number of fruit per plant followed a similar trend (P=0.0293, Fig. 3). 'Corinto' is a parthenocarpic slicing cucumber hybrid with high yield potential and strong vigor, with intermediate resistance to powdery mildew and viruses (Torres and Mazereeuw, 2013), which may explain the outstanding performance. The marketable yield of 'Corinto' was higher than all other varieties, totaling 38,243 kg/ha (P<0.0001, Table 1).

Fruit width was 11% smaller on 'Verdon' (P<0.0001, Table 1). This variety also exhibited the lowest fruit hardness, what can reduce shelf life (P=0.0002, Table 1). Leaf anthocyanin content was 30% higher in Fall 2016 (P<0.0001, Table 1).

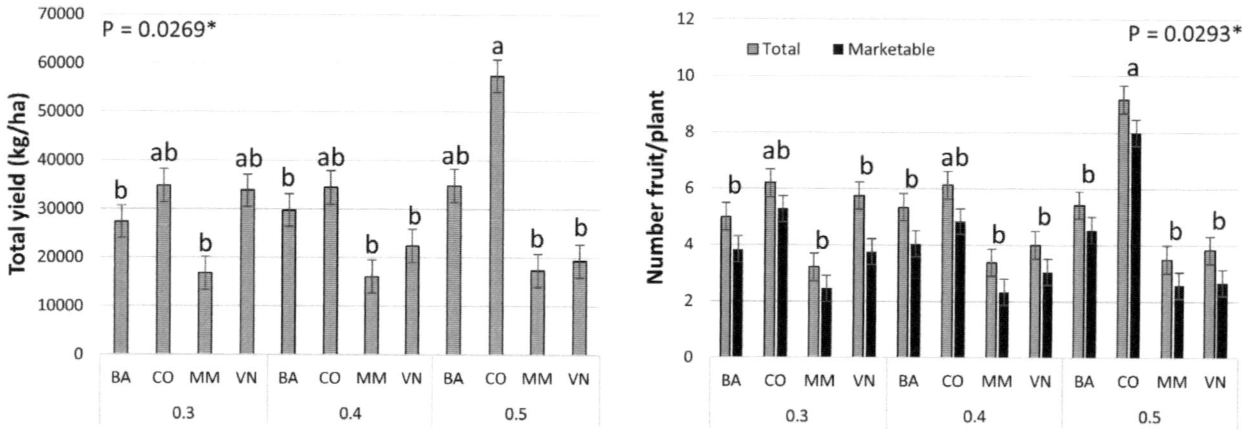


Fig. 3. Total yield (left) and number of fruit per plant (right) in four slicing cucumber varieties ('Boa' [BA], 'Corinto' [CO], 'Marketmore 76' [MN] and 'Verdon' [VN]) subjected to increasing volumetric water contents to trigger irrigation automatically (0.3, 0.4 and 0.5 $m^3/m^3$). Kingshill, U.S. Virgin Islands.

Fruit weight was low on 'Verdon' in Fall 2016 and on 'Marketmore 76' in Spring 2017 (P=0.001, Fig. 4). Fruit were 35% longer on 'Verdon' in Fall 2016 and 75% longer in Spring 2017 compared to the other varieties (P=0.0074, Fig. 4). As 'Verdon' presented the lowest fruit hardness, total yield and fruit weight (Table 1 and Fig. 3), this variety is not suitable for greenhouse production in the U.S. Virgin Islands. Fruit soluble solids content was higher on 'Marketmore 76' and 'Verdon' on both seasons (P=0.0004, Fig. 4). Leaf chlorophyll content decreased in Spring 2017 (P=0.0031, Fig. 4). 'Boa', 'Corinto' and 'Marketmore 76' presented the highest plant growth index in Fall 2016, with all varieties having drastically reduced plant growth in Spring 2017 (P=0.0018, Fig. 4). Reasons for such drop were not identified.

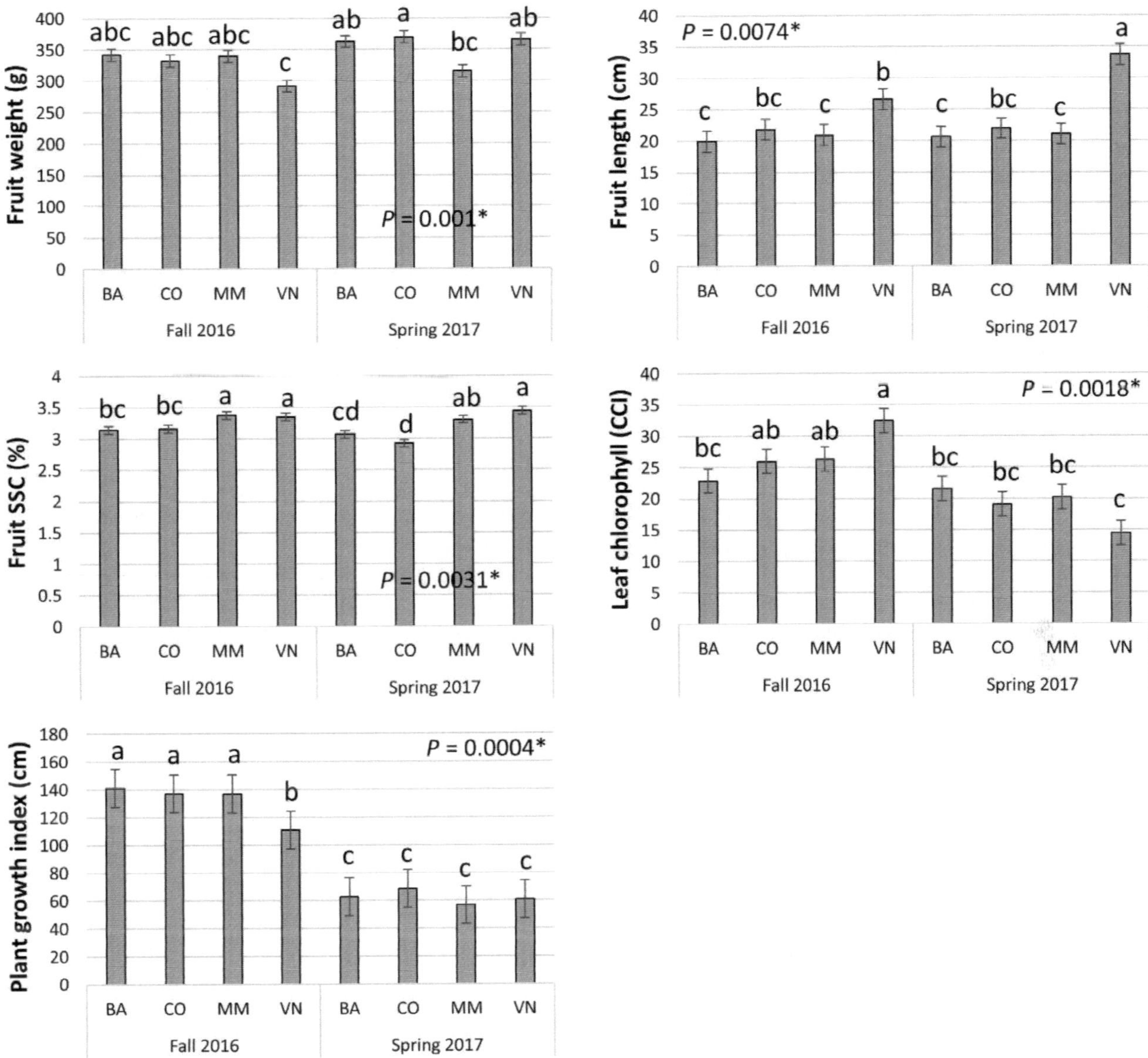


Fig. 4. Fruit weight, fruit length, fruit soluble solids content (SSC), leaf chlorophyll content and plant growth index in four slicing cucumber varieties ('Boa' [BA], 'Corinto' [CO], 'Marketmore 76' [MN] and 'Verdon' [VN]) cultivated in two sequential seasons (Fall 2016 and Spring 2017). Kingshill, U.S. Virgin Islands.

## Conclusions

'Corinto' consistently showed higher yield in all VWCs used to trigger irrigation automatically, producing longer and more fruit than the other varieties, being a promising material for greenhouse cucumber production using sensor-based irrigation in the U.S. Virgin Islands.

## Acknowledgements

We thank Victor Almodovar and Naima Jenkins-El (Horticulture and Aquaculture program), Kalunda Cuffy, Jayar Greenidge and Shamali Dennery for their technical assistance. Funding for this research was provided by USDA-NIFA-Hatch Funds (Accessions no. 227355 and 1010280) and USDA-NIFA-Multi State grant W3128 (Accession no. 1006963).

## References

Ferrarezi, R.S. and Testezlaf, R. 2017. Automated ebb-and-flow subirrigation for citrus liners production. I. Plant growth. Agricultural Water Management 192: 45-57.

Ferrarezi, R.S.; Weiss, S.A.; Geiger, T.C.; Beamer, K.P. 2016. Edible-pod peas as high-value crops in the U.S. Virgin Islands. HortTechnology 26(5): 683-689.

Ferrarezi, R.S.; van Iersel, M.W.; Testezlaf, R. 2014. Subirrigation automated by capacitance sensors for salvia production. Horticultura Brasileira 32(3): 314-302.

Ferrarezi, R.S.; van Iersel, M.W.; Testezlaf, R. 2015a. Monitoring and controlling ebb-and-flow subirrigation with soil moisture sensors. HortScience 50(3): 447-453.

Jones, H.G. 2007. Monitoring plant and soil water status: Established and novel methods revisited and their relevance to studies of drought tolerance. Journal of Experimental Botany 55: 2427-2436.

Nemali, K.S. and van Iersel, M.W. 2006. An automated system for controlling drought stress and irrigation in potted plants. Scientia Horticulturae 110: 292-297.

Nemali, K.S.; Montesano, F.; Dove, S.; van Iersel, M.W. 2007. Calibration and performance of moisture sensors in soilless substrates: ECHO and Theta probes. Scientia Horticulturae 112: 227-234.

Shock, C.C. and Wang, F.-X. 2011. Soil water tension, a powerful measurement for productivity and stewardship. HortScience 46: 178-185.

Torres, L.M. and Mazereeuw, J.P. 2013. Cucumber hybrid 'Corinto'. US Patent US 8,399,740 B2, filed May 8, 2009, and issued Mar 19, 2013.

U.S. Department of Agriculture. 2015. Census of Agriculture: Census of Horticultural Specialties (2014). 14 Oct. 2017. https://www.agcensus.usda.gov/Publications/2012/Online_Resources/Census_of_Horticulture_Specialtes/HORTIC.pdf.

U.S. Department of Agriculture. 2009. Census of Agriculture: Virgin Islands of the United States Territory and Island Data (2007). 14 Oct. 2017. https://www.agcensus.usda.gov/Publications/2007/Full_Report/Outlying_Areas/usvi.pdf.

## My Irrigation System is Plugged: What Do I Do Now?

It is almost inevitable for irrigation emitters to be plugged after a short period of time. Blockage of emitters is one of the most common problems when dealing with micro-irrigation systems. Properly designed and maintained filtration systems generally protect the system from most blockages. Blockages can cause irregular water distribution, which in turn can damage the crop. Our research shows that plugging is random, which means irregular patterns of plugging in emitters. When the plants show excessive stress, it is generally too late to correct the problem. The main causes of plugging include algae, bacterial slime, particles, construction debris, and sediment. With adequate filtration, line flushing, and chemical treatment most blockages can be prevented.

If the blockage is caused by mineral precipitates it is either from the well pumping sand or soil which can be caused by cracks in the piping or low water level. The minerals most likely to cause plugging is silt. Silt can be ultra fine sand, clays, and other insoluble soils. This is a filtration issue that can be solved by using an effective filter. Coarseness of filters, the costs, etc are all variables involved in choosing the right filter for your system. Generally, the best type of filter is a media filter. The sand can be supplemented with DE (diatomaceous earth) for very fine filtration if necessary.

Some growers have been convinced they have a calcium blockage in their system. In the company's 40 years of water treatment, calcium/scale has never been found to cause plugging in emitters. What happens is that the blockage in the emitters grows so thick that water puddles on top of the emitter and when the water evaporates, a small amount of calcium/scale is visible.

A mineral scale will generally not form without heat and pressure. You would find scale forming in cooling towers, cool cells, boilers or something we are more familiar with: a tea kettle. It takes heat, pressure, an imbalance of alkalinity vs. calcium, or a recirculating system with evaporation for scale to form.

How much calcium carbonate is in the water? To put the amount of calcium carbonate in perspective, consider how little is in the water and what it means. At 200 ppm of calcium carbonate (hardness), that translates to 200 pounds of calcium carbonate for every million pounds of water. That is 200 lbs. of calcium carbonate in every 120,000 gallons of water. On a percentage basis this is 0.02%. This is a very small amount and is being spread over a large area. If 200 lbs. of calcium carbonate is spread over 40 acres using drip tape, it would not form a mineral scale thicker than a very thin paper. Mineral scale usually forms on the bottom and sidewalls of tubing as the water generally flows out of a system when it is shut down. At this concentration, the calcium carbonate would not form a thick scale for many years. There is always the debate of which is better well water or surface water. Sometimes there isn't really a choice due to location and availability. We had one customer in Louisiana that was pumping out of a bayou and had problems with plugging. They spent over $10,000.00 installing a DI (deionization) system that produced almost pure water. They still had terrible problems with plugging. The source of the water is not necessarily a factor in the formation of plugging.

Most blockages are caused by some form of microorganisms that are growing. All the factors are there for this to occur: water, warmth, food (fertilizer), and time. If you are using fertigation in your irrigation system, just as the fertilizer makes your plants grow, it will also make algae and slimes grow. Flushing the irrigation system after fertigation would reduce the amount of fertilizer in the system.

In the most general sense, a plant is a member of the lower or vegetable order of living organized things. Thallophyta are the lowliest organized plants and include a great variety of forms, the vegetative portion of which consists of a single cell or a number of cells forming a more or less branched thallus. They are characterized by the absence of differentiation of the body into root, stem and leaf which is a common feature in higher plants. Both sexual and asexual reproduction occurs in these types of organisms. They can be unicellular or complex organisms, lack mobility, have simple processes for digestion and reproduction, have little defense mechanisms, tend to have thinner cells walls, and can either be aerobic or anaerobic. They can survive and thrive in sunlight, darkness, or a combination of the two. Even if they become substantially dehydrated, these organisms will revive when exposed to water again. Types of these include algae (including Seaweeds) which contain chlorophyll, the Fungi which have no chlorophyll and therefore lead a saprophytic or parasitic mode of life, and the Lichens which are composite organisms consisting of an alga and a fungus living together in a mutual parasitism (symbiosis). A study of phylogeny has suggested twelve classes arranged in the following sequence: (1) Bacteria; (2) Cyanophyceae (Blue-green algae); (3) Flagellatae; (4) Myxomycetes (Slime-fungi); (5) Pendineae; (6) Conjugatae; (7) Diatomaceae (Diatoms); (8) Fleteroconteae; (9) Chlorophyceae (Green Algae); (10) Characeae (Stoneworts); (II) Rhodophyceae (Red Algae); (12) Eumycetes (Fungi);

## What to Do to Restore the System

Frequently growers find their irrigation system is plugged and they are baffled as to what to do. Sometimes the only help available is the salesman of the irrigation system or a local extension agent. Most extension services have recommended chlorine at various dosages and applications. Chlorine at low dosages will kill most organisms, but not remove them. Many times, these dead cells become food for the next generation of growth. As these micro-organisms sense an attack, their only defense is to reproduce to survive. Many times, after a dose of chlorine is injected in irrigation systems, 7-10 days later the problem is much worse. Algae which is the most common organisms in water systems has a re-generation cycle of 7-10 days which is readily apparent in a swimming pool. If you go on vacation without having someone treat the pool, after a week at the beach the pool will turn green.

When you consider municipal water systems that are generally treated at 1-2 ppm of chlorine, inspection of most toilet tanks will show a slight film on the inside that is slimy to the touch. In some areas, the growth in the toilet tanks can be excessive and can cause odors and stains in the toilet. The neighbors also had the same problem and the city was baffled and never did find out what caused it to grow so profusely. Most likely, it was like many irrigation systems that have a bio-film growing on the walls of the piping that breaks off and causes the plugging in emitters. In this municipal water system (similar to irrigation systems), the entire system needs to be treated to have a clean water system.

If you consider chlorine, think about the affects with its use in laundry. At low doses, it will remove some organic stains (coffee, food, etc.), but it has no effect on mineral soils such as mud, clay, iron, etc. Mineral soils are generally removed from clothing with water and the physical action of a washing machine. The mineral stains are not generally soluble so it's more of the physical action and dilution that removes dirt, soils, etc. When you use higher dosages of chlorine, the fabrics will have holes in them if they made with natural fibers like blue jeans, kakis, etc. If the dosage is high enough to eat holes in cloth, it will also eat holes in plants.

For years growers have gotten advice from the fertilizer companies to try flushing the system with acid and/or chlorine. Both of these options usually have poor results. Neither acid or chlorine are effective at removing blockages in emitters at low dosages and at higher dosages, they are harmful to plants as well as the damage they cause to metals in the irrigation system. The acid that is generally available is 54% phosphoric acid which is contaminated with impurities. This can make the plugging much worse. Sulfuric acid is also a cheap acid which is generally used to reduce the alkalinity of water to lower the pH that certain plants thrive on like blueberries. Sulfuric acid will not dissolve calcium carbonate. One time the city water department came to our office to buy sulfuric acid drain cleaner. After several trips, an inquiry was made as to what they were trying to achieve. A critter had crawled into a drainage pipe and the bones were working to collect debris. They thought drain opener would be cheaper than digging up the pipe. After selling them a few gallons of muriatic acid, their problem was solved. High levels of hydrochloric acid will dissolve calcium carbonate scale. The only effectiveness of low dosages of chlorine and acid treatments is for the companies to increase their sales.

An old well driller's trick is to dump chlorine tablets, powder, or liquid down the pipe and allow it to sit for a day or two. Without some agitation to move the chlorine over a large area, the chlorine simple drops to the bottom of the well and much of the chlorine either doesn't dissolve or is ineffective. The water is pumped to waste until the chlorine levels drop. Not only does it corrode metal pipes, but it works only for a limited time.

## REMOVING PLUGGING IN MICRO IRRIGATION SYSTEMS

With new technology and a few hours of work, the grower can restore the irrigation system to working order. The discovery of a new safer and more concentrated form of peracetic acid has proven effective at removing blockages in all types of irrigation systems from drip tape, hard pipe with emitters or spaghetti tubing, micro jets, spinners, spitters, overhead sprayers and other irrigation parts. Growers have many different irrigation schedules, water requirements, flow rates, length of irrigation times per zone, the inability to shut down the irrigation system and other factors can determine how the grower can use this new technology. This chemical can be used in several different ways. This compound can adapt to the growers schedule without interrupting the irrigation regimen required.

This compound removes the deposits in emitters with a 2-4 shock treatments at a ratio of 1:6,000 (1 gallon of treatment for every 6,000 gallons of irrigation water). This is frequently use in Tree Farms, Orchards, Row Crops, and Areas with Large Zones that can be Irrigated for Longer Times. Each zone needs to be treated and allow contact time of 4-8 hours (overnight is better). Depending on the amount of plugging, 2-4 shock treatments are required. The cost using this compound at a flow rate of 100 gpm (gallons per minute) is $12.50 per zone.

It can also be used in a continuous treatment at 1:24,000 (1 gallon of treatment for every 24,000 gallons of irrigation water). The continuous injection is ideal for liners/starters and plants that are irrigated for shorter periods of time which can even be for only a few seconds at a time. These irrigation systems usually include switching zones frequently, applying small amounts of water many times a day. Injection time usually is 8 -24 hours depending on the amount of deposits in the emitters. The cost at a flow rate of 100 gallons per minute is $10.69 per hour.

At either of the above injection rates, peracetic acid does not affect the pH, will not affect plants, has no taste, leaves no residue, 100% organic, and is economical to use. Peracetic Acid can also be used in weekly dosages to prevent the blockage from ever occurring. It has been used in greenhouses and has been sprayed on orchids and other plants at a ratio of 1:1.000 with no resulting damage. The only effect during this experiment was the removal of lichen moss that was growing around or on the ground below the plants.

One of our customers had a malfunction with their injector and it pumped an entire 55 gallon drum of treatment over the weekend. The drum would normally last three months and it was an almost full drum. When they called on Monday, they reported there was no damage to the plants and to send another drum.

This new compound is non-specific in that it removes all deposits in all parts of the irrigation system. If possible, injection before the filter helps keep the filter and the system cleaner. This compound cleans every part of an irrigation system safely, without harming any plant or the irrigation system. An injector for precise control has yielded best results in unplugging emitters in drip tape, drip lines, micro jets, and other micro irrigation emitters.

Respectfully submitted,

Jim Wynn
Line Blaster, Inc.
139 Altama Connector #417
Brunswick GA 31525
jim@lineblaster.com
912-554-3700 Office
912-996-4292 Mobile
866-384-5311 Fax

# EVALUATION OF POLYETHYLENE SUBSURFACE DRIP IRRIGATION DRIPLINES AFTER 26 YEARS OF SERVICE IN A CORN FIELD IN NORTHWESTERN KANSAS

**Adriana Velasquez, Mark Jablonka, Freddie Lamm, Eddy Garcia-Meitin**
The Dow Chemical Company / KSU Northwest Research Extension Center

Subsurface drip irrigation (SDI) delivers increased crop yields, improves utilization of farmland, reduces fertilizer run off and more efficiently uses water. One concern commodity growers have when investing in an SDI system is the longevity of the dripline, requiring a long life to be economically practical. The use of high quality raw materials is a key factor in maximizing the value of an investment in SDI.

This paper contains an actual example of how the use of high quality materials enabled driplines to last 26 years in service. A new dripline was installed in Kansas as part of an SDI system in 1989. The system operated with the same dripline until 2015, when it began to fail. The overall dripline held up well and remained ductile over its 26 year life, demonstrating longevity when quality raw materials are used.

Through this real life example, the key inputs to maximizing the longevity of driplines are identified and outcomes of the aging process described.

## INTRODUCTION

A polyethylene Subsurface Drip Irrigation (SDI) system that was installed in a corn field in 1989 in the Northwest region of Kansas was taken out of service due to leaks in 2015. Portions of the dripline were analyzed for characterization. The purpose of this work is to highlight the importance and advantages in the use of high quality materials to manufacture the polyethylene driplines for such application. This report documents the characterization of this dripline and the failure analysis.

## BACKGROUND

Dow was the first company to develop polyethylene grades specifically for the microirrigation segment in the 1970's. The products Dow developed greatly extended the longevity of tubing at the time. As the segment continued to develop, so did Dow's product offering. FINGERPRINT™ are Dow's resins for microirrigation that have been used to improve the way watering of crops is done, which has contibuted to increasing the utilization of farmland and also has help growers save in the usage of fertilizers.

One of the key application areas for growth within the microirrigation dripline segment is referred to as Subsurface Drip Irrigation (SDI). In this application, driplines are buried up to 24 inches below the surface of the soil, water is pumped through the dripline and delivered to the roots of the crops via emitters built into the driplines. Figure 1 illustrates where the dripline is placed relative to the crop. The dripline is indicated by the circle below the surface with arrows indicating soil water redistribution radiating outward from it.

**Figure 1. Image of Subsurface Drip Irrigation system placement for growing corn.**[i]

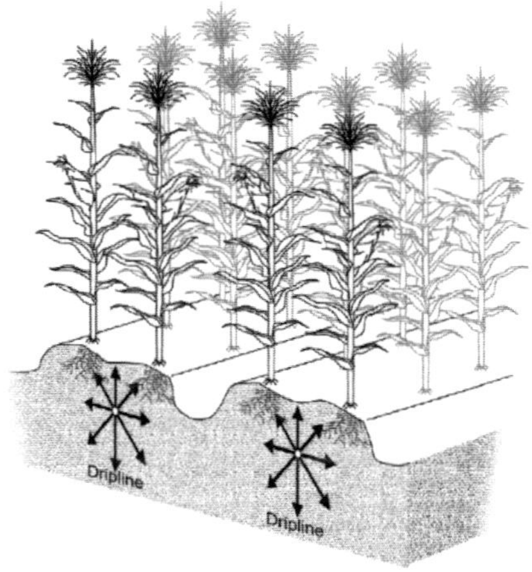


Research in the area of subsurface drip irrigation has been ongoing for over 30 years. Kansas State University began studying and developing subsurface drip irrigation (SDI) techniques for growing commodity crops in 1989.[ii] The driver for this technology is to make more efficient use of water, water conservation is one of the most effective ways to positively impact the environment these days.

For over 25 years the Northwest Research-Extension Center in Colby, Kansas has been focused on subsurface drip irrigation (SDI). The dripline evaluated in this study was installed in a corn field in Colby in 1989. Dripline longevity is a key concern for growers investing in an SDI system, especially when commodity prices are low, which makes the fact that this dripline lasted 26 years very compelling.

## EXPERIMENTAL PROCEDURES

### Materials

The dripline materials are listed in Table 1. Sample 24-1 is a dripline manufactured in the 1989-1990 time frame but was never put in service, sample 24-2, is the in service dripline that failed, had been buried underground during its lifetime. It was used to carry water to the roots of corn plants. Sodium hypochlorite (NaOCl) was used in the dripline as a disinfectant periodically to remove algae/growth. In addition, urea-ammonium nitrate fertilizer was applied through the dripline. Water, sodium hypochlorite and urea-ammonium nitrate are the only chemicals used in the dripline during its lifetime. Sample 24-3 is a standard polyethylene resin produced and sold by Dow into drip irrigation systems today and was used in this study as control.

**Table 1. Summary of materials analyzed.**

| Databook No. | Sample Name | Sample Description |
|---|---|---|
| 201500084-24-1 | Control dripline (unused) | 15mil thick Dripline[iii] purchased in 1990 and stored inside a building at Kansas State University's Northwest Research-Extension Center in Colby, Kansas. |
| 201500084-24-2 | Failed dripline | 15mil thick Dripline installed in 1989 as part of Kansas State University's Subsurface Drip Irrigation (SDI) system in their corn fields at their Northwest Research-Extension Center in Colby, Kansas. |
| 201500084-24-3 | FINGERPRINT™ Resin | Dow's resin used to fabricate driplines. |

## Test Methods

### *Differential Scanning Calorimetry (DSC)*

Since the samples were in the form of dripline, the specimens were punched out and pressed into a pan. TA instrument DSC Q 2000 series was used. The first step is an equilibration step to remove thermal history, then the sample is cooled at 10°C/min to -90°C, isothermal time is 5 minutes, the sample is then heated at 10°C/min to 290°C, isothermal time of 5 minutes. This was done under a nitrogen atmosphere (nitrogen at 50mL/min).

### *Differential Scanning Calorimetry (DSC) – OIT*

Using a DSC instrument, specimens were heated to 200°C in a 100% nitrogen environment, then the nitrogen was replaced by oxygen and the time to full oxidation recorded.

### *Dynamic Mechanical Spectroscopy (DMS)*

Frequency Sweep, viscoelastic properties are measured under controlled strain at 190C, with varying frequencies from 0.1 to 100 rad/s with 10% strain.

### *Thermogravimetric Analysis (TGA)*

A TGA based method was used to quantify carbon black and inorganic residue levels in the dripline. In this method, material from the dripline is heated in an inert environment until it reaches 520°C, at which time the environment is changed to air, the carbon black oxidizes and the weight loss is determined. Residue represents the material left at 800°C.

### *X-Ray Fluorescence (XRF)*

This test was used for determining catalyst residues and relevant additives in all polyolefin samples. The following elements can be requested as an individual test: Al, Ba, Ca, Cl, Mg, Mo, Na, P, S, Si, Ti, and Zn. The XRF is calibrated with polymer standards. Based upon method development data, the error in the accuracy is typically less than +/-10%, but is dependent upon the concentration. The precision (%RSD) of XRF analysis is usually better than +/-5%, but is also dependent upon concentration. The precision and accuracy are also dependent upon sample homogeneity.

### *Antioxidant (AO)*

Resin was extracted using TDM. Extract was analyzed by LC with UV/Vis detector to identify active Antioxidants and oxidized antioxidants as well as their concentration level. Concentration is reported as parts-per-million (ppm). Analysis follows DOWM 102408-I10B.

### *Tensile*

Tensile tests were performed on electromechanical tensile tester. Load cell was 50 lb (~220N). The test was carried out on the full dripline samples with a 2" gauge length using line grips and at a speed of 20"/min.

## RESULTS AND DISCUSSION

### Characterization

The failed dripline was coated with dirt on both interior and exterior surfaces as received. The presence of dirt on the inside of the dripline was likely caused by either the dripline failing or being cut and removed from the field. Figure 2 is a photo of pieces of the materials received. Sample A is a portion of the failed dripline that was split. A sharp linear failure occurred along one of the two edge creases indicated by the red arrows in the photo. This failure is a machine direction split in the dripline. Sample B is a portion of the failed dripline that did not contain a failure and sample C is the control dripline which was manufactured back in 1989 but was never put in service, instead it was stored for 26 years.

**Figure 2. As received dripline samples.**

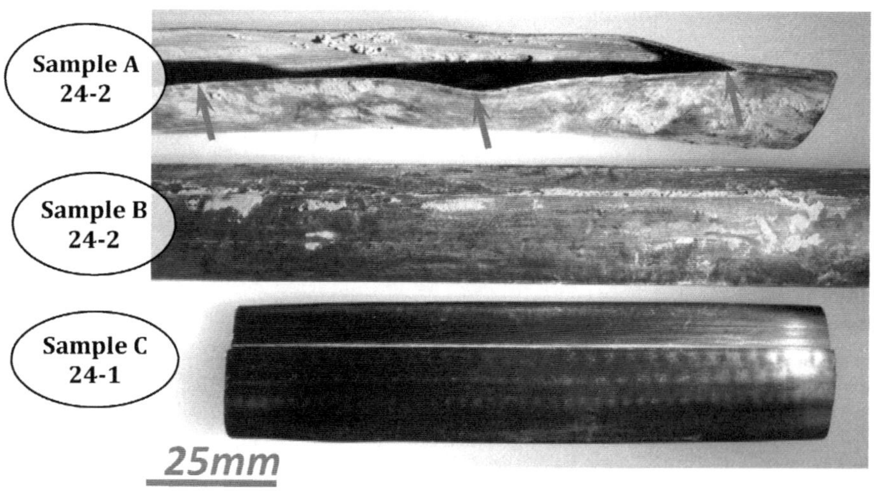


The driplines were confirmed to have been made with 100% Dow resins by identification of the tracer which is added to Dow's microirrigation FINGERPRINT™ products as well as pressure pipe products, in order to positively identify the material as being from Dow in the case of a failure.

Today FINGERPRINT™ resins continue to be the leading resins used in the manufacturing of microirrigation driplines. DSC was used to characterize thermal properties of the driplines and results are summarized in Table 2. The failed and control driplines both had similar melting characteristics which were comparable to FINGERPRINT™ providing evidence that FINGERPRINT™ was used to make this dripline. Usually as PE ages its density slightly increases, but the DSC measurements of the failed and control driplines relative to the FINGERPRINT™ control do not indicate this, suggesting this material aged very well, perhaps due to the quality of the resin, it's AO package and the protection from UV light afforded it by being buried in the soil.

**Table 2. Summary of DSC melting and crystallization temperature measurements.**

| | Tm (°C) | Tc (°C) |
|---|---|---|
| Control dripline (24-1) | 119.0 | 108.3 |
| Failed dripline (24-2) | 119.4 | 108.8 |
| FINGERPRINT™ Control (24-3) | 119.0 | 107.7 |

Figure 3 contains a plot of the DSC second heat melting curve comparing the control dripline, failed dripline and the FINGERPRINT™ resin. This plot illustrates the equivalent thermal characteristics of all three materials.

**Figure 3. DSC Second Heat Melting Curve Comparison.**

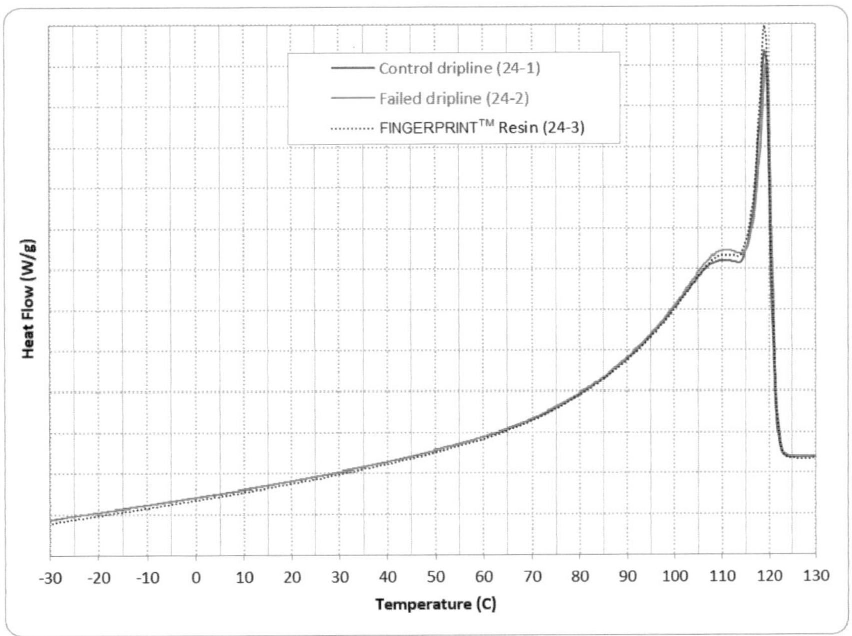


DMS rheology was performed on the materials and the viscosity curves are summarized in Figure 4. The overall viscosity of the material in the failed dripline (24-2) is about 7% lower than FINGERPRINT™ in the low shear region and about 10% lower in the high shear region. Typically the viscosity of carbon black containing polyethylene is greater than polyethylene absent of carbon black. The control dripline (24-1) which was never put in service has a viscosity profile very close to that of FINGERPRINT™. This difference is likely due to normal variation during the process of extruding the dripline, and doesn't necessarily indicate that there is degradation by chain scission.

**Figure 4. DMS Viscosity Overlay at 190°C.**

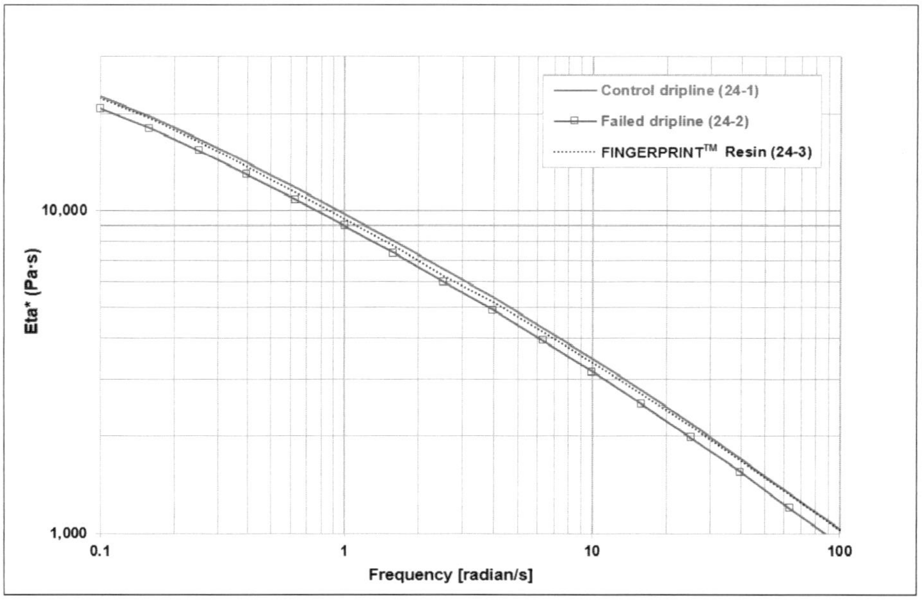


Standard resin characterization tests were performed on the control and failed driplines along with FINGERPRINT™ resin and results are summarized in Table 3. The failed dripline measured density is greater than the Fingeprint™ resin because it contains 2.3% carbon black. The measured melt index of the failed dripline is greater than the FINGERPRINT™, which suggests it has an overall lower molecular weight since typically the addition of carbon black would yield a lower melt index, more viscous, product. Carbon black levels in both the failed dripline and control dripline were within the expected range of 2.0-3.0 wt% and residue levels were extremely low at 0.05 wt%, indicating the carbon black used was very clean and no fillers were used in the manufacturing of the dripline.

**Table 3. Density, Flow Rate and Carbon Black Measurements on the dripline samples.**

| | Density ($g/cm^3$) | Melt Index (dg/min) | Flow Index at 190°C, 21.6kg (dg/min) | MFR (I21/I2) | Carbon Black (wt%) | Inorganic Residue (wt%) |
|---|---|---|---|---|---|---|
| Failed dripline (24-2) | 0.934 | 0.68 | 49.3 | 72.5 | 2.33 | 0.05 |
| FINGERPRINT™ resin (24-3) | 0.922 | 0.53 | 45.8 | 86.4 | 0 | 0 |
| Control dripline (24-1) | Not Measured | Not Measured | Not Measured | Not Measured | 2.37 | 0.05 |

Additive type and levels were determined in the failed dripline indicating that there were still 46 ppm of active secondary antioxidant in the dripline, However all of the primary AO had been oxidized.

Environmental stress crack resistance (ESCR) properties were measured on compression molded specimens created with material from the 26 year old dripline. The method used is described in ASTM D-1693 and specified in ASAE S553.[iv] The ASAE S553 standard requires the material used in the dripline to have an ESCR greater than 1,000 hours using condition A (regardless of the density of the material) at 50°C. In this case, the ESCR of the 26 year old material was measured to be greater than 1,000 hours. Test specimens were taken off test at 1,000 hours and not allowed to go to failure. This result provides evidence that the bulk material in the dripline continued to be of very high quality, even after being in service for 26 years.

Tensile properties of the walls of the dripline were characterized and results are summarized in Table 4. Both the failed dripline and control dripline remained strong even after 26 years. Both driplines exhibited equivalent strain behavior. The control dripline exhibited about 400 psi greater yield stress and 300 psi greater break stress than the failed dripline. In general, the failed dripline exhibited strain at break in the 260 to 350% range. Two of the control driplines exhibited results in the 325% range, but three of them were quite a bit less in the 140 to 200% range. Overall, the failed dripline exhibited relatively consistent tensile performance, suggesting it remained strong even after 26 years of service.

**Table 4. Micro tensile result summary of Control (24-1) and Failed (24-2) driplines.**

| | Failed dripline (24-2) | Control dripline (24-1) |
|---|---|---|
| Stress at Yield (psi) | 2,030 | 2,598 |
| Strain at Yield (%) | 13 | 13 |
| Stress at Break (psi) | 3,185 | 3,520 |
| Strain at Break (%) | 280 | 223 |
| Peak Load (lbf) | 8.5 | 9.6 |

The characterization performed confirms that the failed dripline was fabricated with a Dow FINGERPRINT™ resin. The dripline contained the tracer used in FINGERPRINT™ products at a level consistent with what had been using since the introduction of the products in the market. Thermal characteristics of the dripline are comparable to FINGERPRINT™, with an equivalent melting shoulder at around 110°C and peak at around 119°C. Rheological characterization indicates the material used in the dripline has a lower overall viscosity than FINGERPRINT™. This may be due to the fact that the comparison was made between an extruded dripline and natural resin. Residue levels were less than 0.1 wt%, suggesting the carbon black used was very clean and no foreign material such as recycle was used in the fabrication of the dripline. Additive characterization indicates the failed dripline contained 46ppm of active antioxidant, suggesting the antioxidant package in the dripline had held up extremely well over the years.

## Failure Analysis

Although dripline failures occurred along a crease the overall dripline remained pliable and visual evidence of degradation did not exist[12]. Figure 5 includes a cross section of the failed dripline along with a cross section of the control dripline used, with the red arrow indicating where the dripline failed. This area is where the dripline is creased as it is rolled onto the spool after being fabricated. It remains on the spool in a collapsed state until it is installed in the field.

**Figure 5. Cross sections of failed and control dripline samples.**

Failed dripline Cross Section (24-2) | Control dripline Cross Section (24-1)

Figure 6 contains dripline cross sectional optical microscopy images taken at the failure region of the dripline. Surface striations were present along the inner surface of a control dripline which followed edge creases (blue arrow). These inner surface striations acted as crack initiation sites for the failures as shown by the internal cracks (red arrows) in the failed in-service dripline.

**Figure 6. Cross sectional optical microscopy images of failure region in both Failed (24-2) and Control (24-1) dripline Specimens.**

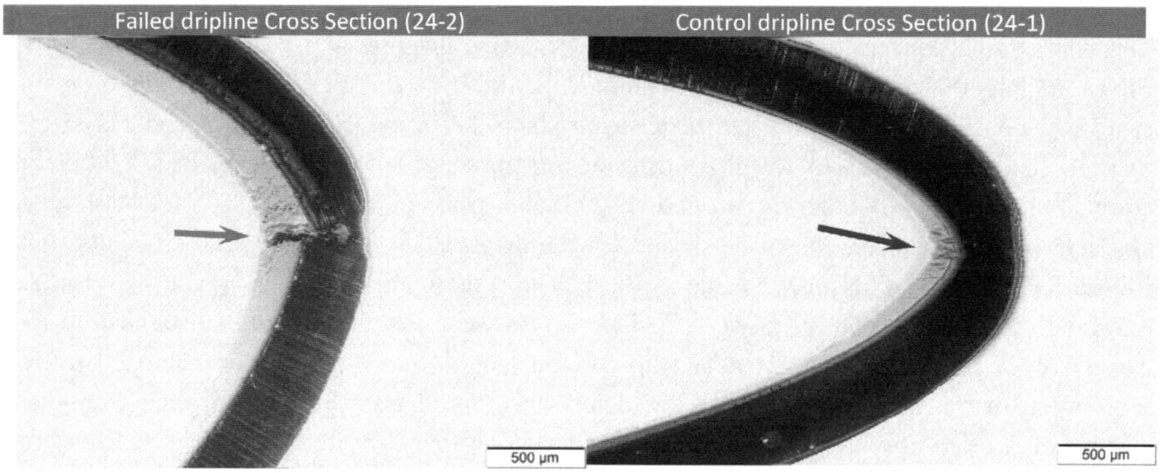


Figure 7 contains higher magnification views of the cross section of the failure region of both the failed and control driplines. Surface striations were present along the inner surface of a control dripline which followed the edge crease (blue arrows). Outer surface deformations were also observed in the control dripline. Sharp internal cracks (red arrows) were present on the inside surface of the failed dripline in the crease region, but not on the outside surface of the dripline. Crack penetration was approximately 40um along the inner dripline surface.

**Figure 7. Cross sectional optical microscopy images of failure region in both failed (24-2) and control (24-1) dripline specimens**

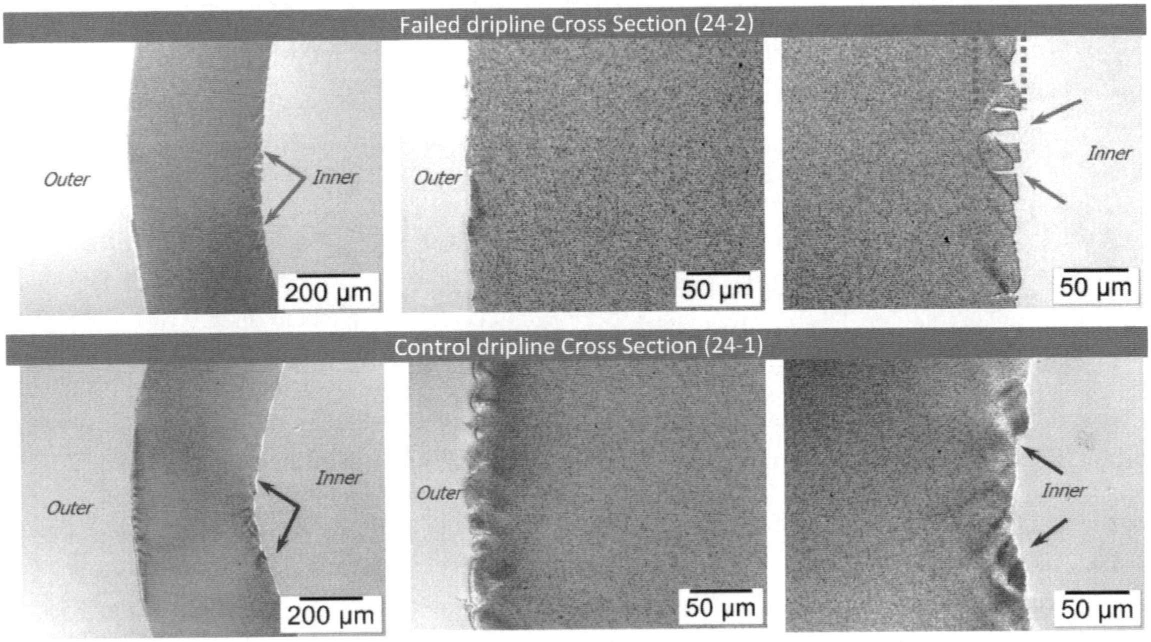


Figure 8 contains cross sectional images of the flat region of both failed and control driplines. The internal surface of both the failed and control driplines in areas away from the crease were absent of any striations, cracks or defects, suggesting the strain in the crease area coupled with exposure to the liquid medium pumped through the dripline over the years led to the failure of the dripline in the failed dripline sample which is expected. Evidence of dripline degradation or cracking was not observed in areas away from the crease. Prior studies have shown an increase in the rate of degradation of polyethylene when a constant strain has been applied.[v] Slight abrasion was observed (indicated by red arrow in Figure 8) along the outer surface of the failed dripline, but is not present on the inner surface of the failed dripline. This abrasion likely occurred during the installation of the dripline because it is not present in the control dripline. There was no evidence of failures initiating from the slightly abraded surface of the failed dripline. This abrasion must have occurred during installation or normal use over time.

**Figure 8. Cross sectional optical microscopy images of flat (non-failure) region in both failed (24-2) and control (24-1) dripline specimens**

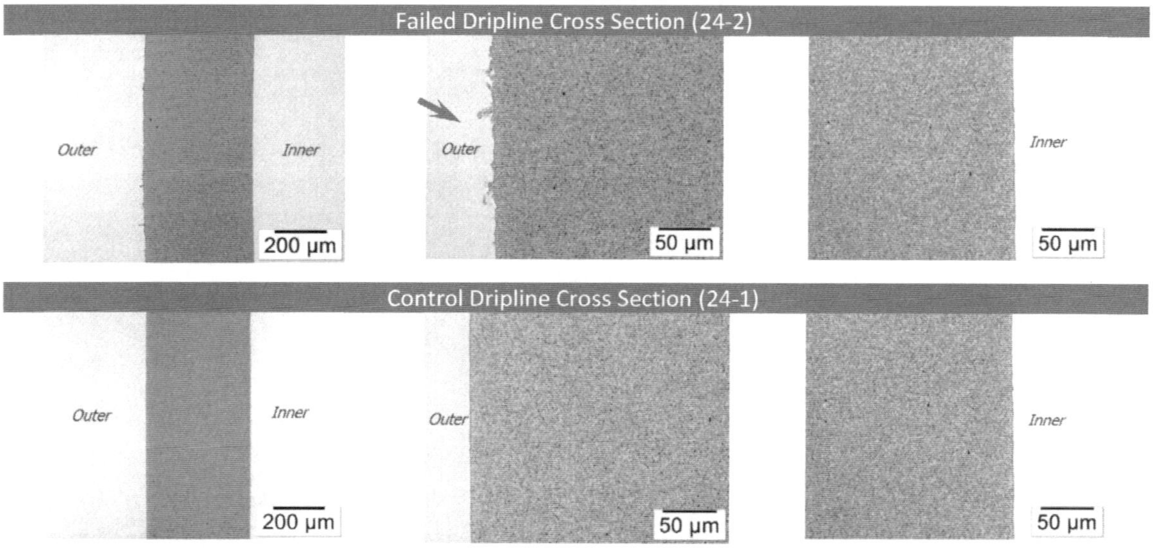


A review of cross section images shown in Figure 7 and Figure 8 indicate remarkable level of carbon black dispersion and homogeneity by today's standards in dripline material used.

An evaluation of the inner dripline surfaces showed parallel striations were present along edge creases, as shown in optical cross sections. Figure 9 contains scanning electron microscope (SEM) images of the inner surface of the crease region in the failed and control dripline specimens. Surface cracks were not observed along creases in the control dripline. Sharp parallel cracks (red arrows) were present along the creases of the failed dripline.

**Figure 9. SEM Images of Inner Surface of Crease Region in both Failed and Control Dripline Specimens.**

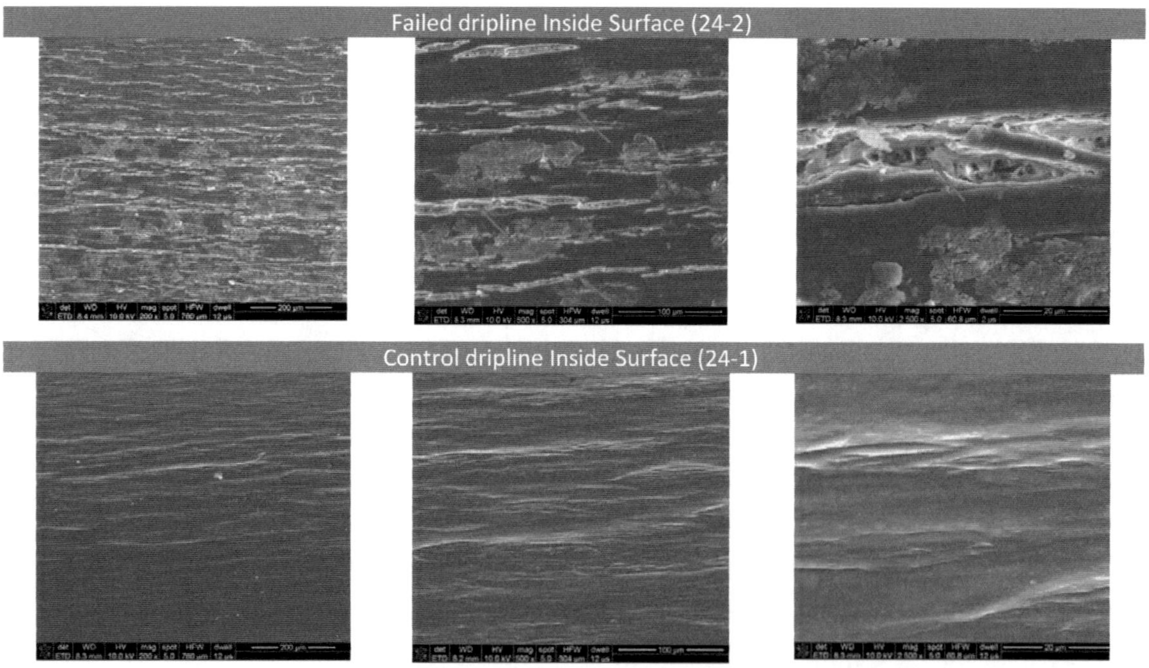

Comparison of the inner surface of the crease region of the dripline using SEM shows the surface striations present in the control driplines, shown in Figure 10 and highlighted with blue arrows. The cracks on the inner surface crease region are also shown in the failed dripline (red arrows), with cracks penetrating approximately 40um into the wall of the dripline.

**Figure 10. SEM cross section images of crease region.**

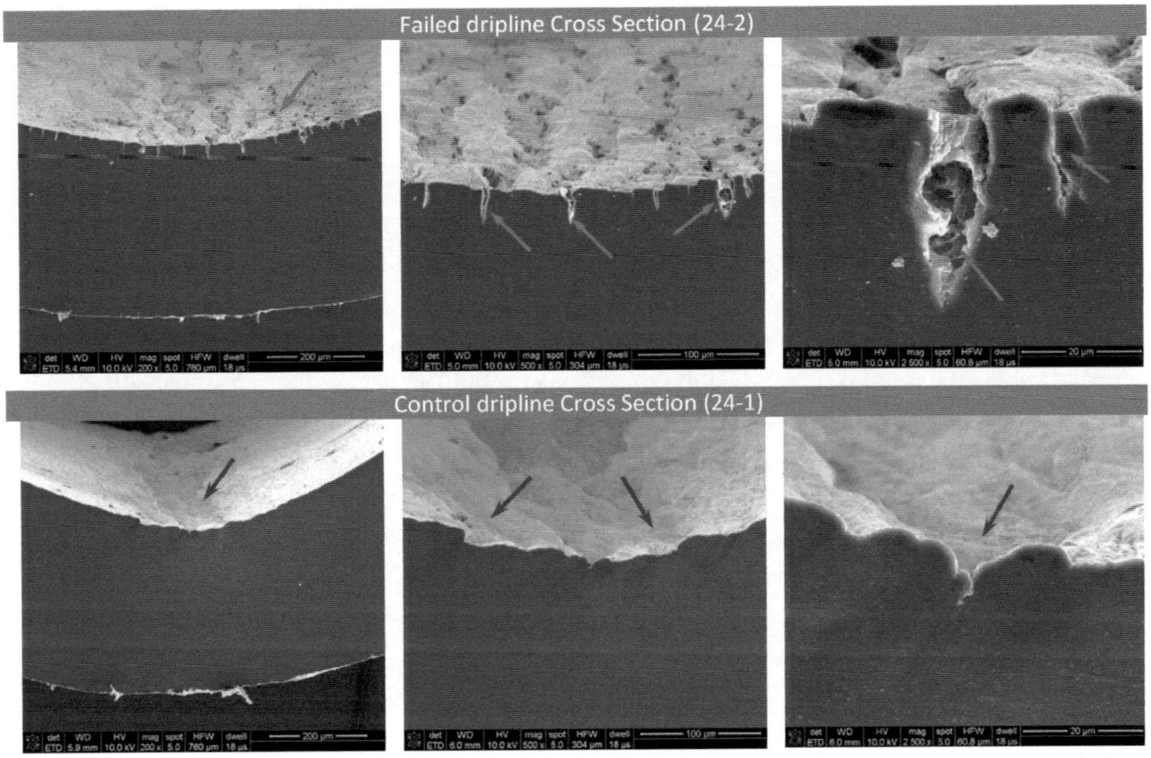


Examination of a fracture surface of the failed dripline showed that failure had initiated at the inner surface of the crease, inside the dripline, and propagated outward. The red arrows in Figure 11 indicate where the fracture initiated. A fracture surface associated with brittle failure was observed to a depth of approximately 125um or 5 mils (indicated by the area between the green dotted lines). This region was associated with parallel striations which formed sharp linear surface cracks along the dripline crease. Beyond the initial brittle fracture, a ductile shear lip was present (indicated by the yellow arrows in Figure 11) which extended to the outer dripline surface. Evidence of ductile tearing was observed beyond the initial brittle fracture zone.

**Figure 11. SEM Images of fracture surface along crease of failed dripline (24-2)**

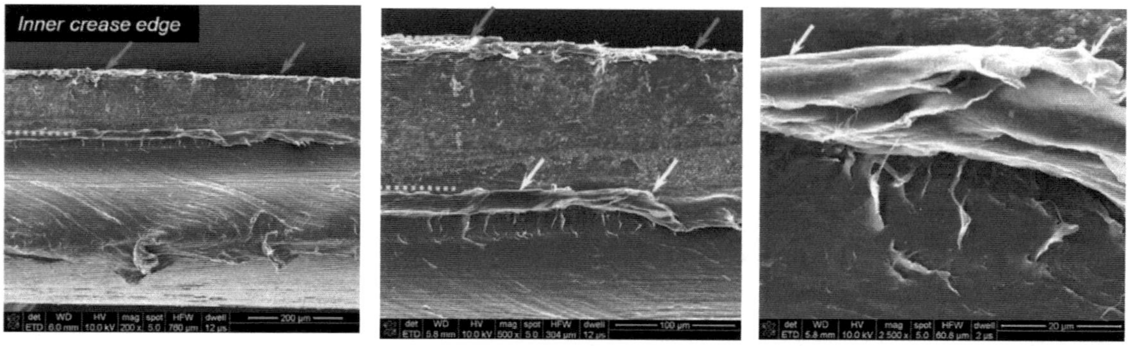


Oxidation induction time (OIT) was measured at the fracture and flat locations of the failed dripline as well as at the flat region of the control dripline. Specimens were carefully cut using scissors. Due to the relatively low thickness of the dripline (15mils) and amount needed for the OIT measurement, the specimen taken at the fracture was a cross sectional specimen representing both the inside, middle and outer areas in the fracture region. A linear direct relationship exists between phenolic concentration and OIT, so this technique can be used to assess the relative amount of active phenolic antioxidant present.[vi] The results are shown in Table 5. A greater concentration of phenolics were present away from the fracture versus at the fracture, suggesting the antioxidants were either consumed or extracted from the dripline wall at a greater rate in the crease region.

**Table 5. Oxygen Induction Time of failed (24-2) and control (24-1) dripline specimens.**

| Description | OIT at 200°C (min) |
|---|---|
| Specimen Near Fracture of Failed Dripline (24-2) | 1.0 |
| Specimen on flat region of failed dripline (24-2) | 1.3 |
| Specimen on flat region of control dripline (24-1) | 3.3 |

It is noted that the fracture surface was found to be extremely brittle, as shown in Figure 11, and it is therefore unlikely any antioxidants were present in this brittle area. The material in the crease region had reached stage 3 of the classical chemical degradation model for polyethylene, the stage at which all antioxidants are depleted, chemical degradation of the polymer is occurring, ultimately leading to an engineering failure.[vii]

It is concluded that the dripline likely failed due to the strain in the crease area coupled with exposure to the liquid medium pumped through the dripline during its lifetime. Evidence of dripline degradation or cracking was not observed in areas away from the crease, nor was any evidence of degradation found on the outside of the dripline. The overall dripline held up extremely well over its 26 year life, suggesting if the dripline had not contained a crease it likely would still be in service today.

Analysis of a dripline manufactured in 2017 with Dow's FINGERPRINT™ resin was done. The results are shown in Figure 12 below. The pictures show that the dripline does not have striations on the crease, there is no evidence of manufacturing deficiencies. Therefore the striations observed on the 26 year old dripline which initiated the failure likely appeared over time due to the depletion of the AO and localized compressive stress in the fold area during its lifetime. That is why the control tape also presented the striations, but didn't fail due to the fact that was not in use. This is the same issue any collapsible/folded tape would have.

**Figure 12. Cross sectional optical microscopy images of Dripline made in 2017 with 100% Dow FINGERPRINT™ Resin**

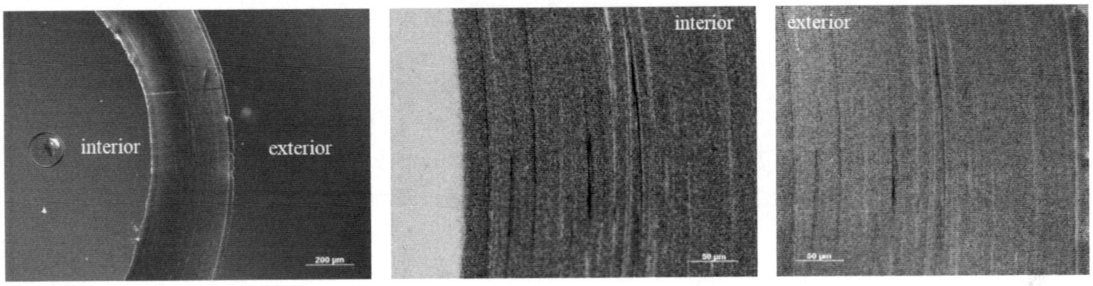


## CONCLUSIONS

The failed dripline was fabricated with Dow FINGERPRINT™ resin. Thermal characteristics of the dripline are comparable to Dow's FINGERPRINT™ resin, with an equivalent melting shoulder at around 110°C and peak at around 119°C. Rheological characterization indicates the material used in the dripline has a lower overall viscosity than Dow's FINGERPRINT™ resin. This may be due to the fact that the comparison was made between an extruded dripline and natural resin. Residue levels were less than 0.1 wt%, suggesting the carbon black used was very clean and no foreign material such as recycle was used in the fabrication of the dripline. Additive characterization indicates the failed dripline still contained some 46 ppm of active antioxidant, suggesting the antioxidant package in the dripline had held up well over the years.

ESCR was measured to be greater than 1,000 hours on the failed dripline. Even after 26 years of service the material used to fabricate the dripline still met ASAE S553 dripline standard.

It is concluded that the dripline likely failed due to the strain in the crease region coupled with exposure to the liquid medium pumped through the dripline during its lifetime. Evidence of dripline degradation or cracking was not observed in areas away from the crease, nor was any evidence of degradation found on the outside of the dripline. The overall dripline held up well over its 26 year life, suggesting if the dripline had not contained a crease it would likely still be in service today.

There is no evidence of manufacturing deficiencies that could have been the cause of the failure of the 26 year old dripline. The striations that initiated the failure likely appeared over time due to the localized compressive stress in the fold area during its lifetime. This is the same issue any collapsible/folded tape would have.

## RECOMMENDATIONS/FUTURE WORK

As next generation products are developed to enable thinner wall dripline for SDI applications, understanding the impact of strain on antioxidant depletion and the degradation of polyethylene in a water environment could be useful. Few studies have been reported and published literature regarding this topic. The importance of using a clean resin and masterbatch is valuable and the advantages include higher performance dripline that will last far longer.

## ACKNOWLEDGMENTS

Thanks to Babli Kapur, Rajen Patel, Cosme Llop, David Gillespie, Teresa Karjala and Kalyan Sehanobish for reviewing this case and sharing insight into the cause of this failure.

## REFERENCES

[i] Lamm, Freddie R., Danny H. Rogers, *"SDI, The Basics of Successful Systems"*, Proceedings of the 24th Annual Central Plains Irrigation Conference; Colby, Kansas; February 21-22, 2012.

[ii] Kansas State University Research and Extension Website: http://www.k-state.edu/sdi/

[iii] United States Patent 4,572,756 titled Drip Irrigation System Employing Adjacently Arranged Flow-Restricting Passages; Richard D. Chapin; January 16, 1985.

[iv] ASAE S553 Standard "Collapsible Emitting Hose (Dripline) – Specifications and Performance Testing" December 2000 (R2012).

[v] Llop, Cosme, Antonio Manrique, *"Cast Silage Wrap Film UV Evaluation"* LAIT-4032; October 22, 1990; CRI 907151.

[vi] Iten, Cicile, Jaime Setien, Carlos Alonso, Bernard Fehr, Cosme Llop, *"Oxygen Induction Time (OIT): What it really measures on Dow's polyethylene?"* LACAT 2006-001; June 22, 2006, CRI 2006004035.

[vii] Ewais, A.M.R., R. Kerry Rowe, "Effect of again on the stress crack resistance of an HDPE geomembrane" Polymer Degradation and Stability 109 (2014) 194-208.

[12] Garcia-Meitin, Eddy, "Failure Analysis of 26 Year Old Drip Tape Made from Union Carbide Resin at Chapin (Jain)", May 13, 2016; Analytical Report AL-2016-001783.

# Corn Irrigation and Fertilizer Use BMPs versus Conventional Growing Practices, a Two-Year Overview

**Maria I. Zamora Re, Ph.D. Student.**

Agricultural and Biological Engineering Department, 1741 Museum Road. Gainesville, FL 32611-0570, mzamora@ufl.edu.

**Michael D. Dukes, Professor and Irrigation Specialist.**

Agricultural and Biological Engineering Department, 1741 Museum Road. Gainesville, FL 32611-0570, mddukes@ufl.edu.

**Abstract.** Can fertilizer be reduced and irrigation scheduling be optimized in corn production without reducing yield? The answer to this question has been studied during 2015 and 2016 on sandy soils in the Suwanee River, Florida region. Due to karst topography, rapid water flow and absence of a natural filtration system, an increased vulnerability to groundwater pollution characterizes this area. Thus, excess nitrogen (N) fertilizer applied with the intention of obtaining higher yields, is a potential threat to waterbodies. The main objective was to improve irrigation and fertilizer Best Management Practices (BMPs) compared to conventional practices. Five irrigation treatments determined irrigation by: (i) GROW, mimicking grower's practices, (ii) SWB, using a theoretical soil water balance, (iii) SMS, monitoring volumetric water content measured by soil moisture sensors and triggered using maximum allowable depletion (MAD) and field capacity (FC) as thresholds to refill the soil profile, (iv) Reduced: irrigation (60% of GROW) representing a low irrigation treatment and (v) NON: non-irrigated plots. Three fertility rates: F1=300, F2=220 and F3=140 lb N/ac were evaluated. The experimental design consisted of a randomized complete block arranged in a split plot with four replicates. During 2015, yield was not significantly different across irrigated treatments (GROW, SMS and Reduced: 193, 191 and 201 bu/ac); however, the non-irrigated treatment had significantly lower yield than all other treatments except SWB (SWB and NON: 178 and 143 bu/ac, respectively). Fertility rates 300 and 220 lb N/ac (196 and 180 bu/ac), or 220 and 140 lb N/ac were not significantly different (180 and 168 bu/ac); however, the 300 lb N/ac treatment was significantly higher than 140 lb N/ac (196 bu/ac vs. 168 bu/ac, respectively). In 2016, no significant differences in yield were found among irrigated treatments, except versus the NON treatment which had significantly lower yield (GROW, SWB, SMS and Reduced: 202, 184, 188, 191 bu/ac vs. NON: 127 bu/ac). Fertility treatments did not differ statistically (F1, F2 and F3: 183, 180 and 173 bu/ac). Irrigation and fertilizer were reduced without reducing yield by using BMPs compared to conventional practices during the first two years of research.

**Keywords.** Irrigation scheduling, soil moisture sensors (SMS), water conservation, BMPs, N leaching, corn.

## Introduction

The Upper Floridan Aquifer (UFA) is the main aquifer in Florida and one of the most productive karst aquifers in the world. Thick carbonate deposits comprise the aquifer and underlay 116,000 square miles among Florida, Georgia and Alabama. The Suwannee River Basin (SRB) is located within the Coastal Plain and it comprises a highly variable hydrogeological system. The southern third of the basin is comprised by a thin layer of highly porous sands which overlies limestone karst and the UFA (Katz and Raabe 2005).

Most of the north-central Florida regions underneath the UFA are unconfined and characterized by a karst topography. By the dissolution of the carbonate rocks, karst features develop zones with enhance porosity causing a highly heterogeneous aquifer system with rapid rates of groundwater movement and recharge (Katz et al. 2009; Tihansky and Knochenmus 2001). As well, direct pathways can be created between surface and groundwater, allowing rapid transport of nutrients or pollutants with little degradation or attenuation (Katz et al. 2009).

Nitrate ($NO_3^-$) is the most common and soluble form of nitrogen (N); thus, it can easily leach from the soil profile by irrigation or rainfall into the groundwater and streams (DeSimone 2009). Background $NO_3^-$ concentrations are considered below laboratory detection limits (i.e. 0.05 mg/L $NO_3^-$, as N; i.e. natural concentrations without human activity intervention). N may originate from natural sources (e.g. precipitation, aquifer materials and organic debri leaching), as well as, from anthropogenic activities (e.g. fertilizer applications, wastewater, animal production) (DeSimone 2009). Nitrate from both sources is possibly the most widespread pollutant in groundwater that can persist for decades (Hallberg and Keeney 1993).

Several studies have shown the increase of nitrate-N in springs from background concentrations (≤0.1 mg/L $NO_3$-N) to above 5 mg/L $NO_3$-N during the last 40 years (Katz et al. 1999; Katz 2004). Steady increments of nitrate-N have been found in several spring waters within the SRB in Florida (e.g. Convict spring with 8 mg/L $NO_3$-N) (Upchurch et al. 2007). During the period 1940-1998, continued studies in the SRB area reported evidence that the greatest nitrate-N pollution found in the lower basin comes from fertilizer sources (Katz et al. 2009).

The north-central regions are estimated as the most pollutant vulnerable regions due to the potential groundwater degradation (Arthur et al. 2007). Some of the consequences are excessive algal growth in aquatic systems, a reduction of the available dissolved oxygen, as well as, death to other organisms (FDACS 2015). Thus, the implementation of Agricultural Best Management Practices (BMPs) is imminent in areas susceptible to pollutants.

### Total Maximum Daily Loads (TMDLs) and Best Management Practices (BMPs)

The maximum amount of a pollutant that can be discharged to a waterbody and meet state water quality standards is referred as a TMDL. Florida Department of Environmental Protection (FDEP) must identify impaired surface waters and establish TMDLs entering waterbodies (e.g. TMDLs established for total N) (FDACS 2015).

BMPs are defined as "practical, cost-effective actions that agricultural producers can take to conserve water and reduce pollutant loads (i.e. pesticides, fertilizers, animal waste and others) entering water resources. BMPs are designed to benefit water quality and quantity while maintaining or even enhancing agricultural production" (FDACS 2017).

Irrigation BMPs address the irrigation scheduling methods; whereas nutrient BMPs focus on the crop nutrient needs, sources and applications to reduce potential water and nutrient losses to the environment (FDACS 2017). Examples of nutrient BMPs consist of fertilizer timing (i.e. apply it in small amounts just before the uptake occurs), performing a soil test prior any fertilizer application, as well as, monitoring crop nutritional status. Thus, fertilizers are only applied when needed to avoid potential losses to the environment (FDACS 2015; Simmone 2006). Then, irrigation BMPs are for example implementing an irrigation schedule to determine how much and when to irrigate supplying the plant's requirements while reducing water and nutrient losses. If the amount of water applied exceeds the amount of water required by the crop, deep percolation and nutrient leaching from the rootzone to the environment occurs. Greater irrigation efficiency is achieved when based on weather and evapotranspiration (ET) parameters or soil moisture content (Irrigation Association 2011).

**Typical corn production vs. new strategies of production**

Corn is a high water and nutrient demanding crop. Traditionally, a calendar-based irrigation schedule is used by growers. This method uses growers' general knowledge of the crop and local weather conditions and applies water according to the crop stages (i.e. the amount of water applied increases as the plant grows). Commonly, nitrogen (N) fertilizers are applied exceeding the N uptake by the plants allowing the excess N to potentially leach increasing the N load to groundwater sources.

The use of soil moisture sensors (SMS) to continuously monitor the volumetric water content (VWC) in the soil, provides information about the status of the soil and the water available to the plants. Therefore, irrigation is applied only when it is needed avoiding overirrigation or deep percolation, while reducing potential N leaching.

Irrigation and N BMPs could be a management strategy to target load reductions for water quality improvements. The objective of this paper is to compare conventional irrigation practices and common applied fertility rates with new BMPs that might increase irrigation efficiency and reduce N leaching to the groundwater.

## Materials and Methods

### Experimental site and design

The experimental site is located at the Suwannee Valley Agricultural and Extension Center (SVAEC), near Live Oak, Florida. Predominant soils were identified as sandy Blanton-Foxworth-Alpin complex, Chipley-Foxworth-Albany, and Hurricane-Albany-Chipley soils (USDA 2013).

The experimental design was a randomized complete block arranged in a split plot. This design included four replicates (i.e. blocks) per treatment, five main plots (i.e. irrigation treatments) with three subplots (i.e. nitrogen rates) for a total of 60 plots. Each plot was 40 ft long x 20 ft wide separated by 20 ft alleys. An alley of 40 ft was included between the blocks and an alley of 20 ft separated the plots (Figure 1).

Corn (hybrid Pioneer 1498 YHR/Bt) was planted on 3 April 2015 and on 22 March 2016. A row spacing of 30 inches and a plant spacing of 6.5 inches were used for a total plant population of 32,500 plants/ac. Corn was harvested on 18 August 2015 and on 3 August 2016 on the 6th and 7th planting rows starting 20 ft inside each plot to avoid boarder effects, for a total of 100 $ft^2$ of each plot.

**Figure 1.** Aerial view of experimental site located at Suwannee Valley Agricultural and Extension Center, near Live Oak, Florida.

Crop evapotranspiration ($ET_c$) was calculated using phenologically based crop coefficients ($K_c$ values: $K_{c-ini}$=0.25, $K_{c-max}$=1.05 and $K_{c-end}$= 0.55) (K-State Research & Extension Mobile Irrigation Lab 2014) and reference evapotranspiration ($ET_o$) as:

$$ET_c = K_c ET_o$$

$ET_o$ was calculated using weather data (daily minimum and maximum temperature, solar radiation, relative humidity and wind speed) from an onsite Florida Automated Weather Network (FAWN 2017)) with the FAO-56 Penman-Monteith equation (Allen et al. 1998).

## Treatments

### Irrigation

The irrigation system consisted of a two span Valley Linear End feed 8000 (Valmont Industries 2015), Valley, NE) with a Variable Rate (VRI) package. This machine is capable of irrigating a field area of 16.6 acres, using a flow of 300 GPM (18.03 GPM/acre) providing an application rate (AR) of 0.96 in/day at maximum capacity. Senninger (Senninger Irrigation, Inc., Clermont, FL) LDN-UP3 Flat Medium Grove ¾ M NPT nozzles were attached to drops at a 10 ft sprinkler spacing. To maintain a constant flowrate, Valley 10 psi pressure regulators (PSR-2 10 10(PSI) 3/4 F NPT) were installed on each drop.

The irrigation treatments consisted of:

1. I1 (GROW): irrigation mimicked grower's irrigation practices which consisted of zero irrigation for the first 30 days after planting (DAP) unless severe dry or windy conditions occurred. Beginning

on 31 DAP, a target amount of 1"/wk was established and could be made up of rain or irrigation but rain events had to be 0.25" or larger. For 40-59 DAP a 1.5"/wk target was established with irrigation events of 0.4". Irrigation was skipped if 0.5-0.75" rainfall occurred and two irrigations were skipped if >0.75" of rain occurred. For 60-105 DAP a 2-2.5"/wk irrigation target was used unless 0.5-1" of rain occurred the day prior to a scheduled irrigation. Two irrigations were skipped if >1" of rain occurred. Finally, around 105 DAP at full dent stage, weekly irrigation targets were reduced to 1.5"/wk until finally irrigation was terminated at physiological maturity. Irrigation was skipped if 0.5-0.75" rainfall occurred and two irrigations were skipped if >0.75" of rain occurred. This treatment used a fixed application rate of 0.4".

2. I2 (SWB): irrigation was determined using a theoretical soil water balance. Weather data (i.e. rainfall, ET, temperature) was obtained from the on-site FAWN weather station located in Live Oak, FL. Calculated crop evapotranspiration ($ET_c$) was used to estimate daily soil moisture and schedule irrigation when 50% of the available moisture was depleted during vegetative stages and when 30% moisture depletion during reproductive stages.
3. I3 (SMS): soil moisture content was monitored using the SENTEK probes (Sentek Pty Ltd 2003). Irrigation was determined using the maximum allowable depletion (MAD) and field capacity (FC) points to refill the soil profile with irrigation according to guidelines proposed by (Zotarelli et al. 2013). Volumetric water content (VWC) was monitored by plots; thus, when VWC at any replicate decreased below MAD, all replicates were irrigated to satisfy the needs of that plot.
4. I4 (Reduced): it applies only a 60% of I1 treatment but using the same frequency with fixed application rates of 0.4". It represents a lower irrigation treatment scenario.
5. I5 (NO): non-irrigated plots or control.

**Fertilization**

Three N fertility rates were evaluated: F1=300 lb/ac; a high scenario commonly applied in corn production, F2=220 lb/ac; representative of the UF-IFAS recommended rate (Mylavarapu et al. 2015), and F3=140 lb/ac; a low N scenario. The low and high treatments were 36.4% lower or higher than the recommendation.

During both years, a pre-plant soil sampling analysis was performed to determine initial soil conditions. At planting an initial liquid application of 30 lb/ac of 16-16-0 was applied on the soil surface across all treatments. The N fertility rates started 14 DAP with the first granular application (i.e. V3 corn growth stage) where 8, 22 and 30 lb N/ac of 33-0-0 (16.49% Ammoniacal N and 16.51% Nitrate-N), were applied on F3, F2 and F1, respectively.

The second granular application (33-0-0) took place on 1st May 2015 and 27 April 2016 (at V6 corn growth stage). A total of 10, 24 and 40 lb N/ac were applied on F3, F2 and F1, respectively. Afterwards, split liquid side-dress applications (28-0-0) took place at between V8 and VT- (tasseling) corn growth stages. At each liquid sidedresss application the fertility rates F3, F2 and F1 had 23, 36 and 50 lb N/ac, applied respectively. In 2016, a supplemental granular application of 30 lb N/ac (21-0-0 at 145 lb/ac) was performed on 19 April to compensate the lost fertilizer due to leaching rain, following the BMP protocol recommendations (FDACS 2015).

Phosphorus and potassium applications were performed based on soil analysis results and equally applied across all fertility rates. Therefore, 75 lb P/ac of 0-46-0 (Triple Superphosphate) were applied during the first granular application in 2015; whereas no phosphorous was required in 2016. In terms of potassium,

98 lb K/ac and 77 lb K/ac of 0-0-60 were applied during the first and second granular applications in 2015, respectively; whereas 35 lb K/ac of 0-0-60 were applied on both granular applications in 2016.

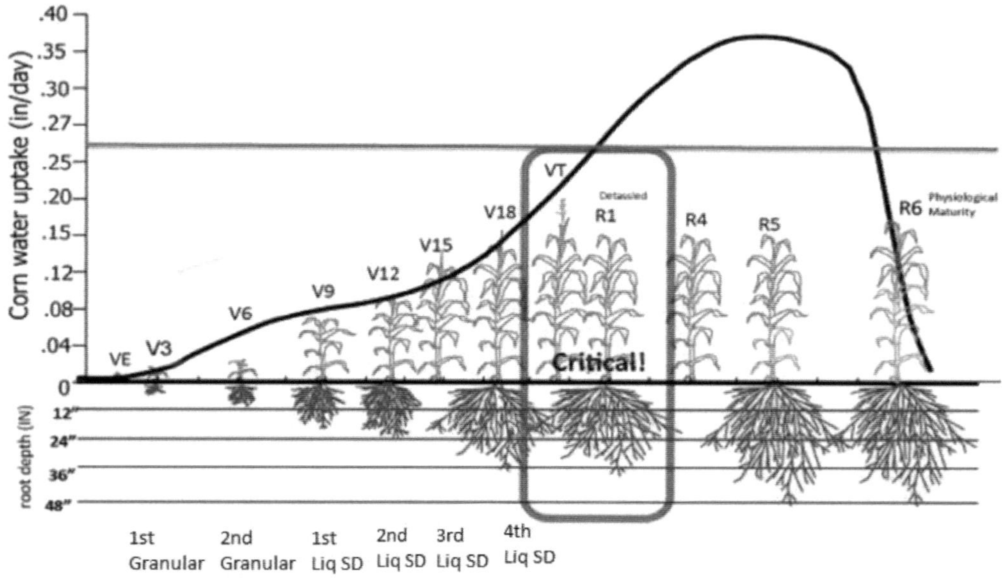


**Figure 2.** Fertilizer applications: initial pop-up (16-16-0), two granular applications and four liquid sidedress (Liq. SD) applications performed in corn experiment 2015 and 2016. An additional N application was performed in 2016 due to a leaching rain (see details in fertilization section) (Cropmetrics 2014).

### Statistical analysis

The statistical program SAS (SAS Institute Inc. 2013, Cary, NC) was used to analyze final corn yield. Further, ANOVA and least squared means (LSD) differences with normal p-values and Tukey's adjusted p-values for multiple comparison were used in this experiment.

## Results

### Irrigation

In 2015, the corn growing season spanned from 3 April to 18 August. Cumulative rainfall was 21.9 inches and estimated crop evapotranspiration ($ET_c$) was 19.5 inches. During this period the GROW, SWB, SMS, Reduced and NON treatments applied 12.6, 7.3, 6.0, 8.3 and 0.6 inches of water, respectively. Irrigation treatments achieved 42%, 53%, 34% and 95% water savings in comparison to the GROW treatment, which represents conventional irrigation practices.

In comparison, 2016 was a drier year with the growing season from 22 March 22$^{nd}$ to 3 August. Cumulative rainfall was 14.6 inches and estimated $ET_c$ was 16.6 inches. Larger amounts of irrigation were required due to the non-uniform rainfall distribution through the growing season. Irrigation treatments applied 20.0, 12.2, 11.5, 12.6 and 1.0 inches for GROW, SWB, SMS, Reduced and NON, respectively. Water savings achieved were 39%, 43%, 37%, 95% compared to GROW, correspondingly.

## Yield

**By irrigation treatment**

In 2015, the GROW, SWB, SMS, Reduced and NON resulted in 193, 178, 191, 201 and 143 bu/ac. Yield obtained by irrigated treatments did not differ statistically versus the non-irrigated, except the SWB. During the first year of evaluation, this treatment initially had a 50% maximum allowable depletion (MAD) through the whole growing season and fixed application rates of 0.75”. These conditions were limiting the available water during the reproductive stages, causing water stress to the plants which negatively impacted yield (Figure 3).

By contrast, in 2016 the MAD threshold for SWB treatment was 50% during vegetative stages and it was reduced to 30% in reproductive stages. These conditions allowed greater moisture in the soil during reproductive stages, which have a greater sensitivity to water stress. Average final yields resulted in 202, 184, 188, 191 and 127 bu/ac for GROW, SWB, SMS, Reduced and NON treatments, respectively. No significant differences were found between irrigated treatments, only versus the non-irrigated (NON) treatment, which resulted in a significantly lower yield (Figure 3).

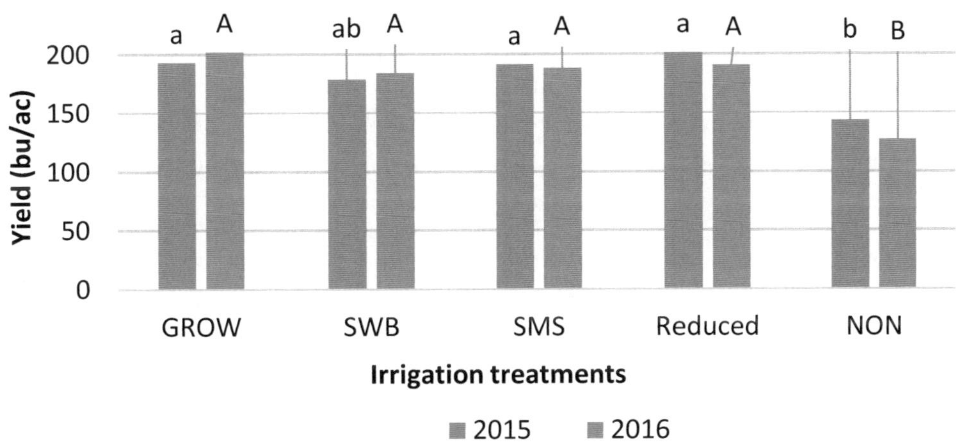


**Figure 3.** Average corn yields resulted from irrigation treatments in 2015 and 2016. Lowercase and uppercase letters indicate 95% CI for irrigation treatment means in 2015 and 2016, respectively.

**By fertility treatment**

Average final yields for the three fertility rates resulted in 196, 180 and 168 bu/ac for F1 (high), F2 (medium) and F3 (low), respectively during 2015. No statistical differences in yield were found between the high and the medium fertility rates. Neither the final yields obtained from the medium and the low fertility rates. Only the high N rate (F1=300 lb N/ac) resulted in a statistically higher yield (196 bu/ac) in comparison to the low N rate (F3=140 lb N/ac, 168 bu/ac) (Figure 4).

In 2016, F1, F2 and F3 final yields averaged: 183, 180 and 173 bu/ac, respectively. No statistical differences were found among the three fertility rates. Due to a leaching rain after the first granular application, an application of 30 lb N/ac extra were applied following the BMP recommendations.

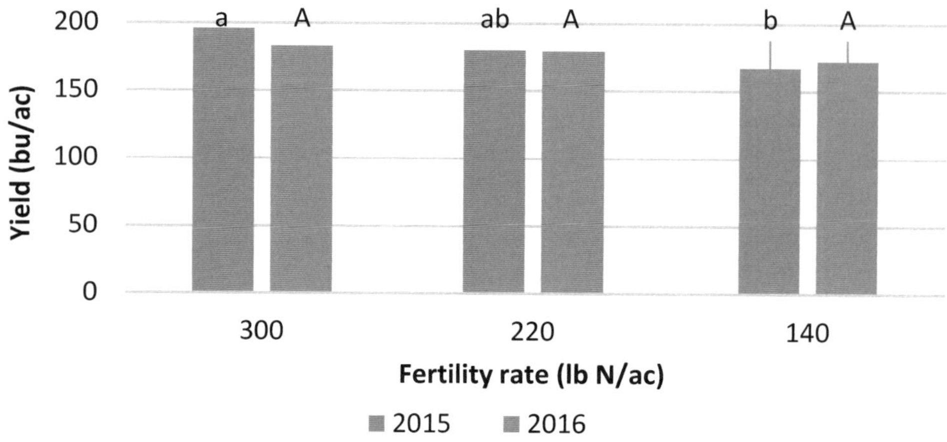


**Figure 4.** Average corn yields resulted from evauated N fertility rates (F1, F2 and F3= 300, 220 and 140 lb N/ac, respectively) in 2015 and 2016. Lowercase and uppercase letters indicate 95% CI for irrigation treatment means in 2015 and 2016, respectively.

## Summary and Conclusions

In 2015, the irrigation strategies evaluated (i.e. SWB, SMS and Reduced) resulted in total water savings of 42%, 53%, 34% compared to the GROW treatment, which mimics common irrigation practices in corn. Reducing the irrigation amounts did not have a negative impact on yield. No significant differences in yield were found between the irrigated treatments, only the SWB treatment resulted in statistically lower yields.

In 2016, the SWB, SMS and Reduced treatment achieved 39%, 43% and 37% water savings vs. GROW treatment. The difference between years was attributed to the variation in rainfall amounts and distribution patterns between 2015 and 2016. Greater amounts of irrigation were required in 2016; however, none of the irrigation alternatives resulted in negative impacts on yield. Average yields did not differ statistically compared to the conventional irrigation practices.

In 2015, following the UF/IFAS corn N fertilization recommendations (F2=220 lb N/ac) resulted in average yields with no statistical difference than average yield obtained with conventional N applications (F1=300 lb N/ac). Thus, same yield could be achieved with 220 lb N/ac while reducing the N applications by 26.6%. In 2016, no statistical differences were found between the three N fertility rates evaluated.

The implementation of BMPs focused on different irrigation strategies (i.e. SWB, SMS and Reduced) resulted in yields without statistical differences than the conventional irrigation practices and with substantial water savings in both years of evaluation. As well, no reductions in yield were found when following UF/IFAS recommendations (220 lb N/ac). These represent potential solutions to reduce N leaching and increase irrigation efficiency in corn production in Florida.

## Acknowledgments

The ongoing research is supported by FDACS, FDEP and the SRWMD. Special thanks to M. Gutierrez, D. Hensley, M. Thomas, C. Harchick from ABE Department, as well as, Ben Broughton, Mike Boyette and staff from SVAEC and UF/IFAS staff for their assistance with field operations and project design.

# References

Allen, R., Pereira, L. S., Raes, D., Smith, M. (1998). "Chapter 2 - FAO Penman-Monteith Equation ." *Crop Evapotranspiration - Guidelines for Computing Crop Water Requirements - FAO Irrigation and Drainage Paper 56* , Paper 56 Ed., FAO, Rome, .

Arthur, J. D., Wood, A. R., Baker, A. E., Cichon, J. R., Raines, G. L. (2007). "Development and Implementation of a Bayesian-Based Aquifer Vulnerability Assessment in Florida." *Nat Resour Res,* 16(2), 93-107.

Cropmetrics. (2014). "Maximizing root zone utilization: A key to precision irrigation management." <http://cropmetrics.com/2014/04/maximizing-root-zone-utilization-a-key-to-precision-irrigation-management/> .

DeSimone, L. A. (2009). *Quality of Water from Domestic Wells in Principal Aquifers of the United States, 1991–2004,* 2008-5227 Ed., U.S. Geological Survey Scientific Investigations, U.S. Geological Survey, Reston, Virginia.

FAWN. (2017). "Florida automated weather network: Data access." <http://fawn.ifas.ufl.edu/data/reports/> .

FDACS. (2017). "Agricultural best management practices." <http://www.freshfromflorida.com/Business-Services/Water/Agricultural-Best-Management-Practices> .

FDACS. (2015). *Water Quality/Quantity Best Management Practices for Florida Vegetable and Agronomic Crops,* 2015th Ed., Florida Department of Agriculture and Consumer Services, Tallahassee, Florida.

Hallberg, G. R., and Keeney, D. R. (1993). "Nitrate." *Regional Ground-Water Quality.* Van Nostrand Reinhold, USA, 297-322.

Irrigation Association. (2011). *Irrigation,* 6th Ed., Irrigation Association, Falls Church, VA.

Katz, B. G., and Raabe, E. A. (2005). *Suwannee River Basin and Estuary: An Integrated Science Program-White Paper,* Open-File Report 2005-1210 Ed., U.S. Department of the Interior and U.S. Geological Survey, .

Katz, B. G., Hornsby, H. D., Bohlke, J. F., Mokray, M. F. (1999). *Sources and Chronology of Nitrate Contamination in Spring Waters, Suwannee River Basin, Florida,* Report 99-4252 Ed., U.S. Geological Survey, Tallahassee, Florida.

Katz, B. G., Sepulveda, A. A., Verdi, R. J. (2009). "Estimating Nitrogen Loading to Ground Water and Assessing Vulnerability to Nitrate Contamination in a Large Karstic Springs Basin, Florida." *J. Am. Water Resour. Assoc.,* 45(3), 607-627.

Katz, B. (2004). "Sources of Nitrate Contamination and Age of Water in Large Karstic Springs of Florida." *Environ. Geol.,* 46(6-7), 689-706.

K-State Research & Extension Mobile Irrigation Lab. (2014). "KanSched." (v3.1.5), K-State Research & Extension Mobile Irrigation Lab, Kansas.

Mylavarapu, R., Wright, D., Kidder, G. (2015). "UF/IFAS Standardized Fertilization Recommendations for Agronomic Crops." *Soil and Water Science Department, UF/IFAS Extension,* (SL129), 10/1/2015-8. <https://edis.ifas.ufl.edu/pdffiles/SS/SS16300.pdf> .

SAS Institute Inc. (2013). "SAS for Windows." (SAS 9.4), SAS Institute Inc., Cary, NC, USA.

Sentek Pty Ltd. (2003). *TriSCAN Manual Version 1.2a,* 1.2a Ed., Sentek Pty Ltd, Stepney, South Australia.

Simmone, E. (2006). "Soil and nutrient management: Best management practices." <http://ipm.ifas.ufl.edu/pdfs/Soil_Nutrient.pdf> (Chapter 3, 2017).

Tihansky, A. B., and Knochenmus, L. A. (2001). *Karst Features and Hydrogeology in West-Central Florida—A Field Perspective,* U.S. Geological Survey Karst Interest Group Proceedings, Tampa, FL.

Upchurch, S. B., Chen, J., Cain, C. R. (2007). *Trends of Nitrate Concentrations in Waters of the Suwannee River Water Management District, 2007,* SDII Global Corporation Project Number 3017076 Ed., Suwannee River Water Management District, Live Oak, Florida.

USDA, N. (2013). "Web soil survey." <http://websoilsurvey.sc.egov.usda.gov/App/WebSoilSurvey.aspx> (3/31, 2015).

Valmont Industries, I. (2015). "Valley variable rate irrigation." <http://www.valleyirrigation.com/valley-irrigation/us/control-technology/variable-rate-irrigation-%28vri%29> (03/02, 2015).

Zotarelli, L., Dukes, M. D., Morgan, K. T. (2013). "Interpretation of Soil Moisture Content to Determine Soil Field Capacity and Avoid Over-Irrigating Sandy Soils using Soil Moisture Sensors." *Agricultural and Biological Engineering Department, Florida Cooperative Extension Service, Institute of Food and Agricultural Sciences, University of Florida,* AE460, 1-4. <https://edis.ifas.ufl.edu/ae460> (December 2009).

# STAMP Decision Tool for Agronomic Crops in Louisiana

**Stacia L. Davis, Ph.D. E.I.T.**

Assistant Professor, LSU AgCenter, Red River Research Station, 262 Research Station Drive, Bossier City, LA 71112, (318) 741-7430 ext 1105, sdavis@agcenter.lsu.edu

**Daniel Fromme, Ph.D.**

Associate Professor, LSU AgCenter, Dean Lee Research Station, 8105 Tom Bowman Drive, Alexandria, LA 71302, (318) 473-6520, dfromme@agcenter.lsu.edu

**Abstract.** In Louisiana, irrigation efficiency can be improved by determining irrigation events based on plant water requirements. The objective of this project was to develop a decision tool to determine when to trigger irrigation based on plant water requirements for agronomic crops. The decision tool relies on a soil water balance to keep track of water movement in the root zone. This simplistic tool was developed using a spreadsheet for ease of access and availability without internet. Calibration of the tool was conducted by using irrigation data collected from research plots in 2015 and 2016 that included soil moisture measurements to determine actual water movement in the soil. In addition to the benefit of knowing when an irrigation event should occur, this spreadsheet can also act as a descriptive record that keeps track of water application and calculates irrigation efficiency for each irrigation event.

**Keywords.** Cotton, irrigation, scheduling, sensors, soil moisture, soybean

## Introduction

In Louisiana, furrow irrigation is the most common method of water application to row crops. Generally, there is very little control in the applied volume per event. Irrigation volumes depend on pump efficiency, available head pressure, pipe-riser system design, hole size selection in the lay-flat tubing, and infiltration characteristics of the soil. Most of these dependencies require considerable investment and effort to change, which is only likely to occur by producers when required (such as replacing an end-of-life pump) and not just for improving irrigation efficiency. However, using tools to determine when to apply irrigation can delay an application, eventually skipping an irrigation event, or increase irrigation during critical growth stages that can lead to increased yield. The objective of this project was to develop a decision tool to determine when to trigger irrigation based on plant water requirements for agronomic crops.

## Materials and Methods

The Smart Technologies for Agricultural Management and Production (STAMP) Irrigation Scheduling Tool was developed as a first step to scientific irrigation scheduling in Louisiana. This excel-based tool uses the soil water balance to estimate when irrigation should be applied. Since it was designed to be used by row crop farmers, it was pre-populated with agronomic information that best fit the available data. However, there's very little agronomic information related to irrigation available regionally, so each site-specific selection can be customized. The tool is currently in the testing phase and has not been released to the public at this time. This paper presents the testing results to date. It is anticipated that the tool will develop into a more sophisticated product as producers become aware of its usefulness through education and demonstration.

A field study was designed to measure irrigation application and soil moisture in cotton and soybean based on the following treatments: A) soil matric potential sensor system, B) volumetric water content sensor system, and C) weekly irrigated treatment. Each treatment was replicated three times with at least six rows on 40 inch spacing and a minimum row length of 300 ft. Irrigation application was measured using volumetric flow meters (McCrometer, Inc., Hemet, CA) assuming equal application across treatments when more than one treatment received irrigation. Yield was used as the primary response variable to determine whether differences in irrigation application resulted in negative impacts to the crop. Yield was harvested from the two middle rows of each plot in a 100 ft portion. None of the field plots were irrigated ideally during any crop season due to logistical restrictions; however, this resulted in a wide range of moisture conditions and crop responses for testing the STAMP tool.

The GS-1 (Decagon Devices, Pullman, WA) and the Watermark (The Irrometer Company, Riverside, CA) were chosen as the volumetric water content and soil water potential sensors, respectively. Translation of soil water potential measurements to volumetric water content for the Watermark sensors required soil sampling and long-term analysis that is in process and will not be finished until next year. As a result, only the plots with GS-1 sensors were evaluated here. The GS-1 was new to the market and meant for agricultural situations. It was chosen primarily due to its comparability to the Watermark based on size and installation style. Decagon RM50G telemetry loggers were used to access the soil moisture data. Each logger can support five sensors thus these sensors were installed every 6 inches up to 30 inches.

The study was conducted in 2015 and 2016 at Louisiana Agricultural Experiment Stations across northern Louisiana to test the treatments on three distinct soil types. The Red River Research Station (Bossier City, LA) was located on sandy clay loam, part of the Red River Alluvial soils inherent to the region. The field at the Macon Ridge Research Station (Winnsboro, LA) was predominantly silt loam representing the Macon Ridge soils. The final location, at the Northeast Research Station (St. Joseph, LA), had cracking clay soils that dominate the Mississippi Delta region. Soybeans were grown at the Macon Ridge and Northeast Research Stations whereas cotton was grown at the Red River Research Station. All sensors were installed after planting and fertilization and removed prior to harvest.

## Results and Discussion

As reported last year, the most accurate data in the study for the GS-1 sensors occurred in the sandy clay loam soil where cotton was grown. Cotton was also a beneficial crop for this evaluation because it is the only agronomic row crop with published local crop coefficients (Kumar et al. 2015). Using the 2015 data, the STAMP irrigation scheduling tool provided a fairly acceptable estimation of volumetric water content considering the known limitations to the methodology (Figure 1). The comparison between predicted and observed daily soil moisture values across all timesteps between 7/17 and 9/20 resulted in a Nash-Sutcliffe coefficient of efficiency ($C_N$) of 0.33 and a root mean square error (RMSE) of 0.028 $m^3/m^3$ (Table 1). All six combinations of soil types and years produced similar model results where decreased model performance was attributed to uncaptured hydrological processes and data quality.

There were two irrigation events, occurring on 7/21 and 8/11,that caused an increase in soil moisture estimations in the STAMP tool, but were not measured within the soil. Removing the data for these two events increased the $C_N$ to 0.77 and decreased the RMSE to 0.015 $m^3/m^3$.

This indicated that there were physical processes not captured in the STAMP Tool. It is hypothesized that adding consideration for infiltration rates and compaction would increase the model's performance.

The STAMP Irrigation Scheduling Tool was easily converted to an irrigation prediction model by introducing irrigation events when maximum allowable depletion was reached. Irrigation was restricted to the period of growth between the developmental phase and late growth that represented the end of reproduction and never exceeded field capacity. The STAMP irrigation schedule represents the ideal irrigation schedule when the model limitations previously discussed were not a factor. These limitations were consistent between the two model outputs thus inconsequential in this analysis. Also, it was already known that irrigation wasn't adequate during most of these scenarios, thus variation between the calculated soil moisture and the predicted STAMP irrigation schedule was expected.

In the 2015 cotton scenario previously discussed, irrigation initiation should have been delayed by three days (Figure 2) and one additional irrigation was necessary for the season, totaling six events instead of five. The delay in initiation was carried through the season with early irrigations occurring for the first four events. The fifth event occurred at the predicted time and the sixth event was predicted for the end of the season, about two weeks after the final event.

The STAMP irrigation schedule predicted the same or more irrigation events than what was applied during the study at all locations (Table 2). Also in most locations, the amount of irrigation required per event was less than what was applied. The lack of efficient application inherent in furrow irrigation situations can negate the benefits to applying less irrigation and ultimately result in more irrigation over the season. For example, irrigation applications of 8.53 inches of water occurred due to failing to turn the water off at the appropriate time and wasn't related to need. Unfortunately, this is a common occurrence in the mid-South.

Table 1. Summary of model statistics from all three locations and both years.

| Year of Crop Season | Location | Crop | Soil Type | Nash-Sutcliffe Coefficient of Efficiency[1] | Root Mean Square Error ($m^3/m^3$)[2] |
|---|---|---|---|---|---|
| 2015 | Bossier City | Cotton | Sandy Clay Loam | 0.33 | 0.028 |
| 2016 | Bossier City | Cotton | Sandy Clay Loam | 0.30 | 0.022 |
| 2015 | Winnsboro | Soybean | Silt Loam | 0.39 | 0.034 |
| 2016 | Winnsboro | Soybean | Silt Loam | -1.9 | 0.074 |
| 2015 | St. Joseph | Soybean | Cracking Clay | -0.02 | 0.028 |
| 2016 | St. Joseph | Soybean | Cracking Clay | 0.20 | 0.019 |

[1]This term ranges from $-\infty$ to 1 where 0 to 1 indicates that the model predicted the mean as well or better than the observed mean.
[2]This term ranges from 0 to $\infty$ where 0 indicates that the model predicted the regression line of best fit perfectly.

Table 2. Summary of irrigation application by location based on what was actually applied and what was predicted by the STAMP tool.

| Year of Crop Season | Location | Soil Type | Number of Irrigation Events by Treatment | Predicted Number of Irrigation Events | Average Depth per Event by Treatment (in) | Predicted Average Depth per Event[1] (in) |
|---|---|---|---|---|---|---|
| 2015 | Bossier City | Sandy Clay Loam | 5 | 6 | 3.54 | 3.36 |
| 2016 | Bossier City | Sandy Clay Loam | 2 | 2 | 2.27 | 3.46 |
| 2015 | Winnsboro | Silt Loam | 3 | 5 | 3.06 | 2.90 |
| 2016 | Winnsboro | Silt Loam | 3 | 3 | 8.52 | 2.56 |
| 2015 | St. Joseph | Cracking Clay | 3 | 5 | 4.98 | 2.59 |
| 2016 | St. Joseph | Cracking Clay | 2 | 5 | 4.39 | 2.40 |

[1]These values were increased by an efficiency factor of 0.7 to directly compare to the gross irrigation estimates measured in the field.

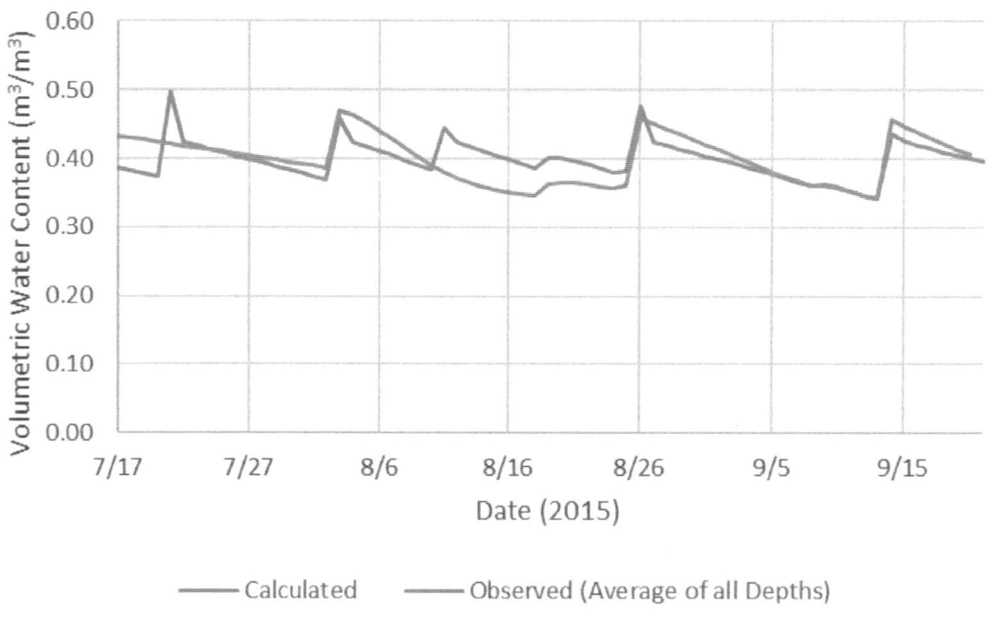


Figure 1. Comparison of soil moisture calculated from the STAMP Irrigation Scheduling Tool and observed from the GS-1 soil moisture data. Irrigation events on 7/21 and 8/11 resulted in a large response in soil moisture using the STAMP Tool, but little to no response was measured in the field.

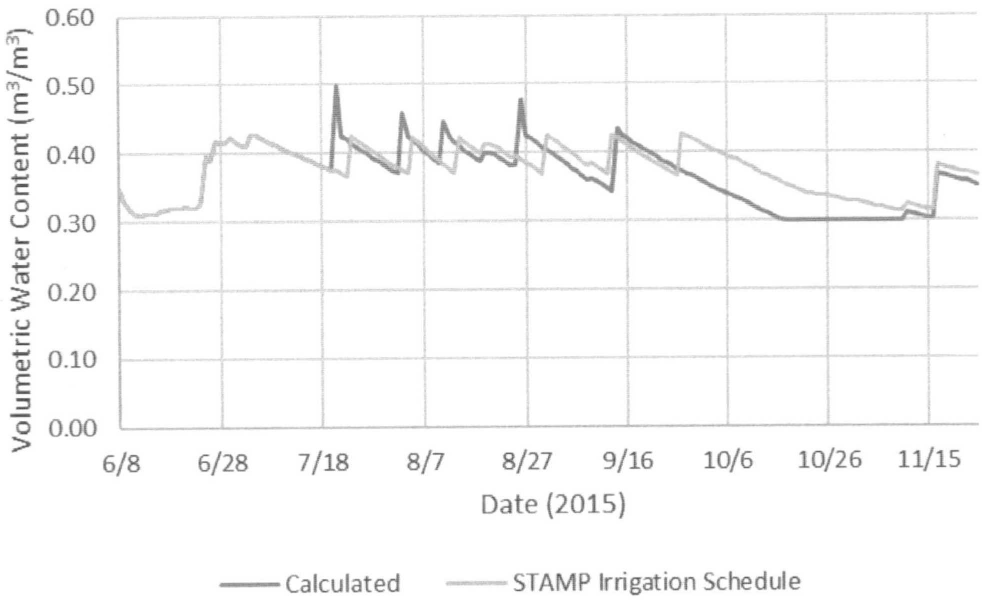


Figure 2. Full season comparison of the soil moisture estimated using the STAMP Irrigation Scheduling Tool. The calculated soil moisture was based on actual irrigation events that occurred whereas the STAMP irrigation schedule was estimated by assuming irrigation occurred when at the predicted times.

## Conclusion

Generally good performance was experienced during the first step to evaluating the STAMP irrigation scheduling tool. Volumetric water content values were related when close in time. Thus, one irrigation event that was measured but produced runoff without infiltrating can not only create an outlier for the day of irrigation, but also for some time after the outlier occurred. Future work will include expanding the data available for analysis, fully calibrating the current model, and exploring more physical processes that can improve the model.

## References

Kumar, V., T. K. Udeigwe, E. L. Clawson, R. V. Rohli, and D. K. Miller. 2015. Crop water use and stage-specific crop coefficients for irrigated cotton in the mid-south, United States. Agric. Water Manage. 156(2015) 63-69.

# Quantifying crop water requirements in the Mississippi Delta using an energy balance approach

Saseendran S. Anapalli, Daniel K. Fisher, Krishna N. Reddy, Sui Ruixui, Prasanna H. Gowda[1]
USDA-ARS, Crop Production Systems Research Unit, Stoneville, MS 38776
[1]USDA-ARS, Forage and Livestock Production Research Unit, El Reno, OK 73036

Timely quantification of consumptive water use of crops (evapotranspiration, ET) from planting through harvest is a critical input for managing irrigations for optimum crop production. Scheduling irrigations based on ET demands of plants can help reduce both water wastage and scarcity by enhancing water use efficiencies in crop production. We developed a residual energy balance procedure (EB) for rapidly quantifying ET from cropping systems based on measurements of the various components of heat energy balance of a land-crop canopy system was developed from components studied previously in the literature. In the EB, ET ($ET_e$) is expressed as the latent heat flux that was computed as the residual energy from a crop-land surface energy balance equation when net radiation and soil heat flux were measured, and sensible heat flux estimated from measurements of plant surface radiative temperature and air temperature, relative humidity, and wind speed at a constant height above the canopy. For a cotton field in Bushland, TX, $ET_e$ compared well with ET measured concurrently in a lysimeter. The EB methodology was then used to quantify ET for irrigated corn in Stoneville, MS. The cumulative seasonal values of alfalfa reference crop ET ($ET_r$) compared better than the grass reference crop ET ($ET_o$) with $ET_e$. Seasonal $ET_e$ was greater than $ET_o$ but less than $ET_r$. From these results, we conclude that the EB procedure presented here, with portable instrumentation, constitutes a viable alternative method to lysimeters for quantifying ET quickly in cropping systems for irrigation water management applications. Additional testing of the EB method is recommended under a broader range of climatic, agronomic and soil conditions.
Keywords: evapotranspiration; energy balance; Irrigation, crop water requirements.

## 1. Introduction

The long-term average annual rainfall received over the Mississippi Delta region is approximately 1300 mm, with only about 30% of it received during the core crop growing periods from April to August (Saseendran et al., 2016a). The crop growing season rainfall is also characterized by large inter- and intra- seasonal variabilities in their amounts and spatial and temporal distributions. To stabilize returns from crops raised in the region, farmers often provide supplementary irrigations, drawing water from the Mississippi River Valley Alluvial Aquifer. In the absence of reliable information on the water needs of the crops, farmers often provide arbitrary irrigations; consequently, agricultural water use from this aquifer far exceed its long-term recharge rates (Powers, 2007). Global warming was also reported to increase pressure on irrigation water requirements in the region (Saseendran et al., 2016b). Accurate, timely quantification of water requirements (or ET) of corn grown in the region is essential for scheduling irrigations for optimizing water use efficiency (WUE) in these cropping systems and to match irrigation withdrawals with the recharge rates of the aquifer.

In this manuscript we present a residual energy balance approach (EB) for rapid estimation of crop evapotranspration. In EB, an energy balance equation is applied to a soil-crop land area using remote or tower-mounted atmospheric boundary layer sensors and near-surface soil sensor measurements of the system variables, and ET (expressed as latent heat flux, *LE*) is estimated as the residual term of the energy balance equation when the soil heat and sensible heat fluxes in the equation are either measured or calculated. Typically, the sensible heat flux (*H*) is quantified assuming an air-diffusion (flow) resistance to heat and water transport across the turbulent atmospheric boundary layer above the plant canopy, and soil heat flux is measured using buried heat flux plates, adjusted to estimate the soil surface heat flux ($G_o$) (Allen et al., 2007; Heilman and Kanemasu, 1976; Su 2002). Heilman and Kanemasu (1976) developed an EB based ET model that uses the diffusion resistance to heat transport in the energy balance equation. They obtained ET estimates within 4% and 15% bias on a seasonal basis of lysimetric measurements for soybean (*Glycine Max* L.) and sorghum, respectively. In general, the values of sensible heat (*H*) in the EB procedure were derived from the measurements of the air and crop canopy temperature differential and modeling the aerodynamic resistance ($r_a$). In vegetated land surfaces, plant-soil surface temperature should represent the temperature of the apparent source/sink of sensible heat flux within in the plant canopy. This apparent temperature, known as aerodynamic temperature ($T_o$) is not a directly measurable variable, so crop canopy surface radiative temperature ($T_s$) is commonly measured using an infrared thermometer and used as a surrogate for $T_o$ in the computations of *H* in cropping systems. For computations of $T_o$ in this study, we used the equation developed by Chavez et al. (2010) for corn and Chavez et al. (2005) for cotton crops. Such empirical relationships linking crop specific characteristics with environmental variables were applied for simulating crop processes in cropping system models across the globe, for example, the CERES-rice and wheat model.

Our objectives were to provide a synthesis of components in the EB approach and (1) develop a state-of-the-science algorithm for computation of ET based on the EB approach, (2) test the ET quantified using this algorithm with cotton ET measured using a large-scale field lysimeter at Bushland, TX, USA, and (3) use the EB algorithm to quantify ET in corn at Stoneville, MS, USA and compare it with grass and alfalfa reference crop ET computed from climatological data for the location.

# 2. Methodology

## 2.1. The energy balance (EB) approach for estimating evapotranspiration (ET)

An energy balance equation for a crop-soil surface can be written as

$$R_n = LE + G_0 + H + \Delta S_{air} + \Delta S_{bm} + \Delta S_{ph} \quad (1)$$

where $R_n$ is the net radiation (positive downward), *LE* is the latent heat flux(positive upward), $G_o$is the soil heat flux (positive downward), *H* is the sensible heat flux (positive upward), $S_{bm}$ is the energy stored in the biomass, $S_{air}$ is the energy stored in the air layer, and $S_{ph}$ is the energy used in photosynthesis, where Δ denotes the change per unit time (s). Units are $Wm^{-2}$ for energy flux and $J\ m^{-2}$ for energy storage. We assume that in annual cropping systems like corn and cotton, $S_{air}$, $S_{bm}$, and $S_{ph}$ are negligible compared with other terms, hence neglected in our calculations. The ET ($mm\ s^{-1}$) is calculated from eq. (1) by dividing *LE* by the latent heat of vaporization of water (λ, $W\ kg^{-1}\ m^{-2}$):

$$ET = (R_n - G_0 - H)/\lambda \quad (2)$$

We employed the resistance to the turbulent exchange of energy and matter between different layers of the atmosphere and the ground surface to compute ET using Eq. (2) (Foken, 2008).

The heat flux at the ground surface, $G_o$ ($W\ m^{-2}$), is estimated by measuring soil heat flux at 8 cm soil depth and accounting for the heat storage in the soil layer above this point (Kimball et al., 1999).

The resistance approach following Triggs et al. (2004) was employed for estimating H.

$$H = \rho_a \, C_p \, (T_0 - T_a)/r_a$$

where $\rho_a$ is the density of air (kg $m^{-3}$) calculated from the ideal gas equation, $C_p$ is the specific heat of air assumed constant at 1005 J $kg^{-1}$ $^{O}K^{-1}$ , $T_a$ is the air temperature at the sensor height above the crop canopy, and $r_a$ the bulk aerodynamic resistance to sensible heat transfer (s $m^{-1}$). $T_0$ is the aerodynamic temperature ($^{O}$K) calculated from Chavez et al. (2010) for cotton and from Chaves et al. (2005) for corn. The methodology used here in computing ET is available in detail elsewhere (Saseendran et al., 2017)

## 2.2. Experimental data

### 2.2.1. Lysimeter and energy balance experiments in cotton (Bushland experiment)

Experiments to estimate cotton crop ET using both lysimeter and energy balance methods were conducted simultaneously in 2008 at the USDA-ARS, Conservation and Production Research Laboratory, Bushland, TX (35$^{O}$ 11'N, 102 $^{O}$06'W, 1170 m amsl) in a Pullman clay loam soil. Cotton crop ET was estimated in a large (3 x 3 x 2.3 m) precision, weighing lysimeter, located in the middle of a 4.7 ha irrigated cotton field. The lysimetric measurement site was also equipped with instruments for measuring net solar radiation, air temperature, relative humidity, wind speed, and canopy surface radiative temperature (view of the ground at 60$^{O}$ zenith angle) measured at 1 m above ground level in the center of the lysimeter field. The site was also instrumented for measuring soil heat flux (Hukse heat flux plate, Campbell Sci. Inc. Logan, UT) at 8 cm depth and soil water and temperature monitored above the flux plate. Both the lysimeter and energy balance components (micrometeorological) data were recorded on a data logger (CR-7X, Campbell Scientific Inc., Logan, UT). Irrigation treatments were applied to refill the soil to field capacity based on weekly neutron probe measurements to maintain the soil water content above the 50% level of maximum plant available water depletion. Cotton was planted on May 21, 2008, and harvested on December 14, 2008. However, continuous measurements of energy balance data were made only from June 7 to August 20, 2008.

### 2.2.2. Corn energy balance experiment (Stoneville experiment)

The experiment in 2016 was conducted on a Dundee silt loam at Stoneville, MS (33.42° N, 90.92° W, 32 amsl) located in the Lower Mississippi Delta region. Corn hybrid DKC66-97 was planted on March 23, 2016, with 102 cm row spacing, at a rate of 33,174 seeds $ha^{-1}$. The crop was furrow irrigated, and irrigation amounts were adjusted to refill soil water contents back to field capacity based on weekly soil water content measurements to maintain the soil water content always above the 50% level of maximum plant available water in the soil. The field size for the experiment was 1.5 ha with dimensions of 200 m in the north-south direction and 75 m in the east-west direction. The tower for measuring energy balance components was located in the middle of the plot. The sensors for measuring air temperature and relative humidity (Vaisala, HMP 155), net solar radiation (Kipp & Zonen Inc., The Netherlands), infrared canopy surface temperature sensor installed to view of the ground at 60$^{O}$ zenith angle (Standard Field of View Infrared Radiometer Sensor, Apogee), and wind direction and speed (Gill 2D-Sonic) were maintained at 1 m above the plant canopy. The sensor heights were adjusted manually to maintain this height whenever there is an increase in crop height exceeding 5 cm. Four soil heat flux sensors (Hukseflux soil heat flux plate, Campbel Scientific Inc. ) were installed at 8 cm depth. Water content and temperature in the 8 cm soil layer above the heat flux were monitored using Stevens HydraProbe (Steven Water Monitoring

Systems Inc.). Phenology observations were recorded every week.

### 2.2.3. Reference crop ET

Alfalfa (0.50 m tall) reference crop ET ($ET_r$) and short grass (0.12 m tall) reference crop ET ($ET_o$) were computed using the ASCE Environmental and Water Resources Institute (ASCE-EWRI, 2005) and FAO Irrigation and Drainage Paper no. 56 (Allen et al., 1998; Pereira et al., 2015), respectively, from weather data collected at the location by assigning fixed resistances for the reference crop surfaces.

## 3. Results and discussion

### 3.1. Evaluation of the EB method for quantifying ET in the Bushland Experiment.

In the Bushland experiment, substantial differences were noticed between the measured $T_s$, and computed $T_o$. On July 9, 2008, a rainy day with contrasting weather conditions , $T_o$ remained above $T_s$ throughout the day with the temperature difference ranging from 2.6 to 10.4 °C (Fig. 1b).

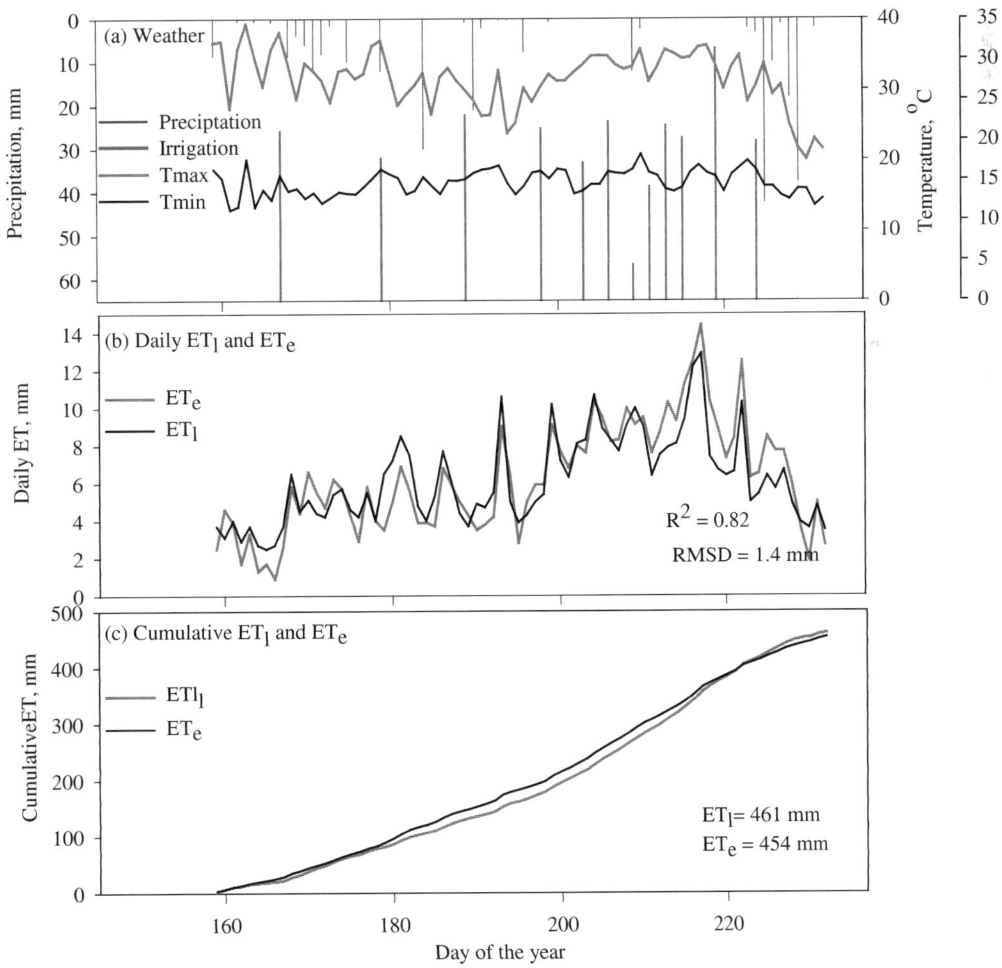


Fig. 1. (a) Daily maximum (Tmax) and minimum (Tmin) temperature(°C) , rainfall and irrigation amounts (mm d$^{-1}$), (b) evapotranspiration (ET) measured by lysimeter ($ET_l$) and energy balance ($ET_e$) methods, and (c) cumulative $ET_l$ and $ET_e$ in the Bushland cotton experiment in 2008.

Daily $ET_e$ computed using the EB method during the cotton growth period matched and correlated well with the lysimeter measured $ET_l$, with an $R^2$ (coefficient of determination computed as the squared value of the Pearson's correlation coefficient, r) value of 0.86 (Fig.1). The total $ET_e$ computed during this period was 454 mm versus a measured $ET_l$ value of 461 mm. In other words, the difference between $ET_l$ and $ET_e$ during this 105 day period was only 7 mm. The root mean squared deviation in (RMSD) in daily $ET_l$ relative to $ET_e$ was 1.2 mm. From these results, we propose that the energy balance procedure developed above is capable of quantifying ET comparable to direct measurements of ET using large-scale field lysimeters. Hence, the EB method for indirectly computing ET from measurements of energy balance components in the cropping system has the potential to provide a viable alternative to more directly measuring ET as a change in mass in large-scale field lysimeters.

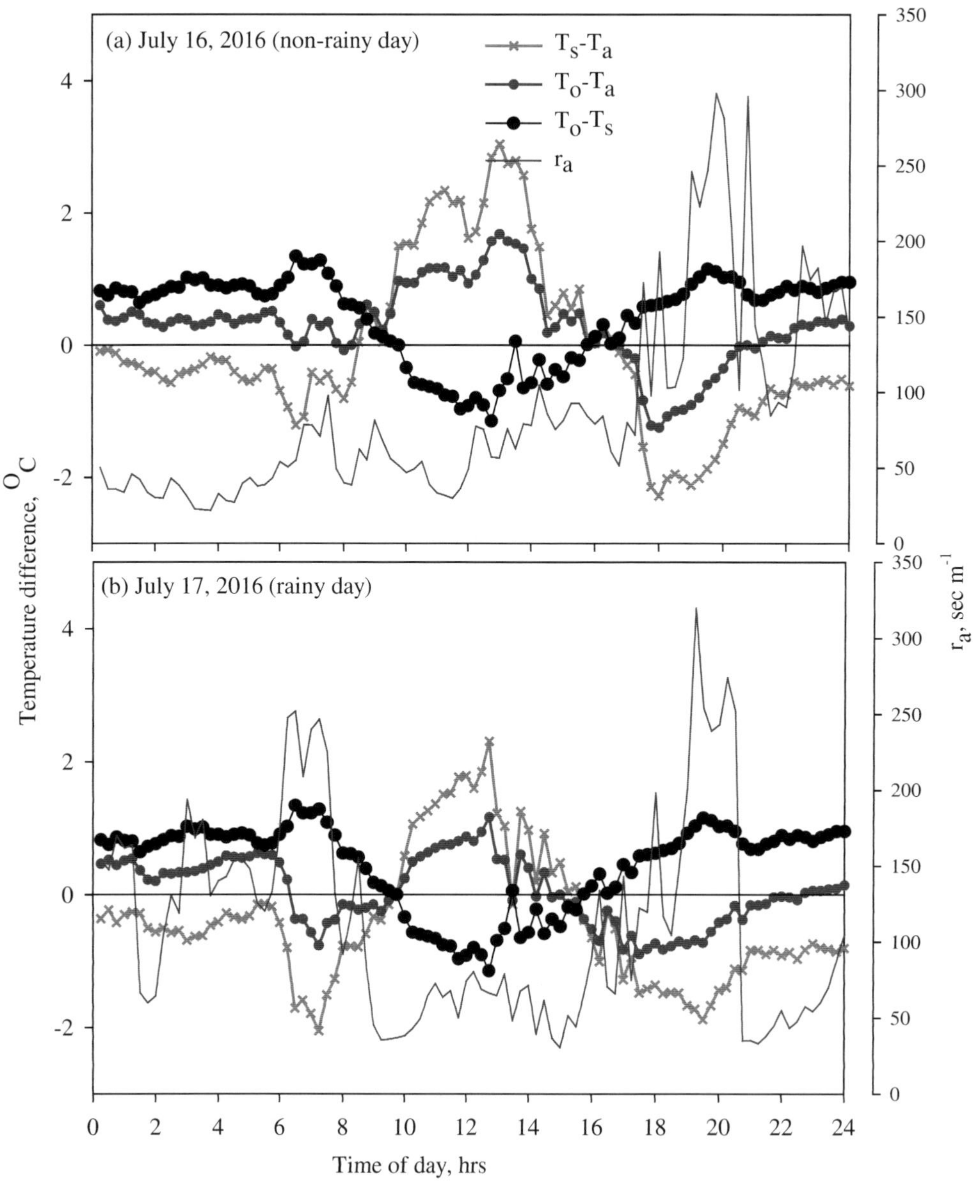


Fig. 2. Measured corn canopy radiative temperature ($T_s$), and computed canopy aerodynamic temperature ($T_o$) and aerodynamic resistance ($r_a$) in the Stoneville corn experiment for representative days (a) without rain and (b) with rain in 2016. The difference between $T_o$ and $T_s$ also presented.

## 3.2. Corn Experiment

The location for the corn crop experiment near Stoneville, MS receives an average annual rainfall of about 1300 mm, of which about 36 % (452 mm) is received during the four months of the corn growth period between April and in July (Saseendran et al., 2016a, 2016b). During this period in 2016, the location received 433 mm of rainfall, characterizing the season as an average rainfall season. The crop was planted on March 23 and it reached physiological maturity on August 2. Maximum crop height averaged 1.9 m with an average maximum *LAI* of 5.5. The $ET_o$ for this season computed from weather data collected on the EB tower was 559 mm, and $ET_r$ was 666 mm. We furrow irrigated the crop with 126 mm of water in three irrigation events – each irrigation event lasted two days. Daily temperatures varied between 8.7 and 28.1 °C in the month of April, between 16.2 and 27.8 °C in May, between 22.5 and 32.7 °C in June, and between 23.5 and 34.1 °C in July. Harvested grain yield in this season was 10,467 kg ha$^{-1}$.

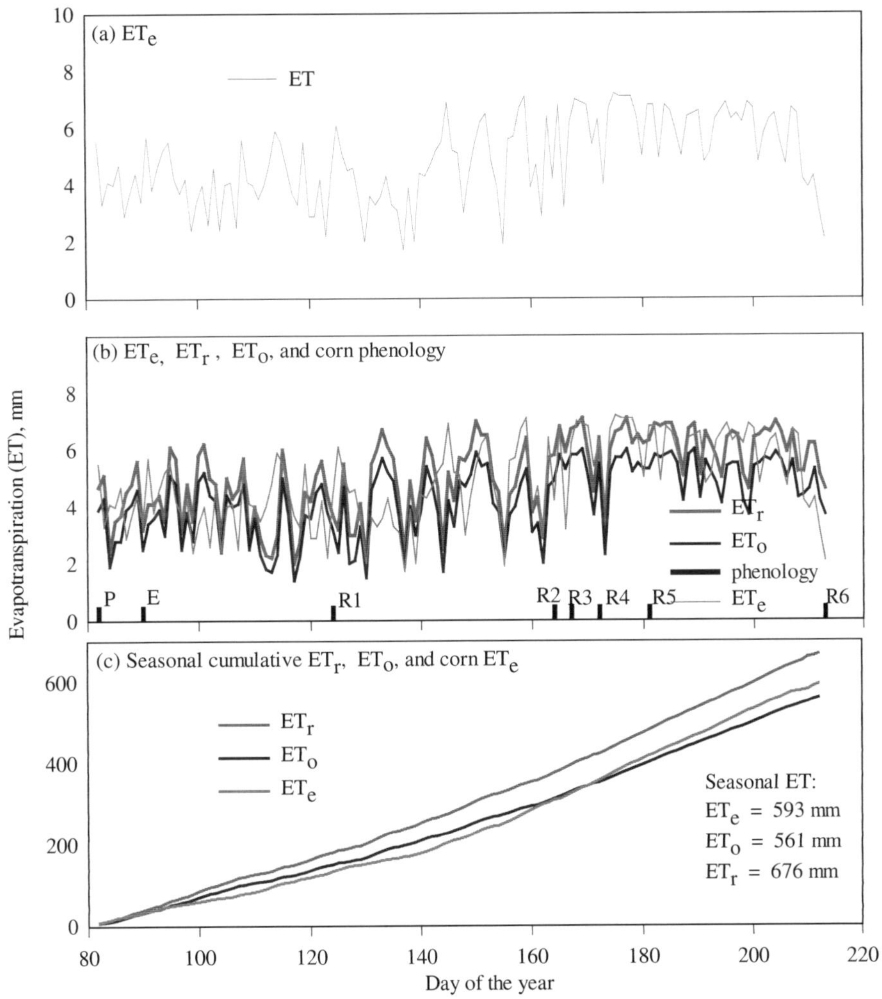


Fig. 3. (a) Corn evapotranspiration computed using the energy balance method ($ET_e$), and (b) grass reference crop evapotranspiration ($ET_o$) and alfalfa reference crop evapotranspiration ($ET_r$) computed from climate data in the Stoneville corn experiment in 2016. Symbols P, E, R1, R2, R3, R4, R5, and R6 represent the dates of occurrences of corn phenological stages: planting, emergence, silking, blister, milk, dough, dent, and physiological maturity, respectively. (c) Cumulative seasonal $ET_e$, $ET_r$ and $ET_o$.

As in the case of cotton, substantial differences were seen between measured $T_s$ and the computed $T_o$. The computed $T_o$ went above $T_s$ at sunset (i.e., $T_o$ - $T_s$ is positive in Fig. 2a) with the maximum value of 1.3°C at 06:30 PM. The general pattern in computed $T_o$ relative to $T_s$ was similar on a rainy day, July 17, 2016, as well (Fig. 2b).

The corn crop was planted on March 23, 2016, reached physiological maturity (R6 stage) on August 02, 2016, and was harvested three weeks later on August 23, 2016. Between planting and physiological maturity, the crop growth duration was 131 days. E During the crop period, the estimated $ET_e$ ranged between 1.7 and 7.2 mm $d^{-1}$ (Fig. 3a, b). The lowest value occurred 56 days after planting (May 17, 2016) due to nearly overcast skies. The values of $ET_o$ and $ET_r$ computed using climatological data collected at the location with values 1.9 and 2.3 mm, respectively, did not deviate substantially from the energy balance computed value.

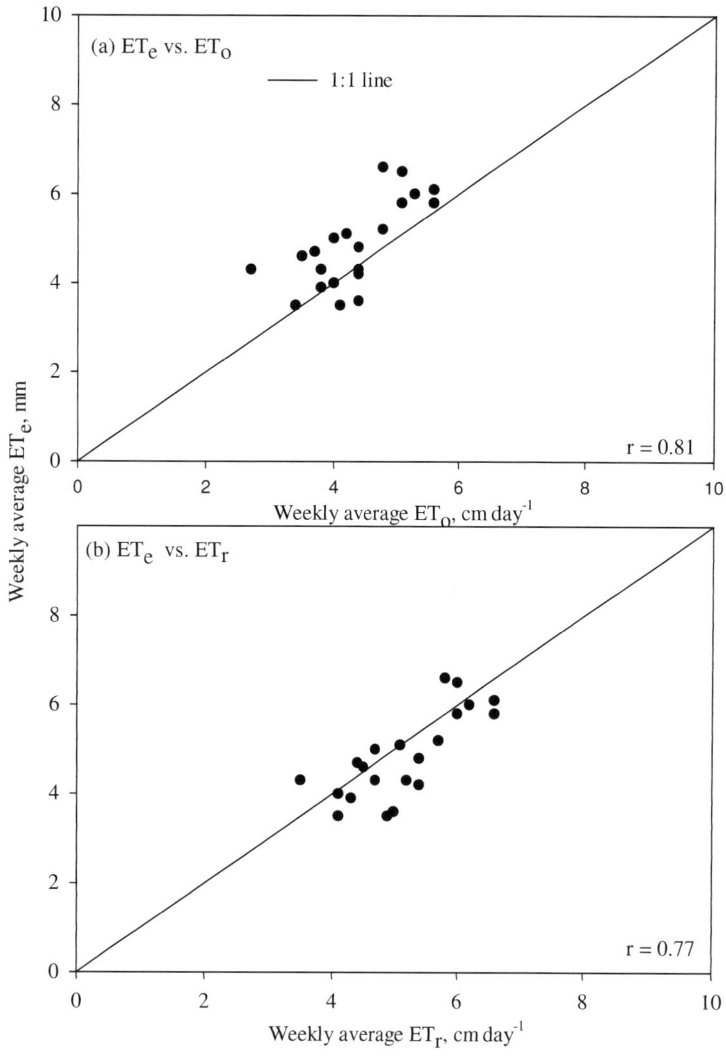


Fig. 4. Comparison between corn evapotranspiration computed using the energy balance method ($ET_e$), grass reference crop evapotranspiration ($ET_o$) and alfalfa reference crop evapotranspiration ($ET_r$) using 2016 climate data in the Stoneville corn experiment. r is the Pearson's correlation coefficient.

On a seasonal average basis, the $ET_e$ values were higher than $ET_o$ and less than $ET_r$ computed from weather data from a Agrometeorological weather station within 2 km from the experiment site. The average daily $ET_e$, $ET_o$, and $ET_r$ values during the corn growth period were 4.8, 4.2, and 5.1 mm $d^{-1}$, respectively. The root mean squared deviation (RMSD) between daily $ET_o$ and $ET_r$ values versus daily $ET_e$ values were 1.4 and 1.5 mm, respectively. The computed weekly total (irrigation decisions in this region are taken mostly on a weekly basis) values of both $ET_o$ and $ET_r$ were correlated with $ET_e$ with Pearson's correlation coefficient (r) values of 0.81 (coefficient of determination, $R^2 = 0.66$) and 0.70 ($R^2 = 0.61$), respectively (Fig. 4a and b). Likewise, though the daily $ET_o$ and $ET_r$ values deviated substantially from $ET_e$ estimates (Fig. 3b), seasonal total values of $ET_e$ and $ET_r$ were close to one other (Fig. 3c). Seasonal cumulative $ET_e$, $ET_o$, and $ET_r$ were 593, 561, and 676 mm, respectively. Therefore, in the absence of direct measurements of crop ET (for example, lysimeters, eddy covariance, or energy balance estimates), $ET_r$ computed from climatological data representing the crop conditions can be the best alternative for irrigation management applications where water is typically not the limiting factor.

## 4. Conclusions

For a cotton field in Bushland, TX, $ET_e$ compared well with ET measured concurrently in a lysimeter. The present EB methodology was then used to quantify ET for irrigated corn in Stoneville, MS. On a weekly total basis, the energy balance computed $ET_e$ was well correlated with reference crop ET for alfalfa ($ET_r$) and grass ($ET_o$) computed from weather data at the location. The cumulative seasonal values of $ET_r$ compared better with and $ET_e$ than $ET_o$. From these results, we conclude that the EB procedure presented here, with portable instrumentation, constitutes a viable alternative method to lysimeters and eddy covariance systems for quantifying ET quickly in cropping systems for irrigation water management applications.

## 5. References


Allen, R.G., Pereira, L.S., Raes, D., Smith, M., 1998. Crop Evapotranspiration: Guidelines for computing crop water requirement. United Nations Food and Agriculture Organization, Irrigation and Drainage Paper 56, Rome, Italy, 300p.

Allen, R.G., Tasumi, M., Trezza, R., 2007. Satellite-based energy balance for mapping evapotranspiration with internalized calibration (METRIC) model. ASCE J. Irrig. Drain. Eng. 133(4), 380-394.

ASCE-EWRI., 2005. The ASCE standardized reference evapotranspiration equation. In: Allen, R.G., Walter, I.A., Elliot, R.L., Howell, T.A., Itenfisu, D., Jensen, M.E., Snyder, R.L. (Eds.), Standardization of Reference Evapotranspiration Task Committee Final Report, Va: ASCE-EWRI, pp. 1-11.

Chávez, J.L., Howell, T.A., Gowda, P.H., Copeland, K.S., Prueger, J.H., 2010. Surface aerodynamic temperature modeling over rainfed cotton. Transactions of the ASABE 53(3), 759-767.

Chávez, J.L., Neale, C.M.U., Hipps, L.E., Prueger, J.H., Kustas, W.P., 2005. Comparing aircraft-based remotely sensed energy balance fluxes with eddy covariance tower data using heat flux source area functions. J. Hydrometeorol. 6(6), 923-940.

Foken, T., 2008. Micrometeorology, Springer, Berlin, Heidelberg, 308 pp.

Foken, T., Wimmer, F., Mauder, M., Thomas, C., Liebethal, C., 2006. Some aspects of the energy balance closure problem. Atmos. Chem. Phys., 6, 4395-4402.

Heilman, J. L., Kanemasu, E. T., 1976. An evaluation of a resistance form of the energy balance to estimate evapotranspiration. Agron. J. 68, 607-611.
Pereira, L.S., Allen, R.G., Smith, M., Raes, D., 2015. Crop evapotranspiration estimation with FAO56: Past and future. Agric. Water Manage. 147, 4-20. DOI:http://dx.doi.org/10.1016/j.agwat.2014.07.031
Powers, S. 2007. Agricultural water use in the Mississippi Delta. CD-ROM. 37th Annual Mississippi Water Resources Conference, Jackson, MS: 47-51
Saseendran, S.A., Fisher, D.K., Reddy, K.N., Pettigrew, W.T., Sui, R., Ahuja, L.R., 2016a. Vulnerability and adaptation of cotton to climate change in the Mississippi Delta. Climate 4, 55, 1-20.
Saseendran, S.A., Pettigrew, W.T., Reddy, K.N., Ma, L., Fisher, D.K., Sui, R., 2016b. Climate optimized planting windows for cotton in the lower Mississippi Delta region. Agronomy 6, 46, 1-15.
Saseendran S. Anapalli, T. Green, P. Gowda, K.N. Reddy, D.K. Fisher, and R. Sui. 2017. Adaptation and application of an energy balance method for estimating evapotranspiration in cropping systems. Agricultural Water Management (in press).
Su, Z., 2002. The Surface Energy Balance System (SEBS) for estimation of turbulent heat fluxes. Hydrol. Earth Syst. Sci. 6(1), 85-99.
Triggs J.M. , Kimball, B.A., Pinter, Jr. P.J., Wall, G.W., Conley, M.M., Brooks, T.J., LaMorte, R.L., Adamb, N.R., Ottman, M.J., Matthias, A.D., Leavitte, S.W., Cerveny, R.S., 2004. Free-air $CO_2$ enrichment effects on the energy balance and evapotranspiration of sorghum. Agric. For. Meteorol. 124, 63–79.

# A crop growth-insect model used in field experimental design

Theodore W. Sammis[1], Ashley Bennett[2] , David R Miller[3]
1. New Mexico State University Department of Environmental Science P.O. Box 3Q, Las Cruces NM 88001
2. New Mexico State University- Agricultural Science Center Los Lunas. 1036 Miller Road Los Lunas NM. 87031
3. Department of Natural Resources and Environment, University of Connecticut, Storrs, CT, (860) 377-3790

Written for presentation the IA Irrigation symposium Nov 8-9 2017 Orange County Convention Center Orlando Florida

## Abstract

A crop growth insect model can simulate the growing of a crop under different climate, water, nutrient, and insect stress conditions. Modeling crop and insect growth has application in the planning of field experiments. An object- orientated model has real world objects with software counterparts and each object consists of encapsulated data (attributes) and methods (behavior, and interactions). In this study, a cucumber spider mite model was developed to evaluate a proposed field research project. The crop insect growth model consisted of seven objects smaller models, which included a growth, water balance, and spider mite population model. Results of the model runs showed that the water irrigation levels of 75% and 50% of none stress Et had statistical difference in yield, but when low and high spider mite infestation were included in the calculation, only the 100% irrigation treatment had statistically lower yields due to spider mite damage. The model runs indicated that the spider mite infestation levels originally proposed for the field based experiments should be increased and the 0.75Et treatment removed.

## Introduction

Combined crop growth and insect models can simulate the growing of a crop under different climate, water, nutrient and insect stress conditions to predict crop yield under different management practices. Consequently, models can help predict fundamental causal relationships in physical and chemical processes related to agricultural production on both small and large farms. Because field experiments are expensive and encounter unexpected abiotic and biotic variables, models can be used to refine experimental design and methodology, improving the results obtained from field based experiments. Once model predictions are field tested and proved correct, model results can be expanded, facilitating field management techniques for specific crops, pests and locations. Thus using the model to test proposed field experiments, reduces the number of experiments necessary to test each important variable. This is a more efficient approach to conducting agriculture research and rapidly applying the results.

A simple process orientated model that simulates a proposed field experiment is a tool that can reduce uncertainty of field experiments. Models can determine whether proposed experimental treatments, such as levels of irrigation, will produce sufficient change in a response variable, for example crop yield.

Model results can guide researchers during the design and planning phase of experiments by suggesting where experimental treatments may produce statically different results.

Cucumber are a short season specialty crop that grows anywhere in the United States. While large scale production of cucumbers is present in states such a Wisconsin, Michigan and Florida (WIFSS, 2017), cucumber are also grown in most backyard gardens and small diversified farms. Cucumbers are a vine crop that can be grown on the ground (bush varieties) or trellises in both fields and greenhouses depending on the variety planted and are hand-picked. The value of cucumber production across the United States was 204 million dollars in 2015 (USDA, 2017). The most common pests on cucumbers are cucumber beetles, spider mites, aphids, squash bugs pickleworms, and squash beetles (Clemson cooperative extension, 2017)

The two-spotted spider mites occur under hot dry weather conditions feeding on the contents of individual cells of the leaves with hundreds of mites per leaf. The damage can develop quickly and appears as pale yellow or reddish brown spots on the upper side of the leaf. The two-spotted spider mite has been reported infesting over 200 species of plants (Perry *et al.,* 1998) causing large economic loss when not controlled by IPM methods. To date, no economic IPM threshold has been developed for when to start spraying field grown cucumbers to control spider mite infestations using either inorganic or organic spraying protocols. Consequently, research is needed to evaluate the interaction of cucumber growth, water stress, and spider mite stress on cucumber yield.

## Objective

A plant growth insect model was developed to assist in the design of an agricultural field experiment. The objective of developing the model is to use the model to estimate the impact of two spotted spider mites infestation levels , *Tetranychus urticae* Koch (Acari: Tetranychidae) on cucumber yield, growth and plant characteristics grown under different irrigation levels. The results of the model simulations will be used to refine the experimental methodology of the field experiment, increasing the probability of a successful experimental design.

## Modeling Considerations

An effective simulation model must predict both abiotic and biotic system variables. Abiotic variables include climate, weather, rainfall, irrigation, while biotic variables may encompass insect, disease, or weed populations

A simulation model is expected to be a user-friendly decision support tool for irrigated or dry land crops and should include all programming “objects” necessary to grow the crop, using either mechanistic or empirical functional relationships (Acock and Reynolds, 1989). The model must contain objects of external stresses caused by lack of rainfall or irrigation, insects, or soil borne diseases. (Reynolds and Acock, 1997). An object orientated model has software counterparts to real world processes. Each model object consists of data (attributes) and methods (behavior, and interactions). Objects and the described variables in the object interact with each other as well as the environment. A benefit of simulation models are users can easily make changes in variables i.e. levels of insect infestation on a crop.

The best known group of plant growth models was developed by IBSNAT (International Benchmarks Sites Network for Agrotechnology Transfer). The model structure is a top down design, modularity, high cohesiveness, loose coupling (Hodges 1991). It was written initially using the FORTRAN language. These initial models developed into a group of sixteen crop growth models with one user interface called DSSAT (Decision Support System for Agrotechnology Transfer) crop models (Jones, et al., 2003).

However, a major problem with DSSAT models is the time and money necessary to develop a new crop and insect module. Consequently, crop modules were developed for the major crop but additional modules may never be available for specialty crops because of the cost and time of development. A second problem with the DSSAT model is the use of a large number of program languages including Python and Fortran, which restrict development of new crop models.

To overcome the need to learn DSSAT modeling languages and facilitate development of crop insect models for specialty crops, we used Excel. Excel is not a programming language; but a spreadsheet program written in C++. It is especially suited for quickly developing object- orientated plant growth models. Object oriented concepts should be separated from implementation in Object oriented languages. Consequently, even though excel is not considered an Object oriented Language the spreadsheet can be structured in the object oriented concept (Schroder, 2017) Each spreadsheet in a excel workbook can be set up in an object programming structure with the transfer of that objects output quickly and easily made to other objects. Graphical display of parameters and output can be made in the object or interfaced with a graphical display object resulting in the developer easily seeing the results of the methods programed in the object.

To use a crop growth- insect model to evaluate proposed agricultural field experiments, the process of designing field experiments must be compatible with the development of the model to simulate the proposed experimental design. The steps in a field design that must be simulated by the crop model are presented by Ganio (1997). They include a specific objective, a plan of action, field operation of the experimental design, and drawing conclusions from the field measurements. Finally, field observations are still needed to verify the management decisions made by the model.

## Description of plant growth water balance spider mite model

The cucumber plant growth water balance spider mite model (flow chart figure 1) was developed to evaluate the plot design and measurements for the proposed research experiment described in the objective. The model was modified from a Pecan tree growth nitrogen model (Andales et al., 2006, Sammis et al., 2013) having a one dimension flood irrigated water balance model replaced with a two dimensional water balance drip irrigation module ( Sammis et al., 2012) because the proposed cucumber experiment was to be drip irrigated. The alternating bearing and pruning, growth and nitrogen balance objects in the Pecan nitrogen model were turned off with a switch in the module. The model retained the soil temperature object from the Pecan model (Sammis, et al., 2013, Sharma et al., 2010) and soil water potential object added from a plant growth phytophthora disease model( Sammis et al., 2012b). The phytophthora disease object was turned off also with a switch because although phytophthora does attack cucumbers, in the proposed experiment phytophthora diseases was assumed to be not present in the soil. After conducting the experiment, if phytophthora disease is observed to be present then this object will be turn on when comparing experimental results to modeled results. A reference Et object described below was develop because the original models acquired the reference Et data from the internet at the location of the experiment. The spider miter growth model described below only needs weather data and calculates the spider mite stress function from that data based on two regression models of low and high spider mite infestation from experiments conducted in the field (Atanassov, 2014), and a population spider mite object that predicts spider mite infestation levels which are higher than field experiments in the literature described by Rabbinge, and Hoy (1980).

The attributes or inputs of each object are part of the parameterization of the object and move with the object when the object becomes part of another plant growth disease model

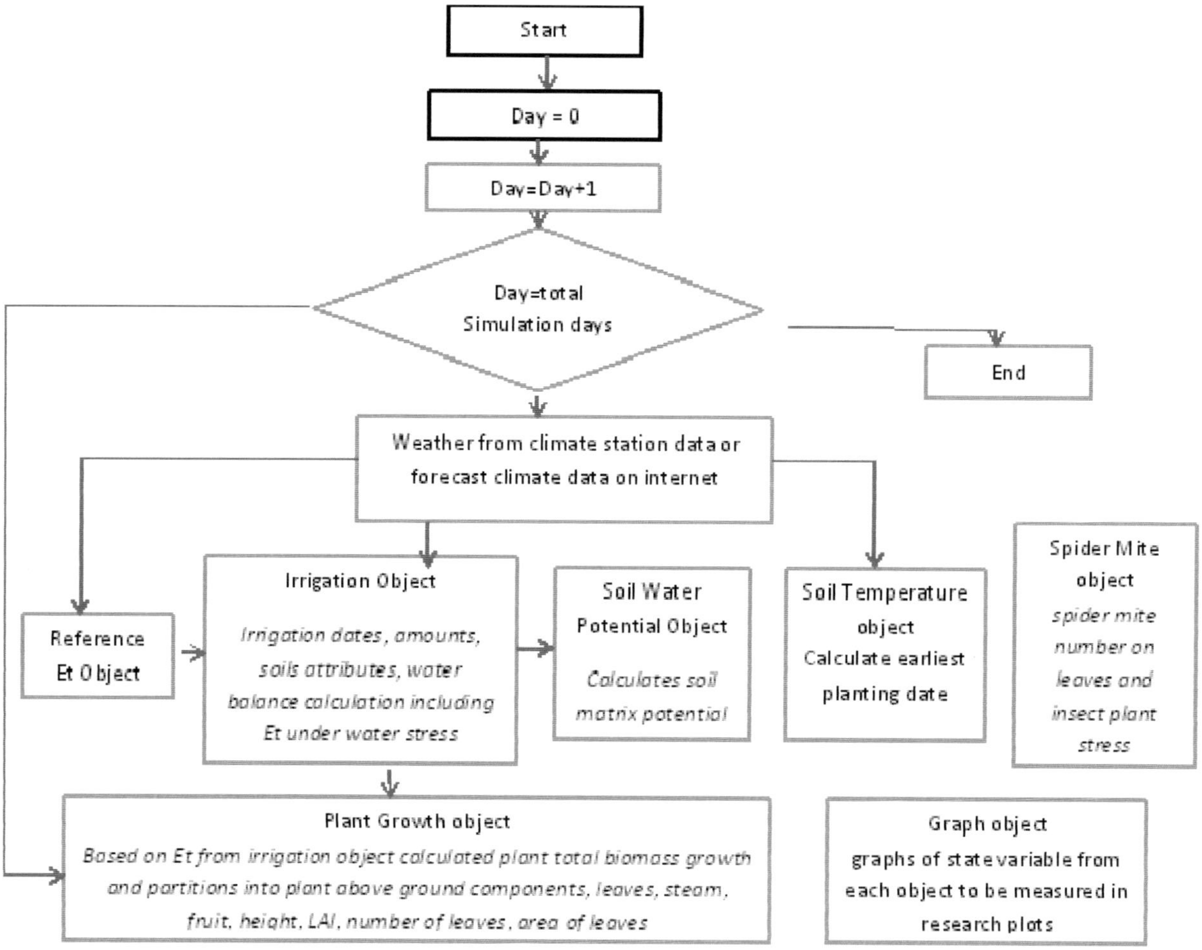


Figure 1 Flow chart for cucumber plant growth spider mite model.

### *Weather Data*

Measured weather data at the proposed research site in Albuquerque NM for the year 2015 was acquired from the internet (Albuquerque Weather Data. 2017). The weather data is needed by the reference Et (Etr), the soil temperature, the irrigation and the spider mite objects. The weather data on the internet was in English units and was converted to metric units for calculation in the Etr object requiring those units. The other objects used metric units in the calculations.

### *Reference Et (ETr) object*

The object was developed by equations described in (Allen et al., 2005) with the required climate input of maximum and minimum temperature and humidity, average wind speed, and solar radiation acquired from automated climate station data on the internet. The output of the object is ETr referenced to a tall crop (alfalfa) for use in calculating Et in the irrigation object. The output is in units of mm/ day but is converted to inch/ day for use by the Irrigation object.

### *Irrigation Object*

The full description of the two dimensional water balance equations in the irrigation object are presented by Sammis (et al., 2012) but several parameters are unique to the experimental design of cucumbers and are listed in Table 1. The third order polynomial describing the crop coefficient (Kc) for non- stress cucumber plants in units of growing degree days (GDD ) over the growing season is described equation 1.

Kc = 7.6 E-3 +3.30E-3*GDD-1.00E-.6 *GDD^2 -2.00E-10*GDD^3 (1)

The coefficient of determination is 0.99 and the GDD with a base of 50 degree F was from Perry (et al., 2016) and the Kc values for specified GDD values came from FAO 24 (Doorenbos and Pruitt, 1977)

In the irrigation object the units of input and calculations are in English units because the original object was developed for use by farmers. Consequently, the GDD and kc has units of F. The Kc equation was derived by Allen (et al, 2017) which gave the kc values for different growing season periods and GDD days for those periods was given by Perry and Wehner ( 2016)

| Object parameters | Parameter Description | Units | Reference |
|---|---|---|---|
| Maximum rooting depth | 24 | Inches | Veggie harvest 2016 |
| Root growth rate | 0.04 | Inches/GDD | |
| Water holding capacity | 2.5 | Inches/ft | Spectrum Technologies 2016 |
| Saturated water holding capacity | 3 | Inches/ft | Spectrum Technologies 2016 |
| Beginning root depth | 4 | Inches | |
| Slope of water stress function | 2 | | Allen et al 1988 |
| Intercept of water stress function | 0 | | Allen et al 1988 |
| Row spacing | 39 | Inches | |
| Management allowed depletion % | 50 | For computer scheduling irrigation | Allen et al 1988 |
| Irrigation depth for scheduled irrigations | 1 | Inch | |
| Management allowed | 99 | For specified | |

| | | | |
|---|---|---|---|
| depletion (MAD) | | irrigation dates and amounts | |
| | | | |
| | | | |

Table 1 Input parameters in irrigation object specified at top of excel spreadsheet in cucumber growth model workbook.

### *Plant Growth object*

The plant growth object was simplified from plant growth object in the pecan nitrogen model and only includes the plant components listed in Table 2.

| Object parameters | Parameter description | Units | Reference |
|---|---|---|---|
| Date | Days during the growing season | m/day/year | |
| Plant Growth (PG) | PG= ET(irrigation object) * Water use efficiency (input in object) | Kg/ha/day | |
| Growth per individual plant (GP) | GP= PG/Plant number/ha( input in object) | Kg/plant/day | |
| Leaf growth per plant (LG) | LG=PG* Leaf allocation (input in object) set to 0 after crop coefficient from irrigation object reaches 1.08 | Kg/Plant/day | |
| Plant stem and branches(PSB) | PSB= PG* stem-branch allocation (input in object) | Kg/plant/day | |
| Plant stem and branch diameter accumulative | Calculated from PSB accumulative and wood density (input in object) | m/day | |
| Plant Height accumulative | Calculated from PSB accumulative and height to radius(input to object) | m/day | |
| Leaf area per plant (LA) | LA= ∑LG* specific leaf area(input to Object) | m^2/plant | |
| Leaf area index (LAI) | LAI= LA/plant spacing | None | |
| Accumulative yield when leaf growth equals 0 (AY) | Ay= ∑PG if LG=0 | Kg/ha | |
| Leaf number (LN) | LN=LA/ leaf size (input to object) | Number | |
| Water Use Efficiency | 15 | Kg/ha/mm (input) | Yaghi et al., 2013 |
| Wood density of stem (cotton) | 185 | $Kg/m^3$ | Gagandeep Kaur Sidhu and Sandhya. 2015 |
| Initial plant radius | 0.0002 | M | |

| | | | |
|---|---|---|---|
| Initial plant height | 0.001 | M | |
| growth to leaves or fruit (cv. Modumoti) | 0.75 | Decimal | Haque et al., 2009 |
| Crop coefficient when allocation of growth goes 100% to yield | 1.08 | | |
| Fruit dry weight (cv. Modumoti) | 0.1 | Kg | Haque et al., 2009 |
| Specific leaf area (22 leaf types) | 21 | m^2/kg | Baret, Frederic and T Fourty. 1997 |
| Harvest Date | 7/27/2015 | | |
| Leaf area size (cv. Modumoti) | 0.037 | m^2 | Haque et al., 2009 |
| Number of plants/ha | 33333 | | |
| Row spacing specified in irrigation object | 1 | M | |
| Height to stem ratio (variety dependent) | 100 | | Haifa 2017 |

Table 2. Description of plant growth parameters and inputs constants in the Object plant growth calculations.

***Spider mite object.***

Spider mite (Acari: Tetranychidae)) infestations on cucumbers cause reductions in biomass, transpiration, photosynthesis ion, dry matter partitioning, chlorophyll reduction, and increases shedding of immature flowers (Park and Lee, 2002). These effects are simulated in the model by the interaction between spider mite number per leaf and the impact on evapotranspiration in the irrigation object which in turn reduces the components of the growth object. To develop the spider mite object, empirical stress function were developed from an experiment reported in the literature (Park and Lee, 2005) on the effect of spider mite numbers per leaf on cucumber yield. Figure 2 shows the spider mite population increase over time for different initial infestation levels on Chun-Gwang Baekdadagi cucumbers (Park and Lee 2005). The time day based data by Park and Lee et al. (2007) was converted to a growing degree time base using a base of 50 degree F (NC extension 2016). Two regression functions based on growing degree were developed for a low and high infestation level at fifth leaf (Figure 3 and 4). A spider mite population model (figure 4) was also develop to generate spider mite counts (Rabbinge, and Hoy. 1980) that were higher than the high infestation level for the field experiment after the initial runs of the model showed that the field experiment spider mite infestation was too low to separate water stress from spider mite stress under large water stress conditions (Et equal to 0.5 Etns). Spider mite infestation levels are reported in the literature as both spider mites per leaf and spider mites per plant, and the spider mite object calculates both spider mites per plant and per leaf.

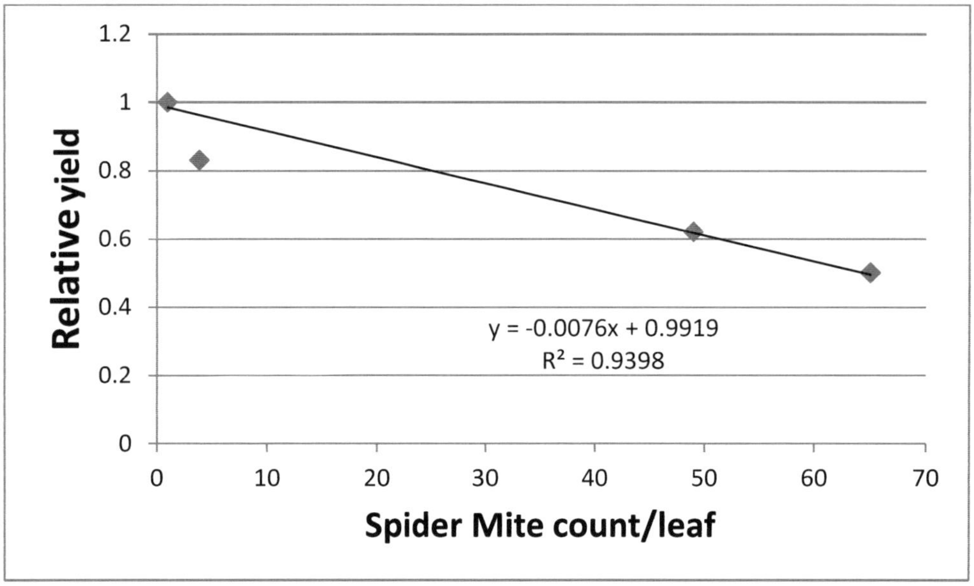


Figure 2. Spider mite count impact on plant evapotranspiration rates and plant growth

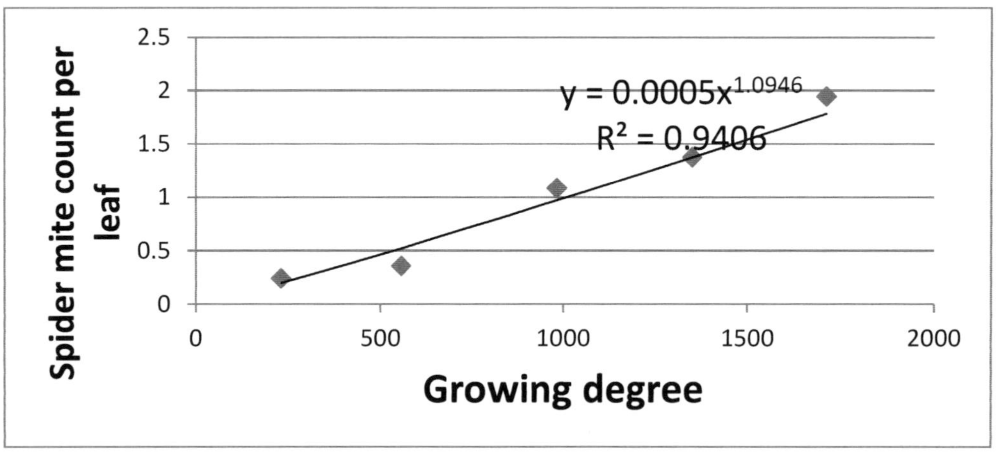


Figure 3 Spider mite count increase with growing degree days with a low initial infestation rate of 98 spider mites per plant at 5 leaf or 20 mites on one leaf at 5 leaf stage

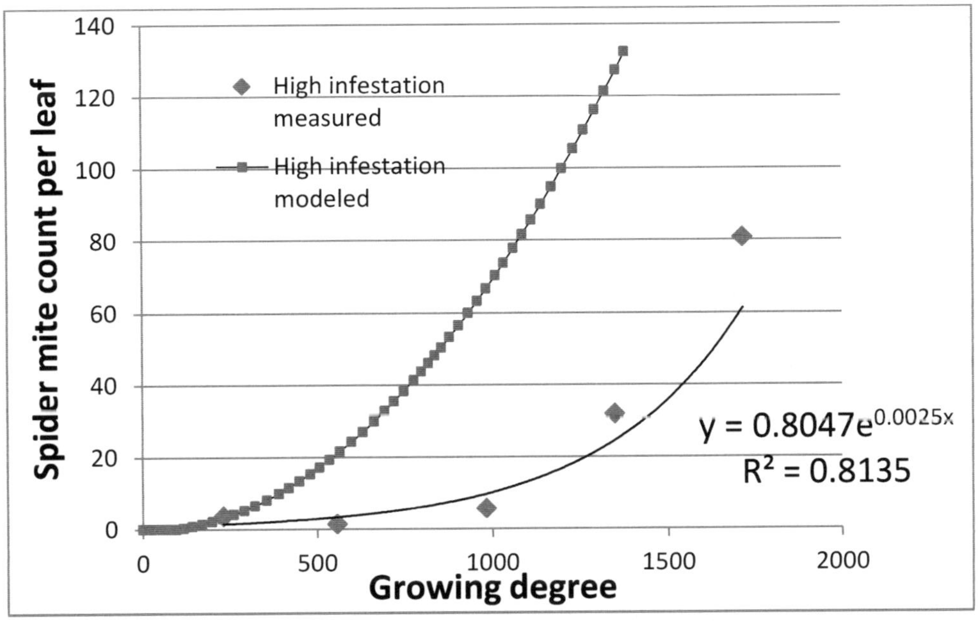


Figure 4. Measured and modeled using a regression equation of spider mite counts increase with growing degree days with a high initial infestation rate of 390 spider mites per plant at the 5 leaf stage and the population spider mite model with an initial high infestation.

The problem with spider mite infestation when a single leaf or two is infested with spider mites is that the population increase rate of the spider mites depends on the survival rate of the infestation. In the greenhouse experiment in the low infestation level the spider mite number per leaf never reached the initial infestation level and for the high infestation level it took the entire growing season to reach the initial infestation level on all leaves.

The literature indicates that with these infestation levels the spider mite number should have been a lot higher than measured because the report spider concentration after one to two generations (15 to 30 days or 250-500 GDD after infestation) on all leaf on a plant should be between 1 to 1.7 time the initial infestation level when two leaves are infested at 5 leaf (Nyoike, and Liburd. 2013, Reisigi and Godfrey, 2007)

An assumptive was made that Et and growth of the cucumbers were affected identical to the reported yield reduction in the field experiment. As second assumption was that evapotranspiration is reduced in a multiplicative model of soil moisture and spider mite stress (equation 2) similar to the pecan nitrogen stress model where Et is reduced as a multiplicative function of soil moisture and nitrogen stress (Sammis et al., 2013). This assumption needs to be verified by comparing the model results to measurements in the proposed experiment.

$$Et = Etns * Ks * Kspider \quad (2)$$

Where: Etns=is the evapotranspiration under non stress conditions
Ks= soil waster stress function (varies between 0 and 1)
Kspider= spider mite stress function (varies between 0 and 1)

$$Etns = Etr * Kc \quad (3)$$

Where : Etr= reference Et for a tall crop (alfalfa) (Allen et al., 2017)
Kc= crop coefficient based on growing degree days (Allen et al, 2017)

The spider mite object calculates the impact of spider mites. It uses a specified coefficient of variation for 0, 1, and 2 standard deviations from the mean spider count predicted by the regression model for high spider mite infestation and the spider mite population model to simulate the spider mite variation expected in the proposed experiment. The model user selects the different coefficient of variations and the model calculates the different spider mite counts. The object input data also has a switch, selected by the user, selecting zero, high regression model or high population spider mite infestation levels in the experiment. If the standard deviation is set to 0 no variation in the spider miter count from the predicted regression and population model will occur.

## Materials and methods

### *Initial Plot design*

The model was used to evaluate an experimental design to determine the interaction of cucumber growth, water, and spider mite stress on cucumber yield. For the proposed experiment, the variety selected was Green Finger cucumber, which is a 60 day maturing slicing variety. The experimental design after modification by use of the model will be a split plot design with irrigation as the main plot and infestation levels of spider mites as the subplot. Three irrigation levels will be tested including full irrigation or the control and 2 levels of deficit irrigation 25% and 50% of the control. Spider mites will be infested at 3 levels including 0, 100 and 400 mites / plant. The research plots will be located in Los Lunas NM, 25 km south of Albuquerque NM on a clay loam soil.

Weekly plant measurements will be collected on plant height, flower number, overall plant area cucumber fruit number, and weight. Mite counts, soil moisture, and canopy temperature will also be recorded on a weekly basis across treatment levels. Plant measurements and mite counts will be made on the three center plants of each plot.

Each plot will be 1 m wide and 2 m long with 4 m between each plot. The distance between blocks (n=4) will be 10 m. The cucumber plants will be planted in the center of the plot on 30 cm spacing containing 5 cucumber plants / plot. Plots will be seeded in mid-May.

The plots will be drip irrigated with an above ground drip system. The irrigation system will be operated every Monday, Wednesday and Friday. The amount of water applied will be determined from the previous week's calculated non-water stress Et applied over the three irrigations times. Irrigation amounts will be 1.0 ET ns for the control, 0.75 Et of the control, and 0.50 Et of the control. Drip irrigation emitters will not be installed in the space between plots.

### *Calibration procedure*

The model parameters for cucumber growth were calibrated by using results from the literature (Tables 1 and 2). The cucumber model was run for the control, no water stress or spider mites, from May 18 to harvest July 20 2015. A frost could still occur after May 18 but the probability would be low (Farmer's Almanac 2016). Cucumber should not be planted until the soil temperature is above 65 degrees F (Veggieharves, 2016) and the soil temperature object predicted that on May 18 2015 the soil temperature would be 77 degrees F. With this planting data, the model predicted a 73 day growing season from planting or transplanting to harvest before the plant started to senesce. Accumulative GDD of 1084 corresponds to an early season maturing slicing cucumber cultivars that require 1,154 GDD. (NC extension 2016)

## Results and discussion

The model predicted a cucumber yield of 2338 kg/ ha of dry matter. Divide the yield by 1.12 to calculate lb/ac. Converting this to wet weight assuming a cucumber moisture content of 96% (HealthyEating, 2017), the yield would be 58400 kg/ha. Commercial drip irrigated cucumber yield in California range from 56000 -89600 kg/ha after unmarketable cucumbers are pulled off the plants and left in the field (Schrader et al, 2017), which is within the range predicted by the model under no water or insect stress conditions.

Plant height was measured for fully irrigated cucumbers (cv. Hoshinokagayaki) in the field at Yamaguchi University, Japan (latitude 34.809 N, longitude 131.827 E and altitude 17 m above sea level) by Yuan (et al., 2016) and compared to the model growth height of cucumbers at Albuquerque, nm (Figure 5). The average air temperature during the spring growing season in Japan was 21 degree C in Japan compared to 23 degree in Albuquerque, NM.

.

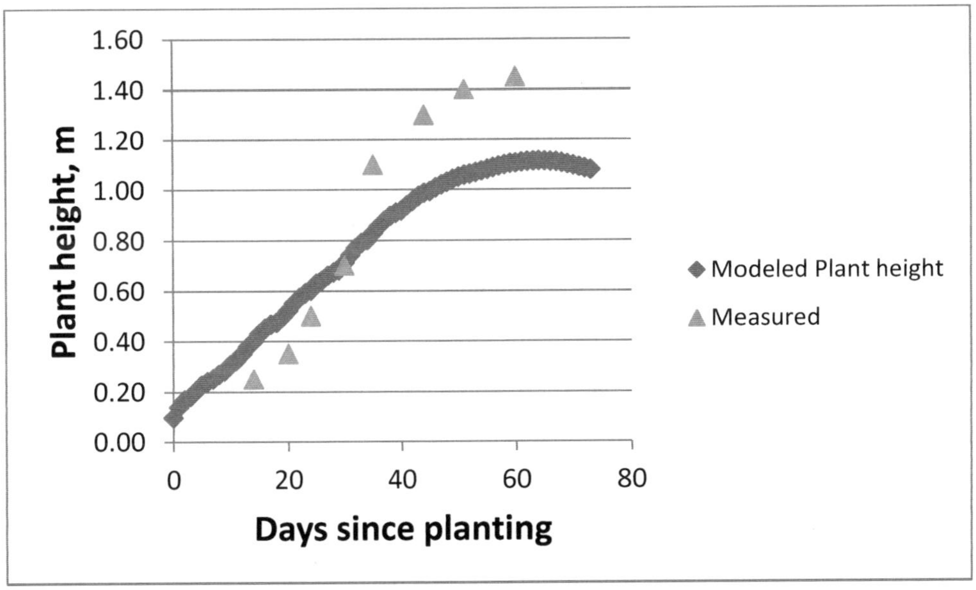


Figure 5. Measured cucumber growth height in the spring by Yuan (et al., 2016) and modeled height under no water or insect stress.

Albuquerque NM is at 35.1 N Longitude, Latitude 106.6 W and at 1614 m elevation, which positions Albuquerque at a higher longitude and elevation compared to Japan. . Despite the difference in longitude and elevation, the modeled and measure heights are very similar indicating that the growth and irrigation objects predicted reasonable results compared to measured cucumber growth data. However, the height to stem ratio is variety dependent (Table 2), meaning the results would have been different with a different height to stem ratio input parameter in the growth object.

The total biomass per plant was modeled as 160 g and the measured value was 100 g (Yuan et al., 2016).The 95% confidence interval was on the high side of the measurement at 140 g. Again different parameterization of the growth object could result in this amount of variability.

The model was run to simulate the application of an irrigation of 1 inch whenever the management allowed depletion in the soil water reservoir reached 50%, for the 1.0 Et treatment. This resulted in irrigation every other or every 3 days. When the 1 inch was applied when the 0.75 Et plots were irrigated at a management allowed depletion level 80% with one inch of irrigation, the time between irrigation increased to one every six days. The amount of water applied for the 1.0Et treatment was 25 inches compared to the 10 inches of irrigation applied to the 0.75 Et simulation treatment.

The relative total biomass for the 0.75 Et simulation produced by the model was 0.74 compared to relative total biomass of 0.65 measured in Japan when the irrigation amount in the field experiment was reduced to 74% of the non-moisture stress plots (ETns).

Yuan (et al., 2016) did not specify the irrigation dates or amounts; therefore the model run is only an approximation of the experiment conducted by Yuan (et al., 2016). However, if the irrigation object interacting with the growth object were not close to conditions for the experiment in Japan, the difference of the total plant growth under full and 75% irrigation would not differ by only 9%.

The difference in literature and model values for cucumber growth response also demonstrates the need to parameterize a cucumber model for the variety that is to be grown in the future field experiment. The need to model the correct variety has been demonstrated by the DSSAT group of models compared to measured value in the field Tsuji (et al., 2017). The plant parameters in the growth DSSAT model are unique for different varieties. The model proposed here will improve the methodology and experimental design for a field trial. The cucumber model as demonstrated by the above comparison does satisfactory predict cucumber growth under different water stress conditions. Verification of the model under spider might stress is needed because only one experiment was available from the literature to develop the spider mite functions in the model.

The literature did not have growth and insect parameters for the selected cultivar to modeled in Albuquerque NM but Table 1 and two used the values derived for varieties grown in those field experiment reported in the literature. Most of the cucumber varieties used in the parametrization of the model were vine types and a vining variety will be was selected for the field experiment that it is adapted to the climate in Albuquerque, NM but the variety will be different that those reported in the literature. Consequently, the model may not predict the same results that will be observed in the proposed future experiment because the model parameters may not be correct for the selected variety. The model simulation represents the average response of vine cucumbers varieties to water stress and spider mite infestation levels.

The model was run for different levels of moisture stress and spider mite infestation with the standard deviation of the spider mite count set to 0,+- 1 and +- 2 to generate the results of the split plot experiment (Table 3 ). The model simulations represent the expected effect on yield and total plant biomass under different levels of spider mite infestation and irrigation at the end of the growing season. The model was run for three levels of spider mite infestation and a control. However, the lowest level of spider mite infestation ( 4 spider mites per leaf at 5 leaf) showed no difference for different irrigation treatments. Consequently the model results are reported for only the high infection level of 390 spider mites per plant at fifth leaf resulting in a final infection level of 60 spider mites per leaf using the regression model in figure 4 and the high infestation level modeled by a population model resulting in a final infestation level of 135 spider mites per leaf (Figure 4). The field experiment spider mite population at the final measurement was 80 spider mites per leaf (Figure 4). However, there is a large standard deviation associated with the measurement.

To model a higher level of infestation because no field experimental results were in the literature for a high level of spider mites population, a spider mite population was generated where the spider mite level at the end of the growing season was double (135 spider mites / plant) the highest field experiment (Figure 4). We used a spider mite population model parameterized and described by Reisig, and Godfrey ( 2007) to model the effects of spider mite stress at this higher level of infestation. The initial model spider mite concentration was 1 spider mite per leaf or 0.64 females per leaf. However, the survival rate is 100 % in the model with no spider mite predators resulting in the high final spider mite count at the end of the growing season.

Consequently, based on the model runs, the field experimental design should have two treatment levels of spider mite infestation: 1. a high 390 spider mites per plant inoculated again at the 5 leaf stage resulting in a final spider mite count at harvest of 60 spider mites per leaf; also 2. a doubled spider mite inculcation level in the field (780 spider mites per plant) which should result in a spider mite count of 135 spider mites per leaf at harvest (Figure 4). The final spider mite counts in the field experiments in the literature were often less than the initial inoculation level. This is due to the problem of determining the sex of the number of spider mite infestations levels. Male's due not produce the next generation. Also, survival of the initial spider mite infestation level is not 100% due to predators.

Because of the poor relationship between initial inoculation levels and final spider mite count at harvest, the spider mite count in any field experiment must be measured after the second generation to determine what final level of infestation will be achieved in the experiment. If the second generation spider mite counts are two low compared to previous field experiments then the experiment predicted by the model will fail.

The model simulation was run for yield and biomass for the high and modeled infestation level at the end of the growing season (Table 3). The model was run for an irrigation water level of 100% of seasonal evapotranspiration (1.0 ET) representing non water stress conditions and for runs where the seasonal Et was 0.75 and 0.5 ET with uniform stress throughout the growing season with no spider mite infestation. The model was then run for a high spider mite infestation level using the regression model) and a spider mite population model set to initial conditions that generate a final spider mite count of 135 spider mites per leaf at harvest.

The model predicted that yield differences would be detectable from the control (no spider mites) at the non-water stress Et level (95 % confidence level based on a Duncan multiple range test) (Table 3) if the two levels of spider mite infestation followed the empirical high infestation level (60 spider mites/leaf) and the population model infestation levels (135 spider mites per leaf (Figure 4). The irrigation treatments of 0 0.75 Et level were statistically the same at the 90 % confidence level with final spider mite counts at 60 and 135 spider mites per leaf. When the irrigation stress was increased to 0.5 Et and the spider mite numbers increased to a final count of 135 spider mites per leaf at the end of the growing season, the yields decreased to 456 kg/ha (Table 3). The results in the model depend on the spider mite stress function presented in Figure 2. If this function is different for different cucumbers varieties then modeled yield would be different if a field experiment was not planted to Chun-Gwang Baekdadagi cucumbers

The cucumber biomass in Kg/ha was the same at the 100 % and 75% Et irrigation treatment for final spider mite counts of 60 and 135 spider mites when comparing biomass yield between the two spider mite levels but the biomass was lower for the two spider mite infestation levels for the 50% irrigation treatment. The model predicted that the spider mites had more impact on cucumber yield than cucumber total biomass. Consequently, it would be more important to measure yield at the end of the field experiment than total biomass given limited manpower resources.

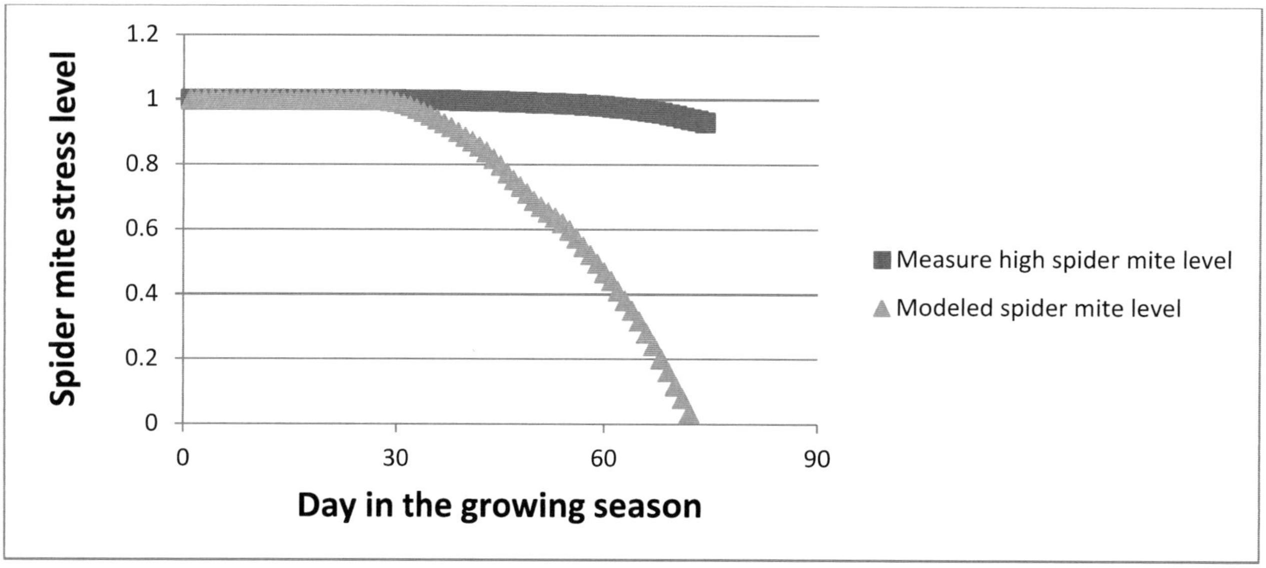


Figure 7 Spider mite stress reduction level in growth and yield for measured and simulated spider mite concentration using the measured and simulated spider mite counts in figure 4.

| Spider mite infestation level | Model prediction of Cucumber Yield kg/ha | | | Model prediction of cucumber biomass kg/ha | | |
|---|---|---|---|---|---|---|
| | Irrigation treatment 100% Et 360mm | Irrigation treatment75% Et 269 mm | Irrigation treatment 50% Et 183 mm | Irrigation treatment 100% Et | Irrigation treatment75 % Et | Irrigation treatment 50% Et |
| None | | | | | | |
| Mean | 2346 a | 1727 abc | 1017 abc | 7772 a | 5773 ab | 3704 ab |
| High infestation 60 spider mite/leaf at harvest | | | | | | |
| Mean | 2266 ab | 1708 abc | 1004 abc | 7556 a | 5595 ab | 3678 ab |
| Standard Deviation | 25 | 7 | 4 | 53 | 100 | 8 |
| Coefficient of variation % | 1.1 | 0.38 | 0.41 | 0.7 | 1.75 | 0.22 |
| Modeled infestation | | | | | | |

| | | | | | | |
|---|---|---|---|---|---|---|
| 135 spider mites/leaf at harvest | | | | | | |
| mean | 744 b c | 719 bc | 456 c | 4721 ab | 3679 ab | 2580 b |
| Standard deviation | 507 | 479 | 329 | 1852 | 992 | 652 |
| Coefficient of variation % | 68 | 67 | 72 | 39 | 27 | 25 |

Table 3 . Modeled cucumber yield and biomass under three irrigation levels and two spider mite infestation levels, one measured (high infestation) and one modeled infestation.

Given limited time and financial resources, the ground truth measurements that should be collected are ranked in Table 4 based on the output of the cucumber growth insect model and the most easily measured parameters based on the simplicity of the measurements, time involved in taking the measurements, and need of the measurements to verify the cucumber growth insect model.

To verify the plant growth insect model, a number of plant growth and mite population measurement would be needed. The frequency of how often field variables should be collected is suggested (Table 4). In addition, the data from field collected variables that are needed to verify the plant growth insect model are ranked in term of their importance: high, medium and low priority (Table 4)

| Object | Value predicted by Model | Original experimental measurements frequency | Proposed experiment measurements based on model output. | Priority of change, high medium , low |
|---|---|---|---|---|
| growth | Plant height (m) | none | Weekly measured by hand or hr by data logger | High, easy to measure |
| Growth | Plant biomass | none | End of the growing season | medium destructive sampling cannot be done during the growing season |
| Growth | Cucumber fruit yield | Weekly | Same as first design | |
| Growth | Cucumber fruit number | weekly | Same as first design | |
| Growth | Leaf number | none | weekly | High, count leaf number |
| Growth | Leaf Area Index | None | End of growing season | Medium time consuming to measure but only one measurement |
| Growth | Plant wood density | none | End of Growing season | Low, but easy to measure |
| Growth | Percent plant | None | End of the | Medium, time |

| | | | | |
|---|---|---|---|---|
| | material as leaves fruit and stems | | growing season | consuming to measure but only one measurement |
| Growth | Specific leaf area | none | Middle of growing season | High, need leaf area meter but only one subsample from plots |
| Growth/irrigation/ spider mite | Leaf damage index due to spider mite | Weekly | Same as first design | High, visual, pictures and hand held spectrophotometer |
| Growth /Irrigation | Plant Spacing | At planting | Same as first design | High |
| Irrigation | | | | |
| Irrigation | Soil texture with depth | none | Before Planting | High, take soil samples to lab |
| Irrigation | Climate data | None | Hourly | High, automated climate station |
| Irrigation | Evapotranspiration | None | Daily | Low, costly and time consuming |
| Irrigation | Soil moisture | weekly | Daily | High ,use soil moisture sensor different depths and data logger |
| Irrigation /spider mite | Crop water Stress Index. stress caused by water and spider mite damage | Weekly, from canopy temperature | Every 1 minutes | High Infrared sensors connected to data logger to measure canopy temperature |
| Spider mite | | | | |
| Spider mite | Spider mite count on leaves | Weekly | Same as first design | High , take pictures and count number using pictures |
| Spider mite | Plant Stress | weekly | Same as first design | High, Derived from leaf damage index |
| Soil Temperature | Soil temperature with depth | None | Hourly | Low, soil temperature sensors connected to data logger. |
| | | | | |

Table 4. Comparison of proposed field measurements in original experimental design compared to additional metrics that could be collected to improve final data set.

## Conclusion

The simulated experiment results shows that the water stress overrides any impact that spider mites might have and that it requires a high spider mite infestation level to detect a change in cucumber yield compared to the non-water stress treatments. The experimental design should drop the 0.75 Et treatment and increase the spider mite infestation levels from the initial design based on previous maximum high infestation level experiments. Based on the simulation model the final infestation level needs to be 2 time the high field infestation level reported in the literature to have an impact on detectable yield at the end of the growing season.

A model has a large number of simulated output measurements that cannot be duplicated in the field experiment because of a limitation of man power, money and time. But as many measurements as is possible should be conducted in the experiment to validate a growth insect model and provide experiment data for future model development and field crop management decisions.

## Reference


Acock B., and Reynolds J.F., 1989. The rationale for adopting a modular generic structure for crop simulators. Acta Horticulturae 248:391-396.

Allen, R. Luis Pereira Dirk Raes Martin Smith, 1988 Crop Evapotranspiration Guidelines for Computing crop water requirements. FAO irrigation and drainage paper 56.

Allen, Richard G., Ivan A. Walter, Ronald L. Elliott, Terry A. Howell, Daniel Itenfisu, Marvin E. Jensen, Richard L. Snyder, 2005 The ASCE Standardized Reference Evapotranspiration Equation. American Society of Civil Engineers /Irrigation Association Reston Va Prepared by the Task Committee on Standardization of Reference Evapotranspiration of the Environmental and Water Resources Institute of ASCE. Available on line at http ://www.kimberly.uidaho.edu/water/asceewri/ascestzdetmain2005. Pdf)

Andales, A., Wang, J., Sammis, T.W., Mexal, J.G., Simmons, L.J., Miller, D.R., 2006. A model of pecan tree growth for the management of pruning and irrigation. Agricultural Water Management 84 (1–2):77–88.

Atanassov, Nikola 2014. Effect of the Spider Mite Tetranychus Urticae Koch (Acarina:Tetranychidae) on Cucumber Yield. Biotechnology & Biotechnological Equipment ISSN: 1310-2818 (Print) 1314-3530 (Online) Journal homepage: http://www.tandfonline.com/loi/tbeq20

Baret, F. and T Fourty. 1997. Estimation of leaf water content and specific leaf weight from reflectance and transmittance measurements. Agronomie, EDP Sciences, 17 (9-10):455-464

Clemson Cooperative extension 2017. Cucumber, squash, melon and other cucurbit insect pests. http://www.clemson.edu/extension/hgic/pests/plant_pests/veg_fruit/hgic2207.html

Tsuji, G. Y. , G. Uehara, S. Balas. 2017 DSSAT V3 http://dssat.info/wp-content/uploads/2011/10/DSSAT-vol3.pdf

Dell, Ralph B, Steve Holleran, Rajasekhar Ramakrishnan 2002.Sample Size Determination ILAR J 43 (4): 207-213. https://www.ncbi.nlm.nih.gov/books/NBK43321/

Doorenbos, J. and W. O Pruitt. 1977. Crop Water Requirements . FAO Irrigation and Drainage paper 24 pages 144

Allen, R. G. , L. S. Pereira, D. Raes, M. Smith 2017. Crop evapotranspiration –Guidelines for computing crop water requirements. Fao irrigation and drainage paper 56. http://www.fao.org/docrep/x0490e/x0490e0b.htm#crop%20coefficients

Farmer's Almanac 2016. The Farmer's Almanac http://www.almanac.com/gardening/frostdates/NM/Albuquerque

Gagandeep Kaur Sidhu and Sandhya 2015.Engineering properties of cotton stalks ( Gossypium hirsitum L.). Indian J Agric Res 4995):456-459

Ganio LM. 1997. Designing a nutrient study. In: Haase DL, Rose R, coordinators and editors. Symposium proceedings, forest seedling nutrition from the nursery to the field; 1997 Oct 28-29; Corvallis, Oregon. Corvallis (OR): Oregon State University Nursery Technology Cooperative. p 88-1 00.

Sandquist RE, Owston PW, Haifa 2017.Nutritional recommendations for: Cucumber in open fields, tunnels and greenhouse. http://www.haifa-group.com/files/Guides/Cucumber.pdf

HealthyEating 2017. List of Fruits & Vegetable With a High Water Content. http://healthyeating.sfgate.com/list-fruits-vegetable-high-water-content-8958.html

Haque, M.M., Mirza Hasanuzzaman and M.L. Rahman 2009. Morpho-Physiology and Yield of Cucumber (Cucumis sativa) Under Varying Light Intensity Academic Journal of Plant Sciences 2 (3): 154-157

Hodges, T., S. L. Johnson and B. S. Johnson. 1991 A modular Structure for crop simulation models: implemented in the SIMPOTATO Model. Agron. J 84:911-915

Jones, J.W, , G. Hoogenboom, C.H. Porter, K.J Boote, W.D Batchelor, L.A Hunt, P.W Wilkens, U. Singh, A.J Gijsman, J.T Ritchie. 2003. The DSSAT cropping system model. European Journal of Agronomy Modelling Cropping Systems: Science, Software and Applications,18(Issues 3–4): 235–265

NC extension 2016. Commercial production of pickling and slicing cucumbers inNnorth Carolina https://content.ces.ncsu.edu/commercial-production-of-pickling-and-slicing-cucumbers-in-north-carolina

Nyoike, T. W, and O. E. Liburd2013. Effect of Tetranychus urticae (Acari: Tetranychidae), on Marketable Yields of Field-Grown Strawberries in North–Central FloridaJ. Econ. Entomol. 106(4): 1757-1766

Park, Y. L. and J.H. Lee. 2005. Impact of Two spotted Spider Mite (Acari: Tetranychidae) on Growth and Productivity of Glasshouse Cucumbers J Econ. Entomol 98 (2): 457-463

Park, Y.L. and J.H. Lee, 2002. Leaf cell and tissue damage of cucumber caused by two spotted spider mite (Acari:Tetranychidae), *J. Econ. Entomol.*, 95: 952–957.

Park, Y. L., and J. H. Lee 2007 Seasonal dynamics of economic injury levels for Tetranchus urtica Koch (Acari, Tetranychidae on Cucumis Saltivus L . Journal of applied Entomology 131(8) :558-592.

Perry, A. S., Y. I., Ishaya, I. and Perry, R. .1998. Insecticides in Agriculture and Environment – Retrospect and Prospects. Narosa Publishing House, New Delhi, India. p.

261.

Perry K. B., and T. C. Wehner. 2016. A Heat Unit Accumulation Method for Predicting Cucumber Harvest Date
http://cuke.hort.ncsu.edu/cucurbit/wehner/articles/art053.pdf

Rabbinge, R. and M. A. Hoy. 1980.
A population model for two spotted spider mite Tetranychus Urticae and its predator Metaseiulus Occidentalis. Entomoby Experimentalls et Applica. 28 (1):63-81

Reisig, D. and L. Godfrey. 2007. Spider Mite– (Acari: Tetranychidae) Infested Cotton: Controlled Studies
Environ. Entomol. 36(6): 1466-1474

Reynolds, J.F. and B. Acock. 1997. Modularity and genericness in plant and ecosystem models. Ecological Modeling 94: 7-16.

Sammis, T., V. Gutschick, J. Wang, D.R. Miller. 2013. Model of water and nitrogen management in pecan trees under normal and resource-limited conditions. Agricultural Water Management 124: 28–36

Sammis, T., P. Sharma, M.K. Shukla, J. Wang, D. Miller. 2012. A water-balance drip-irrigation scheduling model. Agric. Water Manage. (113):30-37

Sammis, T., M. K. Shukla, R. D. French, J. Wang 2012b.
Modeling Phytophthora Disease Development in Chile and Bell Peppers under rain feed and irrigated conditions. Irrigation association proceedings Irrigation show and education conference Orland Florida 2012 Agricultural Track 1:1-18

Sammis, T. V. Gutschick, J. Wang, D.R. Miller. 2013 Model of water and nitrogen management in pecan trees under normal and resource-limited conditions Agricultural Water Management 124:28– 36

Schroder, 2017. What is the advantages of object oriented programing
https://www.researchgate.net/post/What_is_adventages_of_Object_Oriented_Programming

Spectrum Technologies 2016. If plants could talk. http://www.specmeters.com/newsletter/if-plants-could-talkabout-irrigation/

Schrader, W. L., J. L Aguiar, K. S. Mayberry. 2017. Cucumber Production in California Publication 8050 University of California Agriculture and Natural Resources http://anrcatalog.ucanr.edu/pdf/8050.pdf

Sharma, P., M. K. Shukla, T. W. Sammis 2010. Predicting soil temperature using air temperature and soil, crop, and meteorological parameters for three specialty crops in southern New Mexico. Applied Engineering in Agriculture. 26(1): 47-58.

USDA 2017. United States Department of Agriculture Economics, Statistics and market information system . http://usda.mannlib.cornell.edu/usda/current/VegeSumm/VegeSumm-02-04-2016.pdf

Veggieharvest 2016. Cucumber growing and harvest information. http: //veggieharvest.com/vegetables/cucumber.html

Albuquerque Weather Data. 2017. Weather Station Pasitos del Ciero Id.KNMLOSU13. https://www.wunderground.com/personal-weather-station/dashboard?ID=KNMLOSLU13#history/tdata/s20160123/e20170123/myear

WIFSS (Western institute for food safety and security) 2017 Cucumbers https://www.wifss.ucdavis.edu/wp-content/uploads/2016/10/Cucumbers_PDF.pdf

Wortman, S. E. and S. T. Lovell. 2013. Environmental challenges threatening the growth of urban agriculture in the United States. JEQ 42(5):1283-1294

Yaghi, T. , A. Arslan, F. Naoum. 2013. Cucumber (Cucumis sativus, L.) water use efficiency (WUE) under plastic mulch and drip irrigation. Agricultural Water Management128:149–157

Yuan , Bao-Ahong, Jie Sun , Yaohu Kang, Soichi Nishiyama 2016. Response of cucumber to drip irrigation water under a rainshelter. Agricultural Water Management 81(1-2):145-158 ·

# Injectable co-polymers: A tool for soil moisture management

**Bardia Dehghanmanshadi**
Exacto, Inc., 200 Old Factory Rd., Sharon, WI 53585; Bardia@exactoinc.com

**Glen R. Obear**
Exacto, Inc., 200 Old Factory Rd., Sharon, WI 53585; gobear@exactoinc.com

**Frank Sexton**
Exacto, Inc., 200 Old Factory Rd., Sharon, WI 53585

**Abstract.** *Polymers have been used as soil amendments in agriculture for several decades. Some polymer chemistries are specialized for improving soil moisture distribution uniformity, while others are specialized for improving soil water retention. In this study, we combined two distinct polymers to provide both functions, and applied them to newly-transplanted Valencia oranges to evaluate their effects on soil moisture and plant health. Application of the co-polymer resulted in increases in tree height and trunk diameter in 2016 and 2017. Sap flow sensors indicated that a treated tree receiving the same amount of irrigation used approximately 30% more water throughout the 2017 growing season, indicating an increase in irrigation efficiency. Soil moisture data from 2017 show that the co-polymer resulted in more moisture at 12" (30 cm) and 24" (61 cm) depths. The results of this study provide evidence that co-polymers can be useful tools for soil moisture management, allowing growers to either decrease irrigation amounts or increase water-use efficiency with similar inputs. The co-polymer system investigated in this study may be especially useful to growers due to its ease of application through various types of irrigation systems.*

**Keywords.** Chemigation, water-use efficiency, deficit irrigation, irrigation management, water conservation

## Moisture-Retention Polymers in Agriculture

Polymers have been used as soil amendments in agriculture for several decades, and different classes of polymer chemistry have different unique functions (Milani et al., 2017). Some polymers are specialized for decreasing surface tension and improving moisture distribution uniformity in the soil, while others are specialized for absorbing and retaining water in the soil. Moisture-retaining polymers, including polyacrylamide and polyacrylate, have well-proven benefits for soil conditioning and water management, but their application in commercial agriculture has been limited because they are often difficult to apply (Sojka et. al., 2007). The objective of this study was to evaluate soil moisture dynamics and plant health effects from application of a co-polymer (AquiMax®, Exacto, Inc., Sharon, WI) containing an inverse micro-emulsion polyacrylamide formulated with an EO/PO block co-polymer surfactant.

## Materials and Methods

Valencia oranges on Carrizo citrange root stocks were transplanted in October of 2015 in Sanger, CA at a facility managed by SynTech Research (Fig. 1). Trees were irrigated via microsprinkler at 15 PSI with BowSmith Fan-Jet Style-J2 nozzles. The experimental control at 100% grower standard irrigation utilized size 35 nozzles which had an output of 7.3 gal $h^{-1}$ (27.6 L $h^{-1}$). Experimental treatments were evaluated on trees irrigated with size 30 nozzles, which had an output of 5.2 gal $h^{-1}$ (19.7 L $h^{-1}$), approximately 30%

less water than the grower standard (GS) control. In July of 2016, a study was initiated with a randomized compete block design with six replications, where each plot was represented by a single tree. The trees were treated with co-polymer at one of two rates and compared to an untreated control (Table 1) for a total of two applications in 2016 and two in 2017. Treatments were injected directly into microsprinkler lines over a period of 1-3 hours and watered in with approximately 0.25" of water after the applications to clear sprinkler lines and move the product downwards in the soil. The co-polymer was injected neat, and not pre-diluted with water prior to the application.

Figure 1. Valencia orange grove where study area was located in Sanger, CA.

Table 1. Irrigation quantities and application rates for trial on Valencia orange transplants.

| Irrigation Amount | ---------------------- Co-Polymer Rate ---------------------- | |
|---|---|---|
| | 1st Application[a] | 2nd Application[a] |
| 100% GS[b] | - | - |
| 70% GS | - | - |
| 70% GS | 1 gal $ac^{-1}$ (9.3 L $ha^{-1}$) | 1 gal $ac^{-1}$ (9.3 L $ha^{-1}$) |
| 70% GS | 2 gal $ac^{-1}$ (18.6 L $ha^{-1}$) | 1 gal $ac^{-1}$ (9.3 L $ha^{-1}$) |

a. Application Dates: 8/10/16, 8/26/16, 6/6/17, 8/3/17
b. GS, grower standard

Tree height and trunk diameter were measured prior to study initiation, and then 2-3 times per year through 2016 and 2017. Data were subjected to Analysis of Variance (ANOVA) to determine whether there were differences among treatments, and means were separated by Fisher's protected LSD.

In June 2017, sap flow sensors were installed on one tree for the 70% irrigation control, and the 70% irrigation co-polymer receiving two applications of 1 gal $ac^{-1}$ (9.3 L $ha^{-1}$) annually. Sap flow was measured with the "SapIP" systems and sensors (Dynamax, Inc., Fresno, CA), which use heat flux between two locations on the trunk to calculate sap flow every 15 minutes. The sap flow values were normalized to account for differences in tree size. Volumetric water content was measured beginning August 1, 2017 by time-domain reflectometry and logged to a datalogger every 15 minutes at 12" (30 cm) and 24" (61 cm) adjacent to the same trees where sap flow measurements took place.

## Results and Discussion

Four total application of the co-polymer were made in 2016 and 2017. We did not observe issues or difficulties related to product viscosity during any of the applications.

Experimental treatments resulted in numerical differences in trunk diameter and tree height from 7/7/16 through 9/6/17, although the differences were not statistically significant (Table 2). There was a trend towards greater trunk diameter and tree height with application of co-polymer at both 1 + 1 gal $ac^{-1}$ (9.3 + 9.3 L $ha^{-1}$) and 2 + 1 gal $ac^{-1}$ (18.6 + 9.3 L $ha^{-1}$) with 70% GS irrigation, and these increases were similar to those obtained in the 100% GS irrigation. These findings suggest that growers can achieve similar rates of tree growth on new stands with 30% less water with application of co-polymer at 2 gal $ac^{-1}$ (18.6 L $ha^{-1}$) per growing season. The increase in growth in co-polymer treatments at 70% GS irrigation suggest an increase in water-use efficiency.

Table 2. Trunk diameter and tree height as affected by experimental treatments from 7/7/16 to 9/5/17.

| Irrigation | Co-Polymer Rate | Trunk Diameter (in) | | | Tree Height (in) | | |
|---|---|---|---|---|---|---|---|
| | gal $ac^{-1}$ | 7/7/16 | 9/5/17 | % Change[a] | 7/7/16 | 9/5/17 | % Change[a] |
| 100% GS[b] | - | 0.830 | 1.345 | 62.20% | 28.50 | 52.77 | 85.20% |
| 70% GS | - | 0.903 | 1.356 | 50.20% | 33.17 | 54.79 | 65.20% |
| 70% GS | 1+1 | 0.839 | 1.414 | 68.50% | 29.50 | 56.90 | 92.90% |
| 70% GS | 2+1 | 0.845 | 1.395 | 65.10% | 27.25 | 53.56 | 96.50% |
| P<0.05 | | | | 0.280 | | | 0.352 |
| LSD | | | | 17.36 | | | 31.42 |

a. From 7/7/16 through 9/5/17
b. GS, grower standard

Volumetric water content (VWC) of the soil was generally greater at 24" (61 cm) than 12" (30 cm), with the exception of the first several days of measurement between 8/1/17 and 8/3/17. During this stretch, the volumetric water content of the treated plot was over twice as high at 12" (30 cm) compared to 24" (61 cm), whereas the untreated control had more moisture at 24" (61 cm) than 12" (30 cm). Plots had been lightly-irrigated in the preceding weeks, so it is possible that the irrigation events were insufficient to wet the soil profile below a 12" (30 cm) depth for the treated plot. The co-polymer has been shown to decrease saturated hydraulic conductivity and improve lateral water movement, so it is possible that these factors prevented the 24" (61 cm) depth from wetting in the early period of measurement. Deeper irrigation cycles resumed after 8/3/17 and after that point, moisture was typically greater at 24" (61 cm) than 12" (30 cm) for all treatments.

The VWC of plots treated with copolymer was greater than the untreated control for most of the measurement period. To compare the total water content of each plot, the area under the moisture curve (AUMC) was calculated by summing the total area between each set of data points, which was calculated by the following equation:

$$\text{[Eq. 1]} \qquad \frac{((T_2 - T_1)x\ (VWC_B - VWC_A))}{2}$$

where $T_2$-$T_1$ is the different in minutes between two consecutive measurements, and $VWC_B$-$VWC_A$ is the difference in VWC between two consecutive measurements. The combined AUMC for the co-polymer-treated plot at both depths was 224,599 (unitless) compared to 195,497 in the untreated control, which is a 15% increase in AUMC.

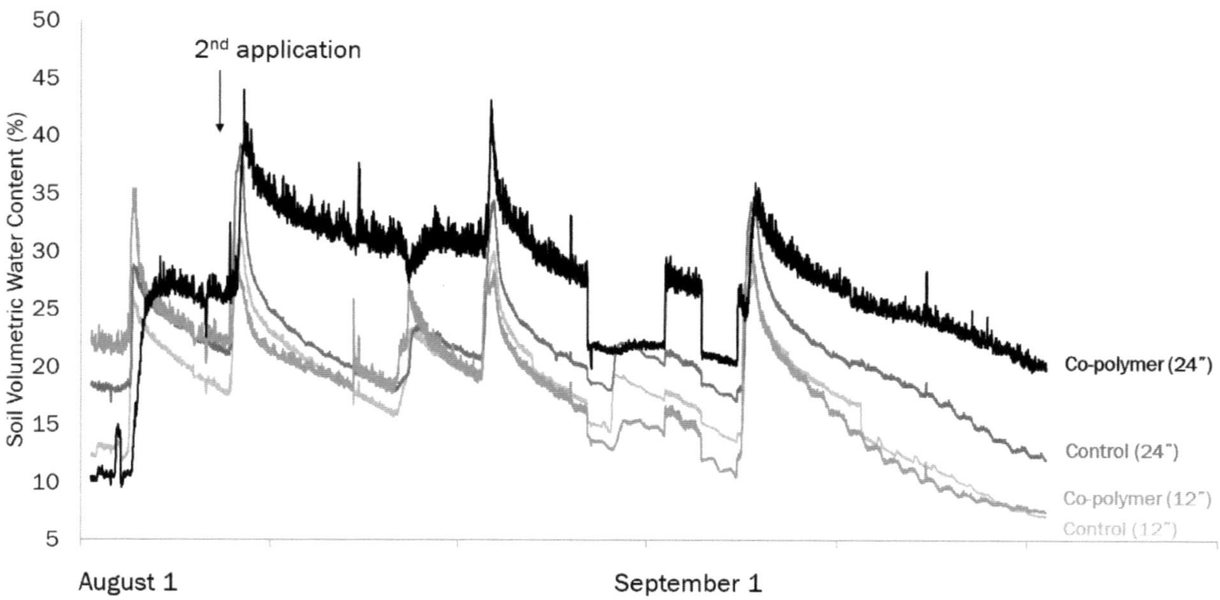


Figure 2. Soil moisture data from August 1 through September 22, 2017.

The sap flow in the co-polymer-treated tree was greater than that of the untreated control on most observation dates in 2017 (Fig. 3). In the month of July, the treated tree used an average of 1.434 gal $day^{-1}$ (5.428 L $day^{-1}$) compared to 1.113 gal $day^{-1}$ (4.213 L $day^{-1}$) in the control, an increase of 28.8%. In Aug., the treated tree had an average sap flow of 1.683 gal $day^{-1}$ (6.371 L $day^{-1}$) compared to 1.496 gal $day^{-1}$ (5.663 L $day^{-1}$) in the control, an increase of 12.5%. The lower difference between co-polymer and the control in Aug. might have been attributed to more frequent irrigation events and higher soil water content during this period. In Sept., the treated tree had an average sap flow of 1.630 gal $day^{-1}$ (6.170 L $day^{-1}$) compared to 0.973 gal $day^{-1}$ (3.683 L $day^{-1}$) in the control, an increase of 67.5%. Between 7/5/17 and 9/22/17, the treated tree transpired a total of 128.73 gal (487.30 L) of water compared to 97.55 gal (369.27 L) in the control, an increase of 32.0%. Given that both trees were irrigated with same amount of water throughout the growing season, this difference indicates an increase in irrigation efficiency. The lower difference during the month of Aug. when soil moisture was generally greater indicates that benefits of the co-polymer will be more pronounced during periods of crop water stress.

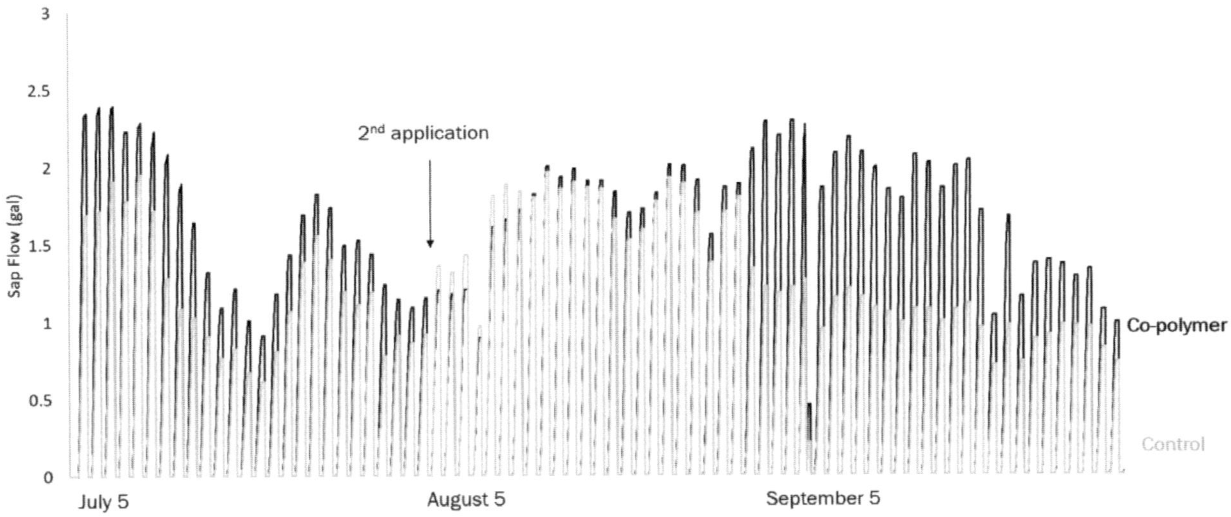


Figure 3. Cumulative sap flow from July 5, 2017 through September 22, 2017. Values were truncated to zero at midnight each day.

The soil moisture and sap flow data show that the co-polymer increased the amount of water held in the soil profile, and this translated to an increase in transpiration. The morphological measurements of the trees indicate that the injectable co-polymer had a positive effect on tree health and increased the rate of growth of these young orange transplants. These findings are relevant for citrus growers who may be replanting orchards after damage from citrus greening or weather-related events. In addition to improving tree health, the results suggest that the co-polymer could allow growers to either cut back on irrigation volumes while achieving similar results, or increase irrigation system and water-use efficiency with the same amount of irrigation. This is particularly useful for regions where water is scarce and periodic droughts limit water reserves.

## Conclusion

The results of this study show that injectable co-polymer systems are a useful water-management tool for tree crops. Injection of the co-polymer provided an increase in trunk diameter and tree height from July 2016 to September 2017, an increase in soil moisture at 12" (30 cm) and 24" (61 cm), and an increase in irrigation system and water-use efficiency. The data from this study suggest that the injectable co-polymer positively affects tree health during non-bearing years and increases the rate of maturation of newly-transplanted orchards. Future research must continue to look at non-bearing trees as they mature and begin to bear fruit to determine whether these improvements in plant health translate to improvements in orange yield.

## References

Milani, P., Franca, D., Balieiro, A.G., and Faez, R. 2017. Polymers and its applications in agriculture. Polímeros. Available at: http://www.scielo.br/scielo.php?pid=S0104-14282017005006108&script=sci_arttext (Verified 19 October, 2017).

Sojka, R.E., Bjorneberg, D.L., Entry, J.A., Lentz, R.D., and Orts, W.J. 2007. Polyacrylamide in agriculture and environmental land management. Adv. Agron. 92:75-162.

# STRATEGIES FOR TARO (*Colocasia esculenta*) IRRIGATION

**Gustavo H. S. Vieira**. Professor. PhD.
IFES *campus* Santa Teresa, Rod. ES 080, km 93, São João de Petrópolis, Santa Teresa-ES, Brazil. 29.660-000. e-mail: ghsvieira@gmail.com

**Guilherme Peterle**. Agronomy undergraduate student.
IFES *campus* Santa Teresa – Rod. ES 080, km 93, São João de Petrópolis, Santa Teresa-ES, Brazil. 29.660-000.

**Jéssica B. Loss.** Agronomist, MS Vegetal Production.

CCA-UFES, Alegre-ES. Brazil. 29.500-000.

**Gabriel Peterle**. Agronomist. MS in reading, Agricultural Meteorology.
Departamento de Engenharia Agrícola – Universidade Federal de Viçosa, Viçosa - MG. Brazil. 36570-000

**Carlos Magno M. Poloni**. Agronomy undergraduate student.

IFES *campus* Santa Teresa – Rod. ES 080, km 93, São João de Petrópolis, Santa Teresa-ES, Brazil. 29.660-000.

**João N. Colombo.** Professor, PhD.

IFES *campus* Santa Teresa – Rod. ES 080, km 93, São João de Petrópolis, Santa Teresa-ES, Brazil. 29.660-000.

**Paola Alfonsa V. Lo Monaco**. Professor, PhD.

IFES *campus* Santa Teresa – Rod. ES 080, km 93, São João de Petrópolis, Santa Teresa-ES, Brazil. 29.660-000.

**Abstract**: This study aimed to evaluate the taro (*Colocasia esculenta*), var. São Bento, in response to different irrigation strategies. The experiment was carried out at Ifes *campus* Santa Teresa, Brazil, at an altitude of 130 meters above sea level, using a randomized block design (RBD) with five treatments for the water availability factor of culture (f factor) equivalent to 0.1; 0.2; 0.3; 0.4 and 0.5, and four replications. Meteorological data were used to estimate the crop water demand, performing daily water balance using spreadsheets. We evaluated the applied water depth, the yield of commercial cormels and the water use efficiency by taro, due to the f factor. The results were submitted to analysis of variance and regressions. Increasing the f factor provided a reduction of applied irrigation depths. Lighter and frequent irrigations improved the development and yield of taro and can be recommended for its management.

**Keywords:** f factor, taro, cocoyam, yield, efficiency

# INTRODUCTION

The taro (*Colocasia esculenta*), also known as cocoyam, is of great economic and social importance in many tropical and subtropical regions of the world, occupying a prominent place in the diet of many people in these regions (Huang et al., 2007). In Brazil, it is grown mainly in the states of the Mid-South. It is present in almost all the municipalities of Minas Gerais State, which is the state's largest producer of the country and it is explored in family farming. Taro presents fleshy cormels, with similar nutritional value of potato tubers (Brasil, 2010).

World production of taro in 2013 reached 9.976 million tons grown on 1.299 million hectares, and Nigeria was responsible for about 3.450 million tons, followed by China with 1.845 million tons and Cameroon with 1.551 million tons (FAO, 2014). However, the FAO estimates that only 145,000 tons are sold and the remainder is used in the food base of these people. Recent data are not found on the production of taro corms on the national scene of Brazil, but it is estimated that this is around 200,000 tons.

Despite the potential that culture has to develop in excess moisture environments, excessive shading and other climatic stress (Heredia Zárate, 1995), there is still a lack of information about it. Even the need for water for the culture is not consolidated, and most often is not linked to technical criteria, resulting in unnecessary expenditure of energy for irrigation and waste of water or under-utilization of irrigation system.

The expansion of irrigated areas, currently held, this tied to concerns about the availability of water for agriculture and energy costs associated with practice. Consequently, the adoption of strategies aimed to reduce the waste of water, without incurring losses in productivity (Coelho et al., 2005), become essential to the efficient use of these water and energy resources.

According to Doorenbos and Kassam (1979), the productivity of a culture is a function of complex biological, physiological, physical and chemical processes, which are determined by environmental conditions and genetic factors of culture itself. Thus, the realization of irrigation tends to raise the productivity of crops. However, only irrigation without an efficient management to consider the cultural and physiological aspects of the species, physical, chemical and physicochemical soil and particularities of the adopted irrigation system (Mantovani et al., 2007), just not provide a significant increase in productivity, but an expressive reduction of the costs associated with the cultivation.

Proper and strategic water management can be done using the water use efficiency (WUE) index for planning and irrigation decision-making, increasing the crop yield (Karatas et al., 2009). Thus, a good management strategy of irrigation is essential to save water without, however, endangering the crop yield (Jalota et al., 2006; Pereira et al., 2009).

In the case of taro, there is a shortage of technical information of Brazilian soils and climate conditions, and most of the work reported in researches comes from other countries (Gondim et al., 2007; Mabhaudhi et al., 2013). As regards, the water requirement of this culture is lack of clear information and it is not being known about water requirements in their phenological phases to conditions of Brazilian agricultural ecosystems. Therefore, the present study was due to evaluate the response of taro to different strategies of irrigation (f factors).

## MATERIAL AND METHODS

The experiment was conducted in the Vegetable Crops sector of the Instituto Federal do Espírito Santo, *campus* Santa Teresa, Espírito Santo state, Brazil, as shown in Figure 1. The area is located in the coordinates latitude 19º48'36" south and longitude 40º40'48" west, has an altitude of 130 m above sea level, with soil classified predominantly as Latosol Yellow Eutrophic clayey, containing 63% of clay in its composition. In the area where the experiment was carried out, the field capacity (FC) is 32%; the permanent wilting point (PMP) is 20.6%; the bulk density is 1.16 g $cm^{-3}$ and the root depth of the culture was 40 cm.

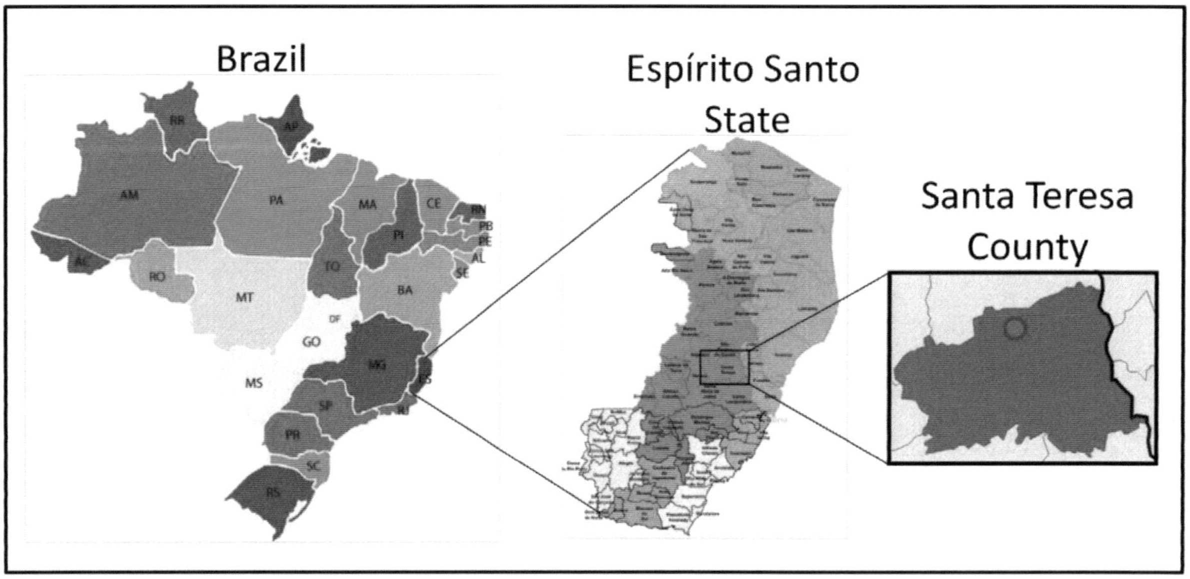


**Figure 1.** Localization of experimental area in Santa Teresa County, Espírito Santo State, Brazil.
**Camaravni (2017); Thinglink (2017); Wikipedia (2017). ** Map assembly made by the author.*

It was installed a drip irrigation system, divided in sectors, to irrigate the plots individually, according to the treatments. The irrigation periods were determined according to the water availability for the culture (f factor), established for each treatment, as shown in Table 1. Thus, irrigation was performed when the water in the soil was depleted equivalent to 10; 20; 30; 40 and 50% of the total water available, respectively, for the treatments T1, T2, T3, T4 and T5, in a randomized block design, with four replications. Figure 2 shows a part of the experimental area.

**Figure 2.** Taro cultivated in the experimental area, localized in the Instituto Federal do Espírito Santo, Santa Teresa *campus*.

The treatments were irrigated at different times, so when the irrigations occurred, the amount of water in the soil was different, which gave different soil water deficits for each treatment, as shown in Table 1.

**Table 1.** Treatments and their respective f factors and water deficits, in mm, at the time of irrigation.

| Treatment | T1 | T2 | T3 | T4 | T5 |
|---|---|---|---|---|---|
| **f factor** | 0.1 | 0.2 | 0.3 | 0.4 | 0.5 |
| **Water deficits (mm)** | 7,3 | 14,6 | 21,9 | 29,2 | 36,5 |

The irrigation levels were determined, individually for each treatment, using spreadsheets, that calculated the water balance with climate measurements (maximum and minimum temperatures and the rain) and sporadic determination of soil water content with the stove method. The calculation of culture evapotranspiration is given by Equation 1 (Mantovani et al., 2009; Vieira et al., 2015). The reference evapotranspiration was determined by Hargreaves and Samani method (Allen et al., 1998). The single crop coefficient (Kc) used for taro was proposed by Fares (2008), which suggests the following values: Kc initial: 1.05 in the first two months planting; Kc mid: 1.15, the second to the sixth month after planting and Kc final: 1.1, the sixth month until harvest.

$$ETc = ET_0 \, K_C \, K_S \, K_L \quad (1)$$

Where:

$ET_C$ - crop evapotranspiration, mm $d^{-1}$;

$ET_0$ - reference evapotranspiration, mm $d^{-1}$;

$K_C$ – single crop coefficient, dimensionless;

$K_S$ – water stress coefficient, dimensionless;

$K_L$ - correction factor due to the drip irrigation, based on the wetted and shaded area, dimensionless.

The irrigation system was assessed following the method proposed by Keller and Karmeli (1975) and determined the Distribution Uniformity coefficient (DU) of the irrigation system. These authors suggest collecting flow at 4 points along the lateral line, i.e., the first dripper, the emitters located at 1/3 and 2/3 of the length of the line and the last dripper. The lines selected within the sector should be: first, those located at 1/3 and 2/3 of the length and the last lateral line, assessing 16 values as a whole.

It was cultivated taro var. São Bento. Before the implementation of the crop, the experimental area was prepared and fertilized. Taro corms were used for the implementation of the experiment and they were distributed in the planting furrows so that the yolks placed face up. Then the corms were covered with soil until about two centimeters above the apex of the yolks.

The yield performance was evaluated from the average productivity of cormels obtained in treatments, and quantified the production of commercial cormels. The classification of the cormels was performed according to the recommendation of Puiatti et al. (1990), from the transverse diameter, being large (> 47 mm), medium (33-46mm), and little cormels (<33mm). For the aggregation of commercial cormels it was considered the sum of large and medium cormels.

The water use efficiency was determined by the ratio of the values of productivity (t $ha^{-1}$) and the respective applied irrigation depth (mm), in each treatment, and the results expressed in kg $m^{-3}$ as cited by Sammis (1980) and Vieira et al. (2015).

The data were submitted to analysis of variance and regressions, and the coefficients were analyzed with 1% level of probability by the F test.

# RESULTS AND DISCUSSION

Figure 3 shows the relationship between the total water depths depending on the soil water depletion. It is observed that the increase of the f factor has provided a reduction of the total water depth applied to the culture. At T1 treatment, with f factor 0.1, the total water depth was 454 mm and at T5 treatment, the largest f factor, 397 mm was demanded by the culture, approximately 12.6% of difference. This occurs because for higher f factor values, the intervals between irrigations (irrigation frequency) become larger, which promotes greater soil water depletion by reducing its moisture and consequent reduction in evapotranspiration and therefore lower replacement water through irrigation.

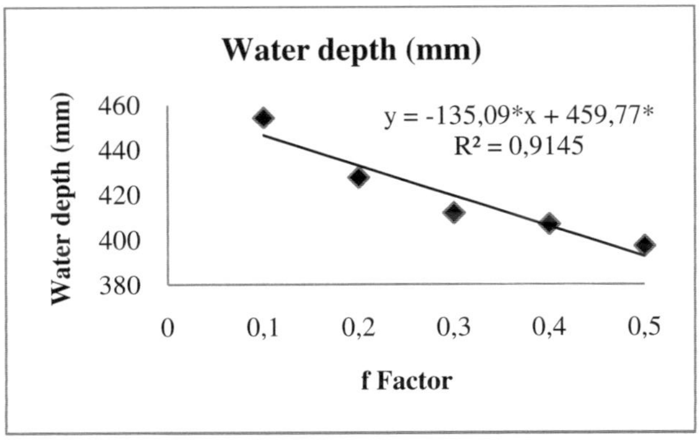


**Figure 3.** Total water applied in different irrigation frequencies (f factor). *Significant at the 1% level of probability by the F test.

In Figure 4 the yield of taro in relation to f factors. It is observed that the increase in the irrigation intervals provided a reduction of yield in commercial cormels.

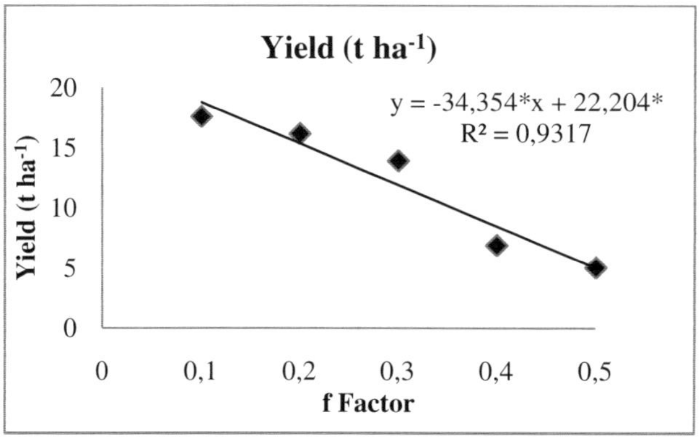


**Figure 4.** Average yield of taro cultivar "São Bento" under different irrigation frequencies (f factor). *Significant at the 1% level of probability by the F test.

The T1 and T2 treatments had the highest productivity among all tested. This occurred because with the high frequency of irrigation, the plants were subjected to less or no water stress, unlike the T4 and T5 treatments that received the largest individual irrigation levels, but in more distant periods, which may be caused drought stress and thus reduced plant yield.

The T5 treatment plants had few or no large cormel, predominantly medium and small cormels, in addition to the mother cormel, occurring cormels damaged by rot, which are not viable to trade. The T1 and T2 treatments presented a number of large and medium-sized cormels in a satisfactory amount for each class, with low number of small and damaged cormels.

It is observed a tendency to increase yield by reducing the irrigation interval. The highest yield of commercial cormels was achieved with the treatment T1, and the observed mean of 17.55 t $ha^{-1}$. For the treatment T5 the average yield achieved was 5.01 t $ha^{-1}$, 71.4% lower than the first. As cited by Caesar (1980), stress conditions such as lack of water led to slow growth and retarded development of cormels, as in this experiment.

In Figure 5, is observed the water use efficiencies in relation to f factors. The greater efficiency can be seen at T1, then T2 with values of 3.87 and 3.77 kg $m^{-3}$, respectively. In these treatments the soil remained wetted between intervals of irrigation, which provided better conditions for the development of culture and to achieve high yield. Even applying greater water depths, it is recommended to apply lighter and frequent irrigations for taro cultivation.

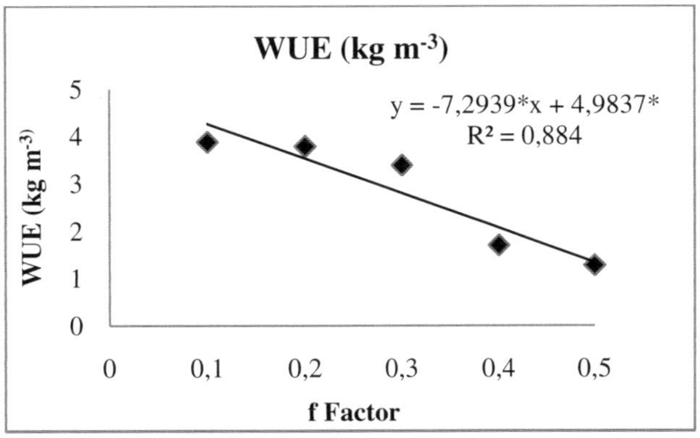


**Figure 5.** Water Use Efficiency (kg $m^{-3}$) of taro cultivar "São Bento" under different irrigation frequencies (f factor). * Significant at the 1% level of probability by the F test.

According to Vicente and Vicente (2004) and Salomão et al. (2014), when the irrigation interval is too long, the friability of the soil immediately obtained after irrigation undergoes a progressive alteration and, depending on the texture, can at the end of the period, present hard consistency, hindering the penetration of roots of cultivated plants. The local soil used has a high clay content, with greater consistency at lower humidity levels. In this case it is important to increase the frequency of irrigation, establishing temporarily lower irrigation frequency, a condition that will keep the moist topsoil for a longer period of time.

# CONCLUSIONS

Increasing the f factor provided a reduction of applied irrigation depth. Lighter and frequent irrigations favor the development and productivity of taro and can be recommended for their management.

# ACKNOWLEDGMENT

To Ifes *campus* Santa Teresa for the financial, operational and infrastructure available for the experiment and for providing scholarship to the students.

# REFERENCES

Allen, R.G.; Pereira, L.S.; Raes, D.; Smith, M., 1998. Crop evapotranspiration: Guidelines for computing crop water requirements, FAO irrigation and drainage paper 56. Food and Agriculture Organization of the United Nations, Rome.

BRASIL. Manual de hortaliças não-convencionais. Ministério da Agricultura, Pecuária e Abastecimento. Secretaria de Desenvolvimento Agropecuário e Cooperativismo. – Brasília: Mapa/ACS, 2010. 92 p.

CAESAR, K. 1980. Growth and development of *Xanthosoma* and *Colocasia* under different light and water supply conditions. Field Crops Research, 3: 235-244.

CAMARAVNI. Espírito Santo State map. Available in: <http://www.camaravni.es.gov.br/mapa_estado.asp>. (Accessed: Oct. 2017).

COELHO, E. F.; COELHO FILHO, M. A.; OLIVEIRA, S. L. 2005. Agricultura irrigada: eficiência de irrigação e de uso de água. Eunápolis: Bahia Agrícola, 7 (1).

DOORENBOS, J.; KASSAM, A.H. 1979. Yield response to water. Rome: FAO, 306p.

FAO. Food and Agriculture Organization of the United Nation. 2014. Estatistics Division: Prodution Taro. Available in: http://faostat3.fao.org/download/Q/QC/E . (Accessed: Aug. 2016).

FARES, A. 2008. Water management software to estimate crop irrigation requirements for consumptive use permitting in Hawaii. Final Report, University of Hawai'i at Mãnoa. Available in: http://files.hawaii.gov/dlnr/cwrm/publishedreports/PR200808.pdf. (Accessed: Apr. 2014).

GONDIM A. R. O.; PUIATTI M.; CECON P. R.; FINGER F. L. 2007. Crescimento, partição de fotoassimilados e produção de rizomas em taro cultivado sob sombreamento artificial. Horticultura Brasileira, 25 (3): 418-428,

HEREDIA ZÁRATE, N.A. 1995. Produção de clones de inhame cultivados no pantanal sul matogrossense. Horticultura Brasileira. 13, 1: 38-40.

HUANG, C.C.; CHEN, W.C.; WANG, C.C.R. 2007. Comparison of Taiwan paddy and upland cultivated taro (*Colocasia esculenta* L.) cultivars for nutritive values. Food Chemistry. 102 (1); 250–256.

JALOTA, S. K.; SOOD, A. G. B. S.; CHAHAL, B. U. 2006. Crop water productivity of cotton (*Gossypium hirsutum* L.) – wheat (*Triticum aestivum* L.) system as influenced by deficit irrigation, soil texture and precipitation. Agricultural Water Management. 84: 137-146.

KARATAS, B. S.; AKKUZU, E.; UNAL, H. B.; ASIK, S.; AVCI, M. 2009. Using Satellite remote sensing to asses irrigation performance in water user Associations in the Lower Gediz Basin, Turkey. Agricultural Water Management. 96: 982-990.

KELLER, J.; KARMELI, D. 1975. Trickle irrigation design. Rain Bird Sprinkler Manufacturing Corporation, 133 p.

MABHAUDHI, T.; MODIA, A.T.; BELETSEB, Y.G. 2013. Response of taro (*Colocasia esculenta* L. Schott) landraces to varying water regimes under a rainshelter. Agricultural Water Management. 121: 102–112.

MANTOVANI, E. C.; SOUZA, J. A. R.; SOUZA, D. O. 2007. Irrigação do tomateiro. In: SILVA, D. J. H., VALE, F. X. R. (Ed.). Tomate - tecnologia de produção. Viçosa: Suprema Gráfica e Editora. 117-157.

MANTOVANI, E. C.; BERNARDO, S.; PALARETTI, L. F. 2009. Irrigação: princípios e métodos. 2. ed. Viçosa, MG: UFV, 355 p.

PEREIRA, L. S.; PAREDES, P.; EHOLPANKULOV, E. D.; INCHENKOVA, O. P.; TEODORO, P. R.; HORST, M. G. 2009. Irrigation scheduling strategies for cotton to cope with water scarcity in the Fergana Valley, Central Asia. Agricultural Water Management. 96: 723-735.

PUIATTI, M.; CAMPOS, J.P.; CASALI, V.W.D.; CARDOSO, A.A.; CRUZ, R. 1990. Sistemas de colocação do bagaço de cana-de-açúcar e do capim gordura, na cultivar de inhame 'Chinês'. Horticultura Brasileira, Brasília. 8 (1): 14-16.

SALOMÃO, L.C.; CANTUÁRIO, F.S.; PEREIRA, A.I.A.; SCHWERZ, T.; DOURADO, W.S. 2014. Influência do turno de rega na eficiência do uso da água de Irrigação e na produtividade de plantas de alface cultivadas em Ambiente protegido. Enciclopédia Biosfera. 10 (18): 2029.

SAMMIS, T. W. 1980. Comparison of sprinkler, trickle, subsurface, and furrow irrigation methods for row crops. Agronomy Journal. 72 (5): 701-704.

THINGLINK. Brazil map. Available in: <https://www.thinglink.com/scene/725367519031853057>. (Accessed: Oct. 2017).

VIEIRA, G.H.S.; MANTOVANI, E.C.; SEDIYAMA, G.C.; LO MONACO, P.A.V. 2015. Lâminas de irrigação em cana-de-açúcar para diferentes condições de disponibilidade hídrica. Irriga, Ed. Esp: 137-148.

VICENTE, A. S. C.; VICENTE, S. E. I. 2004. Proposta para Manejo da Irrigação. Belém. 30p.

WIKIPEDIA. Santa Teresa map. Available in: <https://upload.wikimedia.org/wikipedia/commons/3/3b/EspiritoSanto_Municip_SantaTeresa.svg> (Accessed: Oct. 2017).

# Irrigation technology utilization: A case of central Uganda

Teddy Kizza[1], Winfred Nakyagaba[1], LosiraNasirumbi[1,] Nicolas Kigundu[2], Damalie B. Magala[1], Monica Muyinda[1], Teopista Namirimu[1] and Dan Kirimani[3]

[1] NARO, Mukono Zonal Agricultural Research and Development Institute

[2] Department of Agricultural and Bio systems Engineering, Makerere University

3 Ministry of Agriculture Animal Industry and Fisheries (MAAIF)

## *Abstract*

*Products of irrigated agriculture are at low demand in many Sub Saharan Africa (SSA) countries yet forcing factors for national food self sufficiency are prevalent. This study aimed at understanding the causes for low irrigation adoption and to generate information for improved utilisation. Individual house hold surveys were conducted among 138 vegetable growing households and focus group discussions (FGD) were held with key informants in 4 sub counties in 2012. The results were updated in 2016 among 32 farmers through information sharing meetings. The results indicated that most farmers (77.5%) still used watering cans for irrigating their crops of average 0.5 acres (±0.35). High cost ranked 1$^{st}$ among prohibiting factors to uptake of modern equipment. Lack of water in the uplands coupled with limited knowledge of water harvesting techniques (1% use) restricted commercial vegetable production to low and wetlands.*

**Key words: Irrigation technology, utilisation, Uganda**

## 1.1. Introduction

Informal small scale irrigation practice in Uganda dates way back to the 1940s and formal irrigation development started in 1960s (;fortuneofafrica,*no year....* ). Data from this source indicated that 8 irrigation schemes were established by government between 1970 and 2001. Of these, 7 were in Eastern and 1 in western region. By the year 2006, irrigated land ranged between (0-10 %) of potential irrigable area FAO (2006), with the highest percentage in Eastern region. For most part of central Uganda, irrigated land was reported at (0-1) % of irrigable land. Low irrigation coverage was still identified among the top ten challenges to agriculture in the year 2016, (The statehouse of Uganda, 2016) indicating a persistent low trend of adoption. The government is encouraging farmers to engage in irrigation due to the sporadic droughts that have threatened food security in the country. Irrigation is promoted under the premise that it improves food security and that farmers have been sensitised by the extreme events to its necessity. The factors that have led to the slow adoption rate of irrigated agriculture are however not discussed. Farmers seem to appreciate the fact that irrigation enhances productivity but their reluctance to use it remains a 'paradox'. This study was conducted with aim to document the major irrigation technologies used by small-scale farmers in horticultural production and the socio-economic constraints to use of irrigation technologies. Information generated will be useful in designing user friendly irrigation packages to enhance uptake of the technologies. The information may also be constructive to decision makers in drawing policies and processes geared towards popularizing irrigation among local farmers. The study tried to contribute to research and information Institutional cluster of the (2010-2035) National irrigation master plan for Uganda.

## Methodology

The study was carried out among horticultural farming households (HHs) in_Mpigi and Wakiso districts (Fig.1), located in central Uganda.The two selected districts were the areas predominantly growing vegetables for supply to Kampala city according to un published market survey data on vegetable market chain. Wakiso district is located within: 0° 24′ 0″ N, 32° 28′ 57″ E and Mpigi lies within 0° 13′ 48″ N, 32° 19′ 48″ E longitudes and Latitudes respectvely. Their relative location to Kampala is depicted in Fig.1.

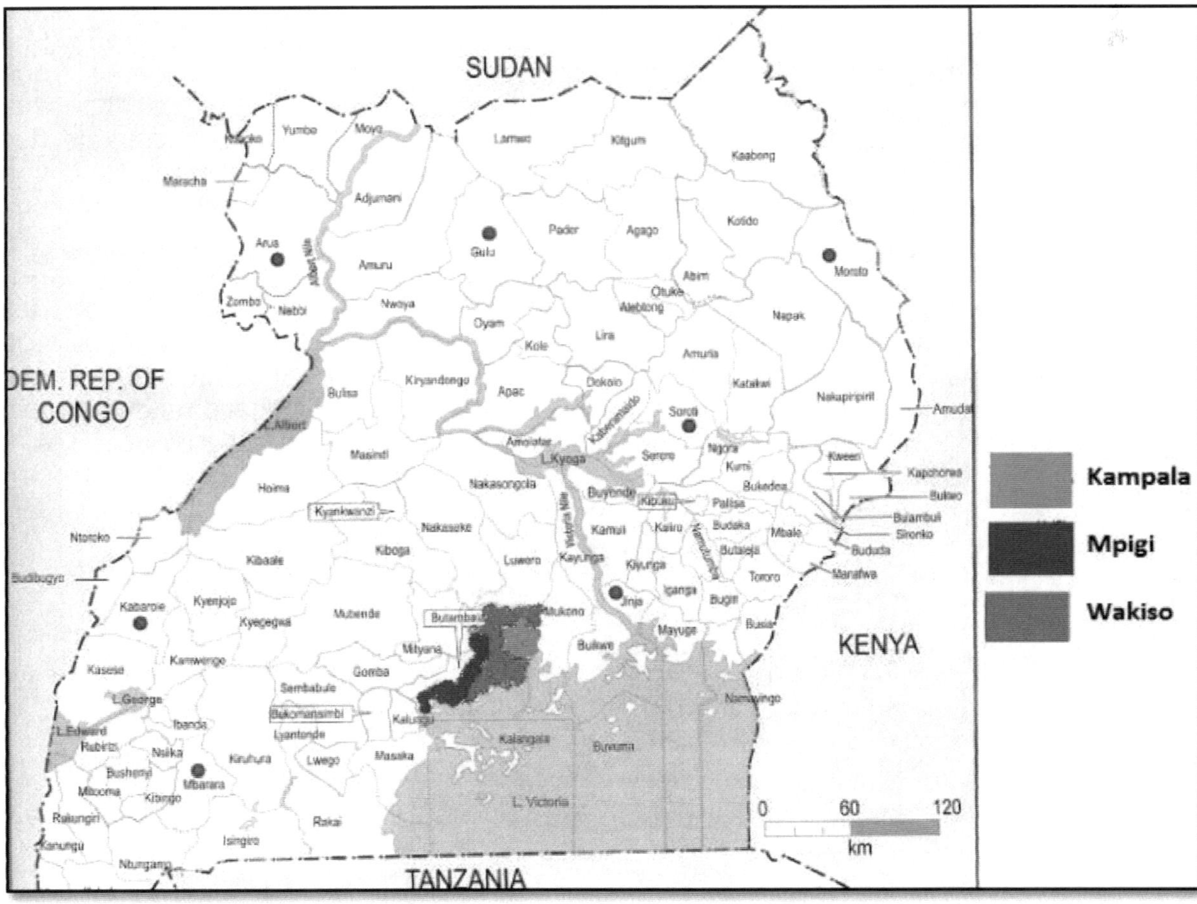


**Figure 1: Map of Uganda presenting study districts and Kampala city area**

## Data Collection and analysis

Individual house hold surveys were carried_out among 138 homes, using pre-tested structured questionnaires and four focus group discussions were held one at each sub county of project implementation area guided by a checklist. The key selection criterion for respondents was those that had practiced vegetable production for atleast two years. Members selected for focus group discussions were those that held leadership positions in the area. The collected data was handled by the socio economist on the team, who had also designed the questionnaire, and was analysed using spss v12. Standard procedures described by Bryman and Cramer(2005) were used for data analysis.

# Results

## Major irrigation technologies used by small-scale farmers in horticultural production

Irrigation practices included use of watering cans,drip kits, sprinklers, bottles and basins (Fig.3.). Majority of farmers were using rudimentary methods of watering their crops. Respondents from Wakiso were more exposed to improved technology (Drip and motorised pumps) than their counter parts in Mpigi. This could be due to proximity of Wakiso to Urban supply markets. The results revealed that 25% of farmers in Wakiso were using wetlands for vegetable production but majority (51%) were getting their water from streams. For the case of Mpigi, 30.8% were sourcing their water from shallow wells and 19.2% were also using streams. The collected water was generally stored in drums (62.5% Mpigi), (31.8%, Wakiso), and Jerry cans (27.5% Wakiso), (33.3% Mpigi). Most irrigated agriculture irrespective of method used was confined to a few meters from the water sources. Within the wetlands, some farmers were still struggling to get water to the plants. As depicted if figure 2, they dug small basins along the channels and then splashed water to the plants using plastic containers. Surface (furrow) irrigation therefore was associated with use of basins.The farmers observed that the quality of vegetables produced using the splash method was poor because at times mud was splashed on the leaves and fruits.

Figure 2 basins dug along water channels for watering crops and (right) land use change just above the wetland

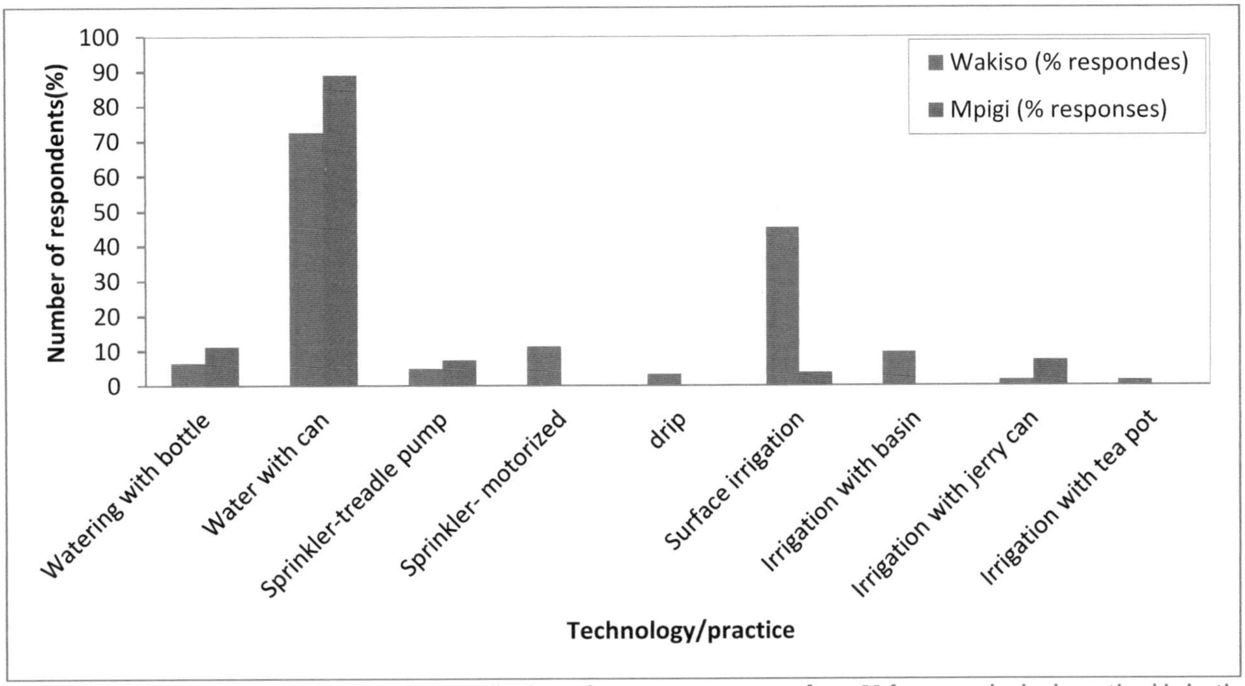


Figure 3. Technology use by farmers in the two districts. These were responses from 89 farmers who had practiced irrigation of any sort.

Farmers noted that with irrigation, they would be able to increase their crop yields (52.3%),grow off season (44.3%), have faster growing crops (18.8%), carryout early planting (3.4%) and have less completion in the market (21.6%). They gave their thought on some of the commonly used technologies as presented in *Table 1*. The number of respondents is shown on the extreme right. Other farmers who did not respond had no idea about the technologies.

Table 1 Farmers' perception on available and familiar technologies

| irrigation technology | Easy to operate | good coverage | Affordable | Knowledge on use | Water saving | Effective and efficient | Less laborious | Total Respondents |
|---|---|---|---|---|---|---|---|---|
| Motorised pump-sprinkler | 8 | 13 | 1 | 0 | 0 | 8 | 2 | **58** |
| Motorized pumps | 7 | 2 | 2 | 0 | 0 | 0 | 2 | **34** |
| pump & pipe | 2 | 1 | 0 | 2 | 0 | 0 | 0 | **17** |
| Drip irrigation | 1 | 1 | 1 | 0 | 1 | 0 | 0 | **14** |
| Watering can | 138 | 0 | 138 | 138 | 42 | 35 | 60 | **138** |
| Treadle -sprinkler | 0 | 0 | 0 | 0 | 0 | 0 | 0 | **2** |

Irrigation was practiced on some crops the farmers thought were of high value. These crops with the corresponding acreages are presented in Table.2.

Table 2 Average acreage under irrigation for the five most irrigated Vegetable crops

| Five top most irrigated Crops | Overall sample | Wakiso | Mpigi | t test |
|---|---|---|---|---|
| Tomatoes | 0.53 (0.47) | 0.51(0.47) | 0.58(0.48) | 0.593 |
| Nakati | 0.49(0.34) | 0.53(0.35) | 0.26(0.02) | 1.668* |
| Cabbage | 0.46(0.34) | 0.36(0.20) | 0.57(0.43) | 1.986* |
| Bugga | 0.48(0.37) | 0.51(0.39) | 0.36(0.28) | 0.925 |
| Green pepper | 0.54(0.51) | 0.53(0.48) | 0.54(0.57) | 0.015 |

*In Parentheses are standard deviations, * significance at 10%*

## Identified gaps and (constraints)

Although farmers acknowledged the benefits of using irrigation, 50% considered most technologies as very expensive cheap ones labour intensive. The level of investment in agriculture was still low although 99.26% (Table 2) of farmers derived their livelihood from it.Poor agronomic practices especially plant spacing increased drudgery and limited mechanisation 8.2% practiced line planting. The farmers requested for training in areas of access to information and knowledge on horticulture. There was a feeling that the rains received were still sufficient to sustain agriculture. The farmers were used to the bi modal rainfall pattern with elevated dry period between December to February with a peak in January as suggested by 79% of the respondents and June-July with peak in July reported by 68.1% respondents. This trend has however been perturbed by the current climate change and the seasons have become very volatile.

## Major Sources of income of farmers

Farmers' sources of livelihood were checked to try and relate with what drove their priorities in investiment. In this study however we did not find straight relationship.

Table 3. Estimated annual income (UGSHs) for the five most reported income sources

| Source of Income | Overall sample | | Wakiso | | Mpigi | |
|---|---|---|---|---|---|---|
| | Mean | Standard Deviation | Mean | Standard Deviation | Mean | Standard Deviation |
| Crop production | 2,804,875 | 1,904,503 | 2,801,036 | 1,939,666 | 2,808,432 | 1,885,777 |
| Livestock production | 1,157,152 | 955,675.9 | 1,170,566 | 1,066,936 | 1,145,893 | 861,062.3 |
| Petty trading-agric produce | 1,540,645 | 111,6262 | 1,539,456 | 1,021,140 | 154,1647 | 1,218,479 |
| Brick laying | 1,032,500 | 438,752.3 | 1,058,750 | 468,781.9 | 986,562.5 | 406,941.9 |
| Petty trading in general merchandise | 836,666.7 | 573,116.9 | 937,333.3 | 722,963.4 | 724,814.8 | 352,546.7 |

## Sources of information on irrigation

Many NGOs and Government Institutions had promoted irrigation practice through training on use and importance. Of the 138 respondents, 49 had undergone training related to irrigation by NAADS ,NARO , Government extension,NGO (BRAC, World vision, Enviromental alert, VOCA, CARITAS, KOFUKAWE, AMFRI), Fellow farmers ,Input supplier/seed companies/Dynapharm , HORT farmers' Association/UNAFE/NOGAM,
Makerere/ Formal education, Marketers/ promotion and JICA. The trainer- farmer coverage was as presented in Fig.4.

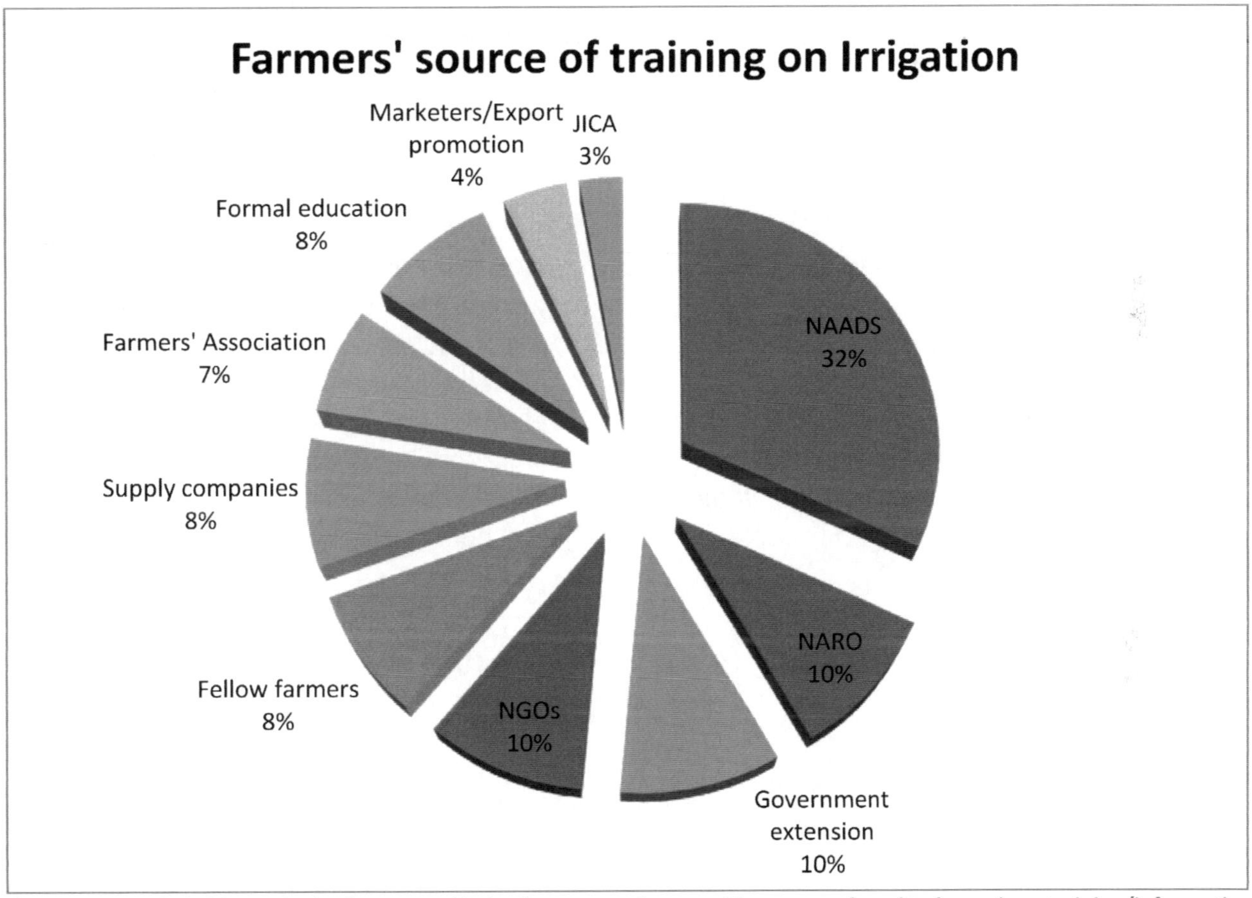


Figure 4 Key stakeholders who had promoted irrigation among farmers. These were found to have given training/information on irrigation to the respondents in the period of five years.

## Support to irrigation

Only 8% of the farmers interviewed reported to have received support on irrigation technologies in terms of funding, technical knowledge and marketing information. The support received however did not reflect much on the practices among farmers. They still indicated need for training and funding. They also expressed need for continuous interaction with trainers and explained that because the interface interval was large, they failed to implement what they were taught.

## Suggestions on management of irrigation demos

Majority (69.6%), of the interviewed farmers opted for group gardens, 24.6% preferred individual farms where as 2.9% suggested management by the sub county (Local Government). We offered two farmer groups with micro sprinkler systems in January, 2017 and planted their selected crops. One group chose three vegetables, Nakati (*Solanum aethiopicum* ),Cabbage and tomatoes while the other opted for two vegetables, *Solanum aethiopicum* and and Bugga (Red amaranth). We observed the farmers' practices and noted their comments as follows: there was tendency for other group members to abandon field activities to the host farmer. Even when they were told that dividends from the harvest would benefit them as a group, they tended to feel that the garden belonged to the funder/Researchers. They lacked ownership of the garden except for the hosts. One group reverted to use of watering cans as a way of saving on fuel for the pump.

## Recommendation(s)

Setting up demonstration plots to show case use of irrigation technologies because many farmers still think these technologies are very expensive and they are not likely to break even. This feeling has created great hindrance to technology uptake and there should be deliberate effort to show that high price of the crop produced with irrigation can offset the cost of irrigation equipment and operation costs. Technologies should be developed to cater for small holder farming because this forms the majority of farmers in this region. A large number of small scale farmers should be targeted as opposed to fewer large scale farmers. Feasibility of renewable energy use in agriculture would be investigated to replace the costly fuel pumps generally used due to limited grid coverage on rural farms.

## Conclusion

Irrigated farms were generally of small holding and close to water sources. There has not been much effort to help small-scale farmers select and utilise appropriate irrigation technologies. The form of assistance given was described by farmers as piece wise and they failed to benefit from it.

## Acknowledgements

MUZARDI staff that carried out data collection are appreciated. The study funds were accessed through ATAAS Irrigation and ATAAS –SLM projects.

## References

FAO. (2006). Demand for products of irrigated agriculture in sub-Saharan Africa. FAO water reports 31. ISBN 92-5-105581-5, P 142.

Fortune_of Africa retrieved through http://fortuneofafrica.com/ug/irrigation-in-uganda/

Republic of Uganda, Ministry of water and environment. (2011). A National Irrigation Master Plan for Uganda (2010-2035). P 218

The state house of Uganda (2016). President names 10 challenges to agriculture in Uganda .Retrieved through http://www.statehouse.go.ug/media/news/2016/06/21

Bryman, A.and Cramer, D.(2005) .Qualitative data analysis using SPSS 12 and 13,A guide for socio scientists

# Application of the Precision Ag Irrigation Language (PAIL)

**Charles Hillyer, PhD**
Assistant Professor & Extension Specialist, Texas A&M AgriLife, 6500 Amarillo Blvd W, Amarillo, TX, 79106, Charles.hillyer@ag.tamu.edu

**Diganta (Dig) Adhikari, PhD**
Director of Engineering, IRROMETER Company, Inc., 1425 Palmyrita Ave., Riverside, CA 92507 DigA@irrometer.com

**Richard Andrus**
Software Engineer, Campbell Scientific Inc, 815 W 1800 N, Logan, UT 84321-1784, randrus@campbellsci.com

**Aaron Berger**
Software & Integration Engineer, AgSense LLC, 259 Dakota Ave. S. , P.O. Box 53 , Huron, SD 57350, aberger@agsense.net

**Dan Berne**
CEO, Next Chapter Marketing, 2752 SW Fairview Blvd, Portland, OR 97205, dan@nextchaptermarketing.com

**Kurtis Charling**
Business Solutions Specialist, Lindsay Corporation, 2301 N. 117th Ave, Suite 200A, Omaha, NE 68164, kurtis.charling@lindsay.com

**R. Andres Ferreyra, PhD**
Manager of Special Projects, Ag Connections, LLC., 1576 Killdeer Trl, Murray, KY, 42071, andres.ferreyra@agconnections.com

**Bart Nef**
Environmental Group Product Manager, Campbell Scientific Inc, 815 W 1800 N, Logan, UT 84321-1784, bart-n@campbellsci.com

**Joe Russo, PhD**
President, ZedX, Inc., 369 Rolling Ridge Drive, Bellefonte, PA 16823,russo@zedxinc.com

**Abstract**: *Over the past six years, a team of industry professionals and Extension researchers has worked in the context of AgGateway's PAIL project to draft a standard for data exchange among irrigation technologies. This draft is currently in process to become an ASABE standard. The North Plains Groundwater Conservation District (NPGCD) has funded development of an integrated irrigation management system. The system is "integrated" in that it combines information from multiple irrigation technologies into a single web application. PAIL is the enabling element of this integrated system: each of the data sources (weather, soil moisture, and pivot control) exchanges information in the PAIL format. Development of NPGCD's system began in December of 2016 and is undergoing testing during the 2017 irrigation season. We present initial results from the development and application of the NPGCD's system and observations relating to how the PAIL standard reduced cost and complexity for the system's software.*

**Keywords**: PAIL, irrigation management, data exchange, standards, systems integration

## Introduction

Irrigated agriculture in the US accounts for 80-90% of the consumptive water use and approximately 40% of the value of agricultural production (Schaible and Aillery, 2012; USDA, 2009). This value, totaling nearly $118 billion, is produced on 57 million acres. Given the increasing challenges in water availability and the likelihood of increased water conflicts from competing users, irrigated agriculture must increase its efficiency without sacrificing a reduction in the value it produces (Schaible and Aillery, 2012). Growers can derive much of this efficiency through application of precision irrigation technologies, and on-farm management systems that facilitate sound agricultural practices. However, less than 10% of irrigated farms use any advanced decision support tools or technologies (USDA, 2009). Improving adoption of these technologies is critical to increasing efficiency.

Recently, a group of companies, industry representatives, academics, and interested parties began collaborating to address the issue of systems integration in irrigation (Hillyer et al., 2014). This project, called **Precision Ag Irrigation Leadership** (PAIL), has the specific goal of producing a set of data exchange standards that enable development of more efficient and easier to use solutions for irrigation management. The PAIL participants represent a diverse group of technologies including companies producing Farm Management Information Systems, Pivot Irrigation Systems, weather and environmental monitoring equipment, and soil moisture monitoring equipment. By having a "common language" for data exchange, manufacturers can collect data from a variety of sources without the burden of developing specialized exchange methods for each different data source. The PAIL standard will improve interoperability of irrigation technologies and, consequently, increase adoption of more efficient irrigation practices.

Details of the purpose, scope, and structure of PAIL have been described elsewhere (Adhikari et al., 2016). In this paper, we present a demonstration project that applies the PAIL data standard. This project includes the development of a "fully integrated" irrigation scheduling program and an on-farm demonstration of the program.

## The argument for PAIL

Consider a generic irrigation management tool that is intended for implementing Scientific Irrigation Scheduling (SIS). For this example, the "system" is a software system that is capable of working in most contexts and is generic in that it is not limited to a particular brand or type of hardware. Figure 1 shows how a software developer might view the system from a very simplified perspective. The system must have some sensors since physical measurements are the basis for SIS. Data from those sensors must move out of the field via some mechanism such as cellular or radio telecommunication. The system must store these data and perform analysis or calculations to produce irrigation recommendations. The bulk of the management system's software resides in this storage & analysis component. The user is also an essential component of the system. Implementing SIS requires knowing how much water was applied. Typically, the user (i.e., the irrigator) must supply this information. Finally, the user is the recipient of any recommendations generated by the system.

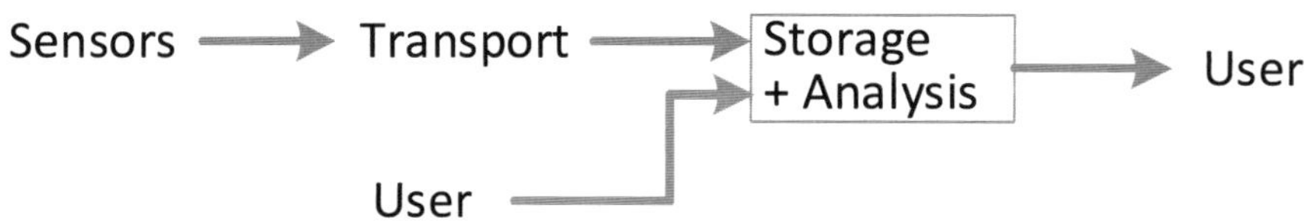


*Figure 1*

Since this system should be generic, the software developers cannot assume that the users have only one sensor type or brand. The three methods of SIS (soil moisture measurement, evapotranspiration, plant sensing) each use different kinds of sensor, and their data differs in structure, units, and meaning. Furthermore, we cannot assume that a particular user has only one brand of sensor. Figure 2 shows how the developer might view the system after taking into account that the grower might use many sensor manufacturers or sensor types might. Conversion components are required to move all those data into the analysis component, and the size of the analysis tool has grown accordingly.

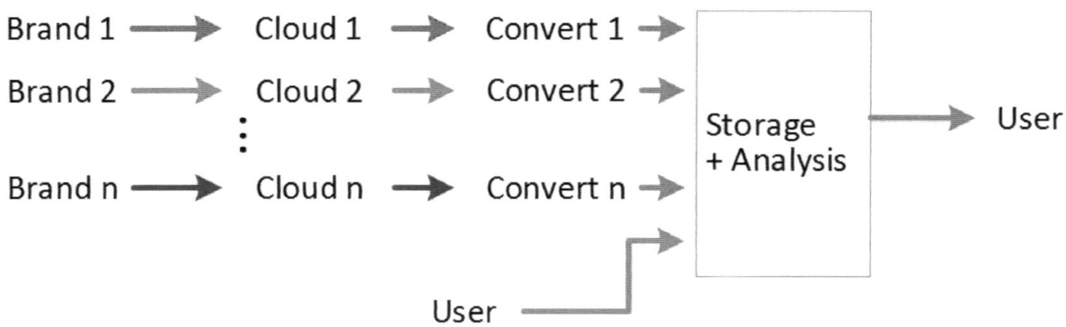


*Figure 2*

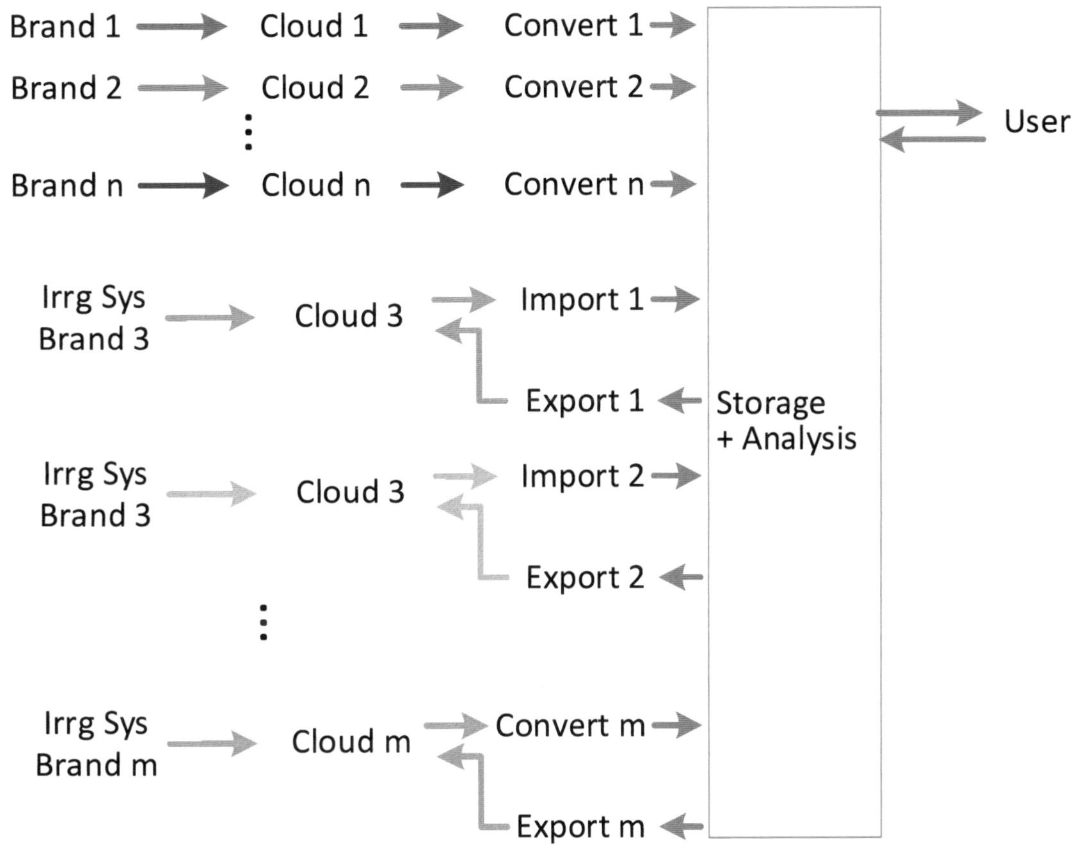


*Figure 3*

The multi-brand sensor view in Figure 2 still shows the user as the source of water application data. Relying on the grower for irrigation data is still a common design decision for scheduling tools and is an additional burden on the user. Most modern pivot control systems have some mechanism to export when and how much water was applied. A fully integrated irrigation management system could take advantage of this data stream and relieve the user of that burden. The obstacle to doing this integration is that each manufacturer uses their proprietary format for the irrigation records. Figure 3 shows how the developer might vie the fully integrated system. In this view, the user is no longer burdened with entering data. The burden has been moved to the software system and manifests as additional import/export/conversion code. Moving this burden to the software is undoubtedly a benefit to the user. However, the additional import/export/conversion are an added cost to development and are a disincentive to the development of generic an fully integrated SIS tools.

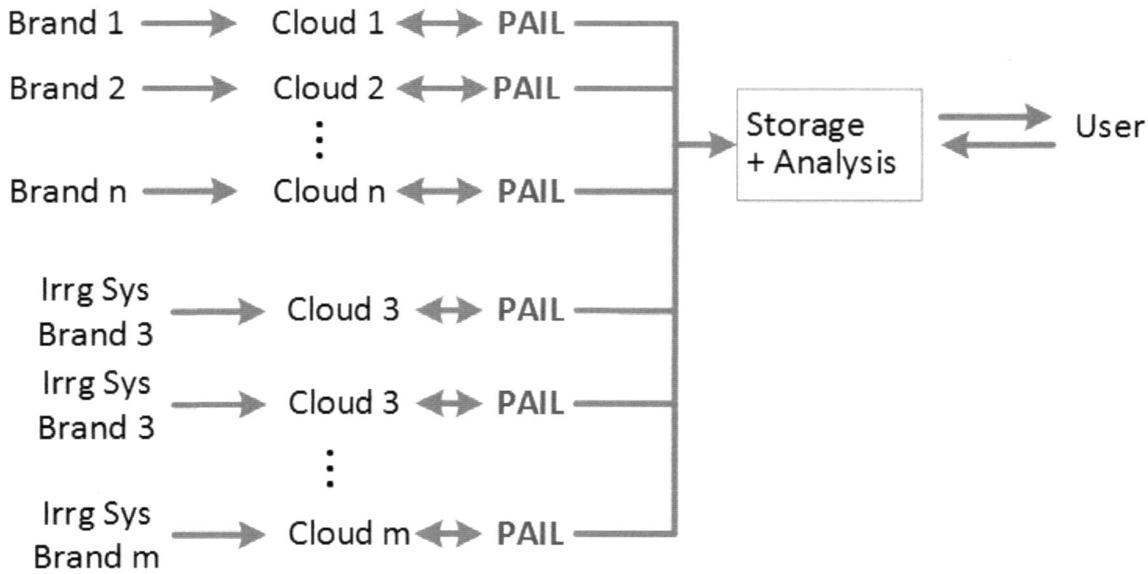


*Figure 4*

The PAIL standard proposes a single format for exchange of all the data relevant for irrigation management. Figure 4 shows a revised view of the system where PAIL is the only exchange format. In this case, the Storage & Analysis component is smaller because it no longer needs additional code to integrate a myriad of data sources.

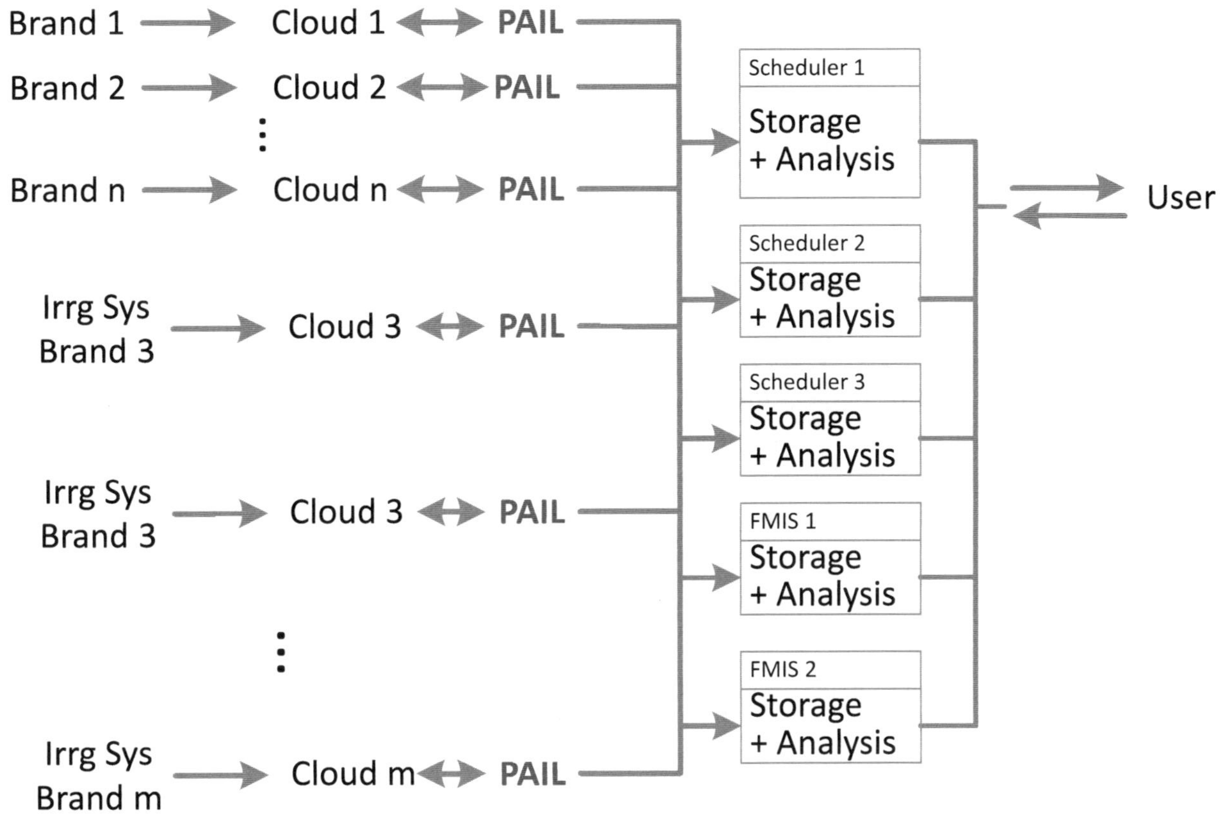


*Figure 5*

One could argue that PAIL is merely pushing the conversion problem back on the manufactures since they would need to implement the conversion code in their products. In fact, the manufacturers face a similar problem as the SIS system's software developer. Many software systems can derive value from sensor data or irrigation system records. Interoperating with each of these systems means the manufacturers face the same problem of converting to a myriad of formats. The PAIL standard has value for the manufacturer since they can build to a single format while still supporting multiple data consumers. Figure 5 illustrates how PAIL benefits both producers and consumers of irrigation data.

The preceding example embodies the basic argument for PAIL. Having a standard format for irrigation data exchange addresses problems for multiple actors in the irrigation space. From the grower's perspective, PAIL addresses the mixed fleet problem where the grower must integrate data from disparate sensing or control systems. From the manufacture's perspective, PAIL addresses the issues arising from having many different consumers of data each with different formatting requirements. In a general sense, PAIL addresses the issue of SIS adoption by making SIS system easier to build and easier to use.

## North Plains Groundwater Conservation District's Irrigation Scheduling Tool

The issues discussed in the previous section are essentially a systems integration problem. Systems Integration (SI) has its origins in military programs (Hobday et al., 2005) and has steadily spread to nearly every business sector. Agriculture has been slow to receive the benefits of SI across many areas of the farm enterprise. This is particularly true in the irrigation sector, and the reasons for slow adoption are varied. Stafford (2000) [1] posited, "data-overload' for the manager has to be overcome by the development of data integration tools, expert systems, and decision support systems." A recent meeting of 44 representatives of the irrigation industry, extension, and academia examined the problems and issues surrounding the adoption of efficient irrigation practices (Two Valleys Roundtable Report, 2015). The group concluded that systems integration (or lack thereof) is one of the things the group cited as a barrier to adopting new technologies. A similar but smaller group of irrigation experts met to examine the future of irrigated agriculture (English, 2015). One of the group's specific recommendations to stimulate adoption of efficient irrigation practices is "making equipment vendors more aware of financial support programs, *management tools*, and outreach sources." The conclusions of these two groups indicate a clear need in the realm of irrigation management: producers need new tools, and the tools must integrate as much data as possible.

The North Plains Groundwater Conservation District (northplainsgcd.org) has endeavored to create a fully-integrated irrigation scheduling tool. This tool is funded by a grant from the Texas Water Development Board (twdb.texas.gov), and Texas A&M AgriLife Extension is building the system (agrilifeextension.tamu.edu). The software is based on the sIMO irrigation scheduling tool originally developed at Oregon State University through an Oregon NRCS CIG grant.

The primary objective of the NPGCD project is to produce an irrigation scheduling tool that is useful to producers in the Texas panhandle region. Growers in this region are progressive, early adopters of practical irrigation technology. Soil moisture probes and pivot controls with remote telemetry have been common in the region for many years. To support these growers, the system should have the following features:

1. The system should be as simple as possible while still implementing SIS.
2. The system should automate all data flows needed for data integration. The data flow is automated in that the user is not required to take any specific action to generate irrigation recommendations. For example, users will not need to manually download ET data or enter irrigation amounts.
3. The system should use a water balance based estimate of soil moisture. This requirement indirectly stipulates that evapotranspiration will drive irrigation decisions however soil moisture measurements will be used wherever possible.
4. The system should mitigate uncertainty associated with estimated soil moisture depletion or recommended irrigation amounts. T this end, the system uses NOAA FRET and QPF forecast products to produce a 7-day forecast of depletion. Additionally, the system includes a user direct correction algorithm to compensate for sensing errors or calibration issues

[1] While citing (Sigrimis et al., 1999)

5. The system will provide reporting features necessary to support the NRCS EQIP medium intensity IWM practice.

The secondary objective of this system is to demonstrate the use of the PAIL data standard in the context of an irrigation scheduling tool. To that end, the sIMO system was modified to accept PAIL formatted documents. These documents include both weather data (ET and precipitation) and water application data (irrigation dates & amounts). The software also generates irrigation recommendations in the PAIL format, but there is as yet no participating consumer for these records. Details of the structure and content of PAIL documents have been presented elsewhere (Adhikari et al., 2016). The goal of this demonstration is to promote adoption of the PAIL standard. The modified sIMO system will use most of the basic functionality supported by PAIL. Once the system is thoroughly tested, the source code will be released under an appropriate open source license.

## System Structure

Figure 6 shows the packages that make up the modified sIMO system. The **ASP.NET Application** contains all the interface code. The sIMO tool is implemented as an ASP.NET web application written in C#. The **Database** is implemented in Microsoft SQL Server, and all the water balance and related calculations are implemented as stored procedures in T-SQL. Some of the API related code (i.e., downloads from NOAA NDFD) and the ASCE Standardized ET equations are implemented in C# as SQL CLR stored procedures. The **PAILlib** package contains a C# implementation of the PAIL object model. This package also contains necessary code to translate from PAIL constructs to sIMO database structure and vice versa. The "**Vendor Specific Adapter**" handles API calls and object translation for those vendors that do not fully support the PAIL standard. Some of the cooperating sites have hardware from vendors that are not participating in PAIL's development (see Table 1). This is treated as an opportunity to demonstrate translation from vendor-specific t formats to/from PAIL native documents. The vendor-specific nature of this code means that it will not be part of the open source version of sIMO unless the vendors explicitly agree to be included.

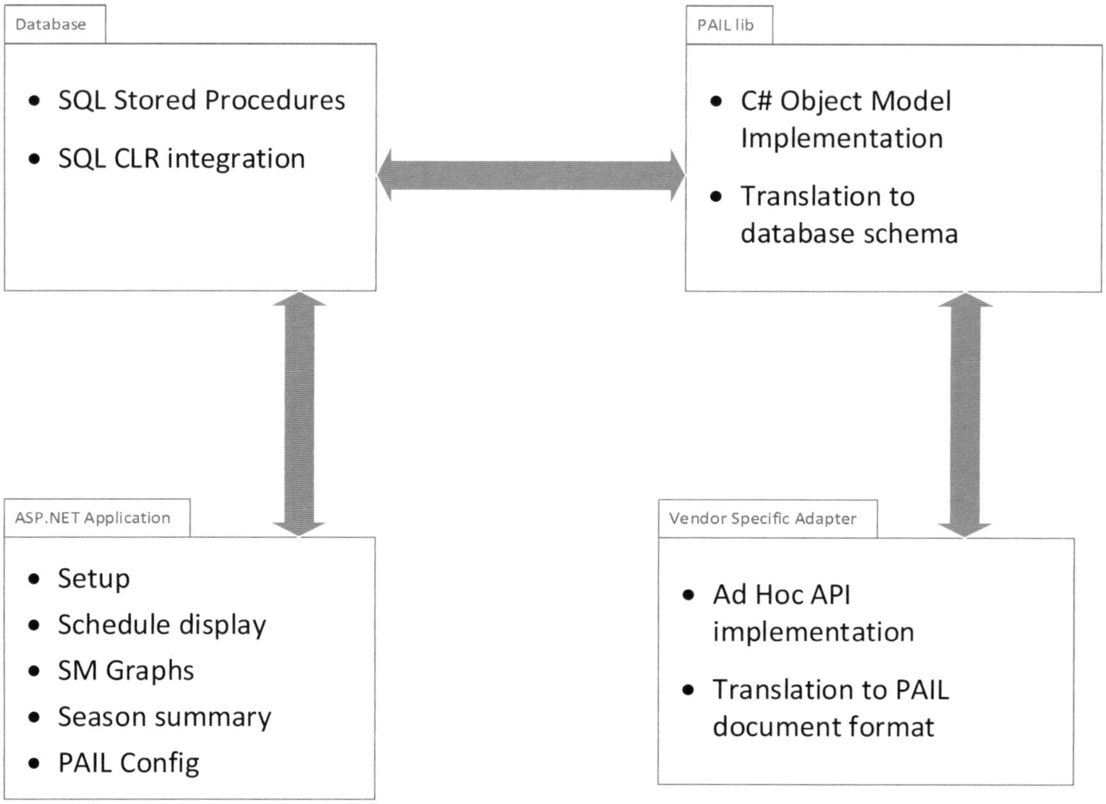


*Figure 6*

## In Season Data Flow

Figure 7 shows a conceptual view of the data flows during the irrigation season. The two primary sources of data mirror the two main scope divisions in PAIL. For the demonstration, the field sensors are primarily weather stations. These stations meet the requirements of ASABE EP505 (American Society of Agricultural and Biological Engineers, 2015) and the stations provide either reference ET or the data needed to calculate ET. The operations side is focused on when and how much water was applied.

Some of the hardware uses cloud-based storage, and others use direct cellular connections. In either case, sIMO obtains the data via HTTP. There is no specific API stipulated by PAIL. Having no standard API means that sIMO must implement separate download code for each vendor. IN nearly all cases, the API code is simple because the PAIL document structure is robust enough to contain any variation that would otherwise require a more complex API interaction.

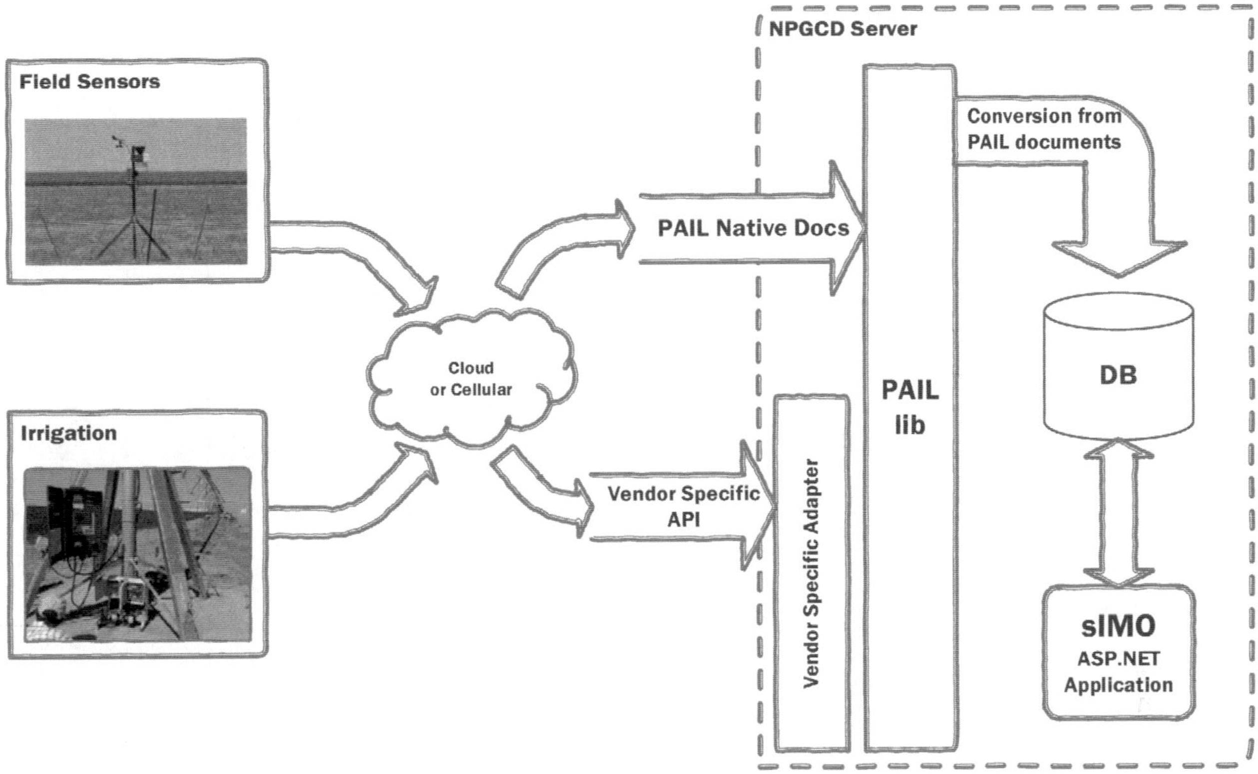


*Figure 7*

## sIMO Interface

The original version of sIMO was constructed to be "as simple as possible." The system achieves this goal via three features: 1) only require information needed to calculate a simple water balance, 2) use mouse-based input for as much user interaction as possible (i.e., minimize typing), and 3) the smallest possible interface. The sIMO interface consists of four primary pages. Figures 8 – 11 show screenshots of each page.

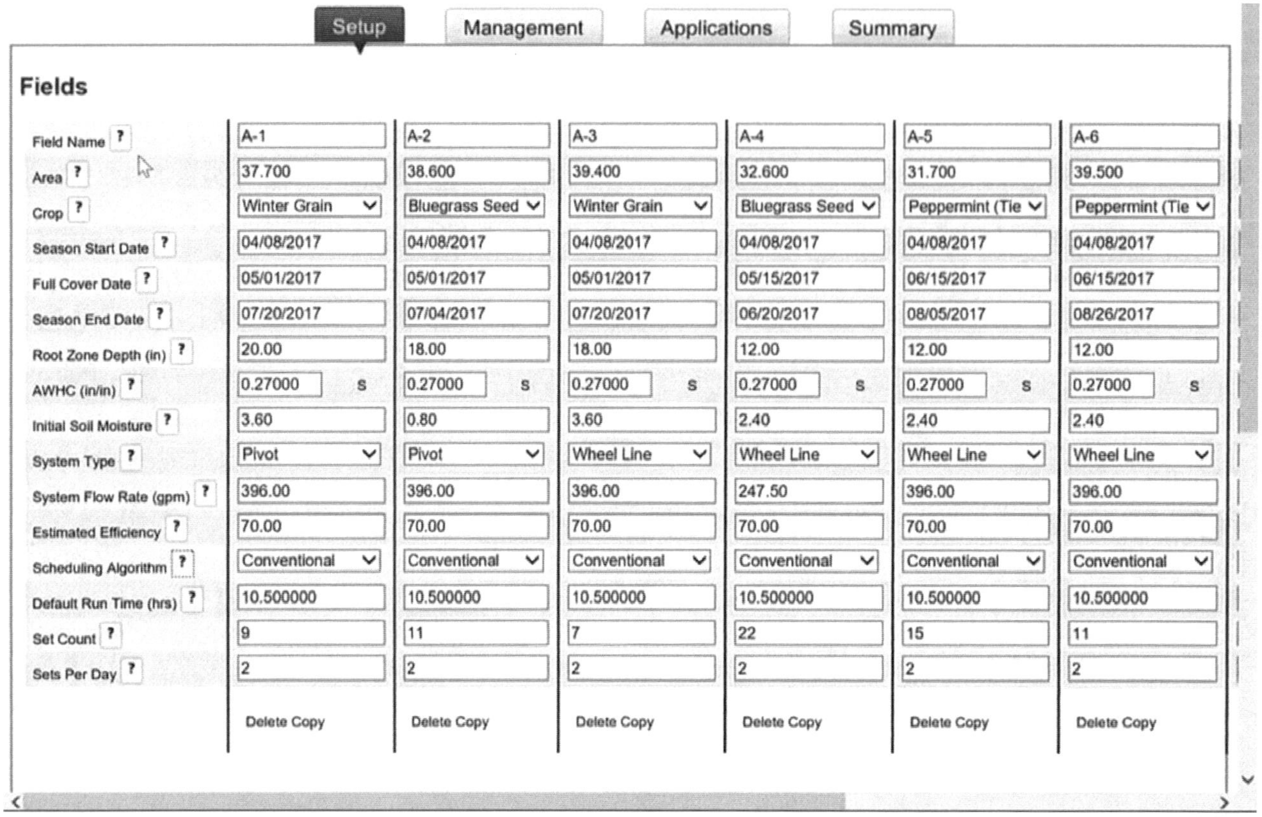

Setup | Management | Applications | Summary

**Fields**

| | | | | | | |
|---|---|---|---|---|---|---|
| Field Name | A-1 | A-2 | A-3 | A-4 | A-5 | A-6 |
| Area | 37.700 | 38.600 | 39.400 | 32.600 | 31.700 | 39.500 |
| Crop | Winter Grain | Bluegrass Seed | Winter Grain | Bluegrass Seed | Peppermint (Tie | Peppermint (Tie |
| Season Start Date | 04/08/2017 | 04/08/2017 | 04/08/2017 | 04/08/2017 | 04/08/2017 | 04/08/2017 |
| Full Cover Date | 05/01/2017 | 05/01/2017 | 05/01/2017 | 05/15/2017 | 06/15/2017 | 06/15/2017 |
| Season End Date | 07/20/2017 | 07/04/2017 | 07/20/2017 | 06/20/2017 | 08/05/2017 | 08/26/2017 |
| Root Zone Depth (in) | 20.00 | 18.00 | 18.00 | 12.00 | 12.00 | 12.00 |
| AWHC (in/in) | 0.27000 S | 0.27000 S | 0.27000 S | 0.27000 S | 0.27000 S | 0.27000 S |
| Initial Soil Moisture | 3.60 | 0.80 | 3.60 | 2.40 | 2.40 | 2.40 |
| System Type | Pivot | Pivot | Wheel Line | Wheel Line | Wheel Line | Wheel Line |
| System Flow Rate (gpm) | 396.00 | 396.00 | 396.00 | 247.50 | 396.00 | 396.00 |
| Estimated Efficiency | 70.00 | 70.00 | 70.00 | 70.00 | 70.00 | 70.00 |
| Scheduling Algorithm | Conventional | Conventional | Conventional | Conventional | Conventional | Conventional |
| Default Run Time (hrs) | 10.500000 | 10.500000 | 10.500000 | 10.500000 | 10.500000 | 10.500000 |
| Set Count | 9 | 11 | 7 | 22 | 15 | 11 |
| Sets Per Day | 2 | 2 | 2 | 2 | 2 | 2 |
| | Delete Copy | Delete Copy | Delete Copy | Delete Copy | Delete Copy | Delete Copy |

*Figure 8. Setup Page*

The Setup page contains all the field-specific information needed to set up the water balance calculations. sIMO presents the information in tabular (pseudo spreadsheet) format, and the user can employ a copy-paste procedure to set up multiple fields. The interface can also download estimated soil water holding capacity from the NRCS Web Soil Service if the user knows the name of the dominant soil. Basic farm-level setup information is also accessible on this page.

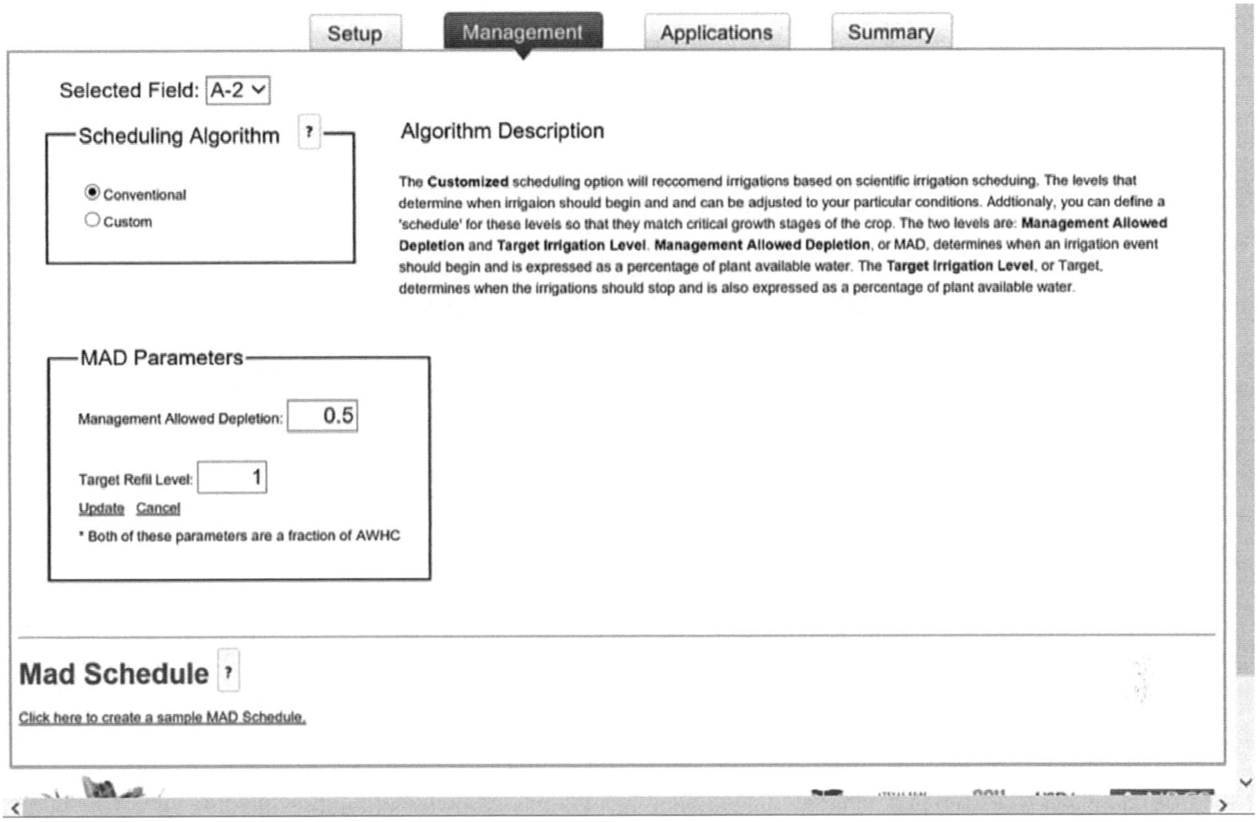


*Figure 9. The Management page*

The Management page is where the user specifies how the system should schedule irrigation. sIMO uses the MAD approach to schedule irrigation (Merriam, 1966). By default, the system uses a MAD of 50% and the system calculates runtimes to refill the soil profile to field capacity. More advanced options enable specific values of MAD and target refill level, including a schedule of MAD & Target levels that change during the season.

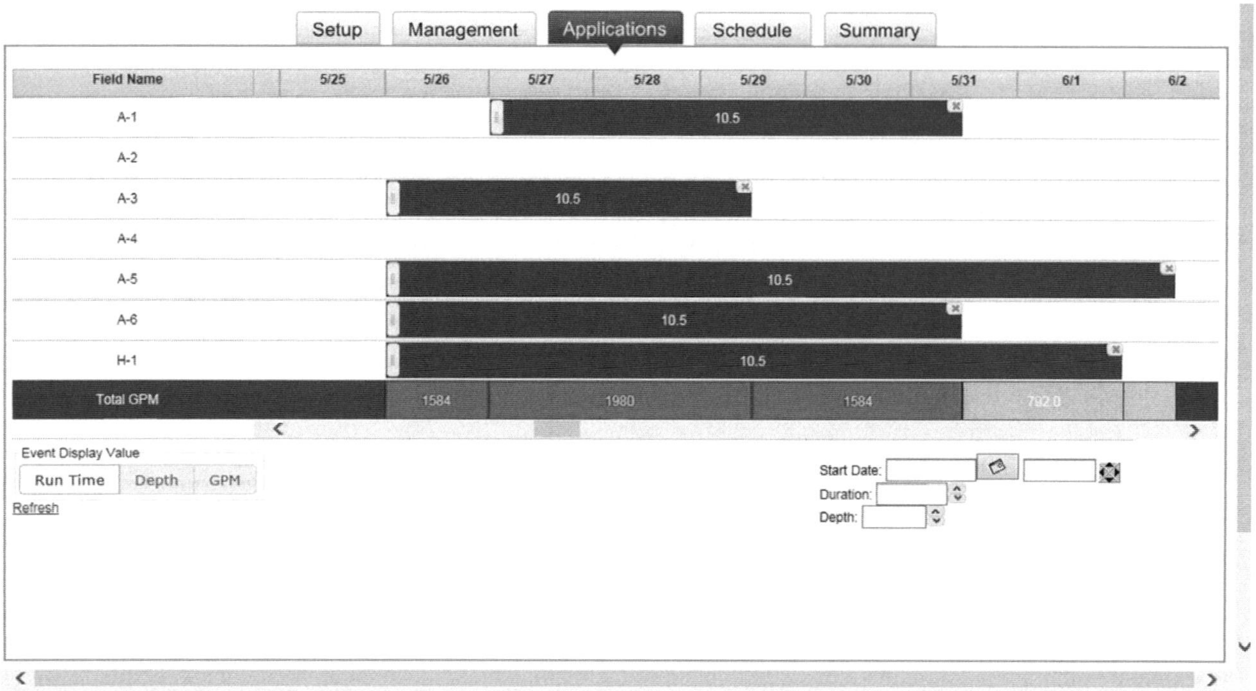


*Figure 10. The Schedule page*

The Schedule page is where the user can see both past irrigation and a recommended schedule for the next 7 days. The interface is a modified Gantt chart where each swim lane represents a single field. The blue bars represent individual irrigation events. The user can click and drag the event to change the start date or duration or double click to create a new event. Clicking on the row causes it to expand and expose a spreadsheet-like interface that shows each of the water balance components for each day. All of the water balance components are editable via spinners.

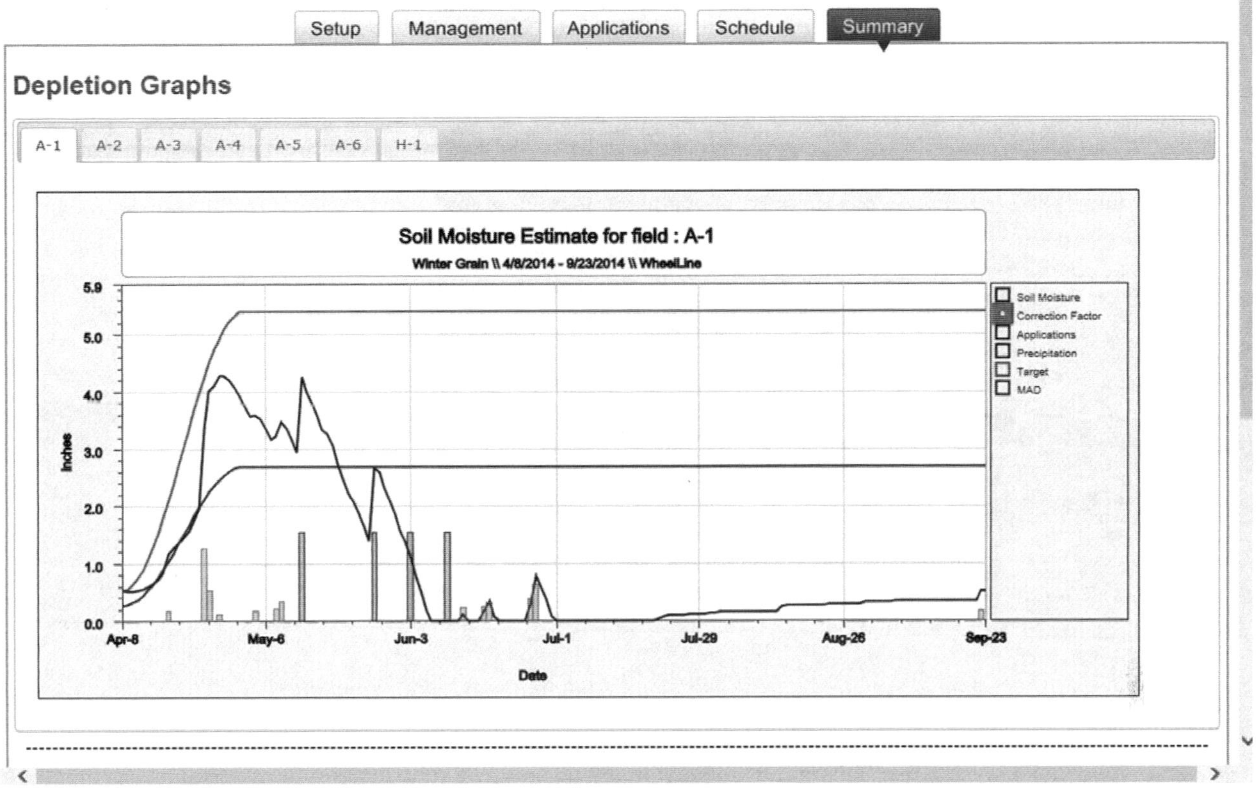


*Figure 11. The Summary page*

The Summary page shows graphs of depletion for the whole season. These are the typical plots soil moisture plots that show available moisture (blue line), irrigation (blue bar), precipitation (grey bar), and management limitations (green/purple lines). This page also has a table of season totals for each field and an option to download the water balance calculation as a CSV file.

## Demonstration Status

A significant component of the NPGCD project involves an on-farm demonstration of the scheduling system. Five sites were selected for the demonstration in 2017. Table 1 summarizes the data sources at each site. The focus during 2017 was on development and testing, so no active scheduling occurred during this season. The demonstration will continue during the 2018 season.

*Table 1 2017 sIMO test sites*

| Site | Field Sensors | Irrigation System |
|---|---|---|
| **NPGCD 1** (Etter) | Campbell Scientific (PAIL) | PivoTrac (PAIL adapter) |
| **NPGCD 2** (Etter) | Campbell Scientific (PAIL) | Lindsay (PAIL) |
| **Cooperator 1** (Dumas) | Ranch Systems (PAIL adapter) | PivoTrac (PAIL adapter) |
| **Cooperator 2** (Texline) | ZedX (PAIL) | PivoTrac (PAIL adapter) |
| **AgriLife (**Bushland, observe only) | ZedX (PAIL) | AgSense (PAIL) |

*Figure 12. Weather station installation at Cooperator #1*

## Conclusion

This paper described an ongoing effort to build a fully-integrated irrigation scheduling tool, sIMO. The management system is designed to be fully integrated so that it can accept data from both field sensors and irrigation control systems and will generate irrigation recommendations in the PAIL standard format. The level and scope of integration are made possible by the PAIL data exchange standard.

Preliminary testing of the system occurred during 2017 and will continue during 2018. We will release an open source version of the scheduling tool after the 2018 irrigation season.

## References


Adhikari, D., Berger, A., AgSense, L., Huron, S., Berne, D., Ferreyra, R. A., Hillyer, C., Melvin, S., Nef, N. B., and Penning, T. (2016). "Precision Ag Irrigation Language (PAIL) Project." *2016 Irrigation Show & Education Conference*, Irrigation Association, Las Vegas, NV.

American Society of Agricultural and Biological Engineers. (2015). "ASAE EP505.1 Measurement and Reporting Practices for Automatic Agricultural Weather Stations." *American Society of Agricultural and Biological Engineers*.

English, M. (2015). *Agricultural Irrigation Initiative: The Future of Agricultural Irrigation*. Northwest Energy Efficiency Alliance.

Hillyer, C., Berne, D., Ferreyra, A., and Smith, A. (2014). "Data Standards for Precision Irrigation Management." *2014 Montreal, Quebec Canada July 13–July 16, 2014*, American Society of Agricultural and Biological Engineers, 1.

Hobday, M., Davies, A., and Prencipe, A. (2005). "Systems integration: a core capability of the modern corporation." *Industrial & Corporate Change*, 14(6), 1109–1143.

Merriam, J. L. (1966). "A Management Control Concept for Determining the Economical Depth and Frequency of Irrigation." *Transactions of the ASAE*, 9(4), 492.

Schaible, G., and Aillery, M. (2012). "Water conservation in Irrigated agriculture: trends and challenges in the face of emerging demands." *USDA-ERS Economic Information Bulletin*, (99).

Sigrimis, N., Hashimoto, Y., Munach, A., and De Baerdmaeker, J. (1999). "Prospects in agricultural engineering in the information age-technological developments for the producer and the consumer." *Agricultural Engineering International: CIGR Journal*.

Two Valleys Roundtable Report. (2015). *Drought, Technology and Agriculture: Increasing Water Efficiency and Yield through Technology*.

USDA. (2009). *Farm And Ranch Irrigation Survey (2008)*. National Agricultural Statistics Service, Washington, D.C.

# Insights into Irrigation from Internet of Things Perspective


## Abstract:

Temecula Valley is turning into high tech hub for winegrowers in Southern California, as it is looking into technology and collaboration to improve the irrigation efficiency. Independent growers and winemakers have come together to serve the best tasting wine and to find mutually beneficial cooperation to conserve water use. In this paper, we will discuss some of the challenges that growers face and technology they are exploring to address these issues. Their observations suggest 20% water savings, enhanced analytics and automation. We will demonstrate tools and technologies that can drive innovation and revolutionize agriculture by (1) providing insight into the farm in real time, (2) analytics of water usage, and (3) irrigation automation. We will also cover a case study on a novel and inexpensive soil moisture sensors called Vinduino Sensors and Stations.


## Introduction

Internet of Things was the next logical steps after proliferation of Internet and mobile devices. As devices that support internet connectivity shrinked in size and power demands new market opportunities and possibilities are a reality. Fueled by Internet's social success, (Facebook, Google … ) engineers believe that Internet can also bridge small engineering systems into larger far more advanced or "smart" solutions.

From this perspective, one of the worlds oldest engineering systems, irrigation systems, can really benefit from the Internet connectivity. When we look at the first engineered irrigation system in Babylon, we see how the moisture content was used as a feedback to determine how much to irrigate. Now, few thousand years later, we have an interent connected world where we can get that feedback everywhere, even on a flight to our favorite, Irrigation Conference. This is the goal, and the desire of the modern world, where decisions are made from far away.

The distance between two machines, machine and a human, or 2 humans is no longer a central problem. We know exactly, which technology to use for communicating plurality modes of information. Thus, the century old bottleneck of communication is no longer there and we can truly communicate information beyond mountains and fields of sight. That's why Internet of Things is the next driver behind agriculture.

In this context, we shall identify challenges for IoT proliferation in Agricultural practice of irrigation and as well as opportunities offered, as well as the most important components of any IoT system first. These essentials can be grouped into 3 areas: communication, computation and usability.

## IoT - Communication:

Internet as we like to think about it is the magical tool that connects devices in a way that you can access information from anywhere. The way it works is just briliant. Everything is organized in layers of abstraction, which utilizing rigorous scientific methodologies. Basic transmission of modulated signals is used to transfer bits of data from point A to B. This layer is called the physical layer (Figure 1). On top of this layer, is the data link layer. Each data is communicated independantly and in isolation in the physical layer. This is the layer that actually transmits information. The layer above is the Network layer. In this layer, information exchange is handled between the nodes on the same small network. However, its in the Transport layer, that one of most important decisions was made, which allows devices to be far closer to each other. It is the hierarchy that allows connection of different small networks to other networks by special Internet Protocol Addressing, a unifying mechanism that allows to make few hopes between devices as depicted in Figure 2. As we move up in the ladder of these communication layers as depicted in the Figure 1, information exchange is used for cohesian and unification of systems into an entity called Internet.

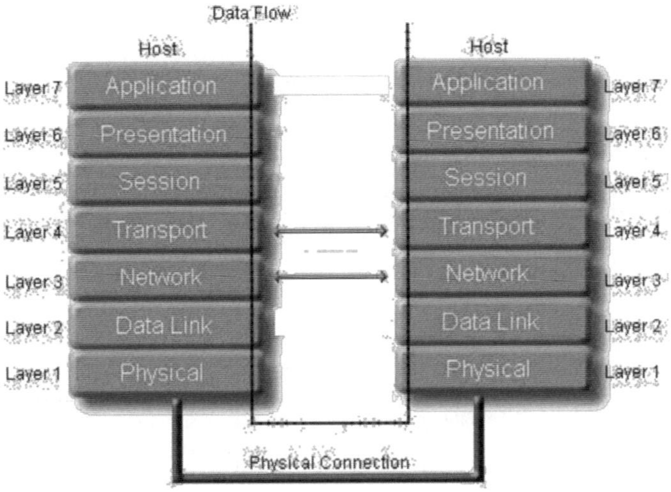


Figure 1: Internet Layers (Nolan)

Figure 2: Internet Model (Pixobay)

In the design of Internet, every to machines is connected to routers, or so caled base stations. These base stations than route, hance the name, pockets of information to next station while effectively routing the shortest path between any 2 devices. This allows having short latencies and elastic throughput. It is important to note that once devices are connected to internet via any base station, they are connected to each other. Thus, Internet of Things simplifies the interdevice communication, given that these devices can connect to any base station.

Originally, Internet was designed to be a wired network, where wires were the hard carriers of signal, but over time it grew into mix of wired and wireless communicaiton systems. Wireless communication is one of the key components in our vision for the Internet of Things for Agriculture as large distance in fields allow require robust communication mechanisms.

Network Structures – Local and Wide Area Networks

Network architecture for networks differs due to their physical connectivity range and type. Local connection thru wired ethernet create a different type of network than wifi connections. However, when we are thinking about agriculture and using sensors and actuators that are connected to one another ranges can exceed few hundred meters. Therefore, the conventional use cases are impractical and longer range solutions beyond WiFi need to be used if we wish not to complicate with labor intensive network modifications.

| Technologies | Frequency Band | Data Rate | Trans. Range | Energy Cons. |
|---|---|---|---|---|
| Wifi | 5-60GHz | 1Mb-1Gb/s | 20-100m | High |
| WiMAX | 2-66GHz | 1Mb-1Gb/s | <50Km | Medium |
| LR-WPAN | 868/915 MHz, 2.4 GHz | 40-250Kb/s | 10-20m | Low |
| 2G | 865 MHz | 50-100kb/s | Cellular area | Medium |
| 3G | 865MHz | 200kb/s | Cellular area | Medium |
| 4G | 2.4GHz | 0.1-1Gb/s | Cellular area | Medium |
| BlueTooth | 2.4GHz | 1-24Mb/s | 8-10m | Medium |
| LoRa | 868MHz/900MHz | 0.3-50Kb/s | <30km | Very Low |

Table 1: Comparison of popular communication technologies (Ray 2016)

Wifi and Bluetooth, which are the two dominant wireles communication protocols use 2.4Ghz band and can support devices up to 100m in unabstracted view. However, in practice this number is far lower considering that quality of communication efficiency tends to get lower causing multiply retransmissions, delays, and higher energy costs. Energy is an important metric particularly for wireless sensor networks, which need to harvest their own energy as wiring them adds manual labor and complicates use. For this reason, lower frequency bands such as 400-900MHz have been proposed for use as they tend to offer a practical alternative for applications such as agriculture. One such new technology is known as LORA. In Figure 3, LORA based Wide Area Network architecture demonstrates an application where a gateway is used as an intermidiatory between the local wide area network and overhauling Internet servers.

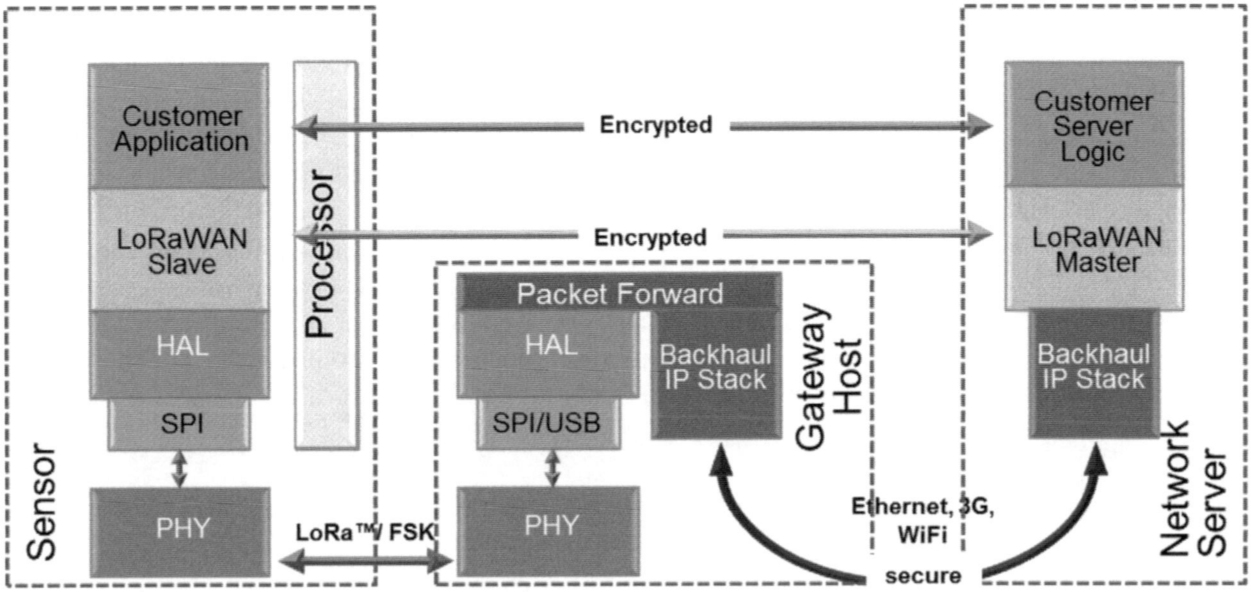


*Figure 3: LORAWAN Structure (LoRa 2017)*

## IoT - Computing

Although, semiconductor industry has yielded plurality computing device, it also created a very complex set of computing devices with parameters that's not easy to use to differentiate. Currently there are many devices ready for use in IoT setting, see Table 2, however, many of these are not meant for industrial and extreme condition use, although they may well be much more powerfull than those in Apollo missions.

Table 2: Comparison of popular IoT Platforms (Ray 2016)

| Platform | CPU | Operating Voltage | Clock (MhZ) | Bus Width |
|---|---|---|---|---|
| Arduino Uno | ATMega328P | 5V | 16 | 8 |
| Arduino Yun | ATmega32u4, | 5V, 3V | 16,400 | 8 |
| Intel Galileo Gen 2 | SoC X1000 | 5V | 400 | 32 |
| Intel Edison | SoC X1000 | 3.3V | 100 | 32 |
| Beagle Bone Black | Sitara AM3358BZCZ100 | 3.3V | 1024 | 32 |
| Electric Imp 003 | ARM Cortex M4F | 3.3V | 320 | 32 |
| Raspberry Pi B+ | BCM2835 | 5V | 700 | 32 |
| ARM LPC1768 | ARM Cortex M3 | 5V | 96 | 32 |

Cloud Platforms

What is cloud? Cloud is the general term used for abstraction of computing architecture into a service that includes a bit more than just computation and storage of information. It's a service provided to users for providing continues and uninterrupted service no matter the scale of the operation in a flexible and scalable manner. Cloud platforms, such as Dropbox offer file storage, while others such as Google Compute Engine offer computational resources. Depending how they charge their users, and what kind of costumer service they provide whether it's a technical expert 24/7 on duty or some of the hardware is located on site of the user, so called hybrid cloud service, there will be lots of choices for the costumers. That's why IoT will always have some sort of cloud service associated as it is just too easy to to integrate these two technologies together and get something more out of it, which is peace of mind that your data, your service is reliable and works all the time.

Table 3: Comparison of popular Cloud services for IoT (Ray 2016)

| Cloud Platform | R.T. Data | Visual. | Cloud Type | Analytics | Cost |
|---|---|---|---|---|---|
| Xively (https://xively.com/) | Yes | Yes | Public | No | Free |
| ThingSpeak (https://thingspeak.com/) | Yes | Yes | Public | Yes | Free |
| Plotly (https://plot.ly/) | Yes | Yes | Public | Yes | Free |
| Carriots (https://www.carriots.com/) | Yes | Yes | Private | No | Pay per use |
| Exosite (https://exosite.com/) | Yes | Yes | Hybrid | Yes | Pay per use |
| GroveStreams (https://grovestreams.com/) | Yes | Yes | Private | Yes | Limited |
| ThingWorx (www:thingworx.com/) | Yes | Yes | Private | Yes | Pay per use |
| Nimbits (www.nimbits.com/) | Yes | Yes | Hybrid | No | Free |
| Connecterra (www.Connecterra.io/) | Yes | Yes | Private | Yes | Pay per use |
| Axeda (www.axeda.com) | Yes | Yes | Private | Yes | Pay per use |
| Yaler (https://yaler.net) | Yes | Yes | Private | Yes | Pay per use |
| AMEE (www.amee.com) | Yes | Yes | Private | Yes | Pay per use |
| Aekessa (www.arkessa.com) | Yes | Yes | Private | Yes | Pay per use |
| Paraimpu (https://www.paraimpu.com/) | Yes | Yes | Hybrid | Yes | Limited |
| Phytech (http://www.phytech.com/) | Yes | Yes | Private | Yes | Pay per use |
| Cayane (www.mydevices.com) | Yes | Yes | Private | No | Mixed |
| EVineyard (www.evineyard.com) | Yes | Yes | Private | No | Mixed |
| WeatherUnderground (https://www.wunderground.com) | Yes | Yes | Private | No | Mixed |

## IoT - Usability – Why now? and How?

Fresh Water is the critical resource for humans, yet, its far more important for life on our planet. Worldwide, roughly 70% of fresh water is used for irrigation, and over 90% in least-developed countries (UN 2015). According to the 2005 US Census Data, total fresh water use in the United States was 355 billion gallons per day (Census 2013). That is over 1000 gallons per day per capita. Out of all fresh water usage in the United States, Irrigation utilizes 38% of all fresh water in the United States, which is often wasteful and highly inefficient (Hsiao 2007). Between 2008 and 2013 total fresh water used for irrigation in farms rose by at least 22% (Maupin 2010). The problem is further exacerbated with extreme climate events such as droughts, which cause environmental disasters. For example, due to drought California implemented a first-ever mandatory water reduction (CNN 2015). Yet, data shows that situation may worsen in 2030 with worldwide 40% fresh water deficit (UN 2012). That said, irrigation control is still largely done by quasi-rational techniques such as feel of soil or condition of crop, respectively, with 78% and 34% popularity among 232K surveyed by US Geological Survey (Maupin 2010).

Figure 4: USGS Irrigation Survey (USGS 2015)

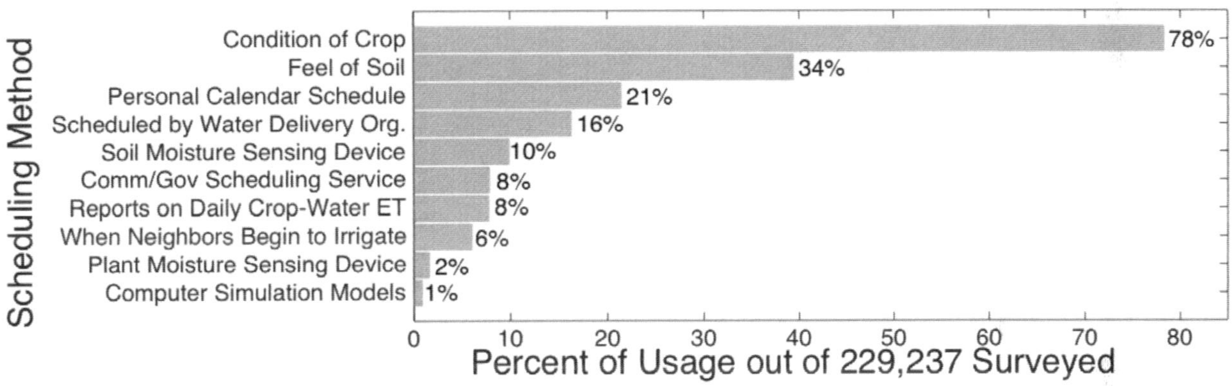


Irrigation systems are cyber-physical systems, because they are composed of man-made systems: irrigation networks and their controllers, and physical world: soil, atmosphere and plants. In cyber-physical systems, all elements involved in the overall picture must be carefully weighted. Moreover, the complete solution for an irrigation system must incorporate every step from design and development to deployment. In other words, there should be means to make design decisions a priori, use the known to engineer tailored or standardized solutions, and finally, recalibrate system settings during or after the final stages of installation on the farm.

Global deficit of fresh water poses challenges just like energy, however, it has not been addressed with the same intellectual investment. Hence, to address water deficit demands with the same level of emphasis as other main stream domains, we have designed set of experiments which will try to examine irrigation scheduling practices and offer new insight to irrigation science. Specifically, our main objectives are in examining relationships between optimal scheduling techniques and yield of crops with respect to state of the art and conventional irrigation techniques and proposed irrigation methodologies.

Irrigation Background

We can write the soil moisture as a differential equation and use commonly used simulation tools to simulate this ordinary differential equation relationship (ODE):

$$\frac{\partial m}{\partial t} = \mathrm{P} - \mathrm{E_T} - \mathrm{R} \qquad (1)$$

or

$$\frac{\partial m}{\partial t} = \mathrm{P} - \mathrm{E_T} - \mathrm{R} + \mathrm{I} \qquad (2)$$

where $m$ is soil water content, $t$ is time, $P$ is precipitation, $E_T$ is evapotranspiration, $R$ is total surface runoff, and $I$ is the irrigation.

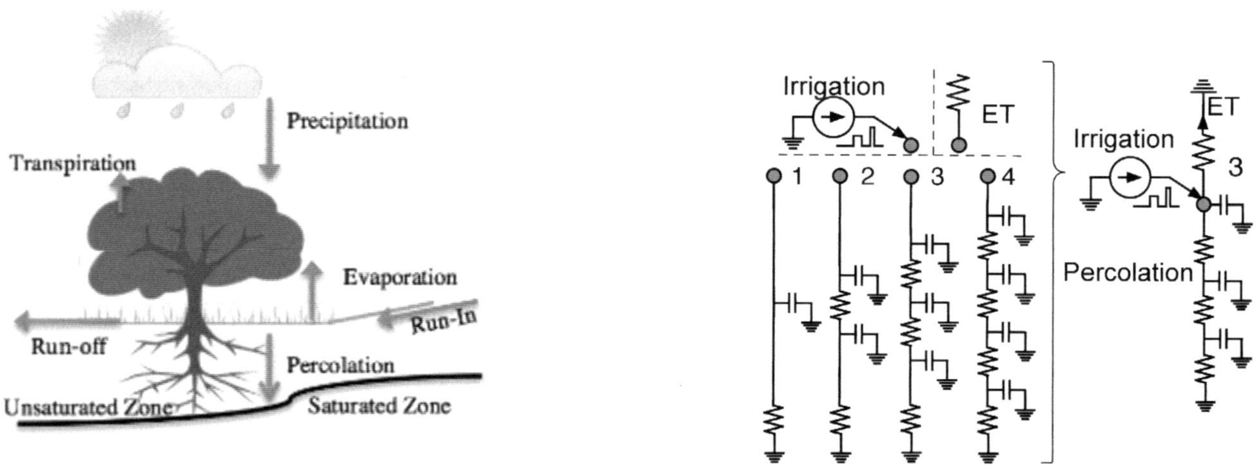


Figure 5: Basic Hydrologic Processes in Irrigation (left) and Equivalent Circuit Model (right) (Hovhannisyan 2016)

Thus, using the analogy of hydroelectric phenomena, we were able to model the soil water percolation as a circuit. The idea of incorporating all the stakeholders in one loop was the main driver for this contribution. On the right is the graphical representation of the model and modeling strategy.

These are the reasons why its important to bringing in hindsight expensive technologies to agricultural proctice. One way to bring IoT into agriculture is the maker community. In the following case study, where we will talk about a maker community in Temecula, CA where not just sensor and weather stations are made for fun, but for reducing community water usage by 5-10% and saving the water district $10mil. Whether they will succeed or fail, only time and math can say, but they won't give up until they do!

## Case Study: Temecula Valley

The traditional irrigation practice for vineyards is a weekly good long soak. However, long irrigation drains deep in the soil, whereas short irrigation achieves majority of irrigated water staying at higher levels. Thus, irrigating once a week, and replacing the weekly amount of water in one irrigation cycle, applies more water than what can be used in one or two days. The surplus of water will drain deeper and eventually become out of reach of the active roots. Unlike popular belief, even for plants with long roots, like grapevines, most of the actual update of water takes place at shallow soil levels (up to 4 feet). Together with the draining water fertilizers also wash away, thus, reducing fertilizer efficiency and polluting aquifer. By irrigating more frequently, like every day, there is more granular dosage of irrigation water, closely following the (daily) evapotranspiration needs. The main goal is to supply the precise amount of water needed and have it delivered only to the soil layers with the active root system, where the plant uptakes it. The expereiments took place in Van der Lee Vineyard as well as other local Vineyards (Figure 7).

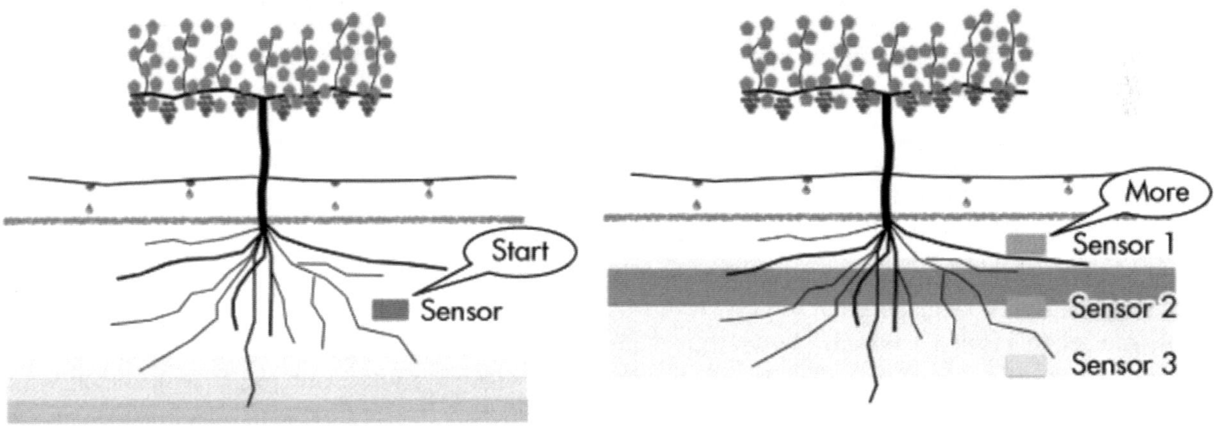


Figure 6: Comparison of single (left) and multi-sensor stations (right)

Figure 7: Van Der Lee Vineyard

We used multiple sensors within the active root zone to monitor the available water to the plants. By placing a soil-moisture below the root zone, we can detect percolation reaching that level (Figure 6: left). Looking at the soil model circuit in Figure 5 we can see that starting from the ground level and moving down, a time-varying signal goes through successive stages of low-pass filters. This is reflected in Figure 6, which were obtained using SPICE simulation of the soil water transport model in Figure 5 which shows that water moisture level becomes more stable at lower soil layers and for the daily irrigation schedule. This behavior predicted in SPICE simulation was verified by actual measurements at different soil levels shown in Figure 9. The moisture levels at deeper and shallower levels track reasonably well, and the deeper sensor moisture levels looks like the moisture level at the shallower level but attenuated and passed through a low-pass filter. The variations in the water levels over the days is due to the different levels of evapotranspiration due to temperature changes within and between days of the experiments. Although, these changes were not captured in the SPICE simulations, they could be easily accommodated into the SPICE models by varying the R value in the ET circuit model based on the daily weather forecast.

Figure 8: SPICE Simulation at different depths.

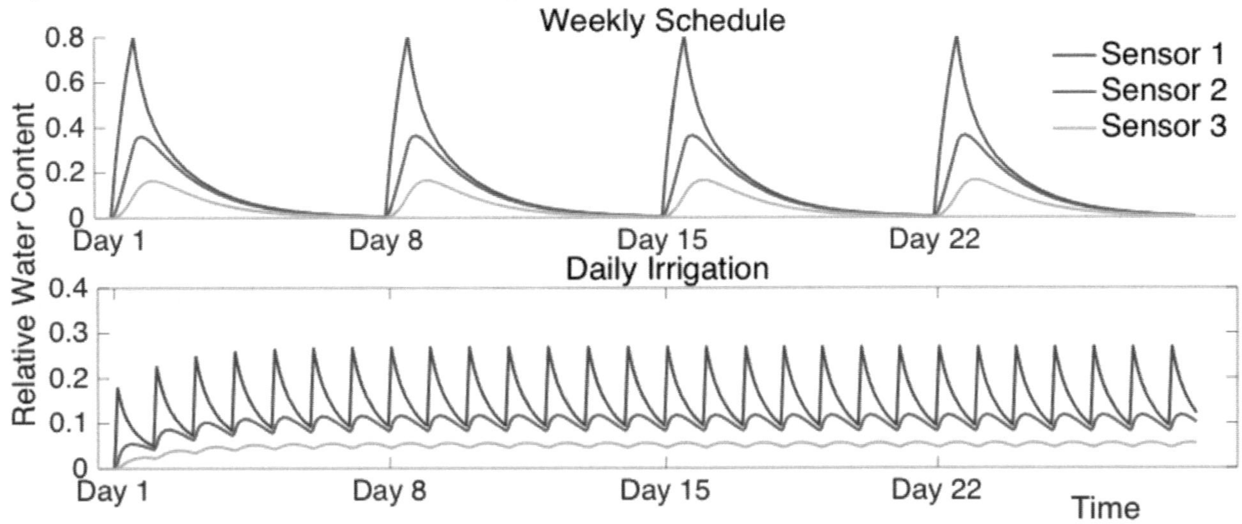


Figure 9: Vinduino Sensor Measurements

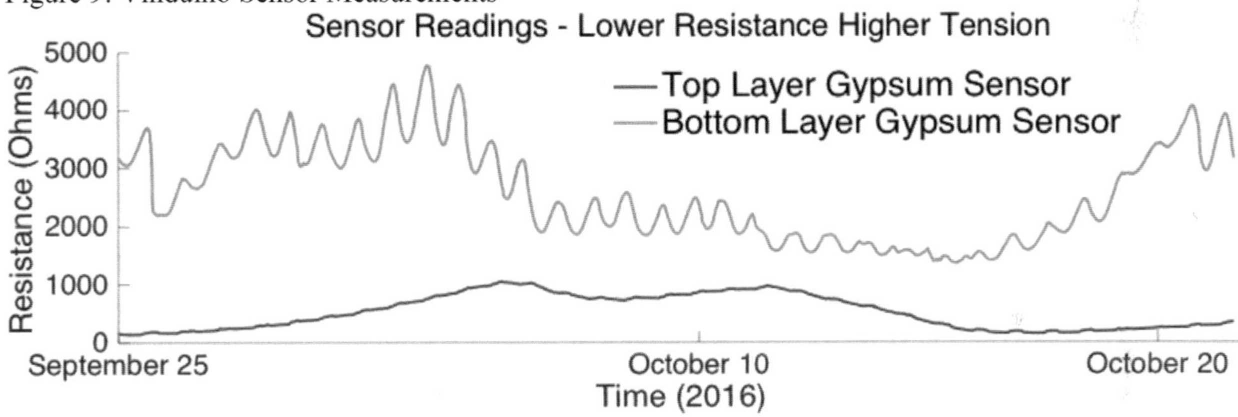


## Conclusion and Future of IoT in Irrigation

In conclusion, we can see that IoT has potential not to just improve efficiency of irrigaiton saving of around 50% for a given period year to year (Figure 10), but also to bring together people and excitement to this very important area, where there is still so much to be learned.

Figure 10: Comparison of Water usage in 100 cu. Ft/month.

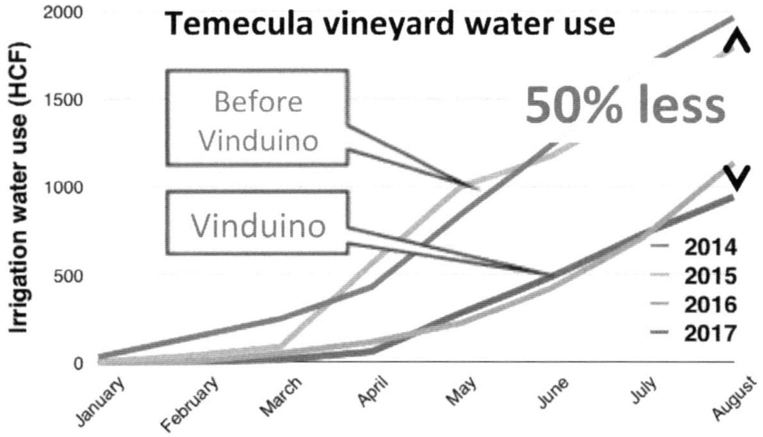


## Future of IoT in Irrigation

Technologies where IoT can be used are:

- Smart Meters: Bidirectional meters for water producers and consumers. Similar to Smart grid for electricity, some farmers might be able to sell water to the water network and their neighbores and use the biderectional metering for that purpose in which case they would be able to utilize IoT metering to be able to get their readings real time without human in the loop. Otherwise, using meters fr internal site-specific metering for irigation zones and division of agricultural (priority) vs other commercial use cases, for flexibile policies from water district which may prioritize day times for normalizing pressure and rates for use cases.
- Irrigation Removte Controlled Self-Sufficient IoT Valves
    - Micro irrigation valves for per site irrigation. IBM has demonstrated in the Galo Vinery that distributed control systems can save signiicant water over traditional single controller.
    - Remote controlled valves for main lines will support emergency shut off conditions that will allow reducing water loss due to busted pipes and animals chewing on tubes.
    - Valves for zones in site specific irrigaiton could control flow as well as pressure and irrigaiton schedule for most efficient water use cases.
- Leak Detection - Major problem for automated tools.
    - Using pressure sensor with valves seems to be possible. In fac we have demonstrated in our lab that we can model water flow and use the model for flow characterization, so nothing gets lost.
    - Using sonic sensors for detection of vibrations across pipes and in areas of breakage.
    - Using flow data and Artificial Intelligence to track the flow areas and leaks.

## Infrastructure Upgrades

- Pipes and Tubes that integrate sensors, can the pipe have the leak detector embedded?
- Valve controllers with internal power generation mechanisms – flow, solar and etc.
- Weather stations – that integrate into one national service.
- Base stations – that allow integration of all parts and components by provide wide area networks.

In the future, these technologies will be avilable and our initial in lab findings suggest that can very well be in market now or in near future.

## References

Ray, P.P. A survey on Internet of Things architectures. Journal of King Saud University – Computer and Information Sciences (2016), http://dx.doi.org/10.1016/j.jksuci.2016.10.003

D. Hovhannisyan, F. Kurdahi, A. Eltawil, A. Aghakouchak and M. A. Al Faruque, "Poster Abstract: Unifying Modeling Substrate for Irrigation Cyber-Physical Systems," *2016 ACM/IEEE 7th International Conference on Cyber-Physical Systems (ICCPS)*, Vienna, 2016, pp. 1-1. doi: 10.1109/ICCPS.2016.7479116
URL: http://ieeexplore.ieee.org/stamp/stamp.jsp?tp=&arnumber=7479116&isnumber=7479059

Vinduino LLC, URL: Vinduino.com

Maupin, M.A., Kenny, J.F., Hutson, S.S., Lovelace, J.K., Barber, N.L., and Linsey, K.S., 2014, Estimated use of water in the United States in 2010: U.S. Geological Survey Circular 1405, 56 p., https://dx.doi.org/10.3133/cir1405.

USDA/NASS QuickStats Ad-hoc Query Tool, https://quickstats.nass.usda.gov/

United Nations, United Nations World Water Development Report 2015: Water. 2015.

T. Hsiao, P. Steduto, and E. Fereres. A systematic and quantitative approach to improve water use efficiency in agriculture. Irrigation Science, 25(3):209-231, mar 2007.

H. Census and N. S. Bureau. Fact sheet. (975):4–5, 2005.

CNN. California drought: Gov. Jerry Brown issues water rules - CNN.com, 2015.

W. W. A. Programme and Earthscan. The United Nations World Water Development Report 3 . Technical report, 2009.

USGS. USGS Water Use in the United States, 2015.

USGS: Howard Perlman. Irrigation: Irrigation techniques, USGS Water-Science School, 2015.

W. W. A. Programme and Earthscan. The United Nations World Water Development Report 3 . Technical report, 2009.

LoRa, Lora Alliance, URL: https://www.lora-alliance.org/technology

Nolan, T., Indian University, http://home.kelley.iupui.edu/notaylor/S305/labs/internetwork.htm

Pixobay, https://pixabay.com/en/internet-network-scheme-154450/

IBM, "How IBM is Bringing Watson to Wine" http://fortune.com/2016/01/09/ibm-bringing-watson-wine/

# Open-source Hardware & Software for Irrigation

Daniel K. Fisher

USDA Agricultural Research Service, Crop Production Systems Research Unit, 141 Experiment Station Rd., Stoneville, MS 38701 USA. daniel.fisher@ars.usda.gov

**Abstract.** *Information is needed in order to study, monitor, or understand a process or event. Data collection efforts often involve compromises related to frequency of measurements, labor requirements, costs, and data access. Advances in electronic technologies, software, and communications infrastructure have resulted in a variety of new and inexpensive monitoring capabilities for agricultural, irrigation, and water-management applications. Programmable microcontrollers and solid-state sensors enable rapid development of unique automated sensing, monitoring, and control systems. Wireless, cellular, and internet infrastructures allow rapid data transfer, and enable transfer of data from remote locations and real-time access to and sharing of data. These emerging technologies are enabled by application of open-source hardware and software, which offer advanced tools and capabilities and are freely available for anyone to use, develop, and modify. Open-source hardware and software resources are discussed, and were used to develop sensing and monitoring systems. Examples of products that have been developed, including a soil-moisture monitor and a mobile plant-canopy monitoring system, are presented to demonstrate the accessibility and usefulness of these development tools.*

**Keywords.** Arduino, sensors, cellular, internet

## Introduction

In order to monitor, study, or understand a process or event, information is needed. Data collection efforts often involve compromises related to frequency of measurements, labor requirements, costs, and data access. Advances in electronic technologies, software, and communications infrastructures have resulted in a variety of new and inexpensive monitoring capabilities for agricultural, irrigation, and water-management applications. Manual measurements can often be automated, reducing time- and labor-intensive activities while increasing the frequency of measurements. Driven by mass-market consumption of electronic devices, costs of electronic sensing and monitoring components have been greatly reduced in recent years, allowing for increased availability and access. As the reach of communications infrastructures, such as wireless, cellular, and internet networks, expands, data transfer options are becoming more accessible for new and unique applications.

While many recent technological advances have been aimed at consumer markets (smartphones, tablet computers, home automation, entertainment systems), the same technologies are available to developers for other markets. Hardware, such as programmable microcontrollers, solid-state sensors, and wireless communications, and the software tools to enable the hardware to function are accessible via the concept of "open-source." Open-source refers to the free and open development of hardware and software, wherein the developer makes all source code, blueprints, and documentation available to all to use, modify, or replicate without limitation. This openness is intended to improve quality, access, and understanding of ideas and products through collaboration rather than competition. Open-source software (Open Source Initiative, http://www.opensource.org) and hardware (Open Source Hardware Association, http://www.oshwa.org) communities have actively advanced software (Linux and Android computer operating systems) and hardware (Arduino) projects, among many others.

The Arduino project (http://arduino.cc) offers options to the agricultural research community, which, due to its small market size, is often underserved and lags in technological progress. The Arduino project consists of hardware and software tools, as well as a large community of collaborators, that can be used to develop unique sensing, monitoring, and interactive devices. Hardware consists of a microcontroller-based development platform, and accompanying software consists of an Integrated Development Environment (IDE) for programming and interacting with the microcontroller. The collaborative community provides programming expertise and assistance, additional hardware and software features to gain increased functionality, and technical support if needed to advance individual projects. Researchers have begun investigating the use of open-source technologies in studies related to irrigation (Bitella et al., 2014; Masseroni et al., 2016; Fisher et al., 2017; Payero et al., 2017), agriculture (Fisher and Sui, 2013; Di Prima, 2015; Thalheimer, 2016; Fisher and Huang, 2017), and the environment (Fisher and Gould, 2012; Mesas-Carrascosa et al., 2015).

The objective of the work presented is to describe and discuss some of the open-source hardware and software options available to the agricultural and irrigation research community. Hardware and software tools for sensing and monitoring, and for data-transmission and access, are presented, followed by examples of open-source monitoring systems that have been developed and deployed under agricultural conditions.

## Open-source hardware

Development of sensing and monitoring systems involves the integration of several main hardware components, including a microcontroller, electrical power supply, sensors, and data communications. Other components can be incorporated to offer additional features if desired, such as timekeeping, data storage, and information display.

### Microcontroller

The microcontroller serves as the basis for sensing and monitoring instruments. A microcontroller is a programmable device that is configured and programmed to accomplish specific tasks. When the Arduino Uno (http://arduino.cc) open-source development platform was first introduced, Arduino-branded and compatible development boards were built around an 8-bit microcontroller which operated at either a 5 V level and a processing speed of 16 MHz or a 3.3 V level and 8 MHz speed. Programming memory was somewhat limited at approximately 32 Kbytes, but the many built-in features, such as 10-bit analog-to-digital converter, multiple communications ports, and multiple digital input/output pins, resulted in a development board with great flexibility that could be adapted to many and varied applications.

A unique feature of the original Arduino development board was its standardized size and arrangement of electrical power and input/output pins. This standardized layout allowed for the development of peripheral devices, called shields, which could be plugged into the microcontroller board to increase functionality. Microcontroller boards and shields offered by third-party developers that were produced in the standardized form were then also compatible, allowing different microcontroller boards and shields to be interconnected.

As the popularity of the Arduino platform increased, the open-source nature of the Arduino project allowed for the development and availability of a variety of compatible devices. The devices use the same 8-bit microcontroller, allowing the use of the Arduino IDE and existing programming libraries, and for the uploading of existing programs. The form factor of newer development boards is often different, however, and additional features are increasingly being built in. Examples of newer devices include the Pro Mini (Sparkfun Electronics, http://sparkfun.com), a smaller, less expensive board with the same features as the original Arduino Uno. The Moteino (LowPowerLab, http://lowpowerlab.com) offers the same features, plus an optional memory chip with approximately 500 Kbytes of data storage capacity, and several low-power, license-free radio options. The Feather series (Adafruit Industries, http://adafruit.com) offer a microcontroller development board with a variety of optional features including built-in micro SD card, Bluetooth and wi-fi radios, and cellular modem. The original Arduino Uno and several newer development boards are shown in Figure 1.

Rapid advances in electronics technologies continue to make more powerful and less expensive microcontrollers available, and the Arduino project continues to add support for many of these microcontrollers. A new generation of Arduino and compatible development boards features a 32-bit microcontroller, which offers an increased processing speed of 48 MHz, increased program memory of 256 Kbytes, 12-bit resolution analog-to-digital converter, and additional communications features. The Arduino Zero and M0 development boards feature the same form factor as the original Arduino Uno, enabling the connection of existing peripheral shields. Other boards maintain the form factors of their respective manufacturer's earlier products, such as the Pro Mini (Sparkfun Electronics) and the Feather (Adafruit Industries), while updating the boards' processing power with 32-bit microcontrollers.

## Electrical power

Sensing and monitoring activities for agricultural, irrigation, and water-management applications often occur in remote areas where access to electrical power is limited. Unattended operation of electronic instrumentation, therefore, requires the use of a self-contained power source, usually in the form of batteries. To enable long-term monitoring, power consumption must be minimized to prolong battery life and extend time intervals between site visits.

To minimize power consumption, software and hardware solutions can be used or combined. Microcontrollers usually have multiple low-power sleep modes that can be programmed to reduce current consumption during periods of inactivity. If peripheral components in the sensing circuit, such as sensors or data-storage devices, remain powered, however, they remain in an

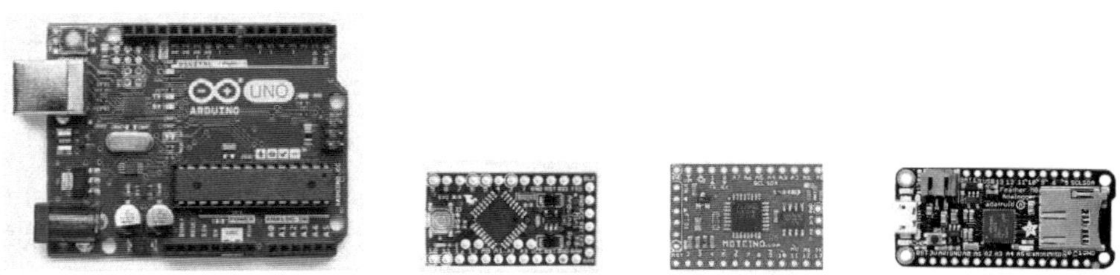

Figure 1. Microcontroller development boards, from left to right; Arduino Uno, Pro Mini, Moteino, and Feather.

active state and their current consumption is not reduced. The circuit can often be redesigned so that power to the peripheral devices is controlled by the microcontroller, enabling the microcontroller to turn the peripherals on during measurement periods and off during low-power sleep periods, greatly reducing drain on the batteries.

Various hardware solutions exist in the form of electronic components designed specifically for enabling low-power circuits. One such component is the TPL5110 Low Power Timer (Adafruit Industries, LowPowerLab), which serves as an interface between an electronic circuit and the battery. The low power timer switches battery power on to the microcontroller-based circuit at a regular time interval, which is set by a resistance value. When power is turned on, the circuit is activated and the microcontroller program executes. After completing the measurement functions, the microcontroller sends a signal to the low power timer instructing the timer to turn power off to the circuit. Battery power is completely disconnected from the microcontroller and circuit, resulting in essentially no current consumption from the battery.

In some applications, the complete shutdown of a monitoring circuit may not be desirable. To maintain long-term operation, a solar-powered circuit can be installed to continuously recharge the battery. The USB/Solar Lithium Ion Battery Charger (Adafruit Industries) and a small Monocrystalline Solar Cell (Newark element14, http://www.newark.com) can be added to almost any circuit to ensure long-term operation.

## Sensors and peripheral devices

Analog and digital sensors have been in use in irrigation and agricultural applications for decades. Soil-moisture sensors such as the Watermark (Irrometer Company, Inc., http://www.irrometer.com) and EC-5 (Decagon Devices, http://www.decagon.com) and meteorological sensors (Campbell Scientific, http://www.campbellsci.com) are well-known to the agricultural and irrigation communities.

Rapid proliferation of sensing and monitoring systems for automotive, consumer devices, and home automation and surveillance purposes has made new generations of sensors available which are designed for integration into microcontroller-based circuits. The often small and inexpensive sensors offer opportunities to expand monitoring efforts for a variety of sensed parameters. Sensors to detect and measure motion (presence, movement, acceleration), location (GPS), temperature (ambient, noncontact/infrared), light/radiation (intensity, RGB/color, multispectral, UV), air quality ($CO_2$, VOC), weather (air temperature, humidity, pressure), and distance (ultrasonic, infrared, lidar) are readily available. In many cases, however, the sensors have been miniaturized to the point where they are difficult to work with other than with specialized surface-mount instrumentation. A number of suppliers, such as Adafruit Industries and Sparkfun Electronics, have recognized this constraint and offer many of these sensors on breakout boards. The breakout boards contain the sensors and other auxiliary electronic components on a circuit board, and make available the connections necessary for interacting with the sensors. Breakout boards provide convenient access to the sensors for prototyping or for inclusion in sensing and monitoring systems.

Sensors are designed to interact with microcontrollers via several standardized communications protocols. Standard two-wire serial ports are common with some components, such as telephone modems and GPS receivers. Newer digital sensors often communicate via Inter Integrated Circuit (I2C) or Serial Peripheral Interface (SPI) protocols, designed specifically for communications between microcontrollers and digital devices. The two-wire I2C interface allows multiple components to connect via the same two microcontroller pins, with each component identified by a unique address. The three-wire SPI interface is similar, allowing multiple devices to share the same communications pins, but each component requires an additional pin for use in identifying that component.

Sensors are often incorporated into a microcontroller-based circuit very simply by connection of appropriate power and logic connections to the microcontroller. Sensor measurements are obtained through the programming of the microcontroller, with the program written to send appropriate addresses or signals to the sensor, wait for the sensor to make a measurement, then read data digitally from internal registers. The microcontroller applies calibration algorithms supplied by the sensor manufacturer to convert the register values to physical measurements. Many programming routines and libraries have been written and are available from the Arduino community, or can be developed from detailed information provided by sensor manufacturers.

Sensors and other peripheral devices are available that can be used with almost any microcontroller. Real-time clocks provide date- and time-keeping functions, allowing events to occur at regular intervals or specific time periods, and as timestamps for data storage. Data can be stored to memory chips or SD cards, which can offer almost unlimited amounts of data to be collected. A number of these sensing and peripheral options have also been developed for specific microcontroller development boards. Shields developed for the original Arduino Uno form factor connect easily to those and other compatible boards. The Feather series (Adafruit Industries) also maintain a standardized form, which while different from that of the Arduino Uno, supports other mating peripheral boards to add increased functionality, such as a GPS receiver, real-time clock, data storage, various radios, and switching relays.

## Communications

An important part of a sensing and monitoring system involves the display or transmission of the data collected. While many commercial dataloggers and monitoring systems contain a display for local viewing of information, in many agricultural and environmental-monitoring applications, monitoring sites are in remote or inconveniently accessible locations. Advancements in communications infrastructures have resulted in several options for access to data from remote locations.

Industrial, Scientific and Medical (ISM) radio bands have been designated internationally for low-power, license-free transmission of periodic data. Due to transmission-power limitations, radios operating in these bands are capable of transmitting data to nearby (usually less than one mile) receivers. Many radios, such as the Xbee (Digi International, http://digi.com), interface with a microcontroller using a standard serial port and are simple to incorporate into a monitoring circuit and microcontroller program. Development boards such as the Moteino (LowPowerLab) and Feather (Adafruit Industries) can be configured with built-in radios, further simplifying development of a wireless monitoring system.

As the popularity of Bluetooth-enabled and wi-fi-connected devices and networks has increased, circuits are increasingly being developed with these capabilities. Bluetooth and wi-fi radios are available for adding on to existing circuits, and available built-in on development boards such as the Feather and Huzzah (Adafruit Industries) and ESP8266 Thing (Sparkfun Electronics), among many others. Bluetooth-enabled devices can be controlled by or interact with portable electronic devices such as tablet computers and smartphones, which can take the place of dedicated data displays or radio controllers, reducing cost and complexity of the monitoring system. Wi-fi-connected devices can use existing wi-fi networks to connect sensing systems and upload data to computers and internet-connected devices.

The rapid expansion of the cellular communications network has resulted in the ability to be connected from almost any place in the country. Cellular modems interface easily with microcontrollers via serial communications, and inexpensive modems and machine-to-machine cellular-data plans make cellular connection an affordable option. Stand-alone modems such as the Fona (Adafruit Industries) can be used to add cellular connection to a new or existing circuit, while other development boards, such as the Feather (Adafruit Industries) and Electron (Particle, http://particle.io), offer built-in cellular capability.

### Open-source software

In addition to the open-source hardware, the microcontroller development board, a key component of the Arduino project is the open-source software used to program the microcontroller. The Arduino Integrated Development Environment provides a programming environment in which programs are written, in a language based on C/C++, to instruct the microcontroller and manage its operation. The IDE is downloaded from the Arduino project website (http://Arduino.cc) and installed on a computer. The IDE allows for external programming libraries to be included in a program. Many libraries have been written to manage specific peripheral devices, such as real-time clocks, sensors, and cellular modems, and allow a user to rapidly and conveniently incorporate advanced devices and features into a program without the need for the user to understand the often complex programming of the peripheral. After a program is written, the IDE compiles the program and checks for programming errors. The program is then converted into the format required by the microcontroller and uploaded. The IDE also includes a serial monitor that can be used to interact with the microcontroller and view output from the microcontroller program.

An increasingly desirable and attainable option for remote monitoring systems is the transfer and access of remote sensor data via the internet. Many "Internet of Things" and "cloud computing" services are available that allow a user to configure a webpage for posting and viewing of sensor data anywhere using a web browser. Services such as Thingspeak (http://thingspeak.com), Adafruit IO (http://io.adafruit.com), and Thinger.io (http://thinger.io), among others, offer data-hosting and viewing at no cost for low-volume use, and are fairly simple to set up and use. Some services, including Thingspeak and Sparkfun IO, are open-source, allowing the cloud-computing software to be downloaded and installed by anyone to implement a similar service on a local or internet-facing computer.

## Example projects

Several projects have been undertaken using some of the components described previously to develop sensing and monitoring systems for irrigation and agricultural applications. Two such systems are described in the following sections. Further details are available in cited publications, and microcontroller programs are available by contacting the author.

### Soil-moisture monitor

A soil-moisture monitoring system was developed to continuously measure soil-moisture status for use in scheduling irrigations. The monitoring system consisted of hardware components to measure moisture status and transmit data, and software components to control the hardware and post data via cellular network to an internet website.

The system was based on a Feather 32u4 Fona microcontroller development board (Adafruit Industries) with a built-in cellular modem. Electrical power was supplied to the microcontroller-based circuit by a rechargeable 3.7 V lithium polymer battery through a TPL5110 low power timer (Adafruit Industries). The low power timer was configured via on-board potentiometer to supply power periodically to the circuit at its maximum time interval of approximately 2.5 h. Female headers were soldered to a prototyping board which mated with male headers soldered to the microcontroller board and low power timer. Spring terminal blocks were soldered to the prototyping board to allow connection of four Watermark model 200SS granular matrix sensors (Irrometer Company, Inc.). Each sensor interfaced with the microcontroller via a half-bridge (voltage divider) circuit, consisting of a fixed resistor on one side of the divider and the variable-resistance Watermark sensor on the other side. The center of the divider was connected to an analog input pin, and the divider's excitation and ground connections to two digital output pins. The completed hardware circuit is shown in Figure 2, and further hardware information and electrical schematic are detailed by Fisher et al. (2017).

A microcontroller program was written and uploaded to manage all sensor measurement and data transmission functions. When the low power timer turned on power to the circuit, the microcontroller program began execution. Each moisture sensor was read sequentially by applying an alternating voltage source to prevent polarization of the sensor. An excitation was first applied to the sensor's voltage divider circuit and the center voltage was measured. The polarity of the excitation was then reversed and the center voltage measured again. The basic voltage-divider equation was used to calculate the electrical resistance of the sensor, and a calibration equation was applied to convert resistance to water potential in units of kPa.

The cellular modem was then initialized and registered on the cellular network, and data-transfer and internet services were established. A unique internet address (URL) was created to post the data to the Thingspeak (http://thingspeak.com) website based on the website and channel credentials and four moisture-sensor data values. The cellular modem transmitted the URL, sending sensor data to the Thingspeak website. Data were then accessible via a web browser on a computer or mobile device, as shown in Figure 3.

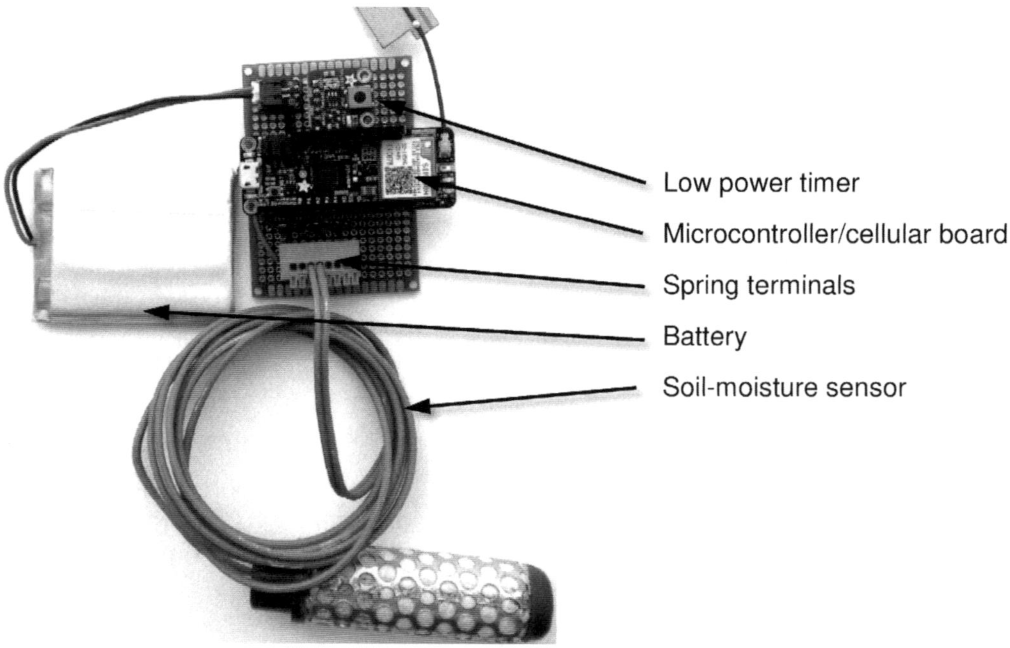


Figure 2. Soil-moisture monitoring system hardware components.

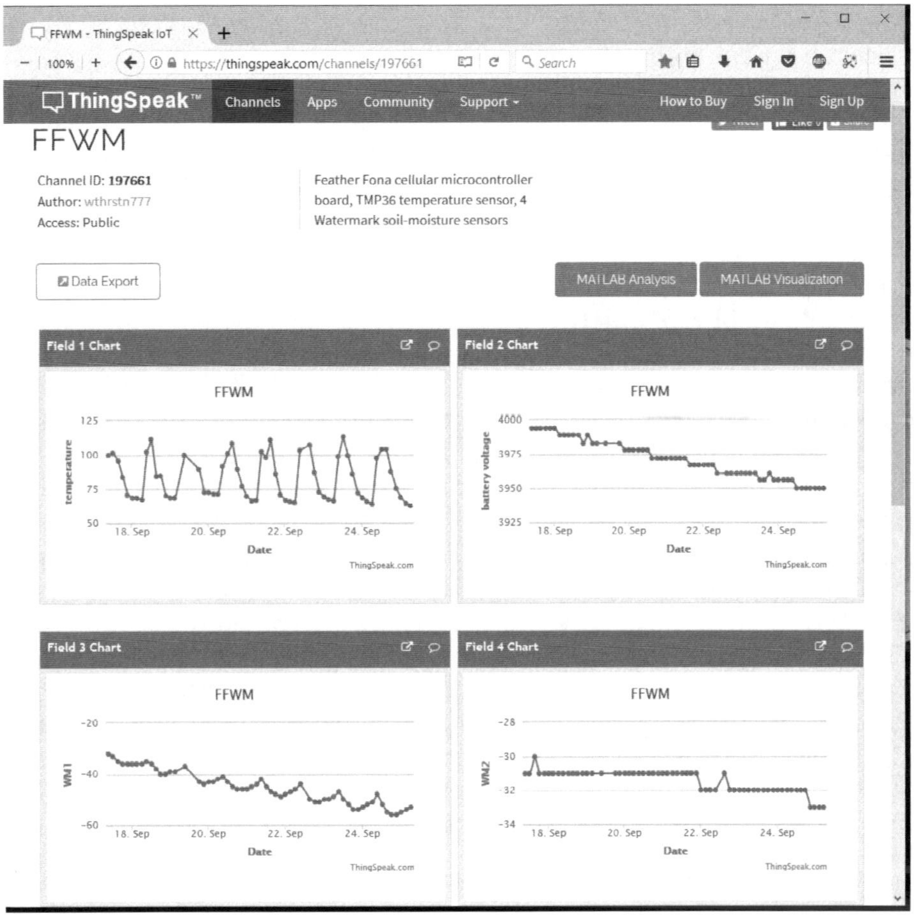


Figure 3. Soil-moisture monitor data posted on Thingspeak website.

Fabrication of the monitoring system circuit board was accomplished in approximately one hour using only basic soldering materials. The final cost of the system was approximately US$85 for the monitoring system circuit, US$30 each for the soil-moisture sensors, and US$30 per year for the cellular SIM card and 1 Mbyte per month data plan (Embedded Works, http://embeddedworks.net).

Monitoring systems were deployed in several fields at the USDA Agricultural Research Service's Jamie Whitten Delta States Research Center at Stoneville, MS, and operated throughout the 2017 crop growing season. Soil-moisture sensors were installed at four depths below the soil surface at each site to monitor water use and determine irrigation requirements. The systems continued to operate the entire season with no maintenance requirements or need to change or recharge batteries, and successfully transmitted approximately 98% of the data via cellular network to the internet website.

## Mobile plant canopy monitor

A project was undertaken to develop a mobile monitoring system for collecting plant-canopy data, including plant height and canopy temperature measurements. The monitoring system was mounted on an agricultural vehicle, and collected and stored sensor measurements, along with concurrent geographic coordinates, as the vehicle travelled across agricultural fields.

Plant-canopy data were collected from four crop rows concurrently and stored to a micro SD card for download to a computer for viewing and analysis.

The system was based on a Feather M0 microcontroller development board with an onboard Bluetooth LE (Low Energy) radio (Adafruit Industries). Additional features, including GPS receiver and micro SD card data storage, were incorporated via FeatherWing peripheral boards (Adafruit Industries) which plugged in to the microcontroller board. The Feather microcontroller and FeatherWing boards were designed to interconnect, and featured similar board dimensions and identical pin configurations, allowing boards to be staked together and simplifying the electrical circuit. Plant height was determined using HC-SR04 ultrasonic sensors (LowPowerLab), each of which interfaced with the microcontroller using two digital input/output pins. Canopy temperature was measured using Melexis model MLX90614ESF-ACF (Mouser Electronics, http://www.mouser.com) infrared temperature (IRT) sensors. The IRT sensors communicate using the I2C two-wire protocol, and the four sensors were connected over the same I2C port's two digital pins. Electrical power was supplied via a 3.7 V lithium polymer rechargeable battery, which was converted to 3.3 V via a built-in voltage regulator on the microcontroller board to power the Feather microcontroller and peripheral boards. The ultrasonic and IRT sensors, however, operated at a 5 V level, requiring a voltage converter (Miniboost 5V Boost Regulator, LowPowerLab) to supply the required voltage. An additional component, a Bi-directional Logic Level Converter (Adafruit Industries), was needed to interface the 3.3 V microcontroller and the 5 V IRT sensors to convert the different voltage levels and ensure that each component was exposed to the proper voltage level. Female headers for mounting the stacked Feather boards and the two voltage-regulating components were soldered to a prototyping board. Male headers were soldered to the prototyping boards to connect the sensors via custom-made cables. The completed prototyping board and electronic components are shown in Figure 4, with detailed information including electrical schematic provided by Fisher and Huang (2017).

The four-row monitoring system was assembled by first housing pairs of ultrasonic and IRT sensors in protective plastic enclosures, with the sensors installed into holes bored through the

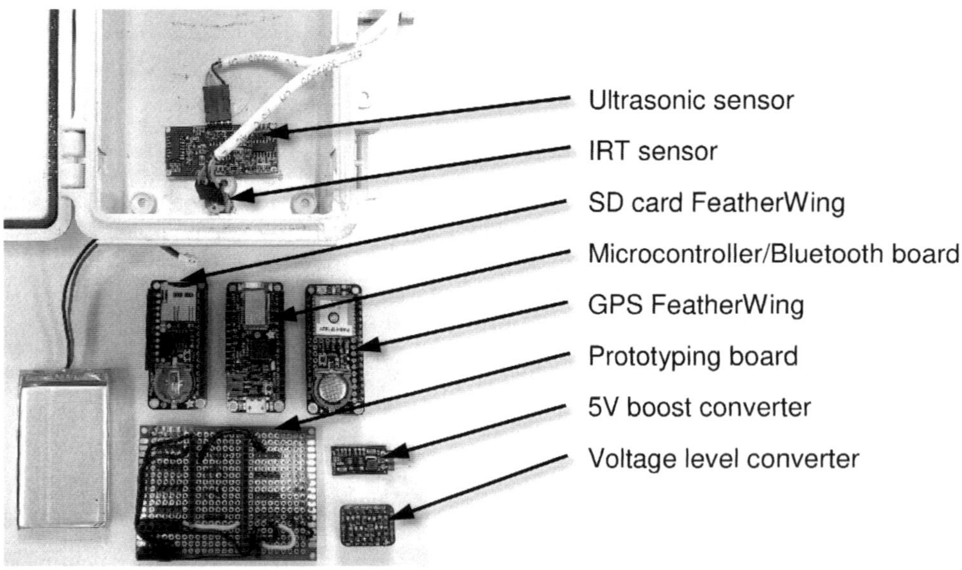


Figure 4. Plant canopy monitoring system components.

Figure 5. Complete plant canopy monitoring system mounted on an agricultural vehicle.

bottoms of the enclosures. Cables were fabricated to connect each sensor to the microcontroller circuit board, which was installed in a separate protective enclosure. The enclosures were attached to a length of aluminum angle stock, with the four sensor enclosures mounted at intervals matching the crop row spacing and the microcontroller enclosure at the center of the aluminum angle. The complete monitoring system mounted on the front of an agricultural spray vehicle is shown in Figure 5.

Construction of the mobile monitoring system, including fabrication of the circuit board and cables, modification of the protective enclosures, and mounting of the completed system on the agricultural vehicle, took approximately four hours. The cost of the hardware components and sensors totaled approximately US$292.

The monitoring system was tested in a field planted to soybean. The monitoring system was powered on and allowed time for the GPS receiver to obtain valid location information. A smartphone was paired to the Bluetooth radio, which was programmed to output GPS information and sensor measurements at 2-sec intervals. Viewing GPS and sensor output in real-time in the field allowed the user to ensure that all components were installed and operating properly prior to field data collection. With the agricultural vehicle stationary on a flat surface prior to entering the field, the heights of each ultrasonic sensor were recorded. Sensor heights were used later to post-process ultrasonic readings and determine plant heights by subtracting the ultrasonic distance measurements (from the sensor to the plant canopy) from the sensor heights.

Samples of plants heights and canopy temperatures collected during one trip across the field are shown in Figure 6. Slight variations in measurements among rows can be observed across the length of the field. Variations in crop characteristics can also be seen down the length of the field. As the vehicle travelled (in the direction from left to right in the graphs), an area of shorter and warmer plants was detected approximately two-thirds down the field. While the monitoring system cannot explain the cause of these differences, the data provide an indication to the user that crop or growing conditions were different in this area. The user could then visit the area to try to determine the cause of the differences.

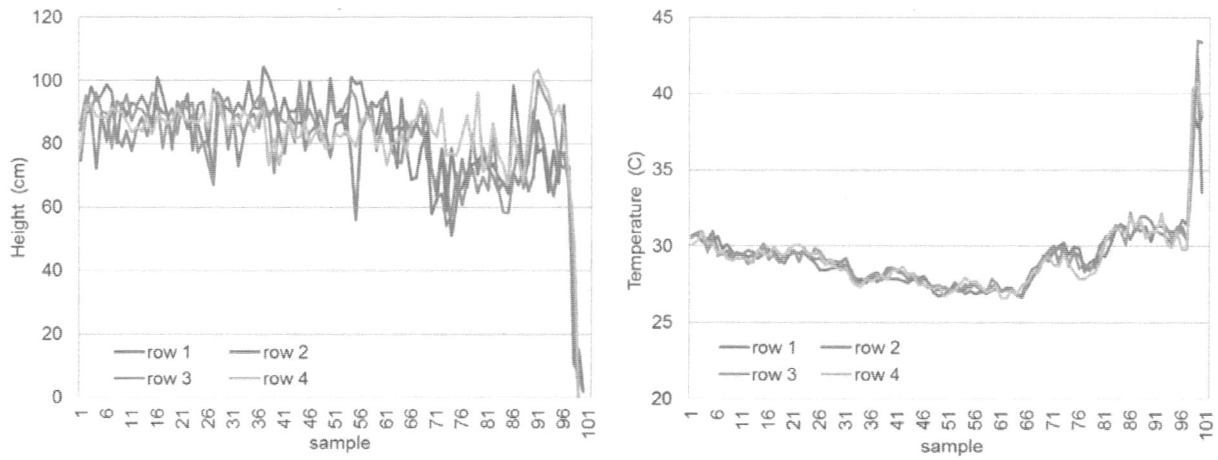


Figure 6. Plant height and canopy temperature data collected during one trip across the field.

## Conclusions

Rapid advances in electronics and communications technologies have resulted in a variety of options for development of sensing and monitoring systems. The open-source concept of free and open collaboration offers additional resources and tools for irrigation and agricultural applications, often underserved and lagging in technological progress. Open-source hardware and software can be used to develop new and unique monitoring systems to collect and transmit data from remote locations. Cellular and internet communications networks and open-source cloud computer services enable remote data to be viewed and shared conveniently in real-time, reducing site visits and time and labor expenses.

## Disclaimer

Mention of a trade name, proprietary product, or specific equipment does not constitute a guarantee or warranty by the United States Department of Agriculture, and does not imply approval of the product to the exclusion of others that may be available.

## References


Bitella, G., R. Rossi, R. Bochicchio, M. Perniola, and M. Amato. 2014. A novel low-cost open-hardware platform for monitoring soil water content and multiple soil-air-vegetation parameters. Sensors 14(10):19639-19659. http://dx.doi.org/10.3390/s141019639.

Di Prima, S. 2015. Automated single ring infiltrometer with a low-cost microcontroller circuit. Computers and Electronics in Agriculture 118:390–395. http://dx.doi.org/10.1016/j.compag.2015.09.022.

Fisher, D.K. and P.J. Gould. 2012. Open-source hardware is a low-cost alternative for scientific instrumentation and research. Modern Instrumentation 1(2): 8-20. http://dx.doi.org/10.4236/mi.2012.12002.

Fisher, D.K. and R. Sui. 2013. An inexpensive open-source ultrasonic sensing system for monitoring liquid levels. Agricultural Engineering International: CIGR Journal 15(4):328-334.

Fisher, D.K., and Y. Huang. 2017. Mobile open-source plant-canopy monitoring system. Modern Instrumentation (in press).

Fisher, D.K., R.S. Fletcher, S.S. Anapalli, and H.C. Pringle III. 2017. Development of an open-source cloud-connected sensor-monitoring platform. Advances in Internet of Things (in press).

Masseroni, D., A. Facchi, E. Vannutelli Depoli, F.M. Renga, and C. Gandolfi. 2016. Irrig-OH: An open-hardware device for soil water potential monitoring and irrigation management. Irrigation and Drainage 65(5):750-761. http://dx.doi.org/10.1002/ird.1989.

Mesas-Carrascosa, F.J., D. Verdú Santano, J.E. Meroño, M. Sánchez de la Orden, and A. García-Ferrer. 2015. Open source hardware to monitor environmental parameters in precision agriculture. Biosystems Engineering 137:73-83. http://dx.doi.org/10.1016/j.biosystemseng.2015.07.005.

Payero, J.O., A. Mirzakhani-Nafchi, A. Khalilian, X. Qiao, and R. Davis. 2017. Development of a low-cost Internet-of-Things (IoT) system for monitoring soil water potential using Watermark 200SS sensors. Advances in Internet of Things 7:71-86. http://doi.org/10.4236/ait.2017.73005.

Thalheimer, M. 2016. A new optoelectronic sensor for monitoring fruit or stem radial growth. Computers and Electronics in Agriculture 123:149-153. http://dx.doi.org/10.1016/j.compag.2016.02.028.

# Continuous Monitoring of Wine Grape Canopy Temperature for Irrigation Management


## Bradley A. King, Research Agricultural Engineer

USDA ARS, Northwest Irrigation and Research Laboratory, 3793 N. 3600 E. Kimberly, ID 83341, brad.king@ars.usda.gov.

## Krista C. Shellie, Research Horticulturalist

USDA ARS, HCRU Corvallis, OR- Parma, ID worksite 29603 U of I Lane Parma, ID 83660, Krista.Shellie@ars.usda.gov.



**Abstract.** *Automated monitoring of plant water status is a prerequisite for precision irrigation and water conservation. The objective of this study was to assess the potential for using a thermal-based crop water stress index (CWSI) as an irrigation management tool to assist with scheduling irrigation events and estimating irrigation amounts for selected wine grape cultivars in the arid Northwest U.S. The temperature of the vine canopy, soil volumetric water content, vineyard environmental conditions, irrigation events and amounts were continuously monitored in field plots of the wine grape cultivars Malbec and Chardonnay at three commercial vineyards in southwestern Idaho during the 2017 growing season. Select measured and calculated parameters were made available in real-time to vineyard managers on a website hosted by the data logger via cell phone modem. At all sites, the daily CWSI rapidly decreased during and following an irrigation event and gradually increased between irrigation events, indicating sensitive and rapid response to changes in available soil moisture. Throughout the growing season, the change in CWSI value reflected the relationship between plant available soil water (PASW) and the water stress coefficient (Ks) of the Penman-Monteith equation for estimating plant water demand. Data analysis suggests that automated calculation of a daily CWSI through continuous remote monitoring of vine canopy temperature and vineyard environmental conditions can be used to guide irrigation scheduling and estimate the reduction in vine water demand when transpiration is restricted by soil water availability.*



**Keywords.** Crop water stress index, wine grape, drip irrigation, automation.


## Introduction

Wine grapes (*Vitis vinifera* L.) are widely grown in arid and semiarid regions where irrigation is used to supplement annual precipitation and maintain a desirable level of vine water stress. Decisions about when to irrigate and how much water to supply during an irrigation event ultimately influence production profitability in terms of input costs, yield and fruit quality. Determining when to irrigate and how much water to supply during an irrigation event can be challenging due to the lack of an easy, reliable method for readily assessing the severity of vine water stress.

Measurements of soil moisture and plant water potential have been used to monitor vine water stress, but each have limitations that restrict their usefulness in an automated system. Williams and Trout

(2005) found that measurement of soil water content to a depth of 3m at nine locations within one-quarter of an individual vine root zone was necessary to accurately determine the amount of water within the soil profile that was available to drip-irrigated grapevines. The low spatial resolution was due to heterogeneous soil attributes, such as texture and depth, spatially heterogeneous irrigation wetting patterns (drip irrigation) and spatially heterogeneous rooting characteristics. Thus, numerous soil moisture monitoring sites would be needed to reliably infer vine water stress status. There is no general agreement as to which measurement of plant water potential (pre-dawn leaf or midday stem or leaf) most reliably indicates vine water status (Williams and Araujo 2002, Williams and Trout 2005, Ortega-Farias et al. 2012). Williams and Trout (2005) found that pre-dawn leaf water potential was unsatisfactory for accurately determining vine water status while midday leaf and stem water potential were linearly correlated and equally suitable for determining vine water status. Midday leaf water potential is the most common method used in California to indicate vine water status (Williams et al. 2012) perhaps because it is less time consuming than either pre-dawn leaf water potential or midday stem water potential allowing more acreage to be covered during optimum midday climatic conditions (Williams and Araujo 2002). Measuring leaf or stem water potential is labor intensive and values can be strongly influenced by environmental conditions (Rodrigues et al. 2012, Williams and Baeza 2007, Jones 2004). Under semi-arid conditions, the influence of vapor pressure deficit (VPD) on midday stem or leaf water potential has been found to differ according to severity of water stress (Williams and Baeza 2007, Williams et al. 2012). Under high evaporative demand, a midday value of leaf water potential less negative than -1.0 MPa has generally been accepted as indicative of well-watered vines (Shellie 2006, Williams and Trout 2005, Williams et al. 2012, Shellie and Bowen 2014, Bellvert et al. 2015).

Thermal remote sensing has been used to estimate drought stress in many crops, including grapevine (Maes and Steppe, 2012). A temperature-based crop water stress index (CWSI), developed by Jackson et al. (1981) and Idso et al. (1981), was found to more reliably indicate plant water status than soil volumetric water content (Jackson, 1982). The empirical CWSI is calculated as:

$$CWSI = \frac{(T_{canopy} - T_{LL})}{(T_{UL} - T_{LL})} \quad (1)$$

where $T_{canopy}$ is the measured temperature of the vine canopy, and $T_{UL}$ and $T_{LL}$ are the upper and lower canopy temperature thresholds when transpiration is completely limited and non-restricted, respectively. The CWSI ranges in value from 0 to 1 where 0 indicates optimum conditions for maximum transpiration ($T_{LL}$) and 1 represents a non-transpiring condition ($T_{UL}$). The need to schedule an irrigation event is signaled when the CWSI value exceeds a desired numerical threshold established by field experiments.

The amount of water to supply during an irrigation event to meet estimated plant water demand is commonly estimated using the Penman-Monteith equation (Allen et al., 1998). The equation used to estimate actual daily evapotranspiration ($ET_{c\ act}$) when environmental conditions, such as drought, limit potential transpiration is:

$$ET_{c\ act} = ET_r \cdot K_{cb} \cdot K_s + K_e \quad (2)$$

where $ET_r$ is the evapotranspiration of a reference crop (mm day$^{-1}$), $K_{cb}$ is a basal crop-specific coefficient, $K_s$ is a stress coefficient that accounts for the decrease in plant water demand due to restricted transpiration, and $K_e$ accounts for soil evaporation from precipitation or irrigation. The soil evaporation coefficient ($K_e$) under drip irrigation was assumed to be negligible in this study. A value for $K_s$ has been estimated from the relationship between percent available soil water (PASW) and the management allowed soil water deficit (MAD) (Allen et al., 1998) or as an asymptotic function of PASW (Jensen et al. 1970). The PASW is calculated as:

$$PASW = 100 \cdot \left[\frac{\theta - \theta_{pwp}}{\theta_{fc} - \theta_{pwp}}\right] \cdot D_{rz} \qquad (3)$$

where θ is current soil water content, $\theta_{fc}$ is soil water content (%) at field capacity, $\theta_{pwp}$ is soil water content (%) at permanent wilting point, and $D_{rz}$ is effective rooting depth (m). The equations proposed by Allen et al. (1998) to estimate $K_s$ are:

$$K_s = 1 \qquad PASW \geq MAD \qquad (4)$$

$$K_s = \frac{PASW}{100 - MAD} \qquad PASW \leq MAD$$

where MAD is a soil water content (%) below which a crop begins to experience a water stress and transpiration is reduced. If $ET_{c\,act}$ is different than 5 mm $day^{-1}$ then MAD can be adjusted as a function of $ET_{c\,act}$. For wine grapes, MAD has a suggested value of 35 to 45% (Allen et al., 1998). A value of 45% for MAD was assumed for analysis in this study. The equation proposed by Jensen et al. (1970) to estimate $K_s$ (Colaizzi et al., 2003) is:

$$K_s = \frac{ln[100 - PASW + 1]}{ln[101]} \qquad (5)$$

Both approaches for estimating $K_s$ are empirical and require knowledge of vine soil water availability (θ, $\theta_{fc}$, $\theta_{pwp}$, $D_{rz}$), which can be challenging due to the spatial heterogeneity issues previously discussed.

A value for $K_s$ has also been indirectly estimated from the CWSI (Colaizzi et al., 2003) and from the ratio of $T_{canopy}$ to $T_{LL}$ (Bausch et al., 2011). In both studies, $T_{canopy}$ in relation to $T_{LL}$ and/or $T_{UL}$ was used to estimate $K_s$, $ET_{c\,act}$ and soil water availability. The relationship proposed by Colaizzi et al. (2003) can be expressed as:

$$CWSI = 1 - \frac{ET_{c\,act}}{ET_{cp}} \qquad (6)$$

where $ET_{cp}$ represents crop evapotranspiration under the same climatic conditions in the absence of transpiration limiting soil water availability ($K_s$=1). This relationship indicates that CWSI = 0 when soil water is not limiting ($ET_{c\,act} = ET_{cp}$) and CWSI = 1 when crop evapotranspiration is zero due to root zone soil water depletion to permanent wilting point. Substituting equation 2 for the numerator and denominator of the right side of equation 6 ($K_e$=0) with a value of $K_s$ = 1 in the denominator ($ET_{cp}$) results in the relationship:

$$CWSI = 1 - K_s \qquad (7)$$

indicating that there is a relationship between the CWSI and soil water content such as that given by equations 4 and 5 or a similar crop specific relationship.

The CWSI has been of limited use with wine grapes due to the practical difficulty of determining values for $T_{LL}$ and $T_{UL}$ while simultaneously measuring $T_{canopy}$ (Jones et al., 2002). Approaches that have been used to estimate $T_{LL}$ include energy balance equations (Sepúlveda-Reyes et al., 2016; Möller et al., 2007) natural or artificial reference surfaces (Sepúlveda-Reyes et al., 2016; Pou et al., 2014; Möller et al., 2007), and the difference in temperature between $T_{canopy}$ and air relative to evaporative demand (Bellvert et al., 2015; Idso et al., 1981). A constant value relative to air temperature has been used to estimate a value for $T_{UL}$ (Möller et al., 2007; King and Shellie, 2016).

King and Shellie (2016) predicted $T_{LL}$ values for the wine grape cultivars Syrah and Malbec using a neural network (NN) model developed from cultivar-specific datasets of measured well-watered vine canopy temperature and environmental variables – solar radiation, air temperature, relative humidity and wind speed. They also estimated $T_{UL}$ as air temperature plus a constant of 15 °C based on the cumulative probability of measured canopy temperature minus air temperature for the study conditions. They showed good correlation of calculated daily average CWSI over a 2 hr period about solar noon with irrigation and precipitation events and amounts. The relationship of the CWSI to other methods of evaluating vine water stress was not evaluated. Given that a vine daily CWSI can be calculated for wine grape, its relation to other common vine water stress measurements and usefulness for irrigation management has not been evaluated. The objective of this study was to evaluate the relationships between the CWSI, midday leaf water potential and soil water content in two cultivars of wine grape to develop an understanding of how a daily average CWSI can be used as a management tool to increase irrigation precision.

## Methods and Materials

Equipment to measure vine canopy temperature, climatic conditions, and soil water content were installed at four sites in three, above-ground-drip irrigated commercial vineyards in southwestern Idaho on June 28th, 2017. Vine canopy temperature was measured using two infrared radiometers (SI-121 Infrared radiometer; Apogee Instruments, Logan, UT) on two vines separated by at least 5 m. The radiometers were positioned approximately 15 to 30 cm above recent fully expanded sunlit leaves located at the top of the vine canopy and pointed northerly at approximately 45° from nadir with the center of field of view aimed at the center of sunlight leaves. The measured canopy area received full sunlight exposure during midday and the radiometers were periodically checked and adjusted as necessary to ensure the field of view concentrated on recently fully expanded, sunlit leaves located on the top of the vine canopy. Environmental parameters; wind speed (034B wind sensor; Met One Instruments, Inc., Grant Pass, OR), air temperature, relative humidity (HMP50 temperature and humidity probe, Campbell Scientific, Logan, UT), and solar radiation (SP-110 pyranometer; Apogee Instruments, Logan, UT) were measured with instruments installed directly above the vine row within 15 m of the infrared radiometers. Soil water content was measured to a depth of 1.2 m in 10 cm depth increments using a Sentek Drill and Drop probe (Sentek Sensor Technologies, Stepney SA, AU) installed in the same vine row within 15 m of the infrared radiometers. The manufacturers calibration of each Drill and Drop probe was used in this study. Canopy temperature and climatic parameters were measured every minute, averaged over a 15-minute period and stored on a data logger (CR6, Campbell Scientific, Inc. Logan, UT). Soil water content was measured every 30 min and stored on the same data logger. The wine grape cultivars Malbec (MB) and Chardonnay (CH) monitored in each of two vineyards are hereafter referred to as MB1, MB2, CH1 and CH2. The upper and lower limits of volumetric soil water content (assumed field capacity $\theta_{fc}$ and permanent wilting point $\theta_{pwp}$, respectively) were estimated in 10 cm increments at each site according to the maximum and minimum values measured throughout the season. When no soil drying was apparent, particularly at deeper depths, the value for $\theta_{pwp}$ was estimated as half the $\theta_{fc}$ value. Irrigation amounts were measured using a tipping bucket rain gauge (RainWise, Inc., Trenton, ME) under a single drip line emitter with irrigation amounts recorded as 15-minute totals.

Cultivar specific neural network models were used to estimate $T_{LL}$ with the four measured climatic variables as model inputs (King and Shellie, 2016). The upper temperature threshold ($T_{UL}$) was estimated as air temperature plus 14 °C for area climatic conditions based on results reported by King and Shellie (2016). Daily CWSI was calculated as the average of 15-minute CWSI values from 13:00 to 15:00 MDT. The CWSI values were calculated in real time by the data logger and stored. Select measured and calculated parameters were made available real-time to vineyard managers on a website hosted by the data logger via cell phone modem. Irrigation decisions were made solely by the vineyard manager, each of which had access to real time values for daily CWSI and soil moisture content.

Vine water status was monitored weekly throughout berry development by measuring leaf water potential at midday ($\Psi_{md}$) using a pressure chamber (model 610; PMS Instruments, Corvallis, OR) following the method of Turner (1988) as described by Shellie (2006). Two, fully expanded, sunlit leaves were measured on each vine monitored with infrared radiometers.

## Results and Discussion

The influence of irrigation events and amounts on daily CWSI for the cultivar Chardonnay at the second commercial vineyard site (CH2) is displayed graphically in Fig. 1. The daily CWSI values were very responsive to irrigation events. The CWSI value rapidly decreased during and following an irrigation event. Larger irrigation amounts resulted in larger declines in daily average CWSI values and vice versa. When irrigation depths were decreased from August 14 through September 1$^{st}$, daily average CWSI values were the greatest and decreased following irrigation to a lesser degree. When an irrigation event was skipped between July 25$^{th}$ and August 1$^{st}$, average daily CWSI continued to increase and rapidly declined to near zero with the relatively large irrigation depth on August 2$^{nd}$. Common irrigation practice by the vineyard manager (personal communication) for study site CH2 is to withhold irrigation until approximately July 1$^{st}$ to develop soil water stress early in berry development and then maintain a mild severity of water stress throughout veraison by applying about 70% of estimated $ET_{c\ act}$ and limiting soil water content deficit to 50% total available water over a 0.9 m soil depth based on neutron probe weekly soil water monitoring.

At the CH2 study site, the measured soil water content at 10 cm increments to a depth of 60 cm is presented in Fig. 2. Active water infiltration and root extraction was apparent only within the 0 to 40 cm soil depth. Soil water deeper than 40 cm was not used to fulfill vine $ET_{c\ act}$ because there was no depletion of soil water below 40 cm during the season. The soil water content from 70 to 120 cm soil depth at the CH2 study site is presented in Fig. 3. The slow gradual decline in soil moisture at depths below 70 cm was negligible and likely due to drainage from 2016 fall irrigation to replenish root zone soil water and winter precipitation. Study site MB2, which was at the same commercial vineyard as study site CH2, also had a limited 40 cm root zone to supply water for vine $ET_{c\ act}$ (data not shown). Study site MB1, which had a much coarser textured soil, had an active root zone of 90 cm to supply water for vine $ET_{c\ act}$ (data not shown). The depth of active root zone at study site MB1 could not be identified because

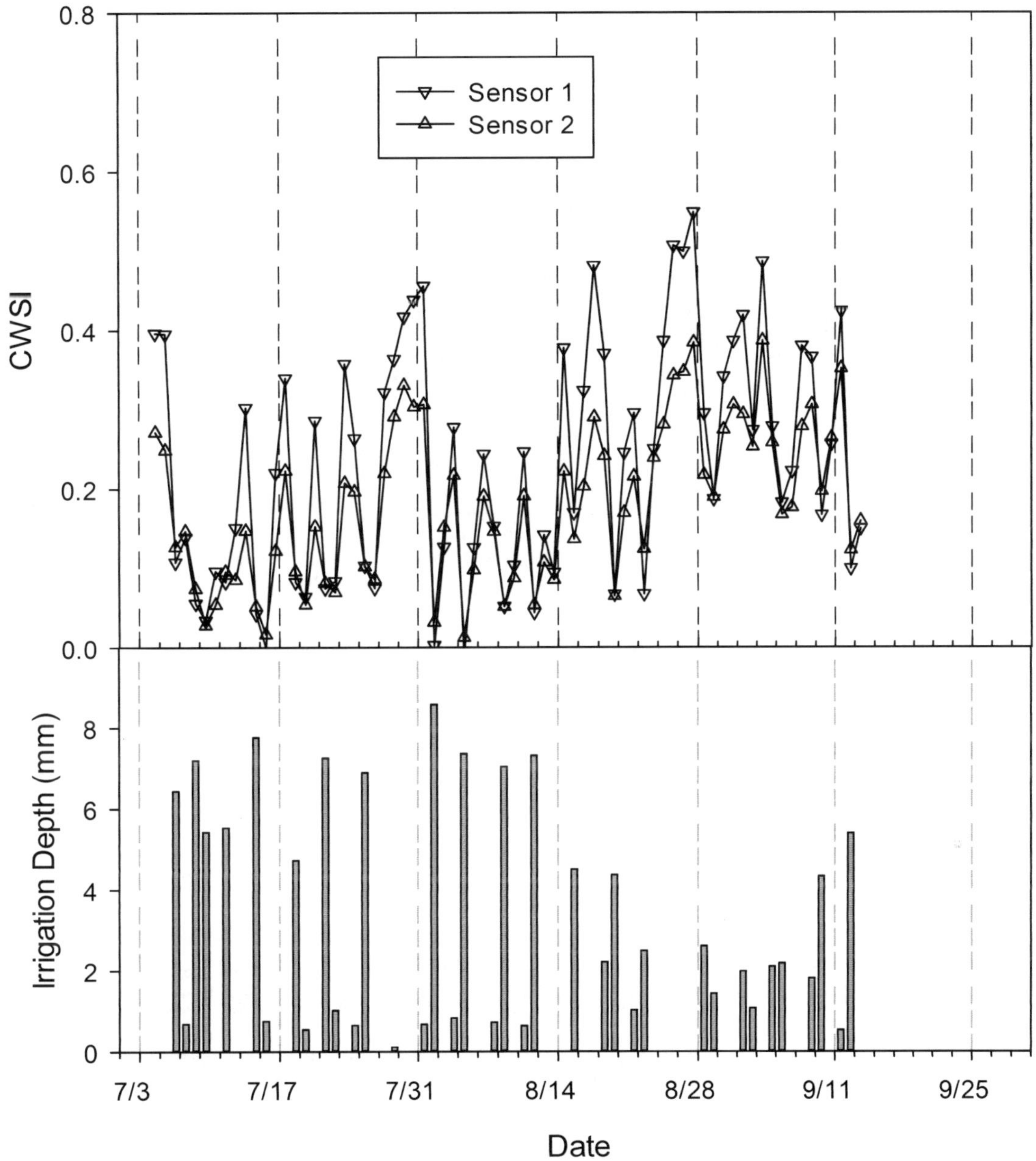


Figure 1. Daily crop water stress index (CWSI) values calculated as the average of 15-min CWSI values ±90 minutes of solar noon (top) and corresponding irrigation events and irrigation amounts per event (bottom) for the cultivar Chardonnay located in the second commercial vineyard site (CH2).

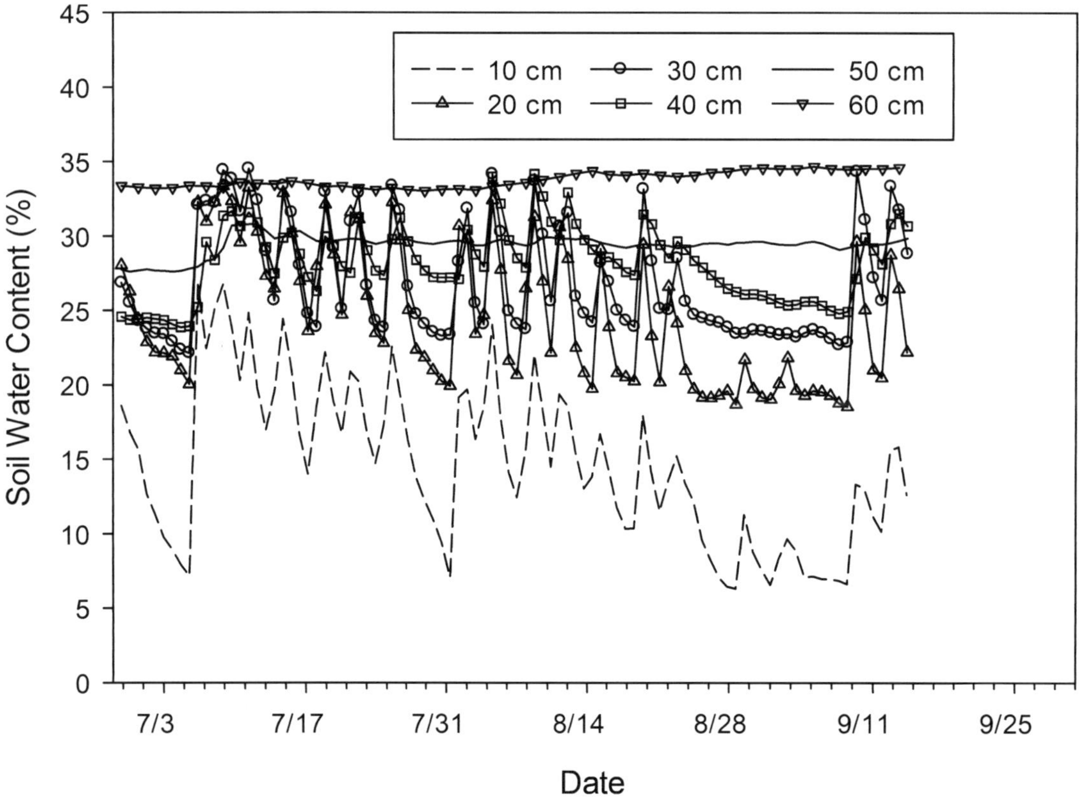


Figure 2. Volumetric soil water content measured at depths from 10 to 60 cm in plots of the cultivar Chardonnay located in the second commercial vineyard site (CH2).

the soil moisture sensor failed to detect a change in soil water content in the upper soil layers (< 20 cm). This may have been due to placement of the sensor relative to the drip irrigation emitters.

The estimated values for $\theta_{fc}$ and $\theta_{pwp}$ that were used to compute percent available soil water (PASW) for the silt loam textured soil at study site CH2, are listed in Table 1. Available soil water throughout the season for the 40-cm root zone at site CH2 (Fig 4) ranged from 23 to 95%. The wide range in available soil water was the result of the limited 40 cm root zone, high evapotranspiration demand, and approximate 4-day irrigation interval.

The relationship between the CWSI and PASW measured daily at 14:30 MST throughout the season at study site CH2 is presented in Fig. 5. The CWSI increased as PASW decreased, in accordance with equations 4 through 7. Also presented in Fig. 5 is the empirical relationships between PASW and the CWSI when the value of $K_s$ in equation 7 is estimated using equation 4 (denoted as FAO $K_s$) and equation 5 (denoted as Jensen $K_s$). Visually, the empirical equation of Jensen et al. (1970) (eqn. 5, Jensen $K_s$) provided a better fit to the data than the FAO $K_s$ piece-wise linear relationship (eqn. 4). Evaluation of each $K_s$ equation fit to the measured data resulted in a mean square error (MSE) value for the Jensen $K_s$

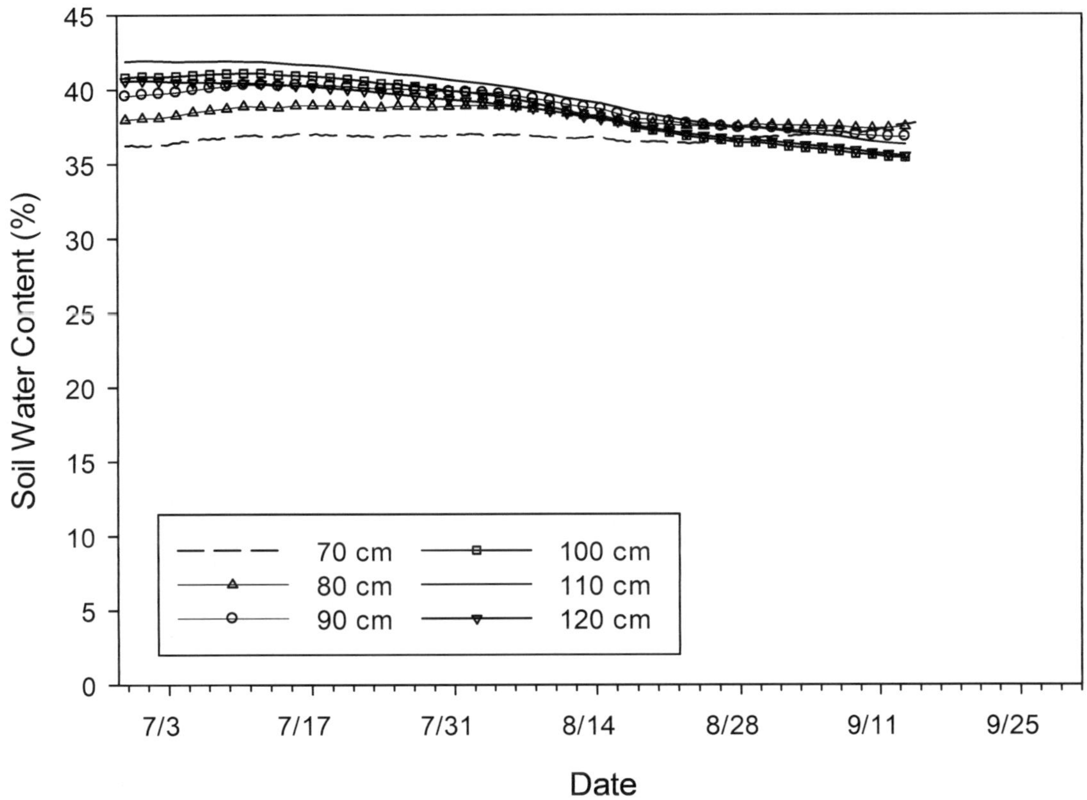


Figure 3. Volumetric soil water content measured at depths from 70 to 120 cm in plots of the cultivar Chardonnay located in the second commercial vineyard site (CH2).

Table 1. Soil volumetric water content (%) at field capacity and permanent wilting point used to calculate percent available soil water (PASW) in a field plot of the cultivar Chardonnay located at the second commercial vineyard site (CH2) in southwestern Idaho.

| | Soil Depth (cm) | | | | | | | | | | | |
|---|---|---|---|---|---|---|---|---|---|---|---|---|
| | 10 | 20 | 30 | 40 | 50 | 60 | 70 | 80 | 90 | 100 | 110 | 120 |
| Field Capacity | 28 | 35 | 34 | 34 | 34 | 34 | 37 | 39 | 41 | 41 | 42 | 42 |
| Permanent Wilting Point | 7 | 17 | 17 | 17 | 17 | 17 | 17 | 18 | 20 | 20 | 20 | 20 |

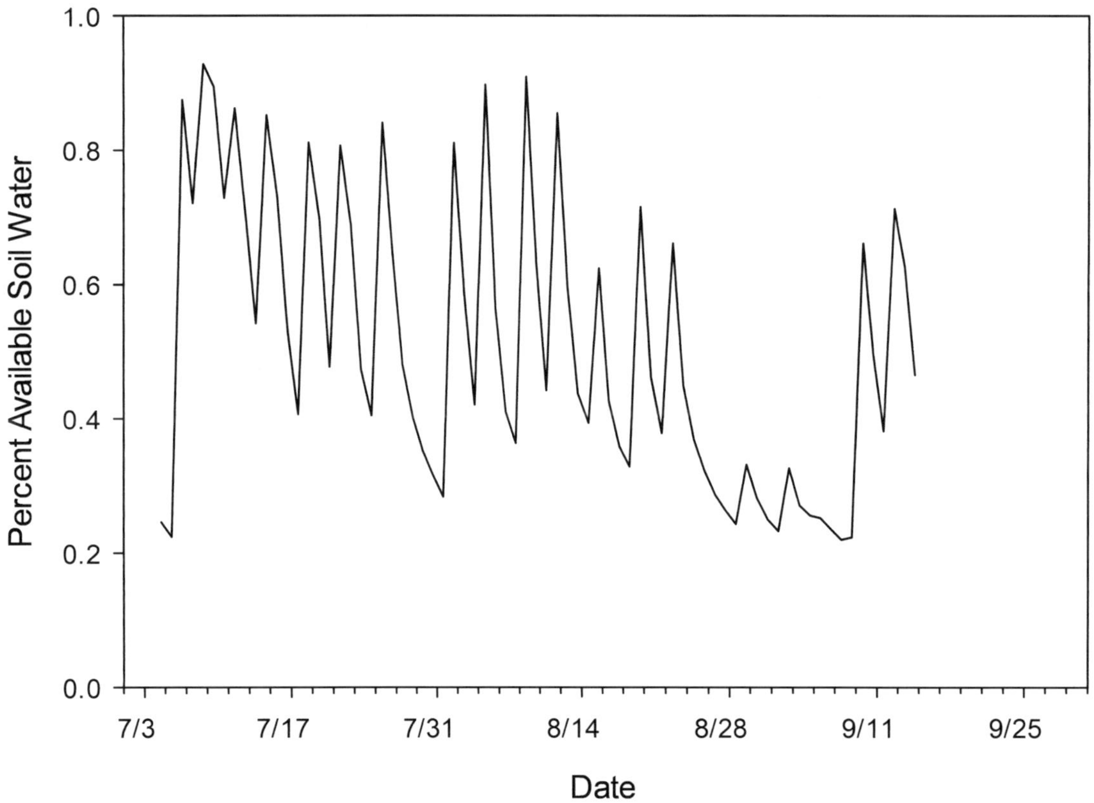


Figure 4. Estimated percent available soil water in the 0-40 cm soil depth in plots of the cultivar Chardonnay located in the second commercial vineyard site (CH2).

of 0.22 and 0.27 for the FAO $K_s$ equation, indicating that the Jensen et al. (1970) equation provided a better fit to the measured data. The relationship between the CWSI and PASW for site MB2 presented in Fig. 6 also shows that the CWSI increased as PASW decreased. The empirical relationship of Jensen et al. (1970) (eqn. 5) visually fit the measured data better than the FAO $K_s$ relationship (eqn. 4). Evaluation of each $K_s$ equation fit to the measured data resulted in a mean square error (MSE) value for the Jensen $K_s$ of 0.08 and 0.20 for the FAO $K_s$ equation, indicating that the Jensen et al. (1970) equation provided a much better fit to the measured data. The relationship between CWSI and PASW for site MB1 is presented in Fig. 7. At the MB1 site, the CWSI increased exponentially as PASW approached zero. Evaluation of each equation fit to the measured data resulted in a MSE value for the Jensen $K_s$ of 0.12 and 0.32 for the FAO $K_s$ equation, indicating that the Jensen et al. (1970) equation provided a much better fit to the measured data. The data for the study sites MB1, MB2 and CH2 are presented collectively in Fig. 8. Again, the Jensen et al. (1970) equation for $K_s$ provided a better fit to the measured relationship between CWSI and PASW having a MSE of 0.14 the MSE of 0.27 for the FAO $K_s$ equation.

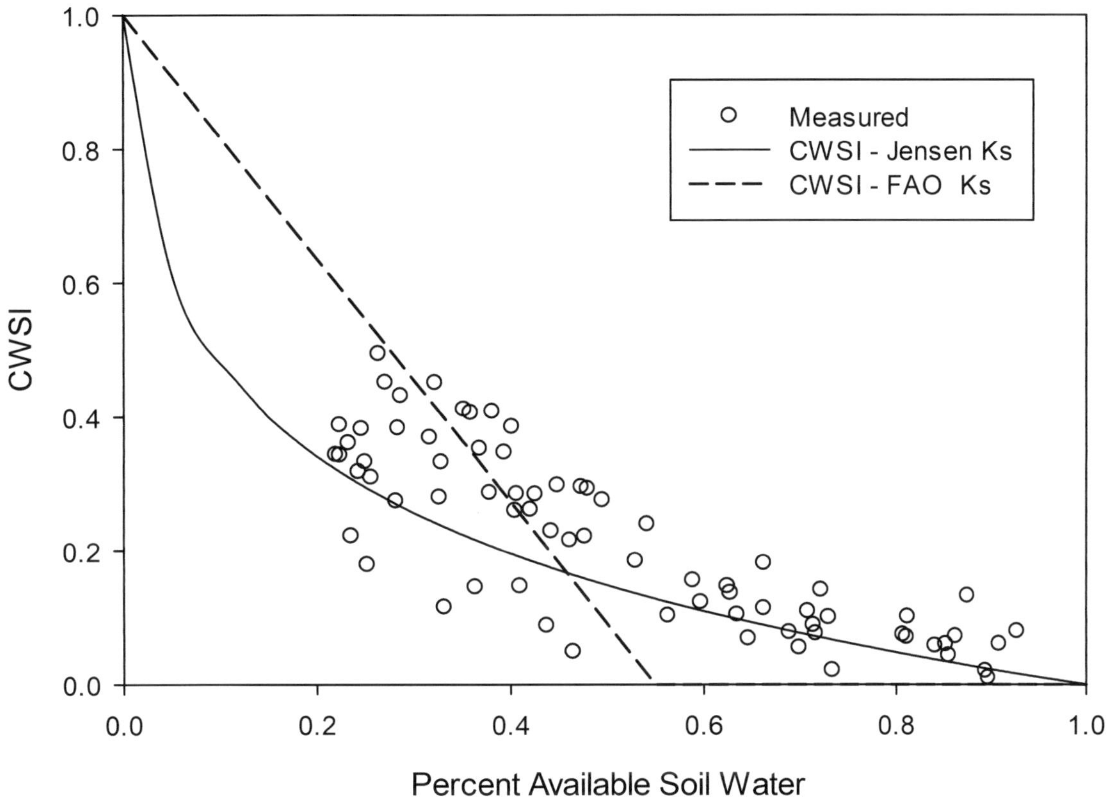


Figure 5. Relationship between crop water stress index (CWSI) computed based on average measured canopy temperature from 14:15 to 14:30 MDT and percent available soil water in the 0-40 cm soil depth based on soil water contents measured at 14:30 MDT in plots of the cultivar Chardonnay located in the second commercial vineyard site (CH2).

There are several sources of inherent variability in the relationship between the CWSI and PASW presented in Figs. 5 through 8. The calculations of the CWSI and PASW required estimation of equation parameters $T_{LL}$, $T_{UL}$, $\theta$, $\theta_{fc}$, $\theta_{pwp}$, and $D_{rz}$ with a level of uncertainty that could account for some of the scatter in the relationship. A single spatial measurement of soil water content was used to estimate PASW when it is well-known that the distribution of soil moisture and roots in the active root zone area of a grapevine is spatially very heterogeneous. At the beginning of an irrigation event, the decrease in vine canopy temperature occurs sooner and is faster than the increase in PASW. This can lead to different calculated values of the CWSI for a given value of PASW. It can also create hysteresis in the relationship between the CWSI and PASW if the soil is in the process of wetting or drying. Measured vine canopy temperature can fluctuate quickly due to variable solar radiation resulting from partly cloudy skies. The CWSI values in this study were calculated regardless of climatic conditions such as clouds and/or rainfall. The neural network models used to estimate $T_{LL}$ also introduces a level of uncertainty in calculated CWSI values. Given the plethora of potential sources of error, it is quite

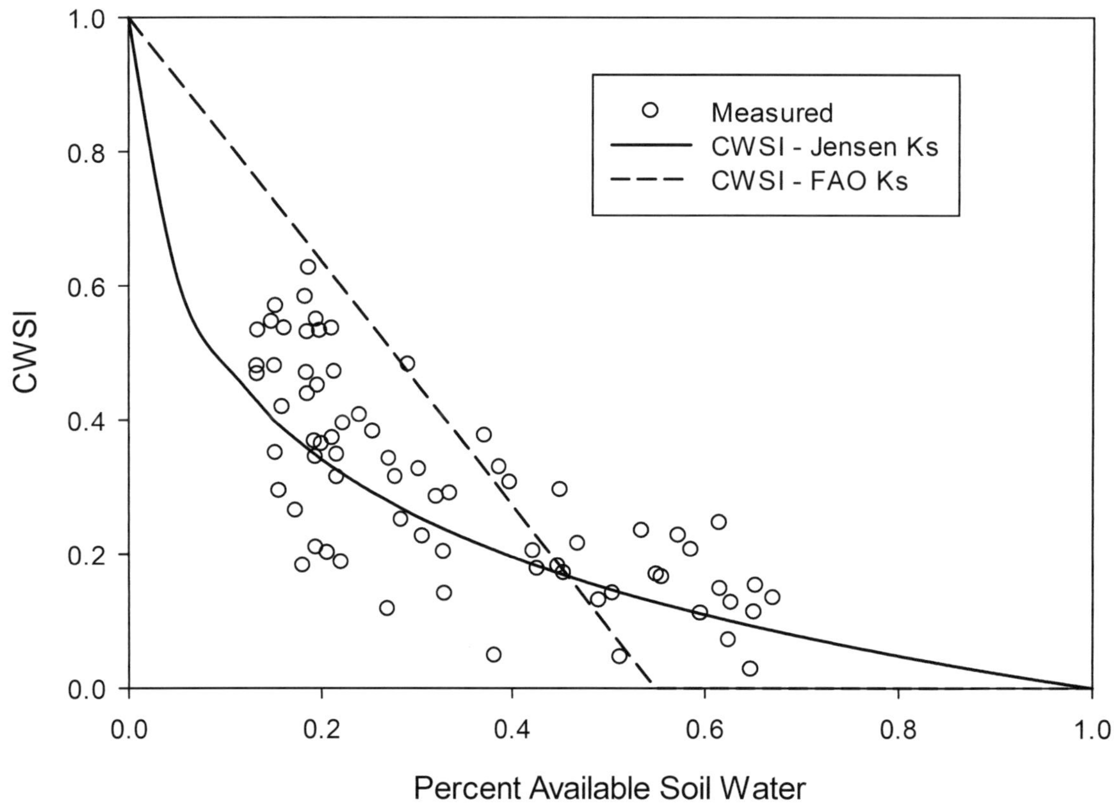


Figure 6. Relationship between crop water stress index (CWSI) computed based on average measured canopy temperature from 14:15 to 14:30 MDT and percent available soil water in the 0-40 cm soil depth based on soil water contents measured at 14:30 MDT in plots of the cultivar Malbec located in the second commercial vineyard site (MB2).

amazing that the relationship between the CWSI and PASW is defined to the degree seen in Figs 5 through 8. Despite this variability, the $K_s$ equation proposed by Jensen et al. (1970) represented the relationship between the CWSI and PASW better than that of Allen et al. (1998).

The relationship between CWSI and $\Psi_{md}$ is presented in Fig. 9. There was a significant ($p \le 0.05$) linear relationship between the CWSI and $\Psi_{md}$ showing that the CWSI increased as $\Psi_{md}$ decreased. However, the low $R^2$ value of 0.21 indicates that there was a large amount of unexplained variability in the relationship. This large amount of variability could be attributed to operator differences (Williams et al. 2012) since the $\Psi_{md}$ measurements were collected by several different personnel throughout the growing season.

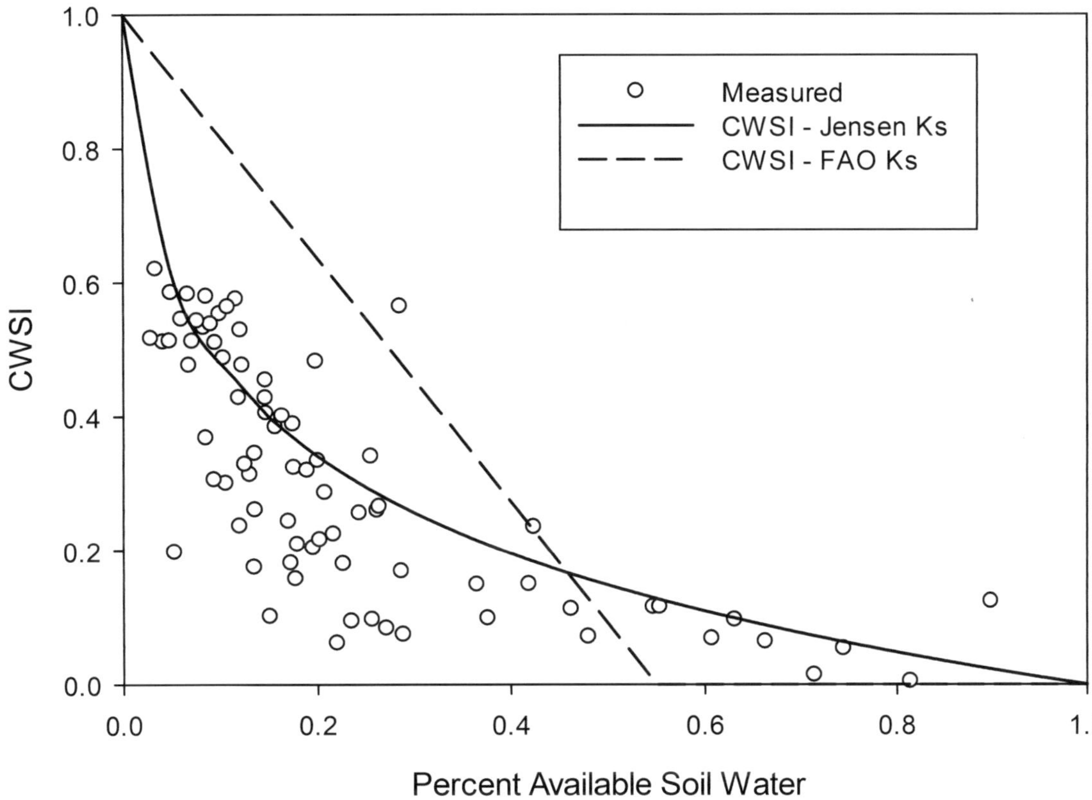


Figure 7. Relationship between crop water stress index (CWSI) computed based on average measured canopy temperature from 14:15 to 14:30 MDT and percent available soil water in the 0-90 cm soil depth based on soil water contents measured at 14:30 MDT in plots of the cultivar Malbec located in the first commercial vineyard site (MB1).

## Conclusions

In this study, canopy temperature, air temperature, solar radiation, relative humidity, wind speed, soil profile water content and irrigation times and irrigation amounts were continuously monitored at four study sites located in three commercial vineyards in southwestern Idaho. A daily average CWSI was calculated using measured parameters ±90 minutes of solar noon. Daily average CWSI was linked to soil water content through the water stress coefficient $K_s$ that accounts for reduced vine transpiration when soil water is limited. This linkage demonstrates that the daily CWSI is a reliable indicator of vine water stress resulting from limited soil water. The equation proposed by Jensen et al. (1970) for estimating $K_s$ provided a better representation of the relationship between daily average CWSI and PASW than the equation proposed by Allen et al (1998). The CWSI was better correlated with PASW than with $\Psi_{md}$. These results demonstrate that a daily CWSI is a reliable method for monitoring grapevine water status under changing soil moisture conditions. The relationship observed in this study between the CWSI and PASW suggests that the daily average CWSI could be used as an irrigation management tool for

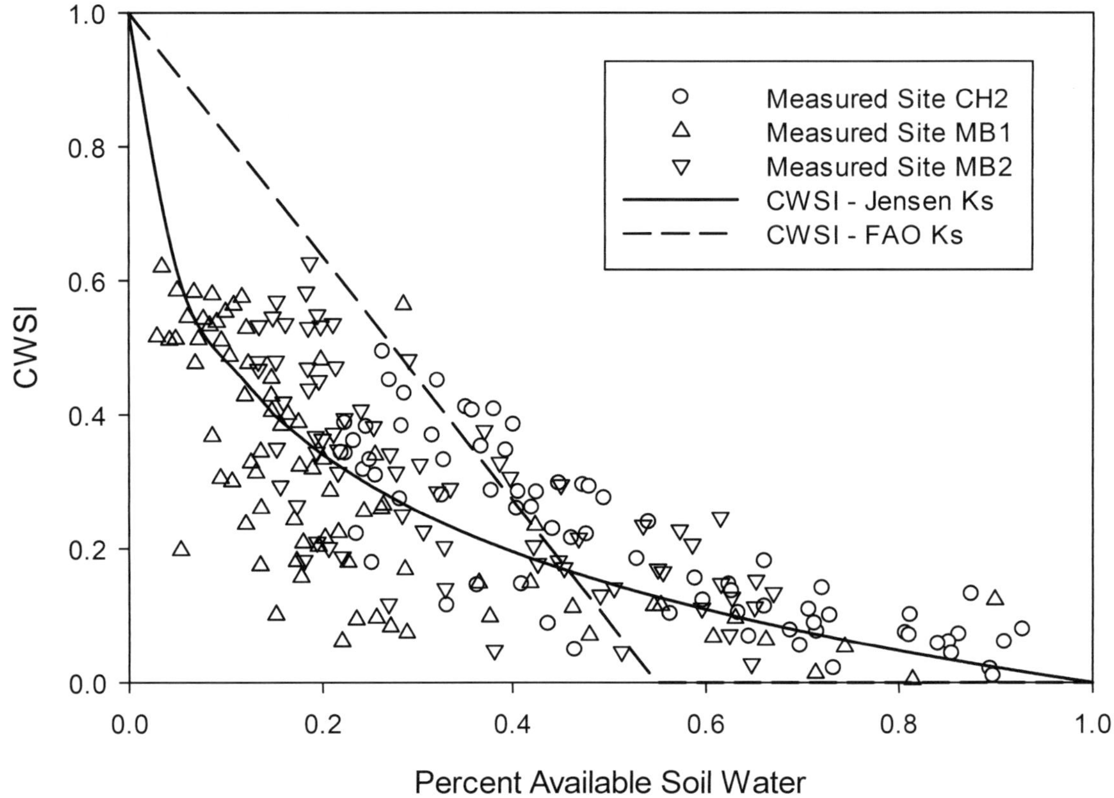


Figure 8. Relationship between crop water stress index (CWSI) computed based on average measured canopy temperature from 14:15 to 14:30 MDT and percent available soil water based on soil water contents measured at 14:30 MDT in plots of the cultivar Malbec at the first (MB1) and second (MB2) commercial vineyard and Chardonnay at the second commercial vineyard (CH2).

irrigation scheduling and potentially also for estimating the amount of water to supply during an irrigation event with further research.

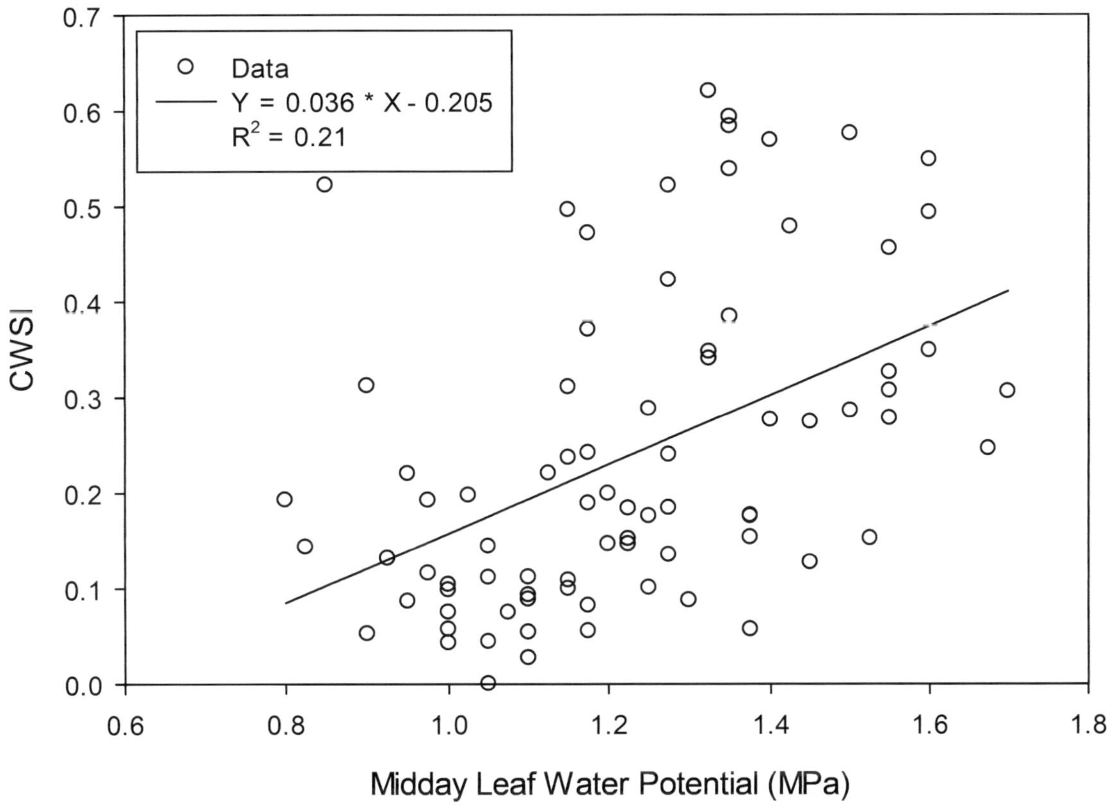


Figure 9. Relationship between the daily CWSI and midday leaf water potential of cultivars Malbec and Chardonnay measured throughout the 2017 growing season at four sites in three commercial vineyards in southwestern Idaho.

## References


Allen, R.G., L.S. Pereira, D. Raes, and M. Smith. 1998. Crop evapotranspiration-guidelines for computing crop water requirements-FAO irrigation and drainage paper 56. Food Agricultural Organization, Rome.

Bausch, W., T. Trout, and G. Buchleiter. 2011. Evapotranspiration adjustments for deficit-irrigated corn using canopy temperature: A concept. Irrigation and Drainage 60:628-693.

Bellvert, J., J. Marsal, J. Girona, and P.J. Zarco-Tejada. 2015. Seasonal evolution of crop water stress index in grapevine varieties determined with high-resolution remote sensing thermal imagery. Irrig Sci 33:81-93.

Colaizzi, P.D., E.M. Barnes, T.R. Clarke, C.Y. Choi, and P.M. Waller. 2003. Estimating soil moisture under low frequency surface irrigation using crop water stress index. J Irrig Drain Eng ASCE 129:36–43

Idso, S.B., R.D. Jackson, P.J. Pinter Jr, R.J. Reginato, and J.L. Hatfield. 1981. Normalizing the stress-degree-day parameter for environmental variability. Agricultural Meteorology 24:45-55.

Jackson, R., 1982. Canopy temperature and crop water stress. Adv Irrig 1:43-85.

Jackson, R.D., S.B. Idso, R.J. Reginato, and P.J. Pinter Jr. 1981. Canopy temperature as a drought stress indicator. Water Resources Research 13:651-656.

Jensen, M.E., D.C.N. Robb, and C.E. Franzoy. 1970. Scheduling irrigations using climate-crop-soil data. J. Irrig. Drain. Eng., 96(1):25–38.

Jones, H.G. 2004. Irrigation scheduling: advantages and pitfalls of plant-based methods. J Exp Botany 55(407):2427-2436.

Jones, H.G., M. Stroll, T. Santos, C. de Sousa, M.M. Chaves, and O.M. Grant. 2002. Use of infrared thermography for monitoring stomatal closure in the field: application to grapevine. J Exp Bot 53:2249-2260.

King B.A., and K.C. Shellie. 2016. Evaluation of neural network modeling to predict non-water-stressed leaf temperature in wine grape for calculation of crop water stress index. Agricultural Water Management 167:38-52.

Maes, W.H., and K. Steppe. 2012. Estimating evapotranspiration and drought stress with ground-based thermal remote sensing in agriculture: a review. J. Exp. Bot. 63:4671-4712.

Möller, M., V. Alchanatis, Y. Cohen, M. Meron, J. Tsipris, A. Naor, V. Ostrovsky, M. Sprintsin, and S. Cohen. 2007. Use of thermal and visible imagery for estimating crop water status of irrigated grapevine. J. Exp Bot 58:827-838.

Ortega-Farias, S., E. Fereres, and V.O. Sadras. 2012. Editorial: Special issue on water management of grapevines. Irrig Sci 30(5):335-337.

Pou, A., M.P. Diago, H. Medrano, J. Baluja, and J. Tardaguila. 2014. Validation of thermal indices for water status identification in grapevine. Ag Water Mgt 134:60-72.

Rodrigues, P., V. Pedroso, J. P. Gouveia, S. Martins, C. Lopes, and I. Ives. 2012. Influence of soil water content and atmospheric conditions on leaf water potential in cv. "Touriga Nacional" deep-rooted vineyards. Irrig Sci 30(5):407-417.

Sepúlveda-Reyes, D., B. Ingram, M. Bardeen, M. Zúñiga, S. Ortega-Farías, and C. Poblete-Echeverría. 2016. Selecting canopy zones and thresholding approaches to assess grapevine water status by using aerial and ground-based thermal imaging. Remote Sens 8:822 doi:10.3390/rs8100822.

Shellie, K.C. 2006. Vine and berry response of Merlot (*Vitis vinifera* L.) to differential water stress. Am. J. Enol. Vitic. 57:514-518.

Shellie, K.C., and P. Bowen. 2014. Ishohydrodynamic behavior in deficit-irrigated Cabernet Sauvignon and Malbec and its relationship between yield and berry composition. Irri. Sci. 32:87-97.

Turner, N.C., 1988. Measurement of plant water status by the pressure chamber technique. Irrig Sci 9:289-308.

Williams, L.E., and F.J. Araujo. 2002. Correlations among predawn leaf, midday leaf, and midday stem water potential and their correlations with other measures of soil and plant water status in *Vitis vinifera*. J. Am. Soc. Hortic. Sci. 127:448-454.

Williams, L.W., and P. Baeza. 2007. Relationships among ambient temperature and vapor pressure deficit and leaf and stem water potentials of fully irrigated, field-grown grapevines. Am. J. Enol. Vitic. 58:173-181.

Williams, L.E., P. Baeza, and P. Vaughn. 2012. Midday measurements of leaf water potential and stomatal conductance are highly correlated with daily water use of Thompson Seedless grapevines. Irrig. Sci. 30:201-212.

Williams, L.W., and T.J. Trout. 2005. Relationships among vine- and soil –based measures of water status in a Thompson Seedless vineyard in response to high-frequency drip irrigation. Am. J. Enol. Vitic. 56:357-366.

# Sustainability of Irrigation with EM Induction Technique

[1]Florence Cassel S. , [1]Dave Goorahoo and [2]Shankar Sharma

[1]California State University, Fresno, CA. USA
[2] University of California, Merced, CA. USA

November 8, 2017
Irrigation Association Technical Session

## Abstract

**Assessing Sustainability of irrigated agriculture using the Electromagnetic Induction technique**

**Short title: Sustainability of Irrigation with EM Induction Technique**

Authors: Florence Cassel S., Dave Goorahoo, and Shankar Sharma
California State University, Fresno
University of California, Merced

Sustainability of irrigated agriculture in Central California is threatened by increased salinization of land and water resources. Intensive irrigation and fertilization, use of saline water and inadequate drainage have resulted in the salinity build up. Increase in soil salinity has been observed in many cropping systems including orchards. Therefore, there is a need to use rapid and cost-effective tools to assess the extent of salinization for developing adequate crop production and land management practices. We applied an electromagnetic induction (EM) technique to map soil salinity through non-invasive and rapid measurements of electrical conductivity. The survey results demonstrated that the EM method could accurately and reliably describe the large spatial and temporal variability in salinity observed across the lands. By identifying the precise locations and levels of salt loading, the EM surveys could help in the development of irrigation and soil management strategies.

Keywords: Salinity, Irrigation, EM survey, Conductivity

## OUTLINE

- Background/Research Activities
- EM Salinity Mapping
- Case Studies

## Research activities

- **Use of remote sensing technologies: EM, satellite/aerial imagery, GIS**
  - Soil and water resources management
  - Precision agriculture
- **Agronomic studies**
  - Fertilizers
  - Barrier crops
  - Seeds
- **Environmental studies**
  - Land application of effluent and drainage waters

## Evidence of Soil Salinity

## Field Variability – Salinity/Sodicity

Near Lemoore, CA

## Soil salinity mapping

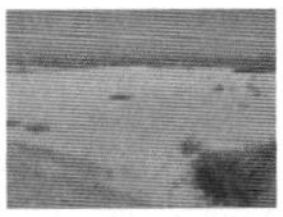

- **Objective:** Assessing soil salinity in agricultural fields
- Use of **Electromagnetic induction** (EM) technique
- Develop salinity **maps using GIS**

## References on Theory

- **EM technology and GIS**

8 Soil Salinity Mapping Using ArcGIS

11 Mapping Soil Salinity Using Ground-Based Electromagnetic Induction Technique*

## Survey equipment

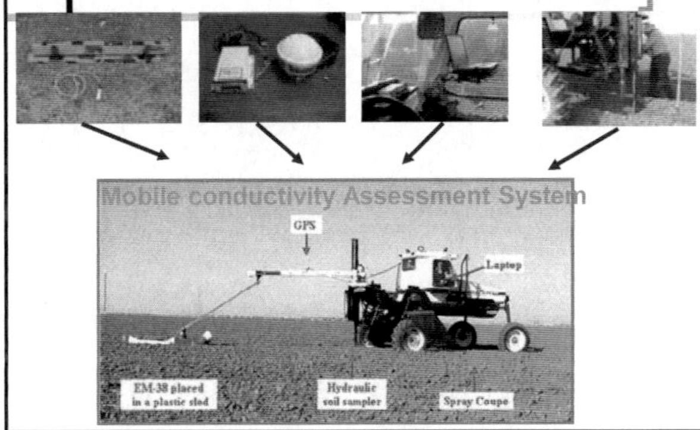


## Steps in EM salinity mapping

- ***Data acquisition***
- ***Data calibration***
  - ***ESAP software***
  - ***Ground truthing***
- ***Mapping soil salinity from point data using GIS and other software***

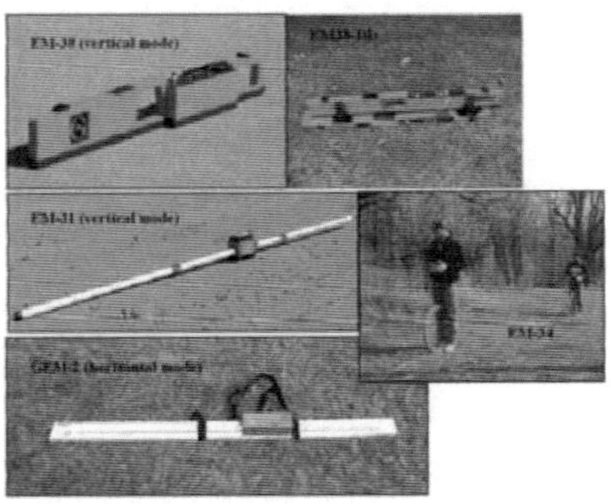

## EM sensor

- Length **(intercoil spacing)= 3.3 ft**
- Weight **= 6.6 lb**
- Operating frequency**= 14.6 kHz**
- Depth of measurement
  - **Horizontal = 3 ft**
  - **Vertical = 6 ft**
- Measurement accuracy **= ± 5% at 30 mS/m**

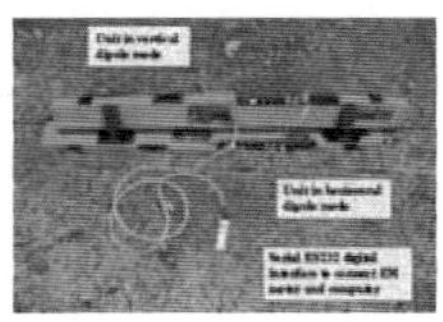

http://www.ars.usda.gov/Services/docs.htm?docid=8918

## Data acquisition - mobile

- EM and GPS data recording along furrows: every 30-40 ft
- 10-15 measurements per acre
- Total of 1500-2250 measurements per field
- Survey time: 2½ - 3 hours (150 ac)
- Calibrate EM data: soil sampling at 6-12 locations

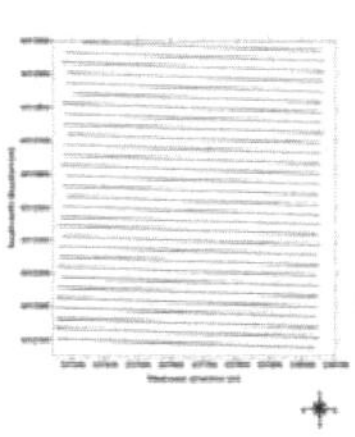

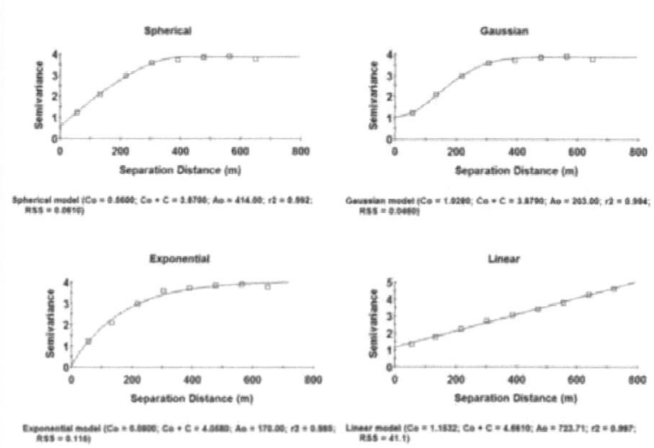

## Soil salinity mapping

- Generate soil salinity maps, based on calibrated EM data
- GIS: Spatial analysis
- Surface maps of soil salinity
- In addition, maps of boron, gypsum, and moisture distribution

## Soil salinity mapping

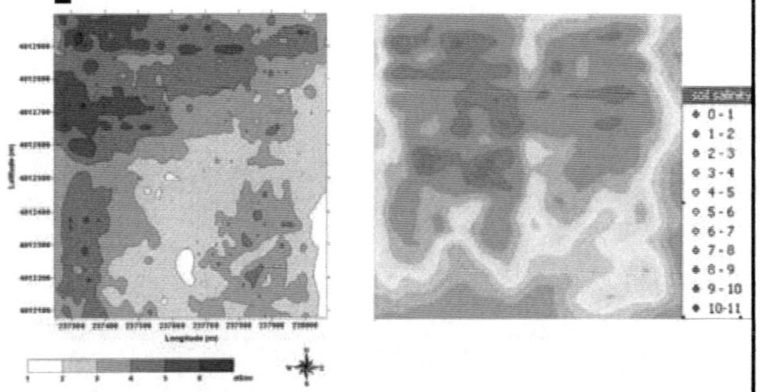


## 1. Effects of Irrigation Drainage

Agricultural fields subjected to IFDM practices (sequential drainage reuse)

- Survey every year
- Salinity variability across farm
- Recommendations on cropping rotation and drainage applications

## 2. Variable rate application

Variable seeding and gypsum application rates based on salinity and boron levels

- Survey before planting
- Variable seeding rate
- Variable amendment applications

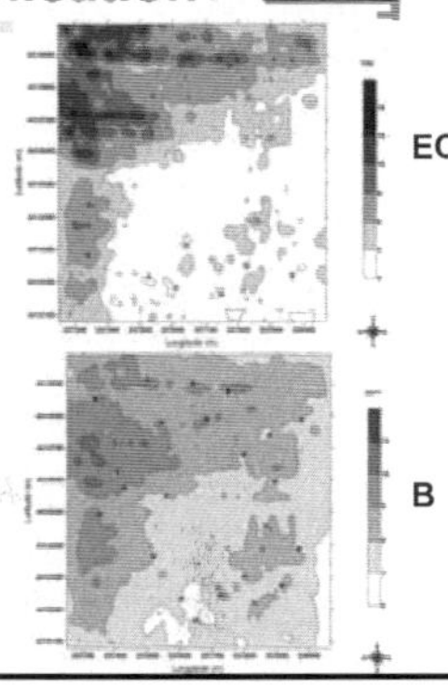


## Variable rate equipment

- Guidance System
- Raven controller
- GPS
- Delivery system

Seed application rate - pixel map

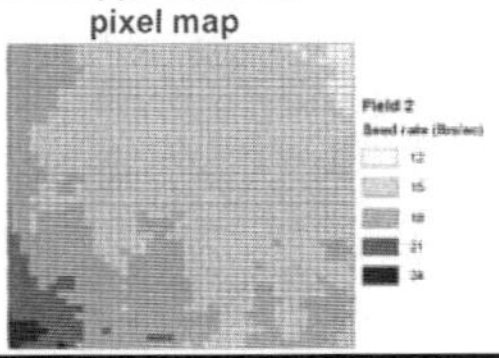


Gypsum mapping

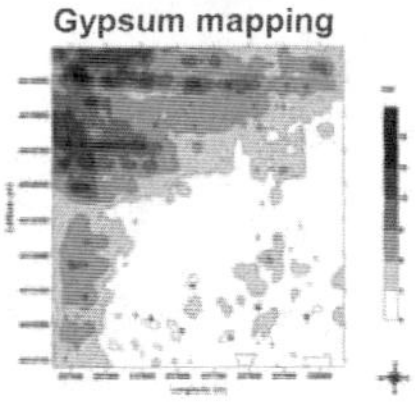

## Data comparison

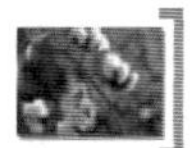

EM survey

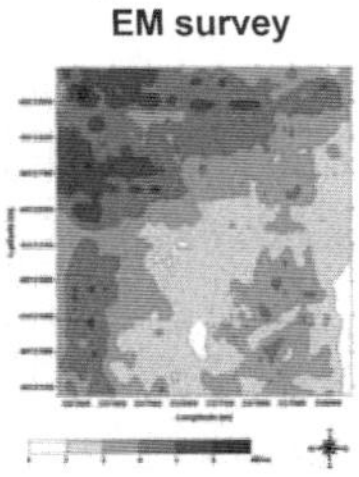

Aerial imagery

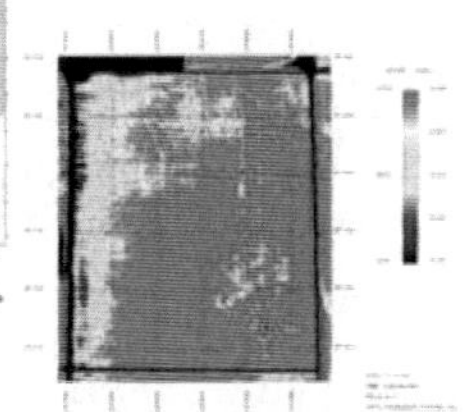

Plant growth (NDVI) map

Yield monitor

## 3. Assessing a bio-filter crop

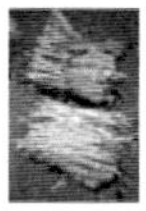

- Objective: Compare effectiveness of elephant grass vs Sudan/Bermuda grass in controlling N and P contamination

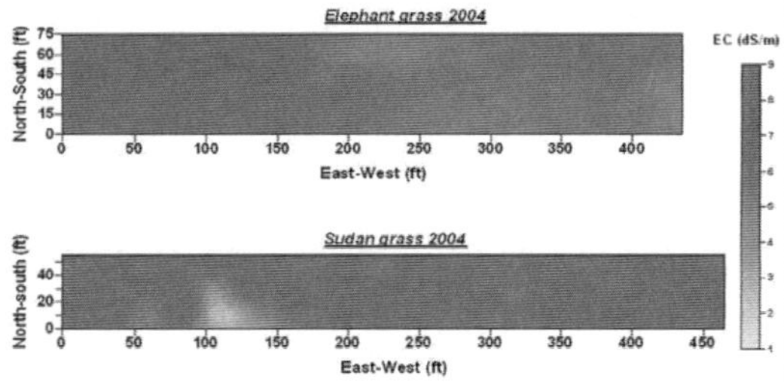


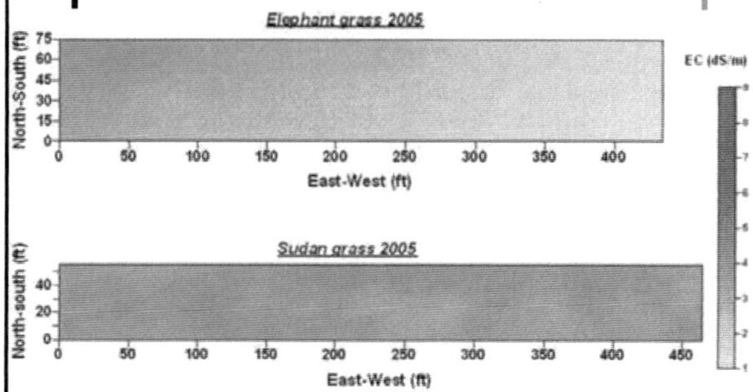


## 4. Assessing yield potential

**Salinization and Yield Potential of a Salt-Laden Californian Soil: an In Situ Geophysical Analysis**

Cassel, F., Goorahoo, D., & Sharmasarkar, S. (2015). Salinization and Yield Potential of a Salt-Laden Californian Soil: an In Situ Geophysical Analysis. *Water, Air, & Soil Pollution*, 226(12), 1-8.

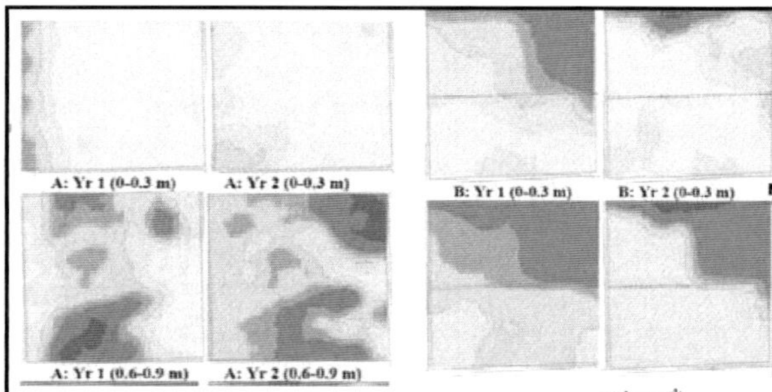


**Fig. 2** Raster maps of conductivity distributions showing spatial and temporal variations in moderately (A) and severely (B) affected salinity zones

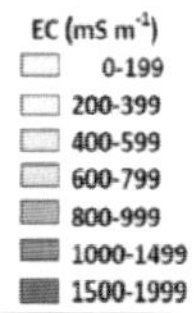


## Concluding Remarks

Electromagnetic induction technique:

- **Precisely assess levels of salinity and moisture across surveyed fields.**
- **Great potential for quick evaluation of soil properties for irrigation over large areas**
- **Cost-effective alternative to extensive sampling**
- **Valuable tool to assess salt problems and effectiveness of irrigation and salt management strategies**

# Drivers and barriers to producers' voluntary adoption of irrigation practices that protect water quality

**Kelly M. Foley, Clinton C. Shock, and Mary V. Santelmann**

Over twenty alternative agricultural practices have been introduced to producers in Malheur County, Oregon, over the last thirty years to protect water quality. Research and outreach were designed to provide technological options for producers. This study sought to better understand the voluntary adoption by producers of practices improving water quality. The Reasoned Action Approach/Theory of Planned Behavior was used as a theoretical framework to identify barriers and incentives to adoption. Study findings suggest that producers primarily consider practical characteristics of practices when making adoption decisions. Some of these concerns include the relative advantage of the practice (derived from the anticipated financial gain or loss, conservation and water quality benefits from adopting a practice), the compatibility of a practice with existing farm operations, the ease or difficulty of implementing a practice, and the ability to observe the success of a practice prior to adoption. These factors varied widely across individual farms because of the diversity in farming operations. Producer age and lack of agency over decision making emerged as barriers to adoption and provide promising areas for future adoption studies. Recommendations are provided for enhanced education and outreach programs and incentive systems that are better suited for the diverse needs of producers operating small to medium sized farms.

## Introduction

The irrigation industry is well aware of the potential negative environmental consequences from irrigation. Groundwater has become contaminated with nitrates and pesticides in the areas of the world with intensive horticultural production. Soil and water erosion losses from farms by irrigation induced erosion transfer sediment and phosphorus to surface waters. The negative environmental effects of inefficient water use are evident. By 1986 the groundwater in Northeast Malheur County was contaminated by nitrate and the residuals of the breakdown of DCPA sold as Dacthal (Bruch, 1986). By 1990 the average groundwater nitrate in the region with intensive horticulture hovered between 18 and 19 parts per million. DCPA residuals were correlated with groundwater nitrate.

Townsend and Howarth (2010) addressed the issue of fixing the global nitrogen problem, and in their manuscript they described the parts of the nitrogen cycle in agriculture that were inefficient and could be improved. They promoted the possibilities of using less nitrogen and applying fertilizer with timing closer to the plants' needs. I n 1985 the Oregon State University Malheur Experiment Station (MES) initiated studies to reduce irrigation induced erosion. In 1990 MES also set out to improve the efficiency of nitrogen fertilizer use and improve irrigation efficiency seeking to improve groundwater quality (Shock and Shock, 2012). Research was conducted to develop locally appropriate technology to reduce nitrogen fertilizer rates on onions, potatoes, sugar beets, and wheat. Research placed emphasis on soil sampling, plant tissue testing, and split nitrogen fertilizer applications. Field trials were intentionally designed to develop locally appropriate technology to result in higher yields using lower total inputs of nitrogen fertilizer.

Nitrogen fertilizer and soil nitrate will only reach groundwater in a semiarid environment when irrigation water is applied in excess of evaporation and crop water use. Numerous trials evaluated potential advantages of sprinkler and drip irrigation compared to the standard flood irrigation. Gated pipe had the potential to increase irrigation efficiency compared to the use of siphon tubes for flood irrigation especially if the gated pipe system was preceded by trash screen filters and if the water was applied via surge irrigation. Additional leveling of fields held potential for improving irrigation efficiency where furrow/flood irrigation continued to be commonly used.

Other production problems compromising water quality were also addressed. Sediment losses could be reduced by adapting filter strips, sedimentation ponds, pump back systems, polyacrylamide, and the use of straw mulch to local conditions. Methods were developed to reduce the application rates of DCPA or eliminate its use completely by adopting other herbicide products and weed control methods. A suite of best management practices was developed, widely advertised, and frequently updated (Shock, 2015). By 2010 some water quality improvements had already been realized. Groundwater nitrate had fallen to 10.5% and DCPA residues were greatly reduced in the most affected wells. Changes that occurred in farming practices happened through a combination of research, demonstration, outreach, and voluntary adoption.

In many parts of the United States water quality has not been improving. It is unknown exactly the extent of the adoption of better practices and why growers adopt certain changes that might improve water quality. We undertook research to understand grower adoption behavior of conservation practices. What motivates growers to change practices and what are the sources of information that they trust as they redesign their farming practices?

Specifically what are the factors that influenced grower decisions to adopt water quality improving practices in northern Malheur County? We sought to understand the drivers and barriers to the adoption of these conservation practices. We sought to investigate the relevant background factors and beliefs and develop recommendations for future study, policy makers, and practitioners of crop production research and outreach.

## Material and methods

Out of 450 growers in the groundwater management area, 29 growers were randomly selected based on a geographically stratified random sample (Foley, 2013). Growers were chosen randomly from five different parts of the intensively irrigated area and from areas representing both flat and rolling topography. Growers were selected from both types of topography since the crops and practice improvements appropriate for the two topographical areas are somewhat different. The 29 growers were each given a semi structured interview which was audio recorded and transcribed. The data analysis consisted of coding the interview for descriptive attributes practices information on the grower in their production, attitudes, norms, and drivers. In general the size of the farms and the age of the growers was similar to the 2007 agricultural census (Table 1). The average age of the grower was 58 years and the average acreage was 649 acres.

Table 1. Characteristics of growers interviewed compared to the 2007 Census of Agriculture (Foley, 2013).

| Characteristic | | 2007 Census of Agriculture Malheur County statistics* | 2013 Respondents |
|---|---|---|---|
| Average age of operator | | 56 | 58 |
| Average acreage | | 937 | 649 |
| Median acreage | | 101 | 550 |
| Farming as primary income (percentage of total operators) | | 62.2% | 82.8% |
| Gender (percentage male) | | 89.6% | 93.5% |
| | *Statistics derived from the USDA 2007 Census of Agriculture | | |

Results

Growers were found to have adopted between two and 12 conservation practices out of 17 practices mentioned in the interviews (Figure 1) (Foley, 2013).

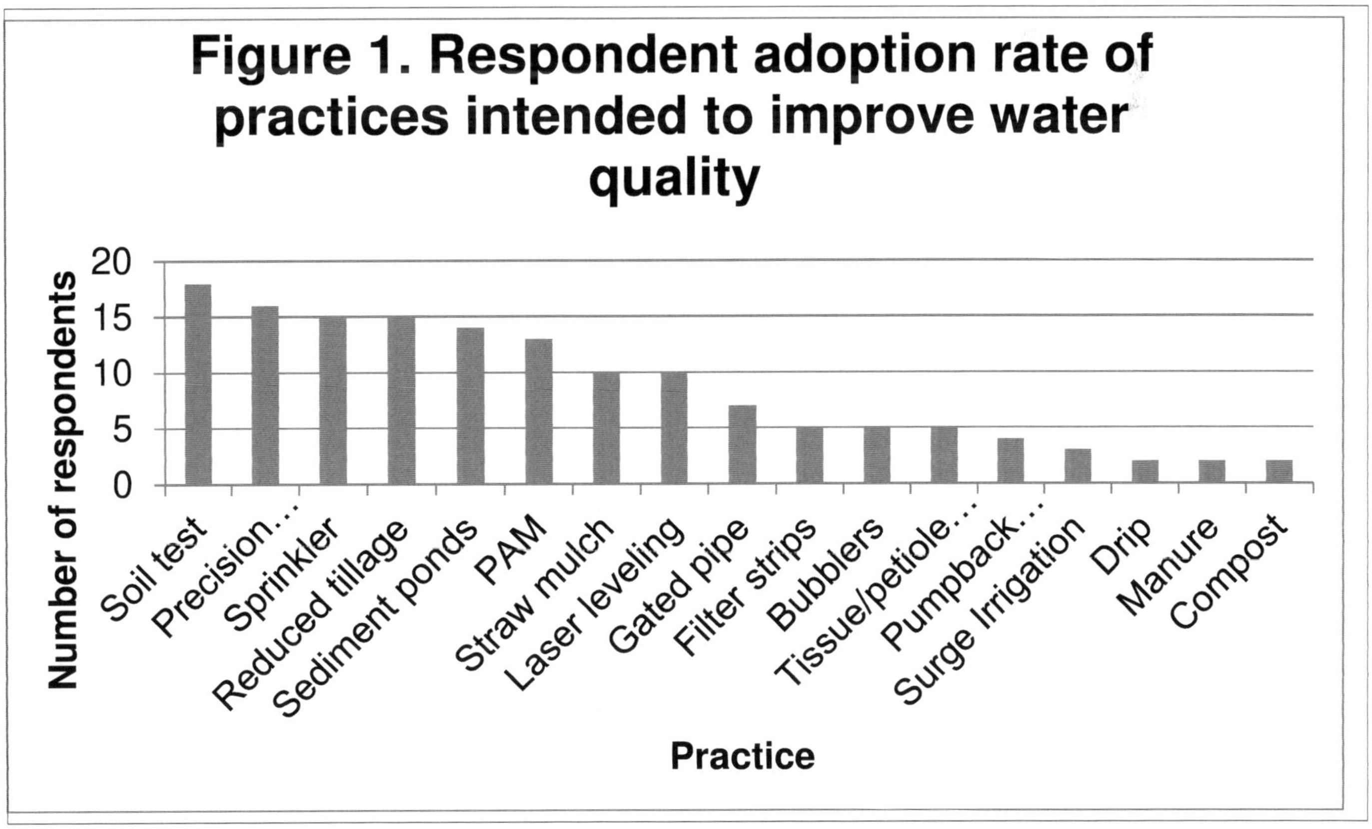

Of the drivers for the voluntary adoption of practices, financial gain was cited by 100% of the growers (Foley, 2013). Other important factors included considering whether a practice compatible with their farm, ease-of-use, and observability (Figure 2). Barriers to the adoption of practices included financial loss, incompatibility with farming practices, difficulty of use, grower's age, and lack of agency. The most important factors affecting adoption were relative financial advantage, lack of complexity, compatibility with current practices, and observability and the most importance barriers were age of the grower and a sense of lack of agency.

Figure 2. Primary drivers and barriers to the voluntary adoption of practices in Malheur County (Foley, 2013).

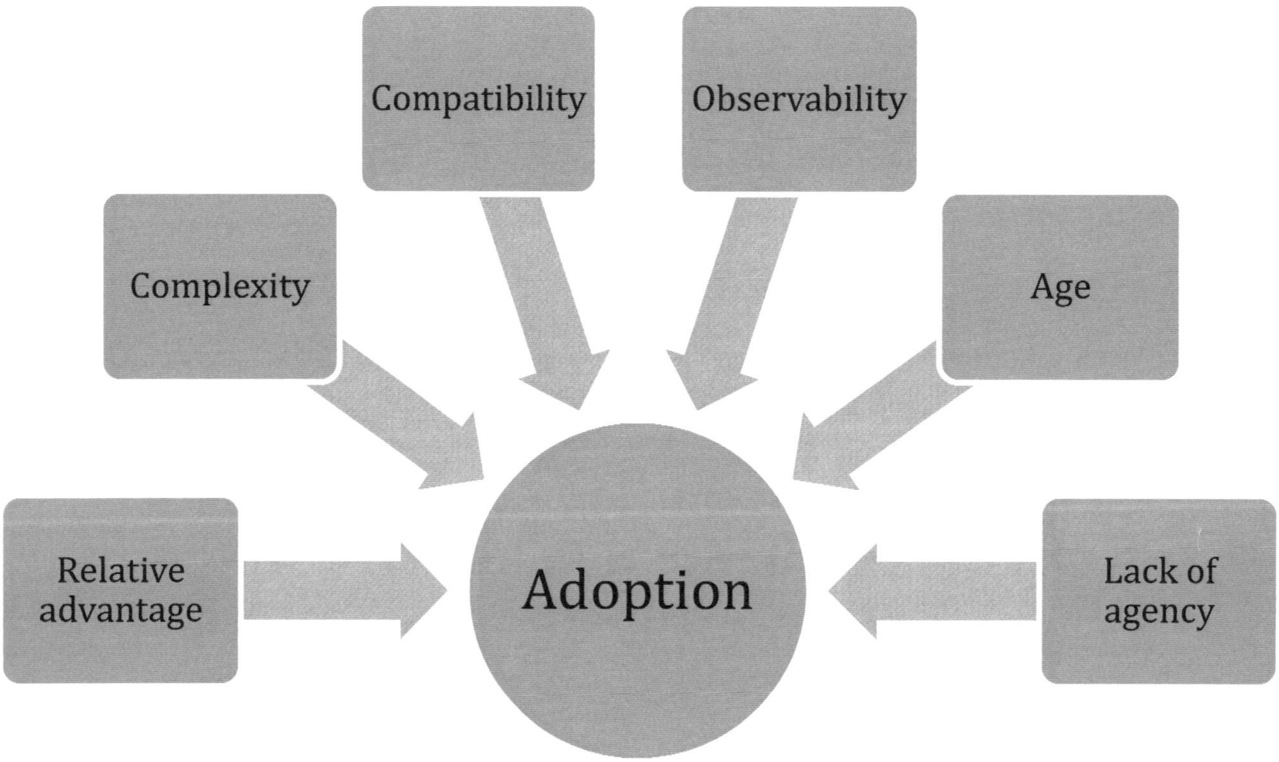


Discussion

Practicality and profitability were uppermost as drivers for consideration of new practices (Foley, 2013). Taking a quote from a grower *"You know, I've always said that the farmers land— the land takes care of you, so you take care of it."*

Observability of a practice was very important for growers. Again taking a quote from a grower *"A lot of people are going that way though. We might have to too if it comes that way. If it looks like my neighbors are doing better with drip then I better [too]."* With regard to observability, the most utilized information sources reported by respondents were information from their neighbors, university research and extension, observation of other farmers in other areas, field men, and written publications (Figure 3).

Figure 3. Most utilized information sources reported by survey respondents (Foley, 2013).

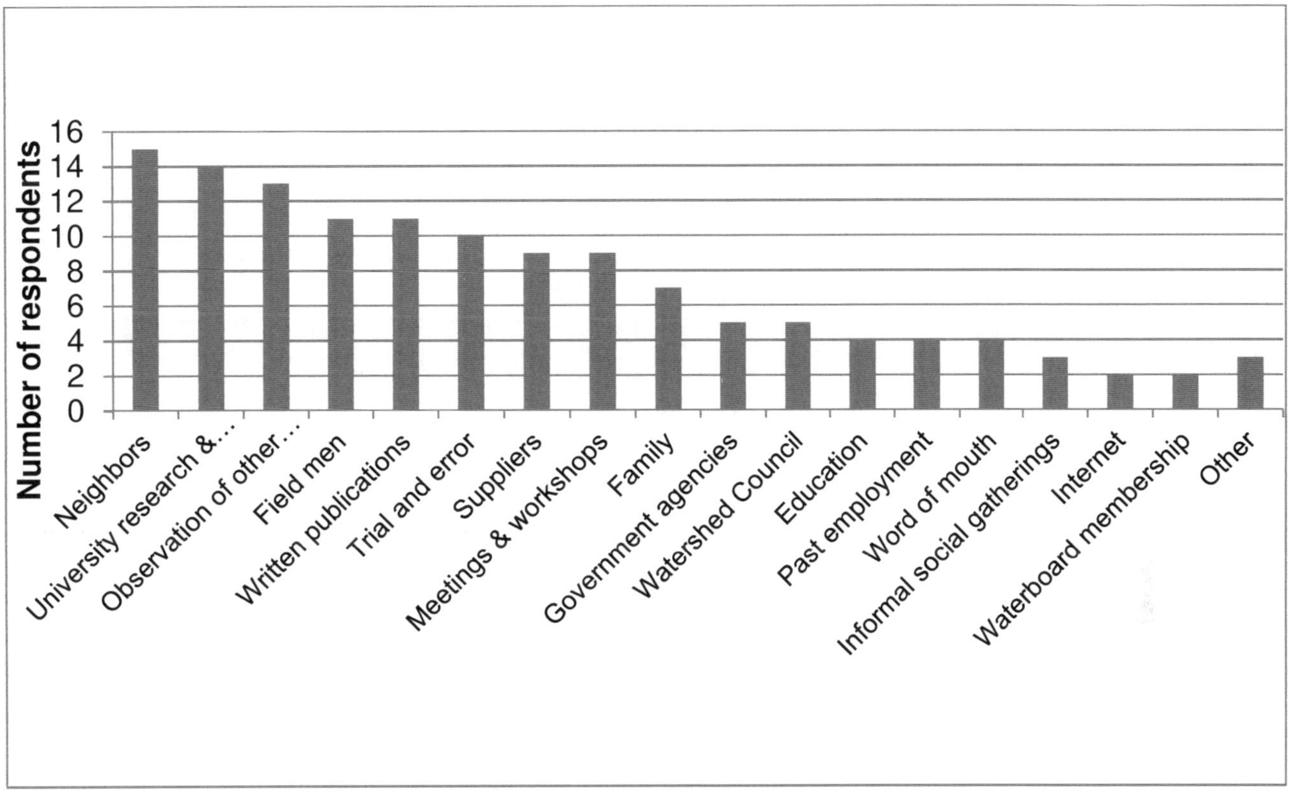


## Conclusions

The relative advantage, compatibility, simplicity, and observation of innovations were important factors in the voluntary adoption of new practices while age and lack of agency emerged as barriers to adoption (Foley, 2013). Growers had very different perceptions of water quality. Defining the adoption of practices can be difficult.

Practitioners of field research and outreach need to creatively imagine alternative practices that could profitably increase the efficiency of input use and that will reduce off-site losses of water, nutrients, and pesticides. Rigorously test and adapt these practices in field trials so that they are viable options for growers.

Outreach agents need to emphasize the practicality and profitability of crop practice options now. They need to design demonstrations that make new practices observable. Extension activities should consistently promote extension as being a reliable trusted source of information.

Policymakers need to provide strong funding and support for applied research and continued use of subsidies and technical support programs for growers. Policymakers can empower growers through education and incentive programs that encourage voluntary adoptions of innovations. There needs to be better communication between growers and policymakers about what the water quality standards are and the goals.

## Literature cited

Bruch, G. 1986. Pesticide and nitrate contamination of ground water near Ontario, Oregon. In: *Proceedings of the Agricultural Impact on Ground Water - A conference.* Omaha, Nebraska; National Water Well Association, Dublin, Ohio.

Foley, K.M. 2013. Examining the Voluntary Adoption of Agricultural Conservation Practices in Northern Malheur County, Oregon. Thesis for the degree of Master of Science in Water Resources Policy and Management, Oregon State University.

Shock, C.C. 2015. Malheur County best management practices. Malheur Experiment Station, Oregon State University. http://www.cropinfo.net/BestPractices/

Shock, C.C. and C.B. Shock. 2012. Research, extension, and good farming practices improve water quality and productivity. Journal of Integrative Agriculture 11(1):14-30.

Townsend, A.R. and R.W. Howarth. 2010. Fixing the global nitrogen problem. Scientific American 32(2):64-71.

# Optimal Efficiency in Gravity Water Distribution Systems

Authors: Jon Peggram - Rubicon Water,
1501 S. Lemay Avenue, Suite 101
Fort Collins, CO 80524
USA
jon.peggram@rubiconwater.com

Damien Pearson - Rubicon Water,
1501 S. Lemay Avenue, Suite 101
Fort Collins, CO 80524
USA
damien.pearson@rubiconwater.com

## Abstract

*New developments in agricultural irrigation technology and management are creating an increased need for on-demand water supplies which can provide water at the precise time and duration needed for optimal efficiency. With most groundwater supplies already being pushed beyond sustainable use, renewable surface water resources must be managed with greater efficiency and timeliness to meet increasingly variable demands.*

*This paper provides examples of improvement in overall customer service and operating efficiency that have been achieved with modernized canal infrastructure. These examples demonstrate how irrigation control gates and flow meters combined with innovative software and advanced control engineering have improved the management of water in canal distribution systems and on the farm.*

*With better management and control of surface water supplies in open channel gravity networks, water availability can be improved for all users thereby enabling food and fiber producers to employ the latest advancements in efficient irrigation technology.*

## Keywords

Water supply network modernization, distribution efficiency, application efficiency, surface water, gravity canal network, automation, flow control, surface irrigation, flood irrigation

# INTRODUCTION AND BACKGROUND

New developments in agricultural irrigation technology and management are creating an increased need for on-demand water supplies which can provide water at the precise time and duration needed for optimal efficiency. With most groundwater supplies already being pushed beyond sustainable use, renewable surface water resources must be managed with greater efficiency and timeliness to meet increasingly variable demands.

Recent examples of modernized surface water conveyance networks which provide a near on-demand water supply powered by gravity can be found in Australia. Extensive investments in irrigation modernization have been made incorporating irrigation control gates and flow meters combined with innovative software and advanced control engineering to improve the management of water in canal distribution systems and on farm.

One key Australian modernization project is the Northern Victoria Irrigation Renewal Project (NVIRP) which is modernizing the delivery assets of the Goulburn-Murray Irrigation District. This investment is improving the efficiency of the conveyance and distribution network from 65% to 85% and will ultimately recover 345,000 acre-feet of water every year for further beneficial use.

## Irrigation Infrastructure in Australia

The majority of Australia's irrigated agriculture is supplied by gravity surface water conveyance networks. In general, Australia does not have the same deep and contiguous ground water aquifers located beneath key agricultural regions as are available in the United States. Surface water represents 74% of Australia's water supply for irrigated agriculture. Only 23% of irrigation water is sourced from ground water[1].

The bulk of Australia's irrigated agriculture is located in the Murray-Darling Basin, which crosses much of southeast Australia. The basin takes its name from its two major rivers, the Murray and the Darling. It is the country's most important river system.

The key irrigation areas in the Murray-Darling basin are the Goulburn-Murray, Murrumbidgee, Murray, and Coleambally irrigation districts. Each of these districts are implementing the modernization programs outlined in this paper. As shown in Figure 1, these districts span the states of Victoria and New South Wales, which straddle the Murray River. The state of Victoria is approximately equal to Colorado in land area, and the Murray River is approximately as long as the Colorado River and has an equivalent discharge.

Victoria's key agricultural region is known as the Foodbowl. The Foodbowl consumes 70-80% of the water used in Victoria, and produces $10 billion dollars of agricultural production including milk, stone fruits, grapes and tomatoes. Approximately $1.7 billion of this produce is exported annually. The Foodbowl region is supplied by a gravity canal network managed by Goulburn-Murray Water. This region covers 840,000 acres of irrigated land supplied by 3,900 miles of open canals. These canals service 14,000 irrigators through 20,000 turnouts. Goulburn-Murray Water has traditionally managed the canal flows through manual operation of approximately 8,000 drop-board regulating structures.

[1] Source: Australian Bureau of Statistics

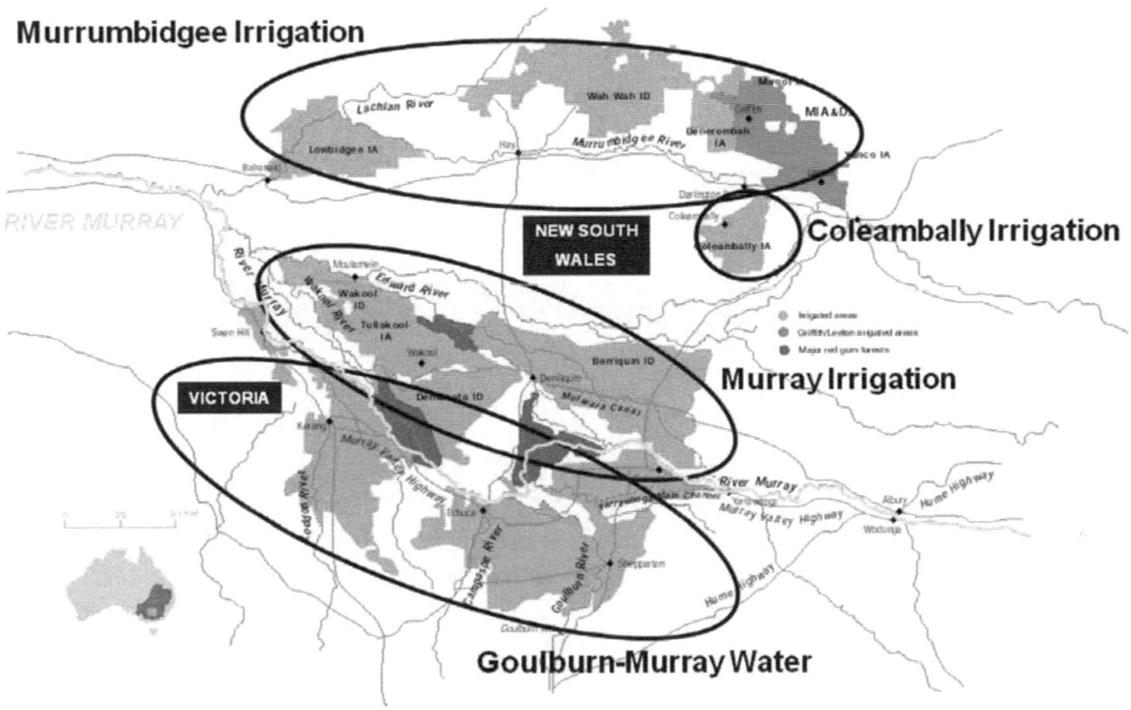


Figure 1 – Modernized Irrigation Districts in the Murray Darling Basin

The farmers that receive this water have historically employed surface irrigation. Surface irrigation (also known as flood irrigation) is used to water approximately 62% of all irrigated farmland in Australia. The land on which the farms have been developed is generally flat and surface irrigation provides an economic means of applying water with no reticulated power and no exposure to energy costs. Approximately 28% of Australia's agriculture is watered by sprinkler irrigation and the remaining 10% is watered by micro irrigation. Due to the widespread use of surface irrigation, many of the on-farm efficiency improvements since the 1980s have focused on constructing sloping borders to achieve more efficient border check irrigation outcomes. Laser grading has been applied in Australia to create field slopes generally of 1:500 to 1:2000. This work was supported by government funded Land and Water Management plans. Sloping borders are the common configuration for surface irrigation in Australia, the use of level basins is not widespread.

## Pre-Modernization Surface Water Distribution Efficiency in the Murray-Darling Basin

Pre-modernized Australian irrigation districts have been very well managed, considering the difficulties of matching supply to demand in gravity canal networks combined with traditional flood irrigation practices. These difficulties resulted in less than half of the water diverted for agriculture productively consumed by crops.

Due to the difficulties of matching supply to demand in gravity canal networks, irrigation customers generally received water with four-day lead times and often received inconsistent flows, with inequitable supply between farmers and inaccurate measurement onto farms.

Manual system operations limited the achievable efficiencies in these networks. The largest losses were during transportation of water and the largest of these losses were due to outfalls (spills). Even with a highly skilled and trained workforce, efficient manual operation of canals was a big challenge. Inefficiencies in water supply networks were caused by antiquated infrastructure and customer desire for improved service which placed priority on maintaining stable pool supply levels ahead of minimizing operational spills.

Approximately 30% of the water that entered Australia's pre-modernized open canal irrigation systems was unaccounted for or lost before it reached the farm. Some of these losses returned as recoverable fractions to the water system, but most of this water was no longer available for beneficial use.

Figure 2 illustrates the key loss components in Australia's pre-modernized agricultural water supply systems. It can be seen in this figure that less than half of the water diverted from storage was productively used by crops. This figure is typical of pre-modernized irrigation system performance globally. Given the proportion of freshwater used by agriculture, recovery of these losses presented an opportunity to reclaim around one third of Australia's freshwater for further beneficial use.

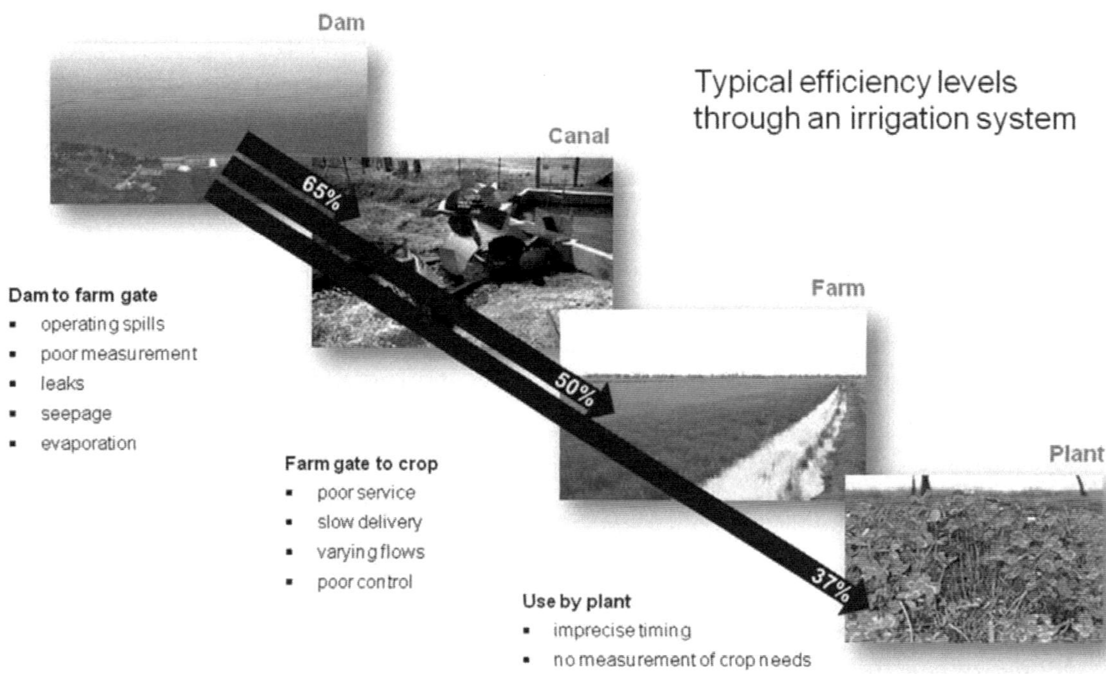


Figure 2 – Typical Efficiency Levels In Pre-Modernized Australian Irrigation Systems

## The Millennium Drought – A Catalyst for Change

Annual precipitation has proven to be highly variable in Australia. It was common throughout the last century for Australia's large surface storage reservoirs to be drawn down over prolonged dry periods. Long-term water availability is governed by the volume of precipitation which is captured and stored. In periods of extended dry, the precise management of stored water for maximum beneficial use becomes a very high priority.

The start of this century witnessed the worst drought in Australia's recorded history. This unprecedented crisis triggered a major overhaul of Australian water law which was underpinned by an investment in one of the world's most advanced modernized irrigation canal networks.

During the Millennium Drought water issues became a top government priority. In response to crisis, new water laws were drafted as part of the National Plan for Water Security, and a new statutory agency – the Murray-Darling Basin Authority- was created to manage the Murray-Darling Basin in an integrated and sustainable manner. The Authority released a major document entitled Guide to the Proposed Murray-Darling Basin Plan, outlining a plan to secure the long-term ecological health of the Murray-Darling Basin. The plan proposed the investment of $10 billion to save the ecology of the Murray-Darling Basin. The specifics included $3.1 billion to buy back water allocations from willing sellers and $5.8 billion for infrastructure investment.

## Modernization Solutions

Prior to 2003 not much had changed in Australian canal management practices in 100 years. As shown in Figure 4, inefficiencies in water supply networks were being caused by antiquated infrastructure and coarse and infrequent regulation adjustments made manually by adjusting traditional drop boards.

Figure 4 – Australian irrigation canal management before modernization – comparison of system operations in 1903 and 2003

An investment in infrastructure modernization was recognized as an opportunity to generate shared benefits. The investment would generate both water savings and improved levels of service to farmers which would accelerate on-farm investment and result in improved irrigation. This system investment would enable improved water use, more agricultural production and less runoff which would in turn result in catchment benefits such as reduced nutrients and salinity.

The cost-performance ratings for various supply modernization options were considered. Water conservation investment decisions were driven by capital costs, operating costs, distribution efficiency and customer service outcomes.

| Distribution system | Gravity channel (manual) | Gravity channel (automated) | Gravity pipeline | Pressurised pipeline |
|---|---|---|---|---|
| Capital cost | ◔ | ◑ | ● | ● |
| Operating cost | ◔ | ◑ | ◔ | ● |
| Distribution efficiency | ○ | ◕ | ● | ● |
| Responsiveness ('on-demand' supply) | ○ | ◕ | ◕ | ◕ |
| **Overall relative cost-performance rating** | ○ | ● | ◑ | ◔ |

○ Very low ◔ Low ◑ Medium ◕ High ● Very high

Figure 5 – Evaluation of Canal Modernization Alternatives

As shown in Figure 5, automated gravity canals were found to provide the best combination of capital costs, operating costs, distribution efficiency and customer service outcomes. Pilot studies of modernized gravity canal systems in the Coleambally Irrigation District and Goulburn-Murray Irrigation District had demonstrated that automated gravity canals could provide distribution efficiencies approaching those of pipelines.

## Northern Victoria Irrigation Renewal Project

One of the centerpieces of the infrastructure investment committed during the Millennium Drought was the Northern Victoria Irrigation Renewal Project (NVIRP). The objectives of this project were to modernize the delivery assets of the Goulburn-Murray Irrigation District. The investment was designed to improve the efficiency of the conveyance network from 65% to 85% and thereby recover 345,000 acre-feet of water every year for further beneficial use. At the same time service levels to irrigators were improved significantly, allowing the full potential of on-farm water savings investments to be realized.

Nearly 12,000 automated gates and meters were installed to automate 1,875 miles of primary canals. The largest agricultural SCADA system in the world was deployed, spanning more than 7,445 RTUs over thousands of miles. A Demand-Integrated Network Control solution was deployed throughout the district to optimize the delivery of water, eliminate spills, provide remote management and data collection and improve farmer service. Works also included the installation of accurate flow meters, and targeted lining of old earthen canals.

Sophisticated management software managed both demand and supply in the entire system to deliver water exactly when and where required. Images of the works undertaken in the project are shown in Figure 6.

In addition to tranforming the canal network from a supply-driven system to a real-time demand-driven system and thereby eliminating operational spills, the modernization project also included the installation of automated flow metering turnouts. These automated flow control turnouts provide accurate and constant flows independent of fluctuations in canal supply levels. Among the many benefits provided by these fully automated flow control turnouts, farmers are now able to easily irrigate at night to reduce their evaporative losses and crop scorching. The new automated turnouts were sized to deliver larger flows than the previous turnouts had delivered, and these higher flow rates were a key part of improving the efficiency of surface irrigation as will be detailed below.

Figure 6 – Modernized Canals in Victoria's NVIRP Project

A farmer survey conducted by Victoria's Department of Sustainability and Environment documented the system benefits shown in Figure 7[2]:

[2]Source: Random survey of 25 farmers, Goulburn-Murray Irrigation District 2008 Irrigation Modernisation Works Post Implementation Review, DSE

| Question | Yes | No |
| --- | --- | --- |
| Do you get a more constant flow onto your farm with your new meter outlet compared with the old one? | 100% | 0% |
| Is having a constant flow onto your farm during irrigation an important factor? | 90% | 10% |
| Was reducing your irrigation order notice time any benefit to your business? | 92% | 8% |
| Was the ability to commence or change your irrigation during non-daylight hours any benefit to your business? | 100% | 0% |
| Was option of upgrading to a remote operating meter outlet any benefit to your business? | 100% | 0% |
| Was option of upgrading to a higher flow rate meter outlet any benefit to your business? | 60% | 40% |
| Is the ability to confirm you requested order instantly via waterline any benefit to you business? | 100% | 0% |
| Overall, how would you describe your support for the modernization projects | Average score: 4 out of 5 | |

Figure 7 – Government Survey of Benefits of Automated Canal Systems

## On-Farm Modernization

In parallel with the investments in conveyance networks, significant investment has also been made in improving on-farm water use efficiencies.

A large proportion of on-farm modernization in Australia is designed to leverage off the significant investments in laser grading of sloping borders made over the last decades, and also to leverage off the in-system canal improvements that have been described above.

Well designed and managed gravity-fed surface irrigation systems have been proven to have the potential to deliver on-farm application efficiencies in excess of 85% and up to 95% on the right soils. Figure 10 below summarizes research by the University of Southern Queensland highlighting the relationship between application efficiency and energy use. Near equivalent water savings can be achieved by high-performance surface irrigation without the increase in energy required by pressurized systems.

Investments in high-performance surface irrigation represent a large proportion of the on-farm water efficiency investments being made in Australia.

The application efficiency of sloping border irrigation is commonly limited through runoff (or tailwater) at the end of the plot or by water infiltrating into the soil below the plant's roots. By applying water at high flow rates, application uniformity can be increased, and both deep percolation and surface runoff can be reduced.

| System | Application efficiency | Water applied (ML/ha) | Water savings (ML/ha) | Energy use (MJ/ha) | Increase in energy use (MJ/ha) |
|---|---|---|---|---|---|
| Traditional surface | 55% | 7.3 | - | 9700 | - |
| Sprinkler | 90% | 4.4 | 2.9 | 17000 | 7300 |
| Micro | 95% | 4.2 | 3.1 | 16000 | 6300 |
| High-performance surface | 85% | 4.7 | 2.6 | 9700 | 0 |

Figure 10 – The Application Efficiencies and Energy use of Various Application Techniques[3]

To irrigate sloping borders using high-performance surface irrigation, water must be supplied onto farms at high flow rates. This is the reason that the turnout flow capacities were significantly increased in the in-system modernization projects such as NVIRP. Secondly, on-farm application systems need to be capable of applying water at high flow rates, which means large bay gates or valves. Thirdly, on-farm systems need to be automated so that gates and valves can be programmed to open and close at a pre-determined time to optimize application efficiency.

With high flow rates, stopping an irrigation even a little late will mean that large volumes of water are lost to surface runoff in a matter of minutes, as well as causing excessive waterlogging, quickly eliminating any efficiency gains up to that point and reducing crop growth. Irrigation duration (or run-time) becomes critical to reaching higher application efficiencies: with high flow rates, the optimal run-time is much shorter and the accuracy and precision required to manage it increases (see Figure 11 below).

The logistics of manually opening and closing gates with short irrigation duration has led to automation of the irrigation to ensure gates or valves close precisely at the right cut-off points and an alert message is sent if they fail to close. The importance of managing irrigation duration (or run-time) is compounded by the fact that the optimal cut-off point in a given bay changes with each irrigation event because conditions such as crop density and soil moisture deficit change. Traditionally, cut-off points have been determined by rules of thumb based on years of experience or by physically visiting a bay to see the progress of the water and then making an assessment. More recently, decision-making is being aided by simple field sensors located halfway along the length of a bay to remotely indicate the progress of the water front.

[3] Kanya L Khatri and Rod Smith (2011), Surface irrigation for energy and water use efficiency, Irrigation Australia Conference.

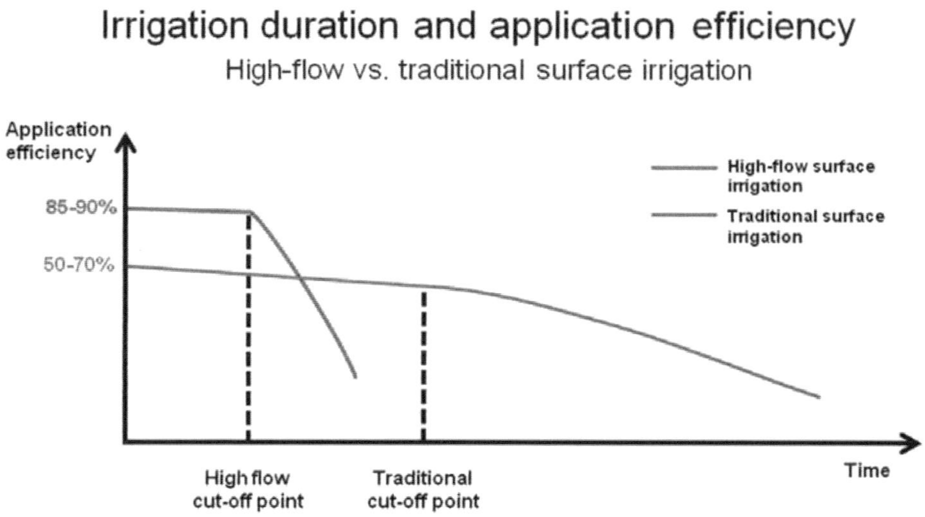


Figure 11 – Increased efficiency and decreased duration of high-flow surface irrigation

Figure 12 – Automated High Flow Surface Irrigation Systems

This technology is achieving application efficiencies similar to sprinkler and drip irrigation, making high-performance surface irrigation a cost-effective on-farm modernization alternative.

## Conclusion

As a result of innovative systems installed within the Murray-Darling Basin's major irrigation districts, canal distribution efficiencies have improved from around 65% to 90%. Farmers can now irrigate more productively with high, consistent flows and near on-demand service. These irrigation districts are now more resilient in the face of reduced rainfall.

An independent review of drought policies endorses the improvements made in the Murray-Darling Basin. A paper released by the Committee for Economic Development of Australia (CEDA) in 2011 states: "The value of these benefits in the Southern Murray-Darling Basin ran into the hundreds of millions of dollars per annum during the drought and represents a major success in water policy. Demands for water will increase as Australia's population grows. The NVIRP and associated water management systems demonstrate how improved efficiencies can ensure this precious resource is managed to meet growing demands and needs now and into the future.

UF | UNIVERSITY of FLORIDA | IFAS

Center for LANDSCAPE Conservation & Ecology

# Long-term Performance of Smart Irrigation Controllers

**Irrigation Association Show**

**Nov., 6-10**

**Orlando, FL**

Michael D. Dukes, PhD., P.E., C.I.D.

Agricultural & Biological Engineering

University of Florida/IFAS

clce.ifas.ufl.edu

# Central Florida Water Initiative

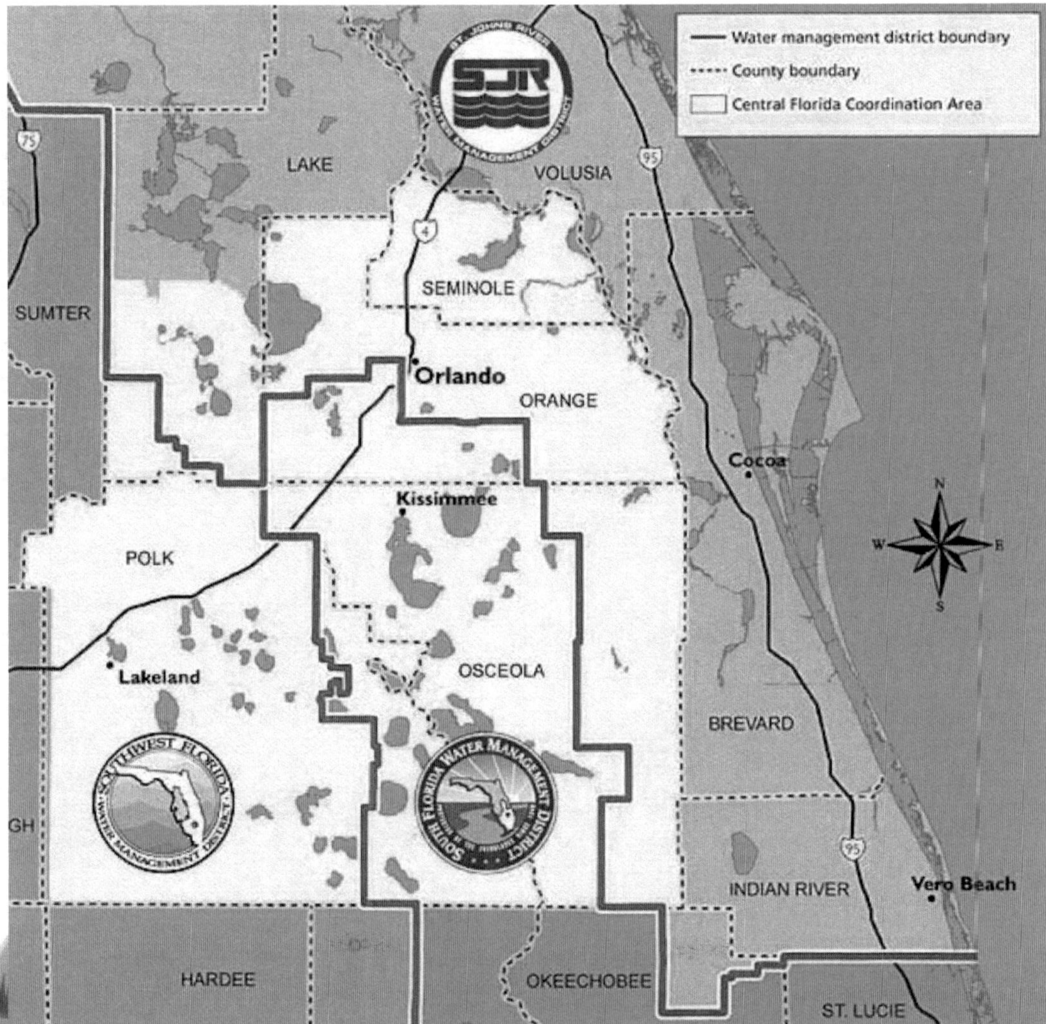


UF UNIVERSITY of FLORIDA IFAS
Center for LANDSCAPE Conservation & Ecology

# Soil Moisture Sensor Controller

# Evapotranspiration (ET) Controllers

- Some can determine runtimes and days
- Programming is key!
  - Soil type
  - Plant type
  - Microclimate
  - Application rates
  - Slope

UF UNIVERSITY of FLORIDA IFAS    Center for LANDSCAPE Conservation & Ecology

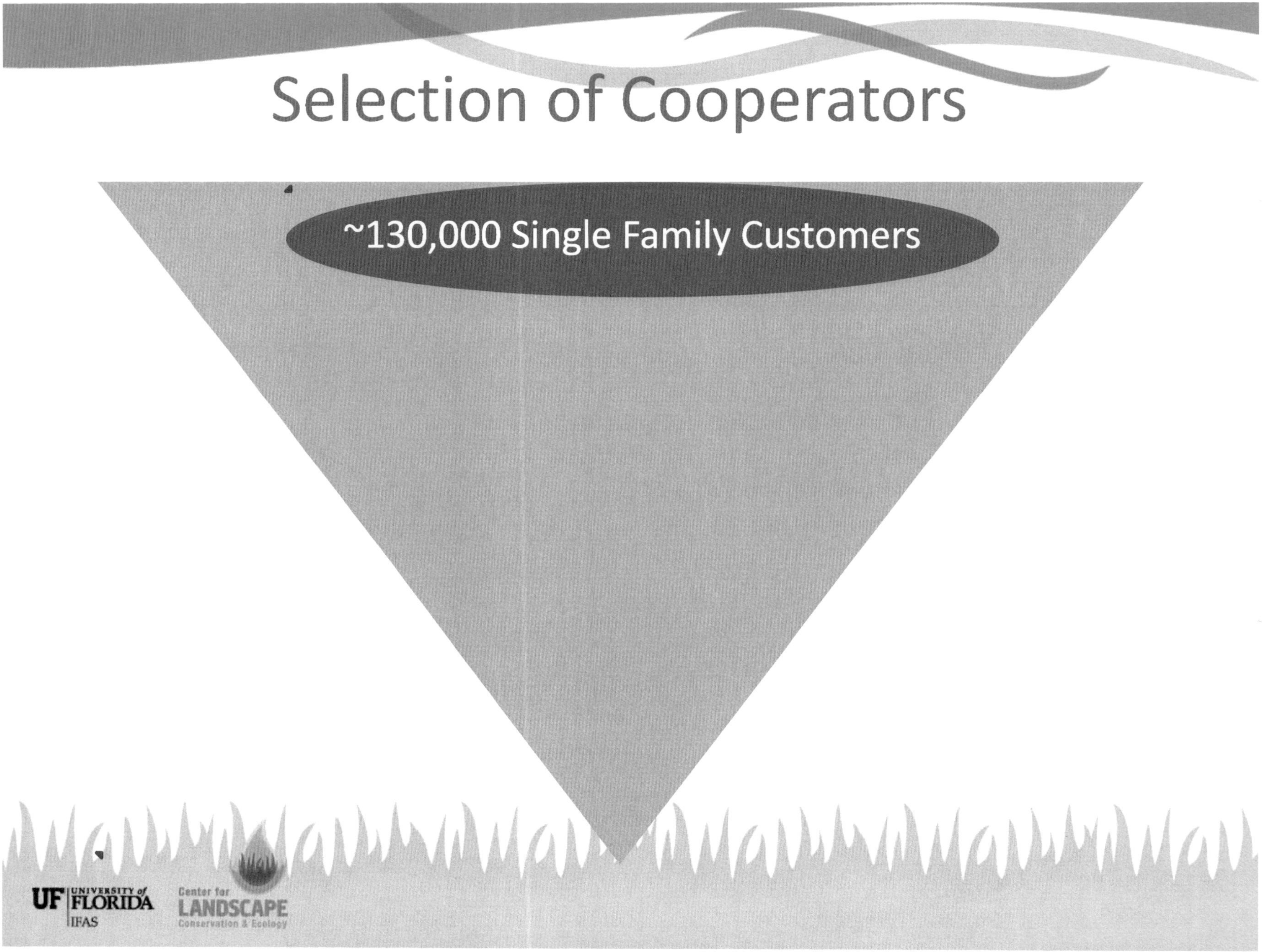
Selection of Cooperators
~130,000 Single Family Customers
UF UNIVERSITY of FLORIDA IFAS
Center for LANDSCAPE Conservation & Ecology

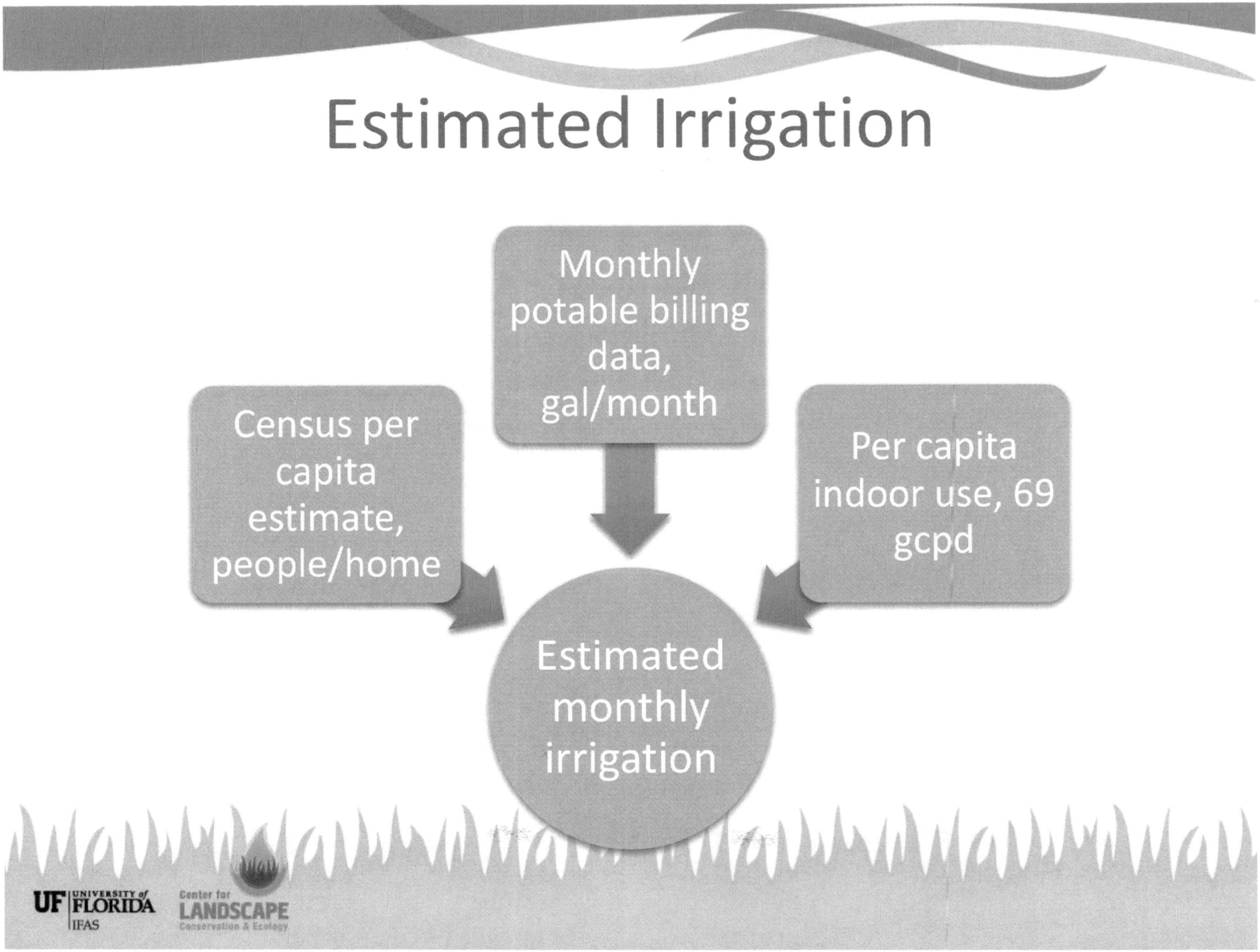
Estimated Irrigation
Census per capita estimate, people/home
Monthly potable billing data, gal/month
Per capita indoor use, 69 gcpd
Estimated monthly irrigation
UF UNIVERSITY of FLORIDA IFAS
Center for LANDSCAPE Conservation & Ecology

# Gross Irrigation Requirements

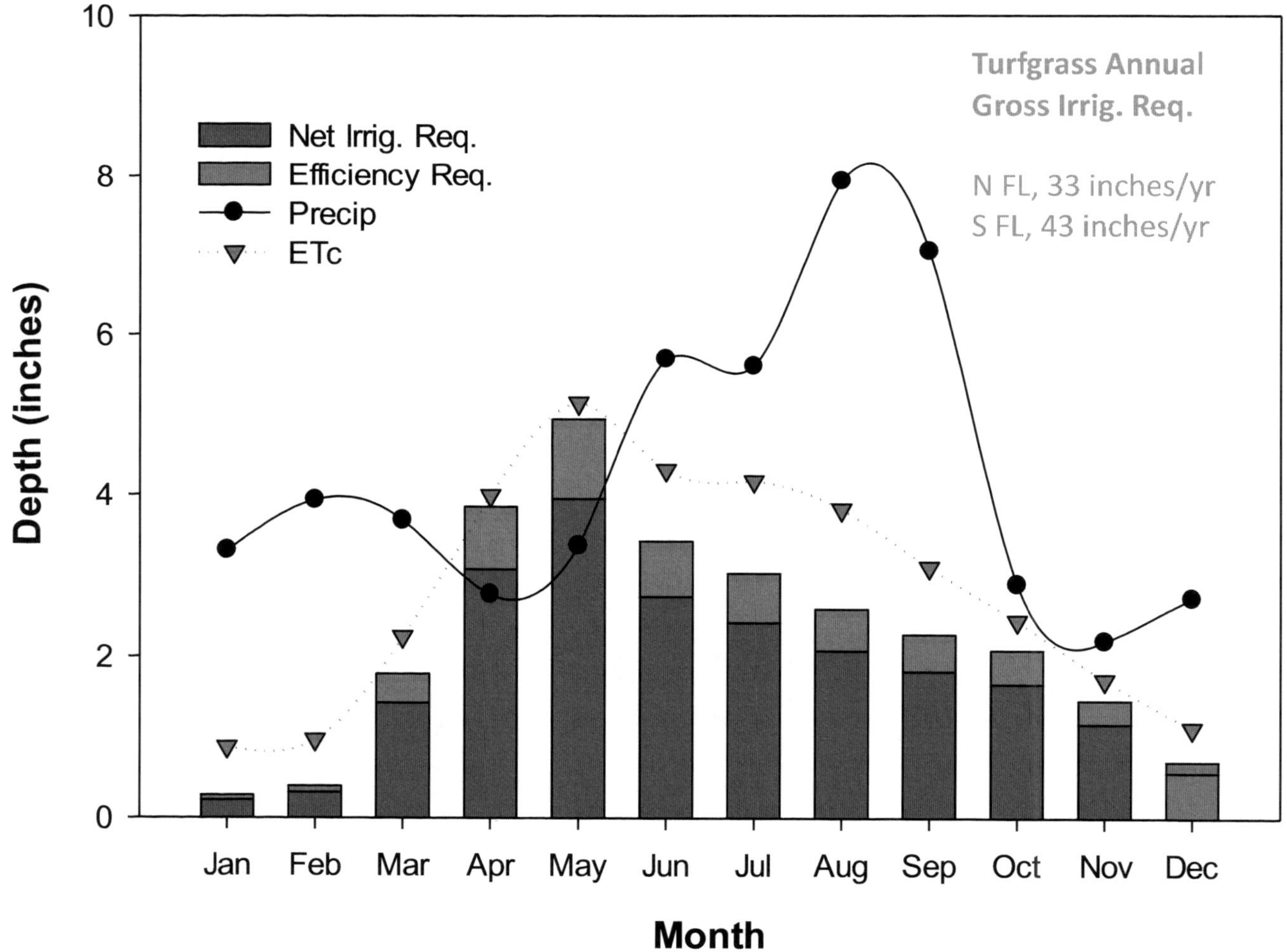


UF UNIVERSITY of FLORIDA IFAS LANDSCAPE Conservation & Ecology

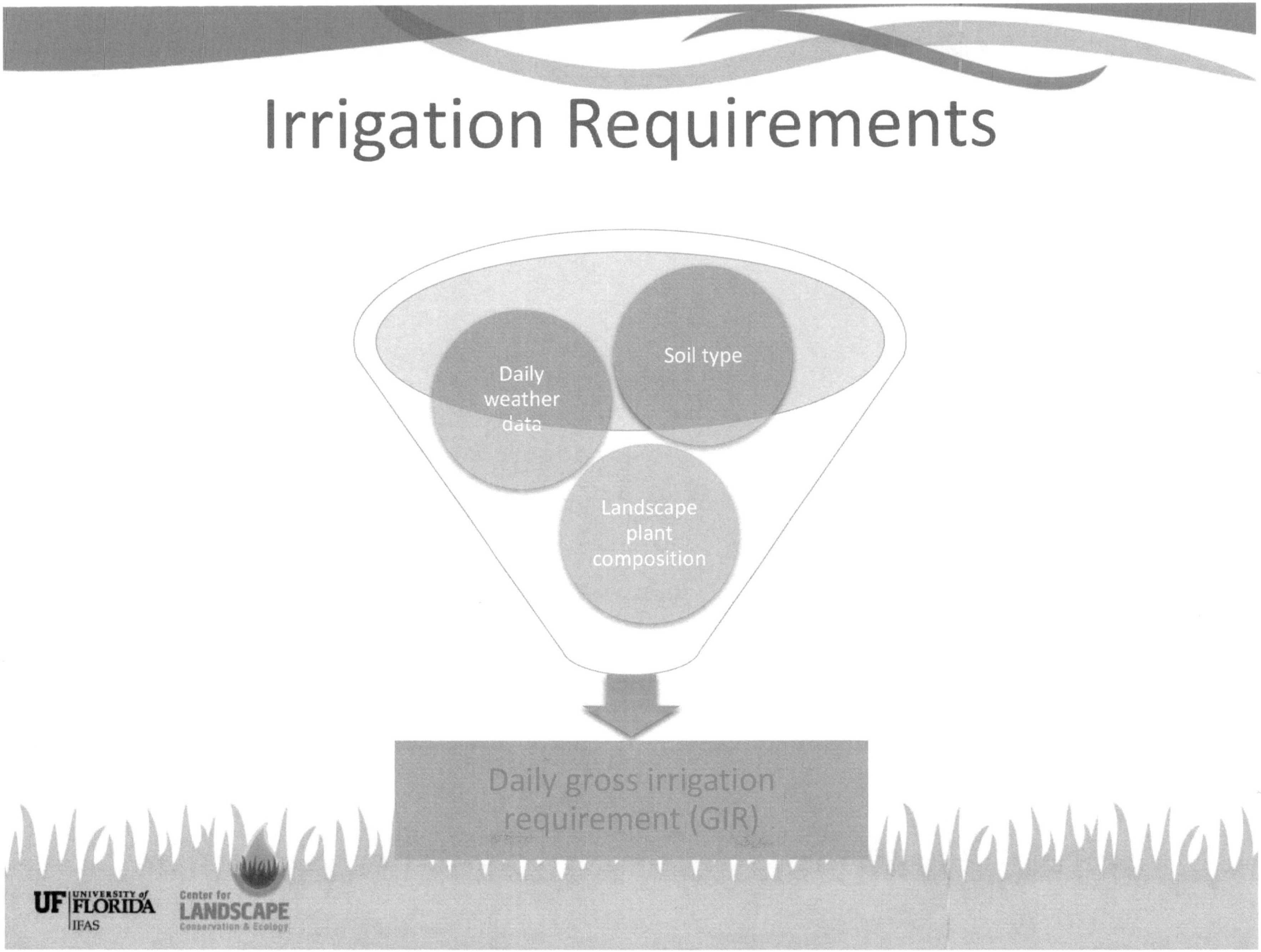
Irrigation Requirements
Daily weather data
Soil type
Landscape plant composition
Daily gross irrigation requirement (GIR)
UF UNIVERSITY of FLORIDA IFAS
Center for LANDSCAPE Conservation & Ecology

# Orange County Evaluation Selection of Excess Irrigators

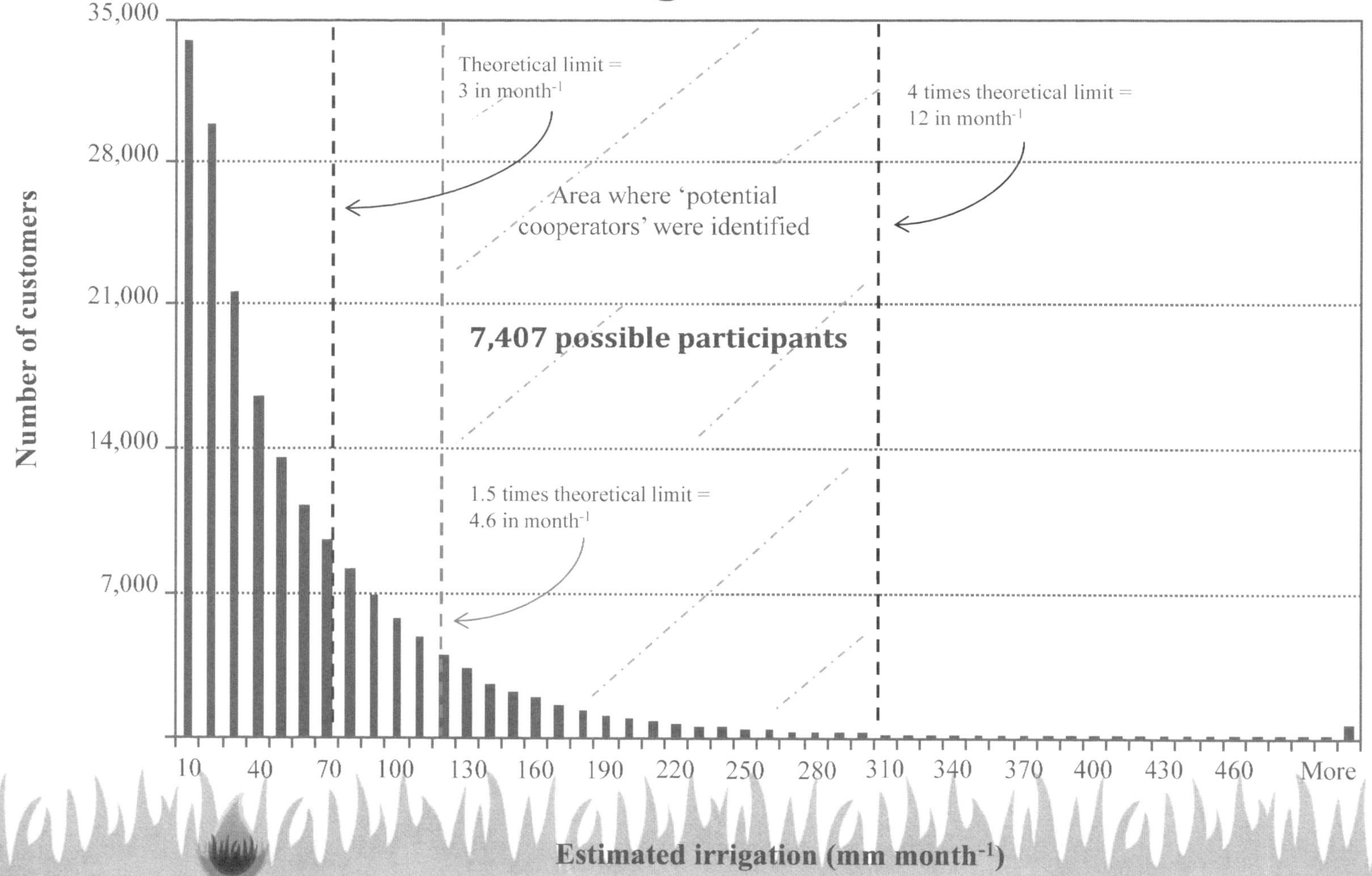


UF UNIVERSITY of FLORIDA IFAS

Center for LANDSCAPE Conservation & Ecology

# Selection of Cooperators

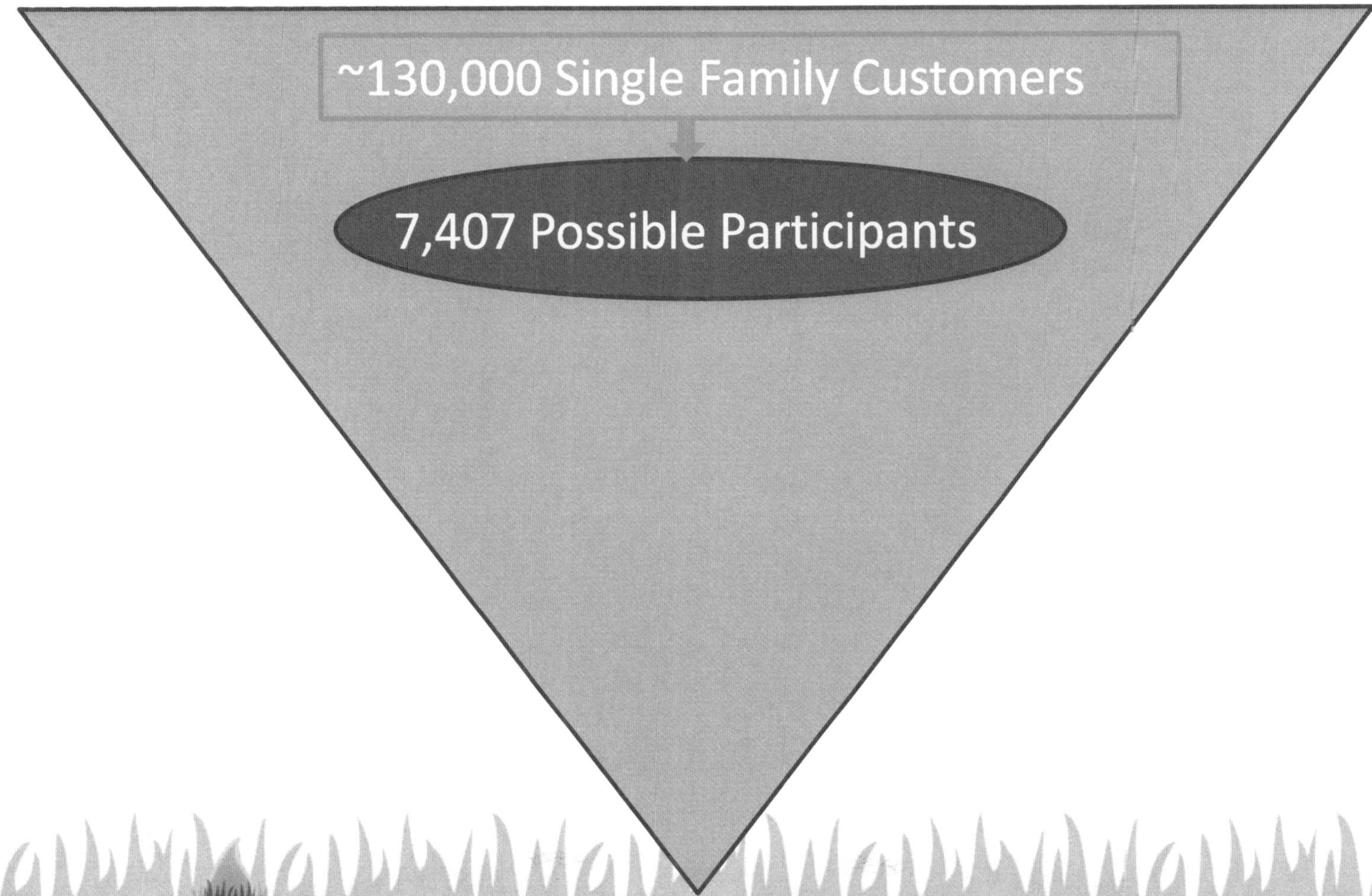

# Cooperator Questionnaire

- Determined homeowner's study interest & Irrigation knowledge
- Irrigation controller or automated irrigation system needed.
- Not a renter
- Intended on living at residence for two or more years

UF UNIVERSITY of FLORIDA IFAS

Center for LANDSCAPE Conservation & Ecology

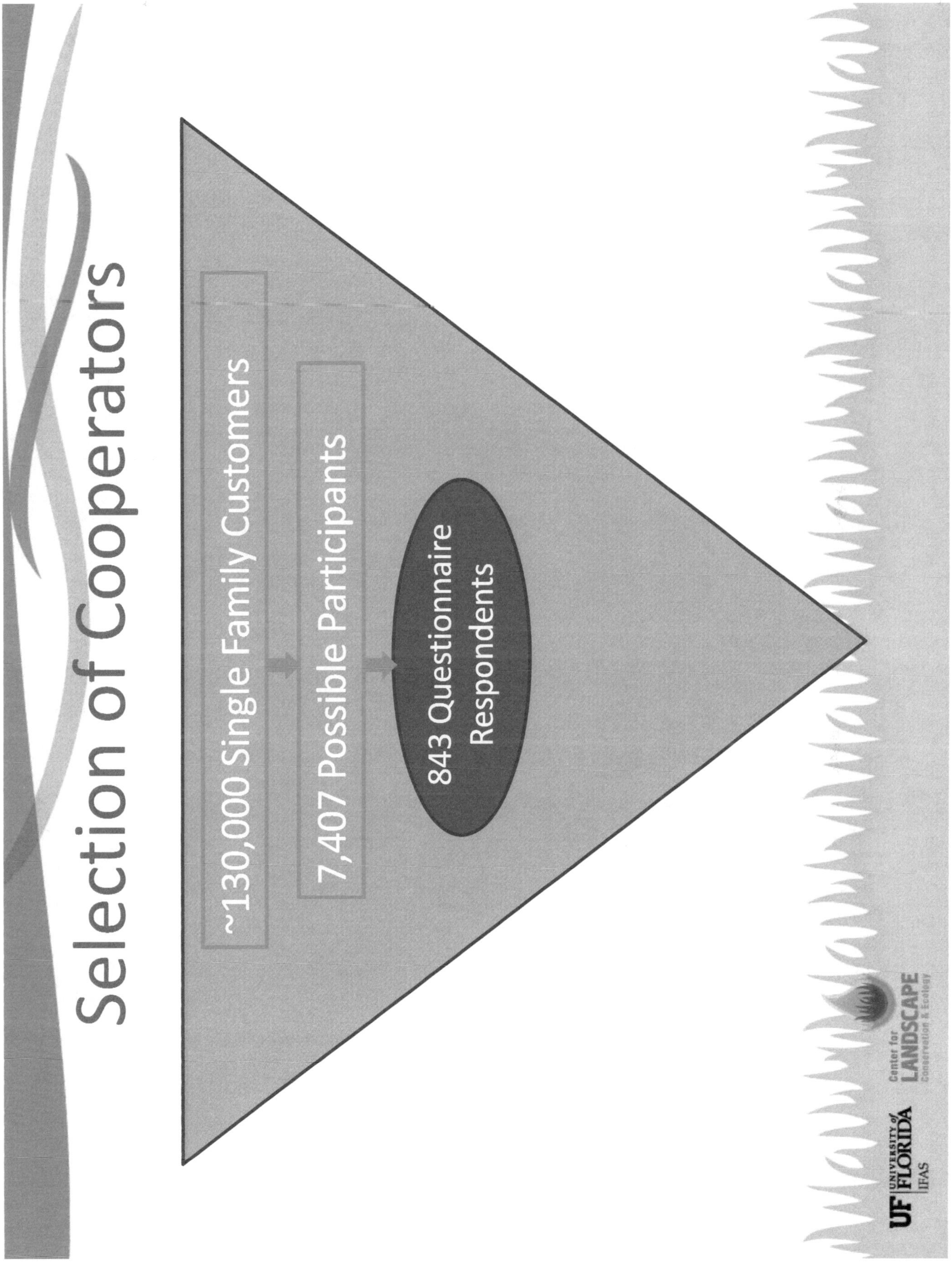
Selection of Cooperators
~130,000 Single Family Customers
7,407 Possible Participants
843 Questionnaire Respondents
UF UNIVERSITY of FLORIDA IFAS
Center for LANDSCAPE Conservation & Ecology

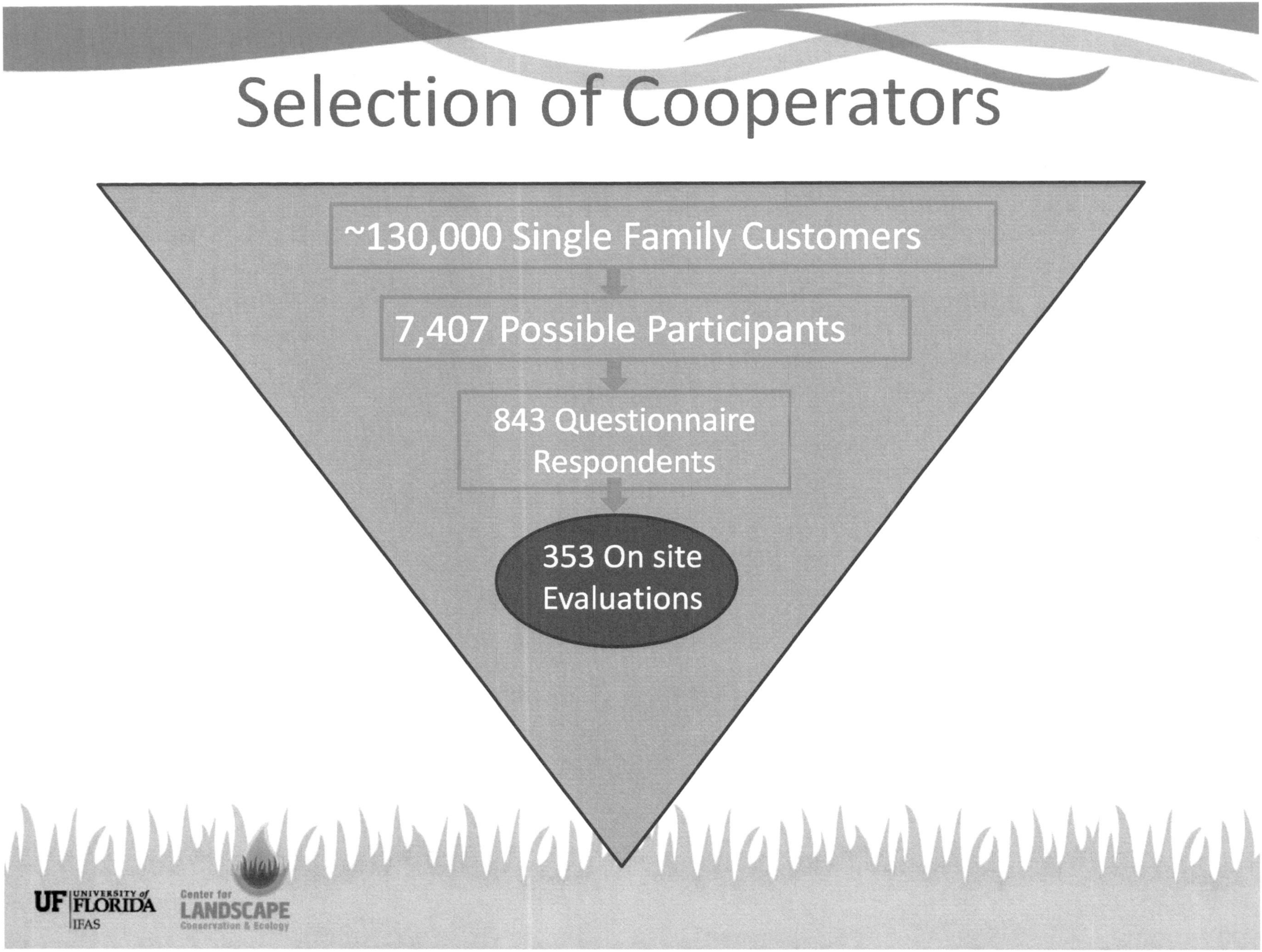
Selection of Cooperators
~130,000 Single Family Customers
7,407 Possible Participants
843 Questionnaire Respondents
353 On site Evaluations
UF UNIVERSITY of FLORIDA IFAS
Center for LANDSCAPE Conservation & Ecology

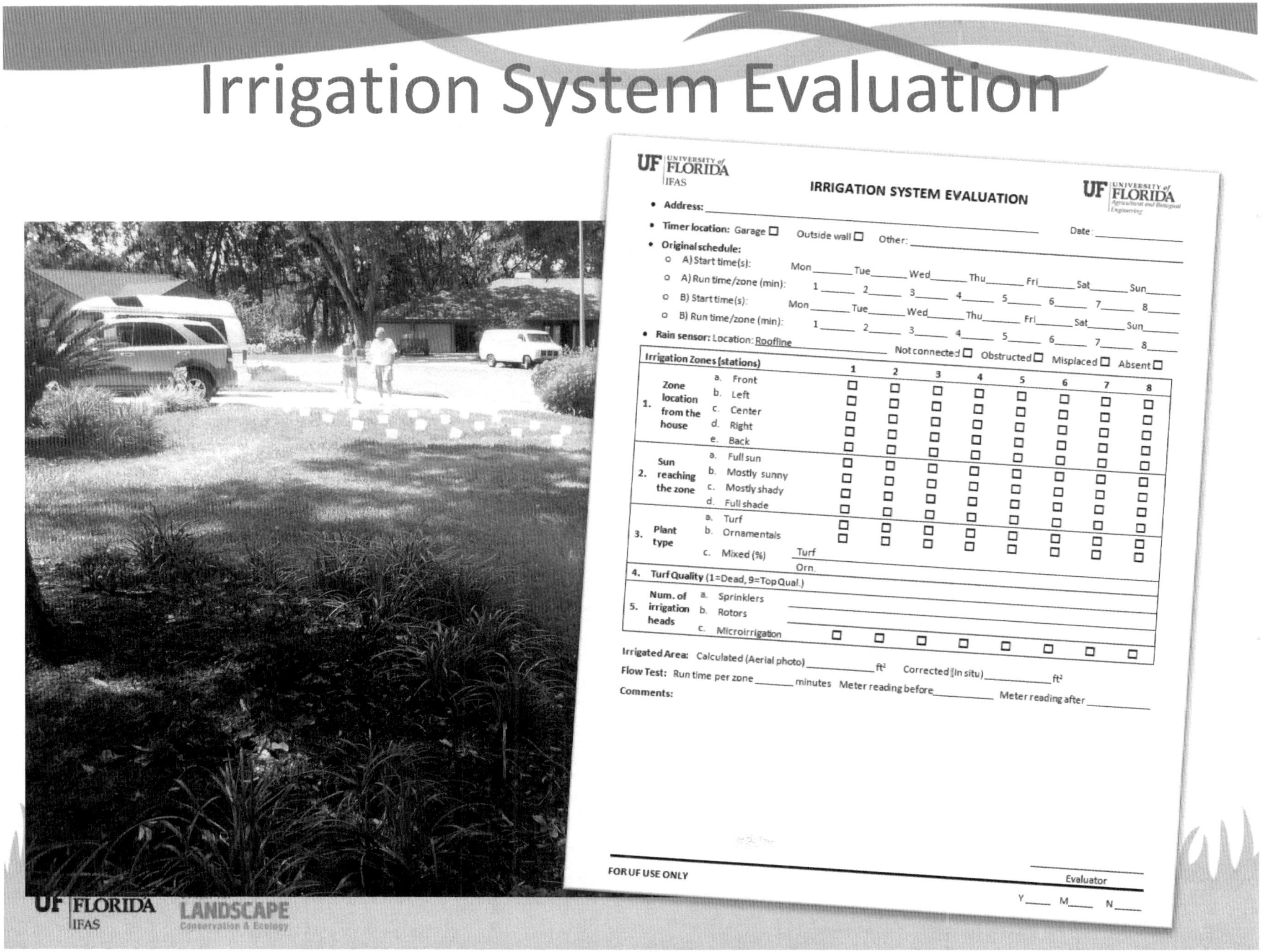

Irrigation System Evaluation
UF FLORIDA IFAS
IRRIGATION SYSTEM EVALUATION
UF UNIVERSITY of FLORIDA Agricultural and Biological Engineering
Address:
Timer location: Garage Outside wall Other:
Date:
Original schedule:
A) Start time(s): Mon Tue Wed Thu Fri Sat Sun
A) Run time/zone (min): 1 2 3 4 5 6 7 8
B) Start time(s): Mon Tue Wed Thu Fri Sat Sun
B) Run time/zone (min): 1 2 3 4 5 6 7 8
Rain sensor: Location: Roofline
Not connected Obstructed Misplaced Absent
Irrigation Zones (stations)
1 2 3 4 5 6 7 8
1. Zone location from the house
a. Front
b. Left
c. Center
d. Right
e. Back
2. Sun reaching the zone
a. Full sun
b. Mostly sunny
c. Mostly shady
d. Full shade
3. Plant type
a. Turf
b. Ornamentals
c. Mixed (%)
Turf
Orn.
4. Turf Quality (1=Dead, 9=Top Qual.)
5. Num. of irrigation heads
a. Sprinklers
b. Rotors
c. Microirrigation
Irrigated Area: Calculated (Aerial photo) ft² Corrected (In situ) ft²
Flow Test: Run time per zone minutes Meter reading before Meter reading after
Comments:
FOR UF USE ONLY
Evaluator
Y M N
UF FLORIDA IFAS
LANDSCAPE
Conservation & Ecology

# Selection of Cooperators

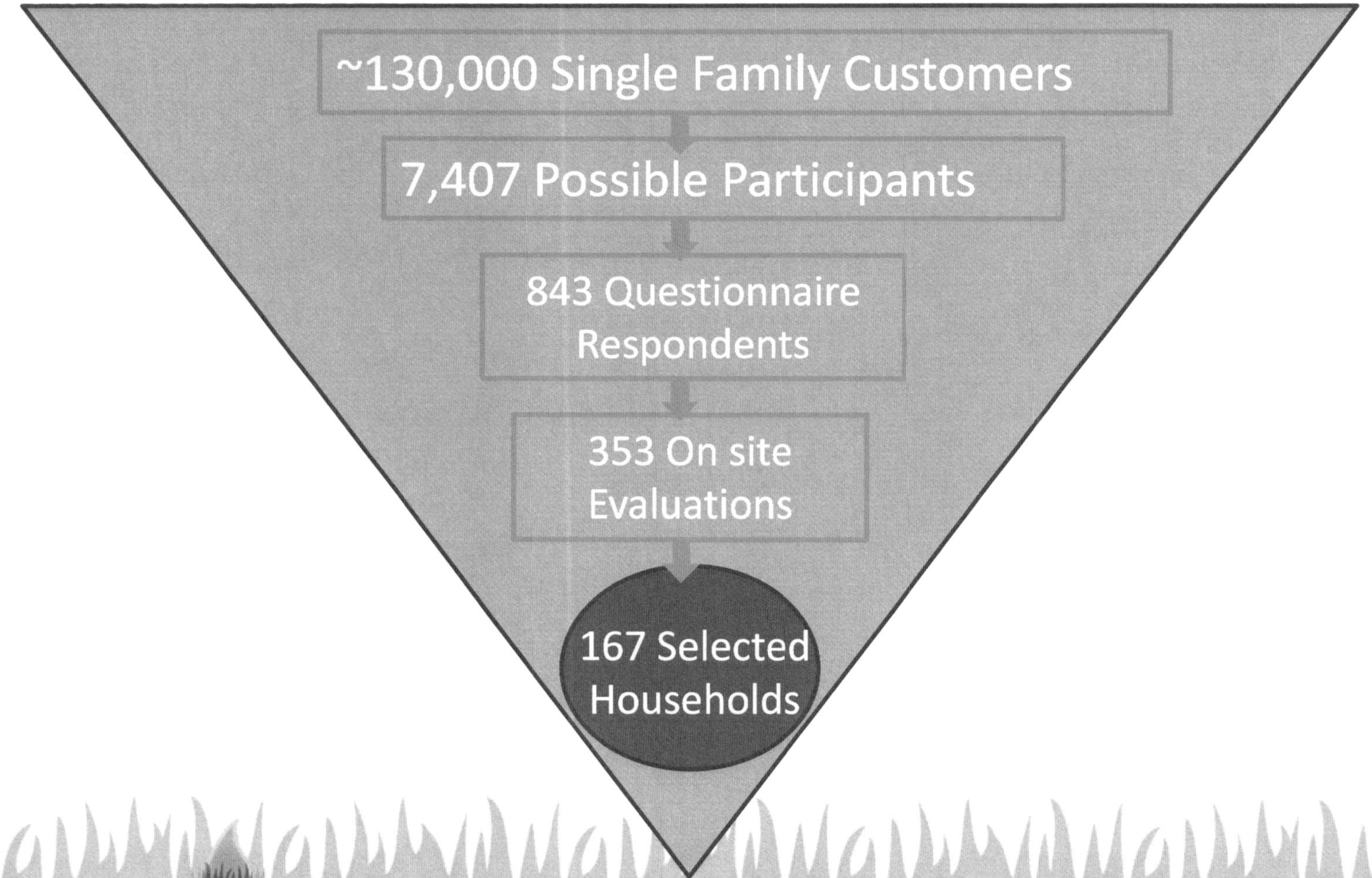


UF UNIVERSITY of FLORIDA IFAS
Center for LANDSCAPE Conservation & Ecology

# Summary of Participants

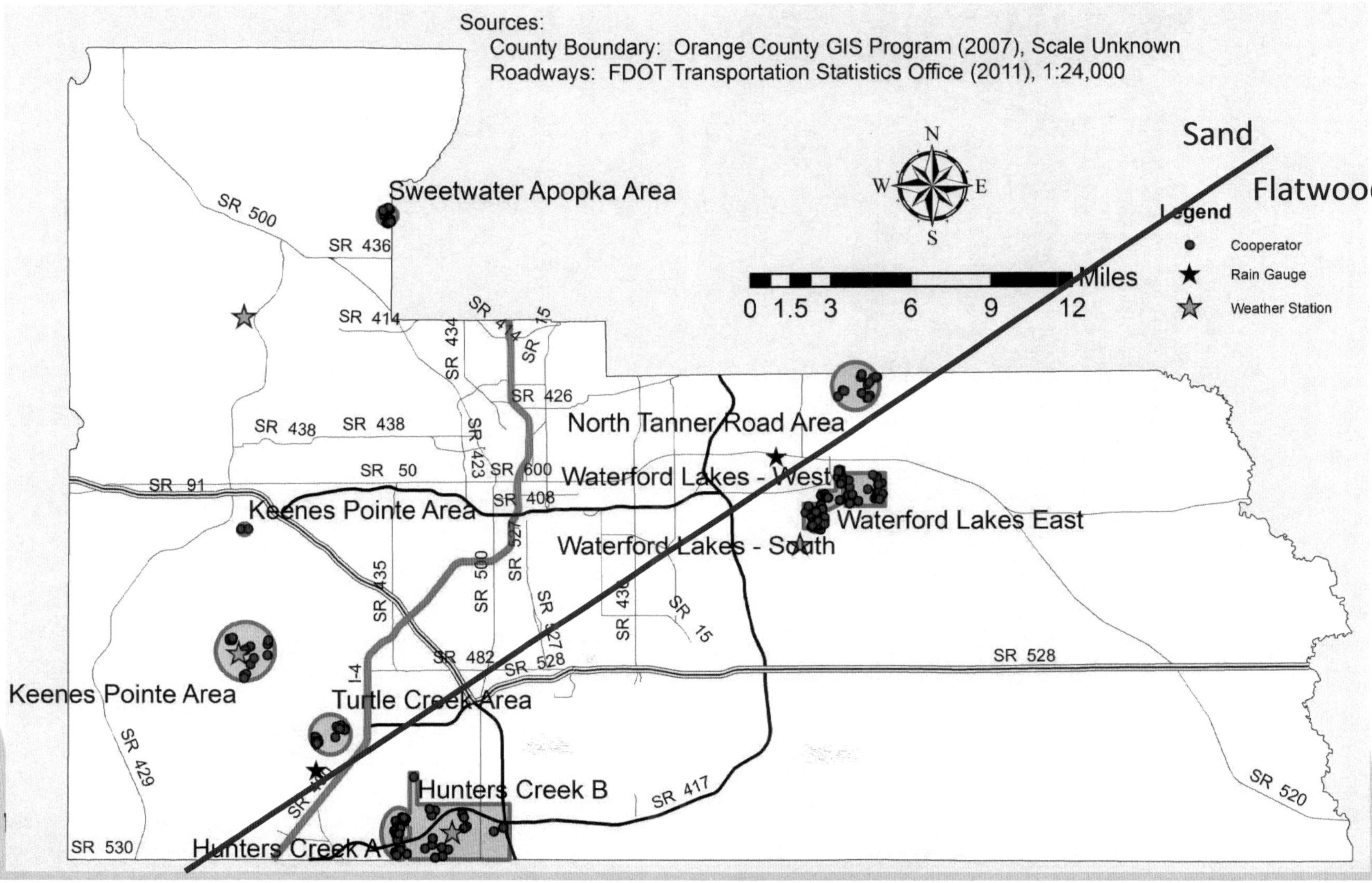

# Two Smart Controllers Evaluated

– Rain Bird ESP-SMT
  - ET treatment

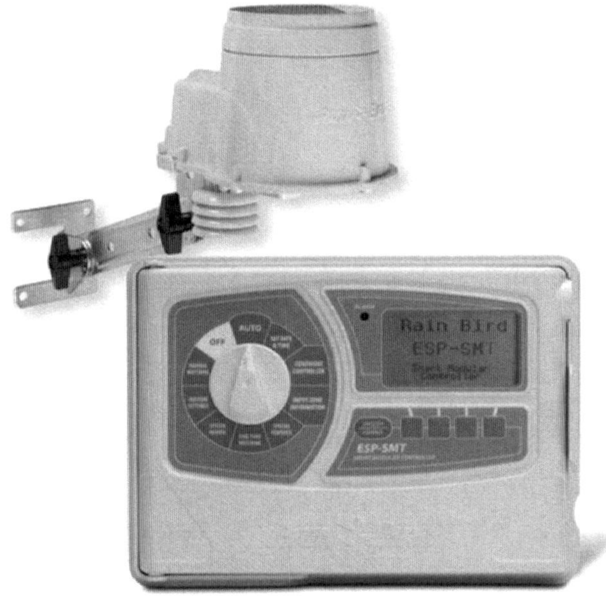


– Baseline WaterTec S100
  - SMS treatment

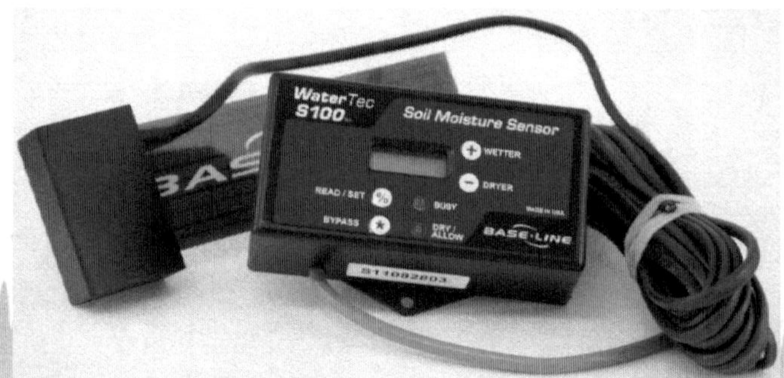


UF UNIVERSITY of FLORIDA IFAS

Center for LANDSCAPE Conservation & Ecology

# Contractor Groups

- ET
  - Contractor programmed with default landscape settings
  - Daily water windows
  - Rare interaction with homeowner
- SMS
  - Buried at 6 inches in minimally compacted soil
  - Re-programmed time clock schedules for daily irrigation:
    - 20 minutes spray
    - 45 minutes rotor
  - Rare interaction with the homeowner

UF UNIVERSITY of FLORIDA IFAS

Center for LANDSCAPE Conservation & Ecology

# "EDU" Groups

- Educational Training
  - ET+Edu treatment
    - Reprogrammed for site specifics
    - 5 minute tutorial
  - SMS+Edu treatment
    - Inserted into soil column at 3 inch depth
    - Reprogrammed for 0.25" per event, 2 events per day, 3 d/wk
    - 5 minute tutorial

UF UNIVERSITY of FLORIDA IFAS
Center for LANDSCAPE Conservation & Ecology

# Automatic Meter Recording devices (AMRs)

- Separated flow meter to measure irrigation only
- Records hourly irrigation volumes
- Monthly downloads

UF UNIVERSITY of FLORIDA IFAS
Center for LANDSCAPE Conservation & Ecology

# OCU Technologies & Expt. Design

| Treatment | ET | ET+Edu | SMS | SMS+Edu | Comparison |
|---|---|---|---|---|---|
| Technology | Rain Bird ESP-SMT | Rain Bird ESP-SMT | Baseline WaterTec S100 | Baseline WaterTec S100 | -- |
| Locations Installed | 7 | 9 | 7 | 9 | 9 |
| Number Installed | 28 | 38 | 28 | 38 | 35 |

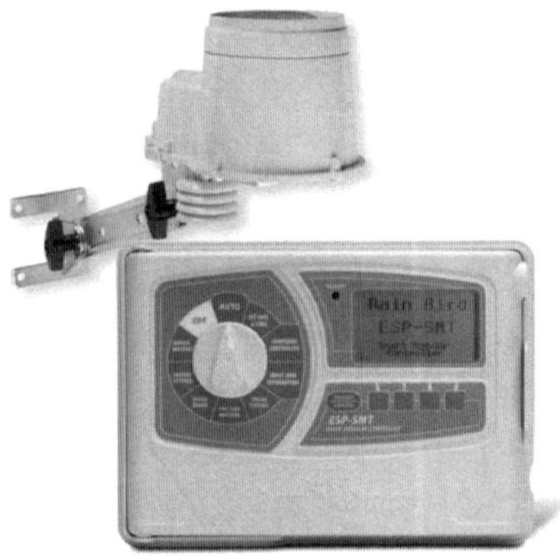


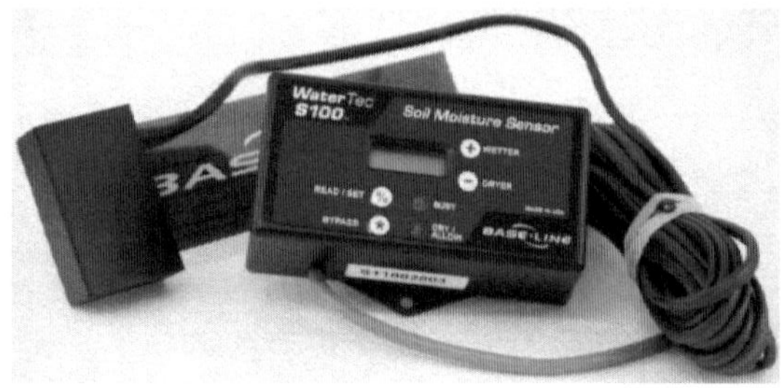


UF UNIVERSITY of FLORIDA IFAS
Center for LANDSCAPE Conservation & Ecology

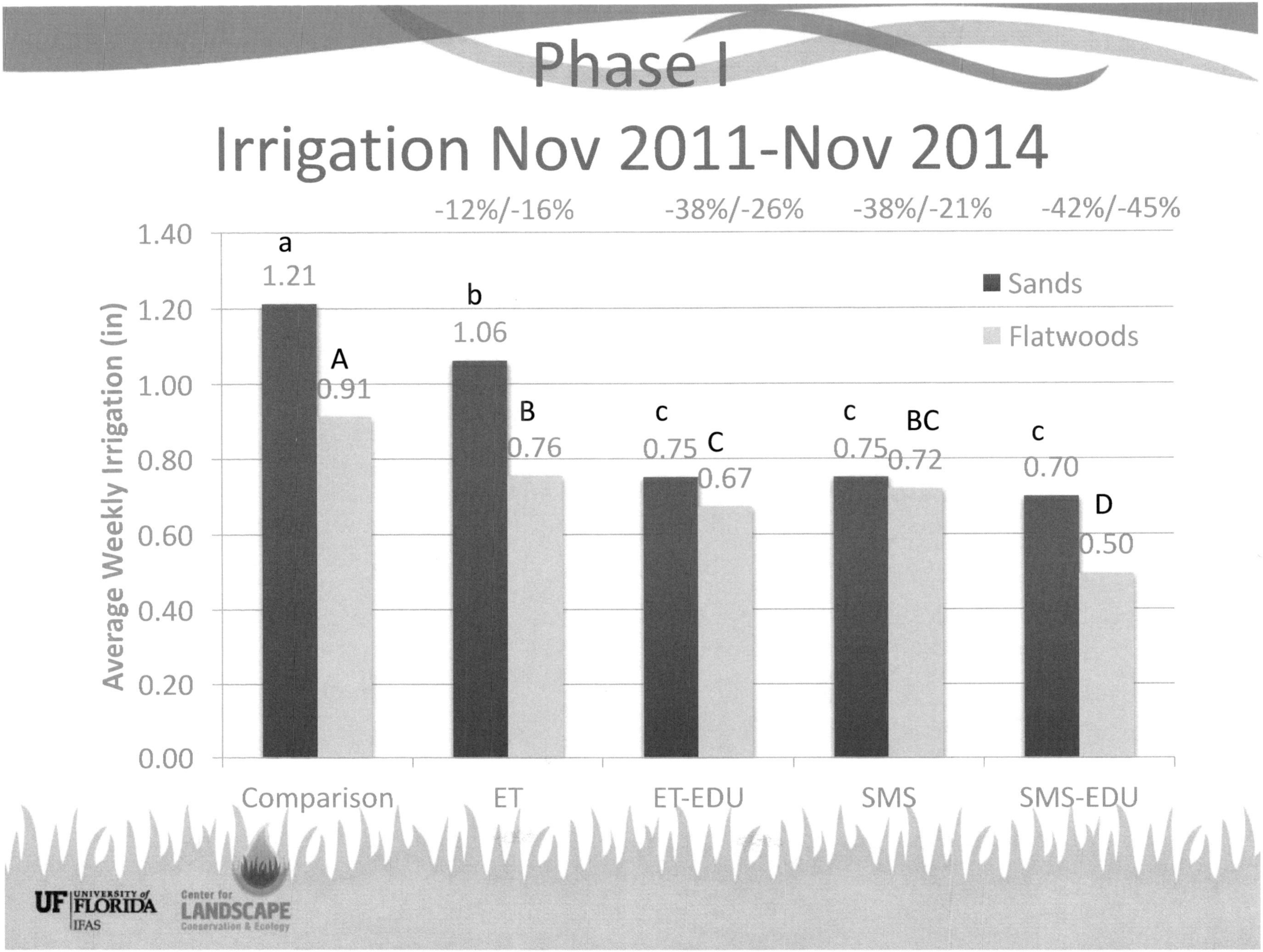
Phase I
Irrigation Nov 2011-Nov 2014
-12%/-16%
-38%/-26%
-38%/-21%
-42%/-45%
Average Weekly Irrigation (in)
1.40
1.20
1.00
0.80
0.60
0.40
0.20
0.00
a
1.21
A
0.91
b
1.06
B
0.76
c
0.75
C
0.67
c
0.75
BC
0.72
c
0.70
D
0.50
Sands
Flatwoods
Comparison
ET
ET-EDU
SMS
SMS-EDU
UF UNIVERSITY of FLORIDA IFAS
Center for LANDSCAPE Conservation & Ecology

# Phase II

## Irrigation Nov 2014-Oct 2015

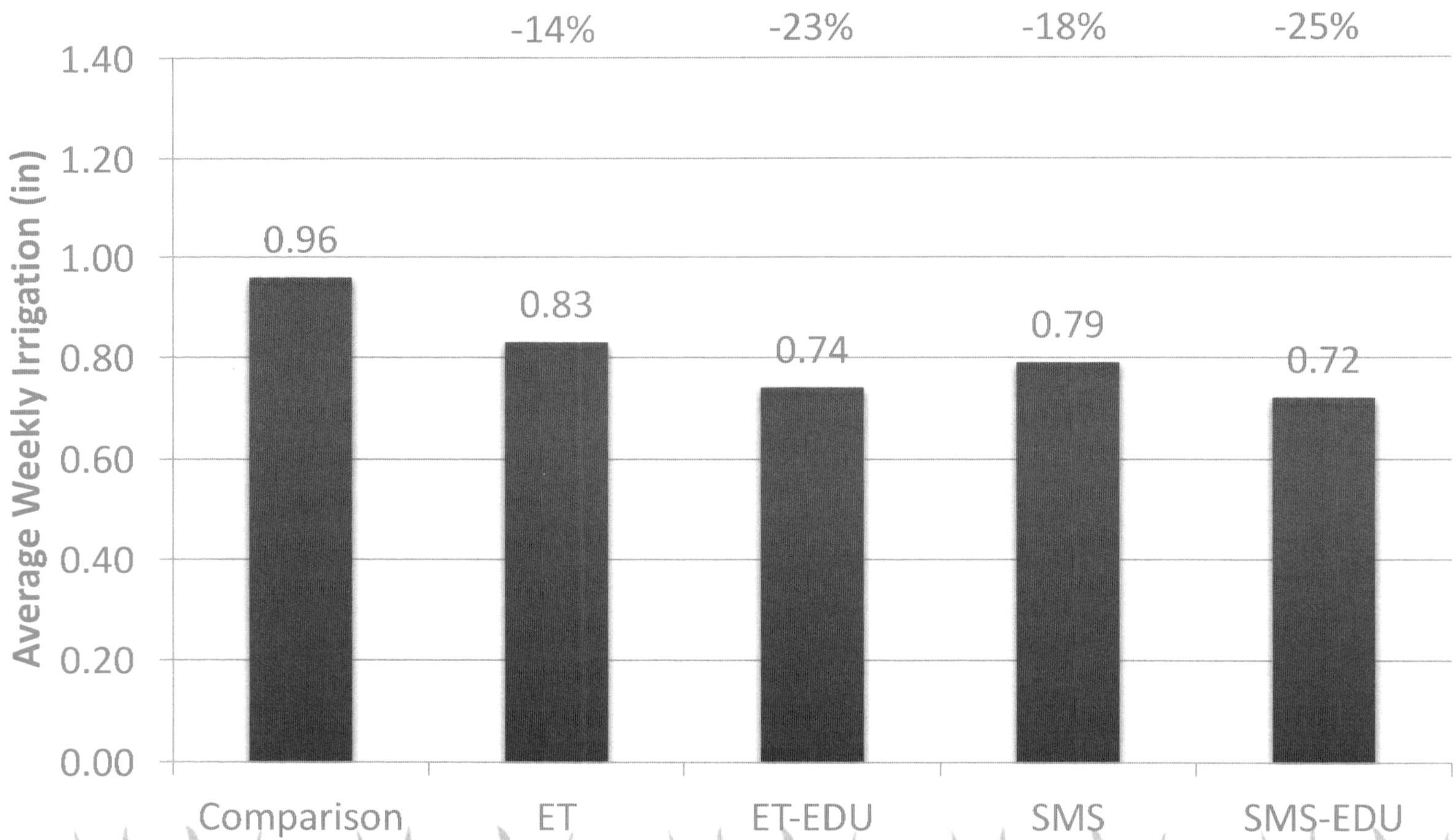


UF UNIVERSITY of FLORIDA IFAS

Center for LANDSCAPE Conservation & Ecology

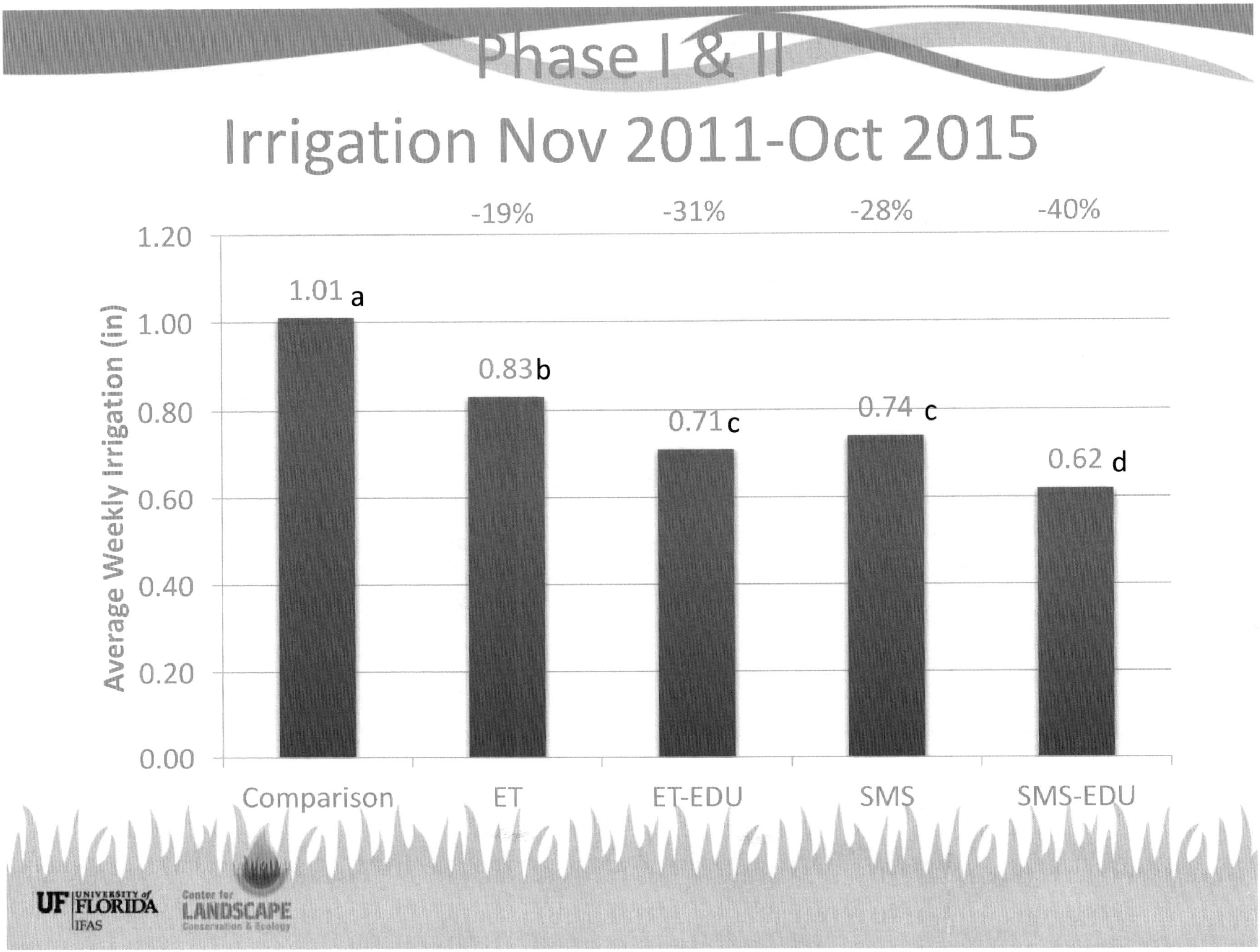
Phase I & II
Irrigation Nov 2011-Oct 2015
-19%
-31%
-28%
-40%
1.20
1.00
0.80
0.60
0.40
0.20
0.00
Average Weekly Irrigation (in)
1.01 a
0.83 b
0.71 c
0.74 c
0.62 d
Comparison
ET
ET-EDU
SMS
SMS-EDU
UF UNIVERSITY of FLORIDA IFAS
Center for LANDSCAPE Conservation & Ecology

# Irrigation/Week Nov 2011-Feb 2016

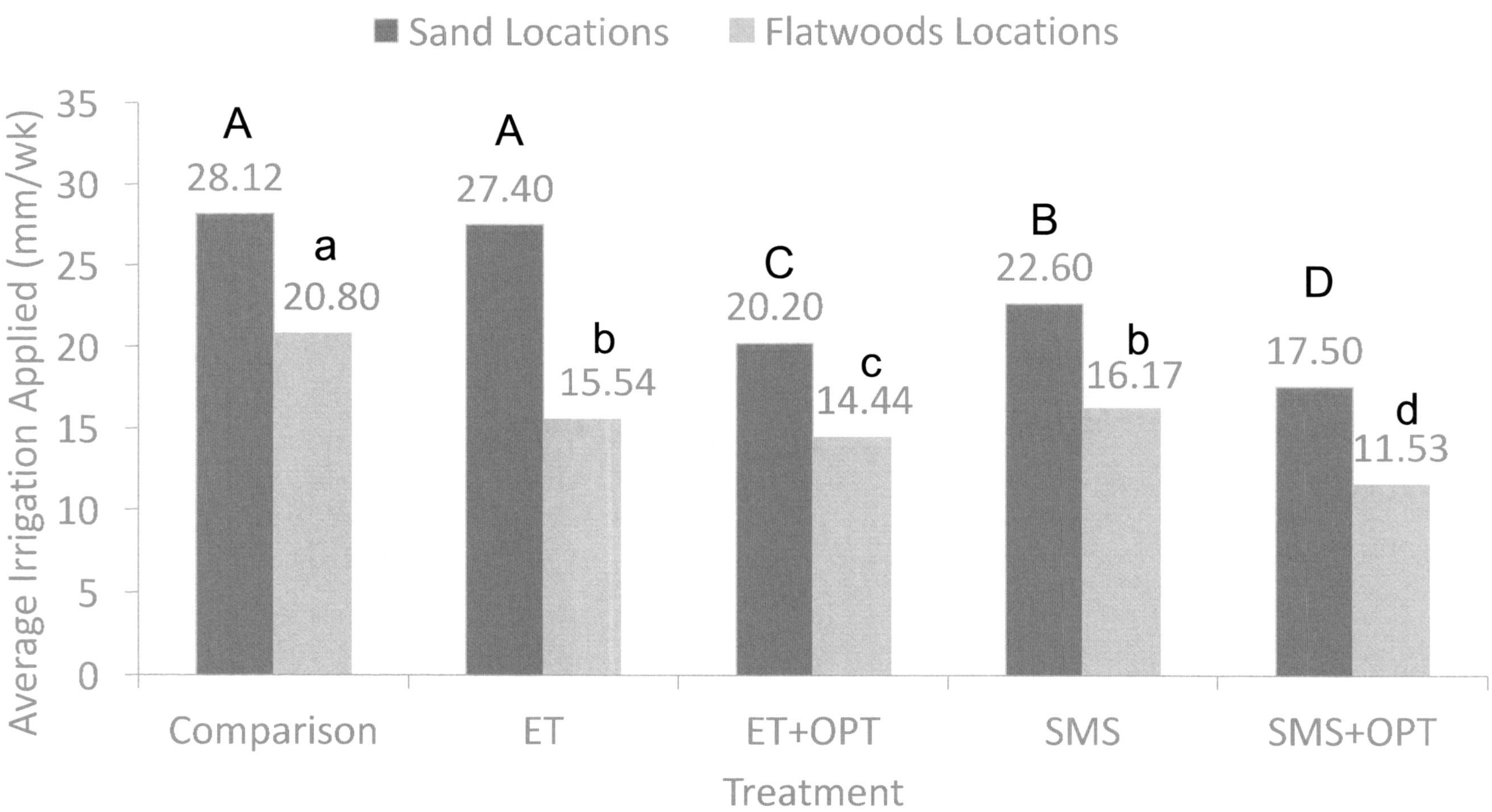


UF UNIVERSITY of FLORIDA IFAS
Center for LANDSCAPE Conservation & Ecology

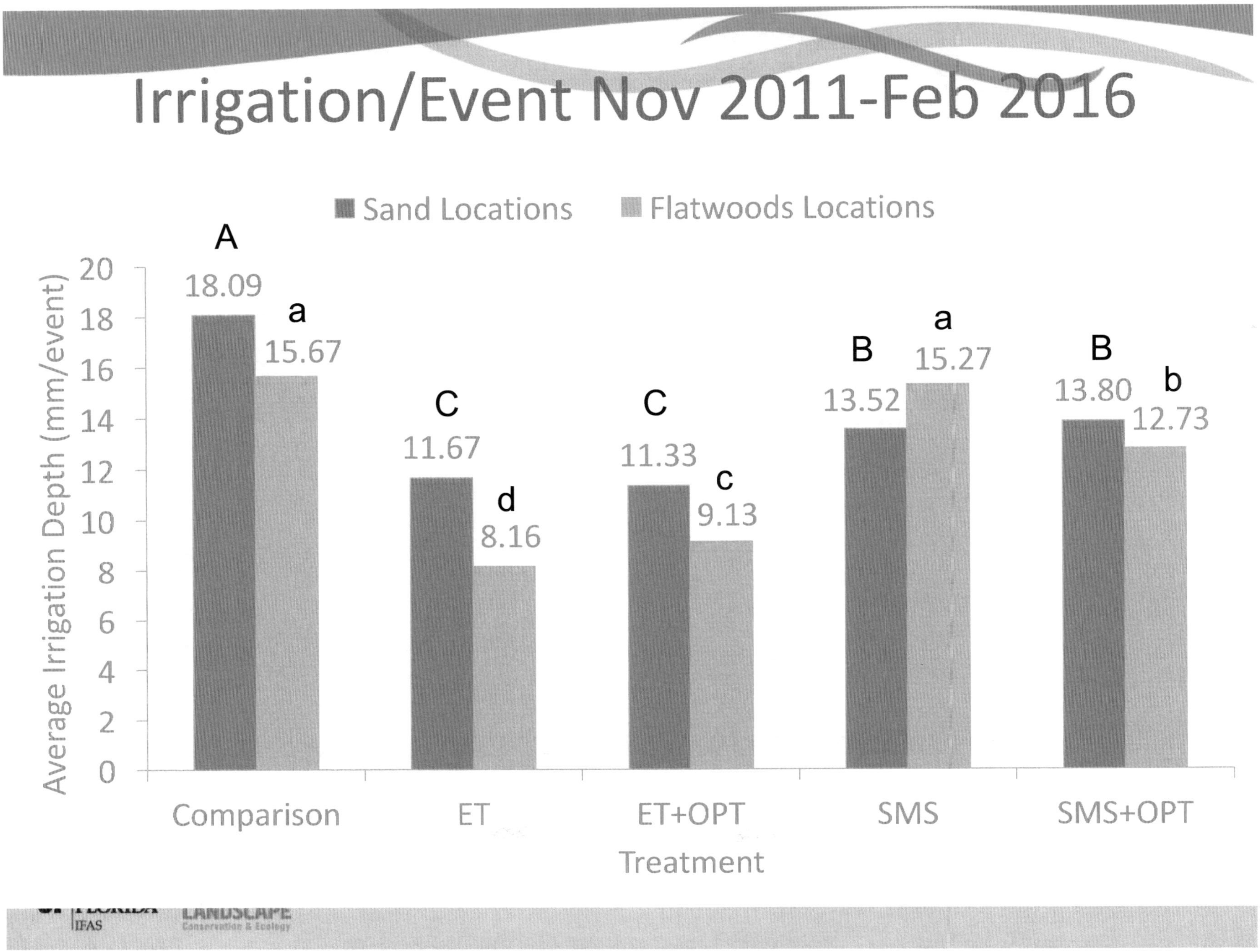
Irrigation/Event Nov 2011-Feb 2016
Sand Locations
Flatwoods Locations
Average Irrigation Depth (mm/event)
20
18
16
14
12
10
8
6
4
2
0
A
18.09
a
15.67
C
11.67
d
8.16
C
11.33
c
9.13
B
13.52
a
15.27
B
13.80
b
12.73
Comparison
ET
ET+OPT
SMS
SMS+OPT
Treatment
IFAS
LANDSCAPE
Conservation & Ecology

# Irrigation Events/Week Nov 2011-Feb 2016

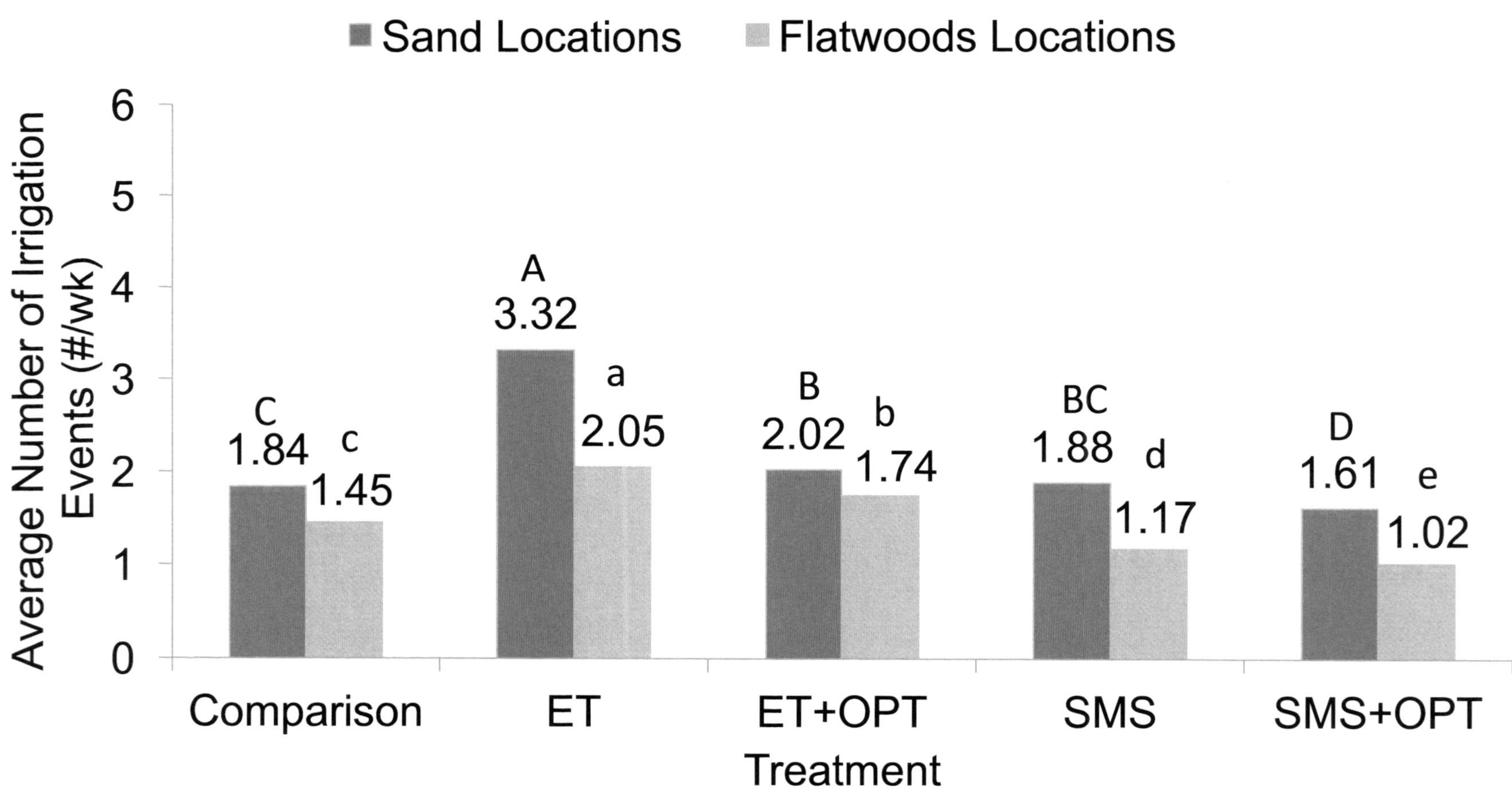


UF UNIVERSITY of FLORIDA IFAS

Center for LANDSCAPE Conservation & Ecology

# Comparison Results

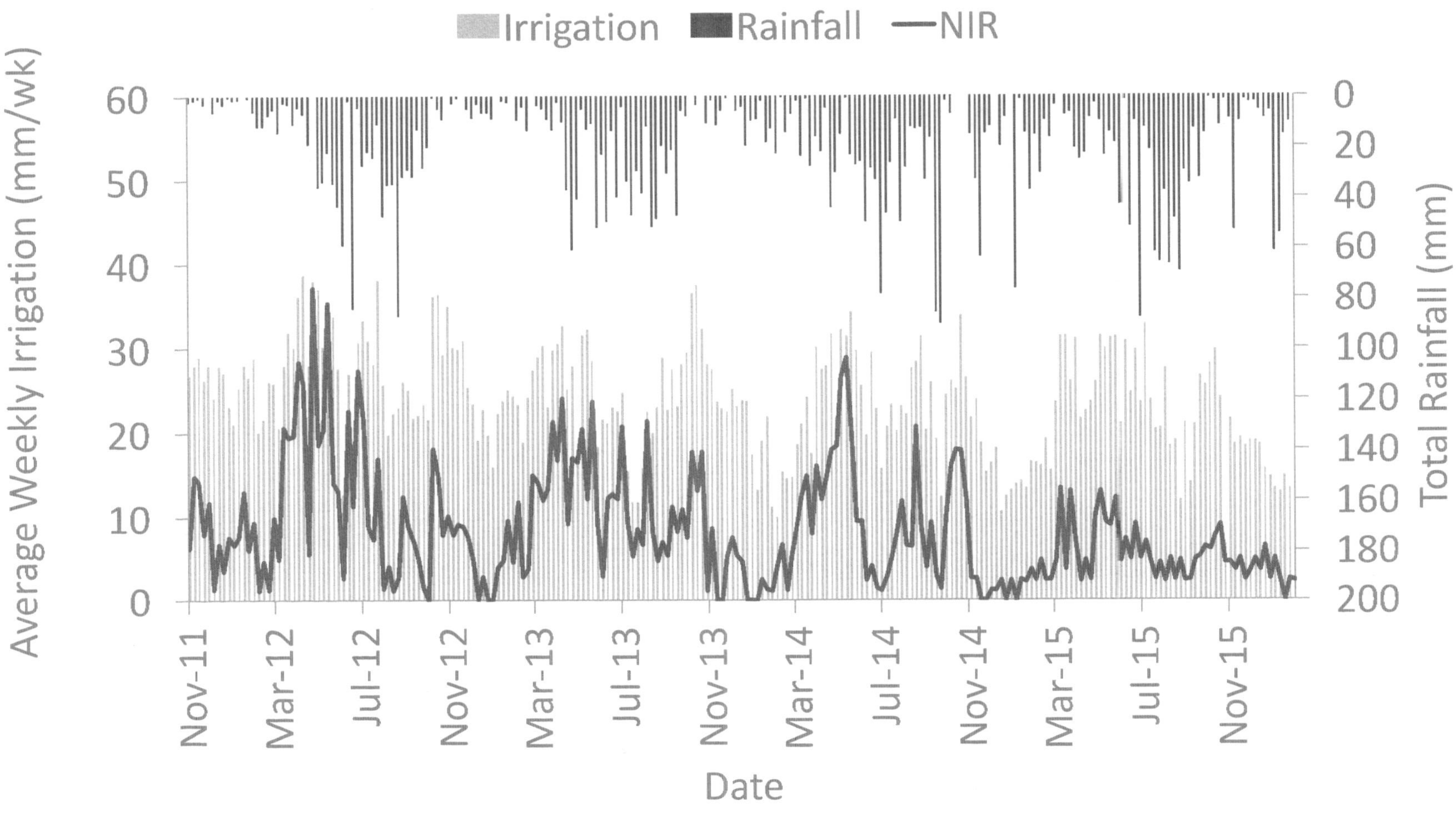


UF UNIVERSITY of FLORIDA IFAS

Center for LANDSCAPE Conservation & Ecology

# ET+OPT Results

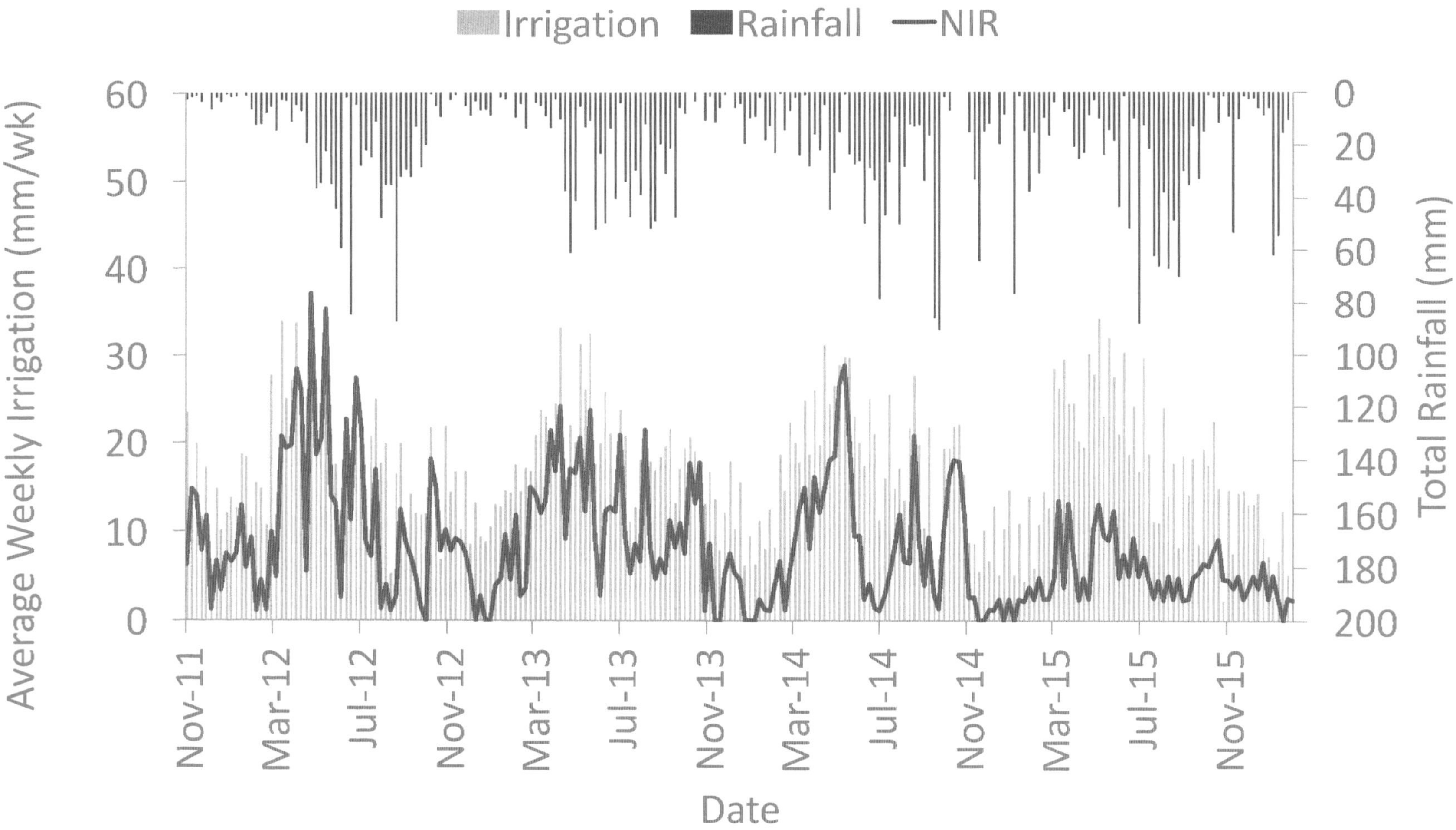


UF UNIVERSITY of FLORIDA IFAS

Center for LANDSCAPE Conservation & Ecology

# SMS+OPT Results

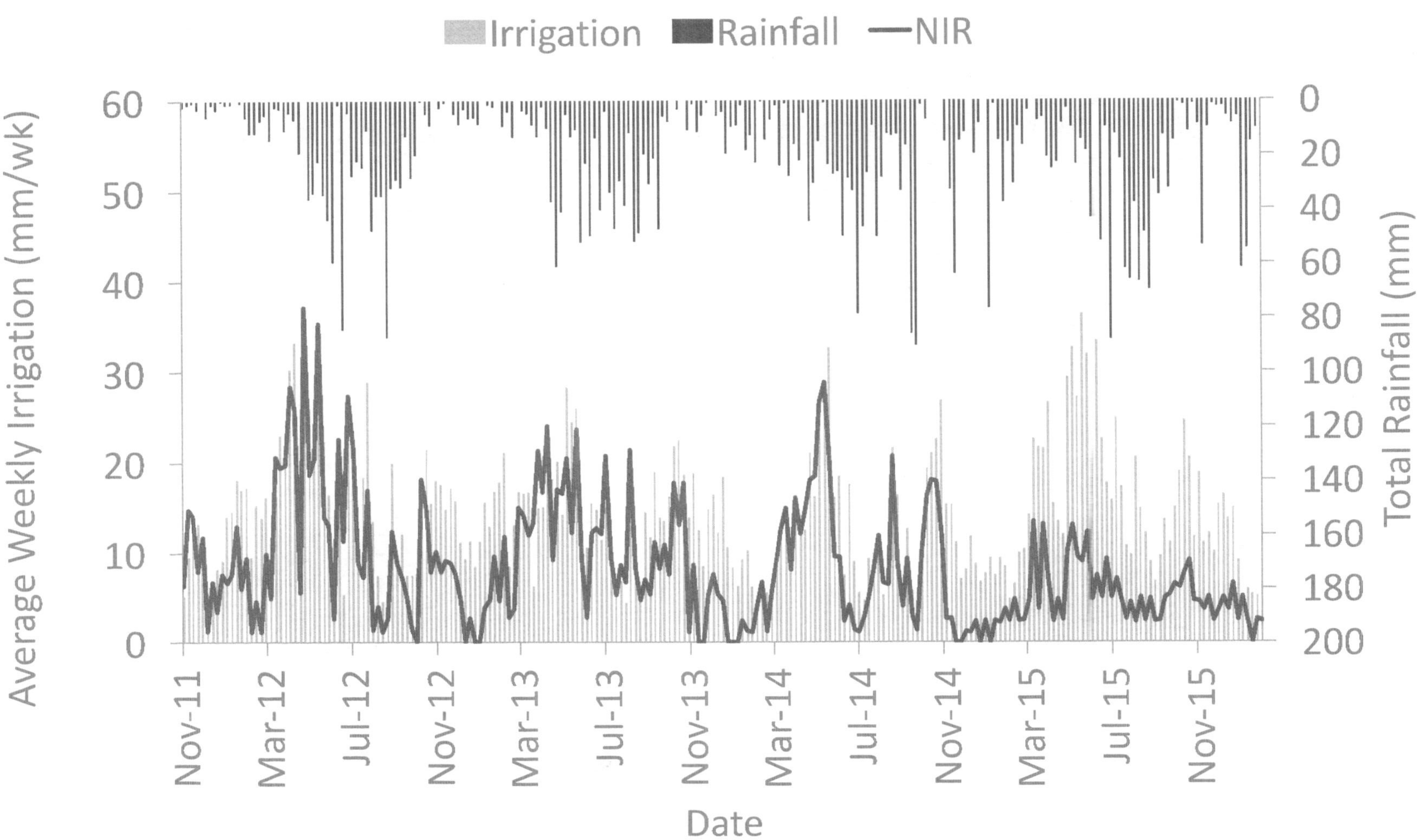


UF UNIVERSITY of FLORIDA IFAS

Center for LANDSCAPE Conservation & Ecology

# Turfgrass Quality

5 9 1

UF UNIVERSITY of FLORIDA IFAS

Center for LANDSCAPE Conservation & Ecology

# Turfgrass Quality

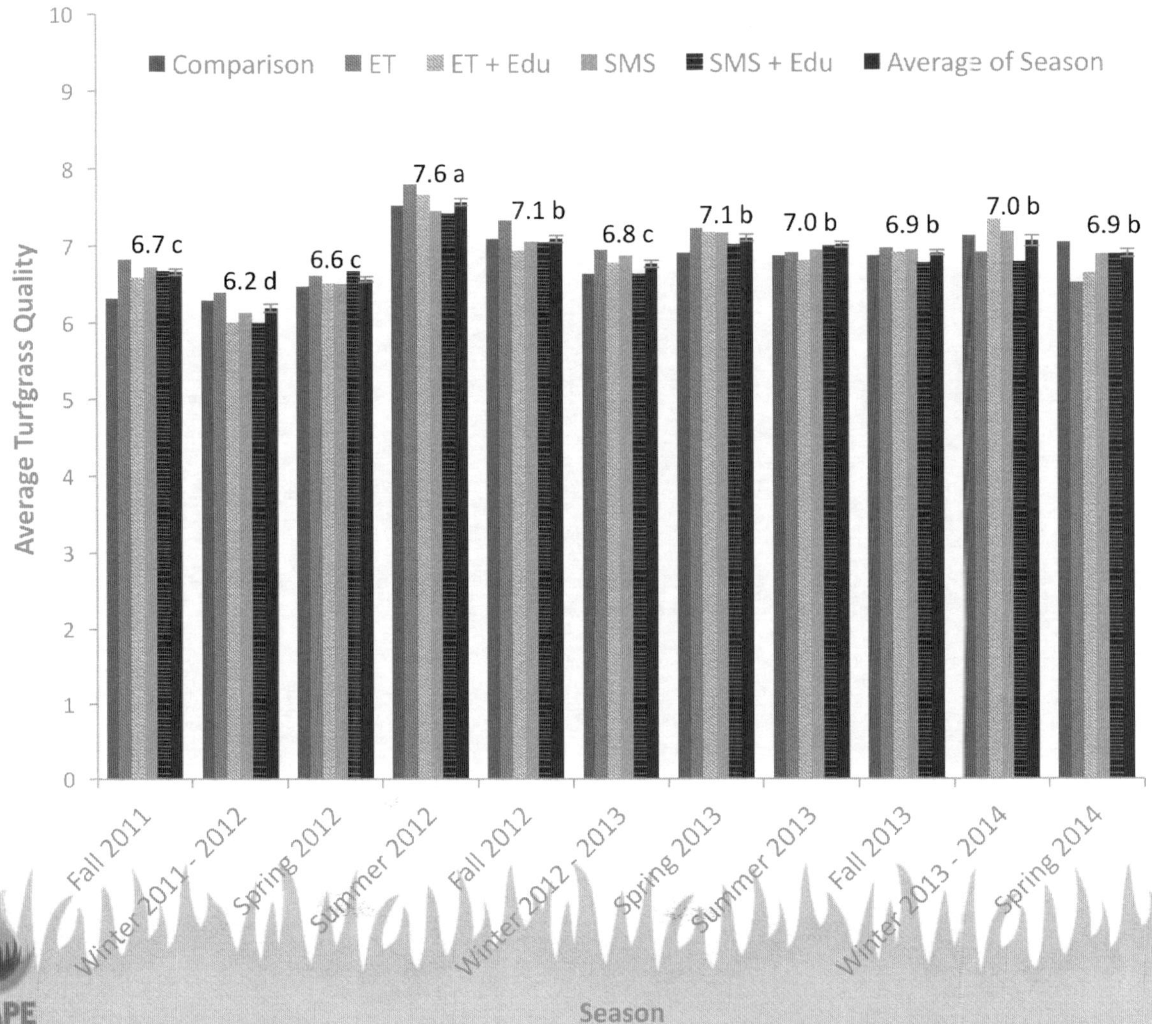


UF UNIVERSITY of FLORIDA IFAS

Center for LANDSCAPE Conservation & Ecology

# Smart Controllers – Bottom Line

- ET/SMS significantly reduce over-irrigation
- ET controllers must be targeted to sites with savings potential
- Proper installation enhances savings
- Longevity of savings?

UF UNIVERSITY of FLORIDA IFAS

Center for LANDSCAPE Conservation & Ecology

__Acknowledgements__: Water Research Foundation, Orange County Utilities, St. Johns River Water Management District, Southwest Florida Water Management District, Stacia Davis, Eliza Breder, Michael Gutierrez

UF UNIVERSITY of FLORIDA IFAS | Center for LANDSCAPE Conservation & Ecology

mddukes@ufl.edu http://abe.ufl.edu/mdukes/

# Increasing Water-Use Efficiency Using Block-Zone Design on Tees

**Russell A. Geiger, CGIA, CGCS, MS Ag. Eng.**
**Hydrologic Irrigation Design, Consulting and Engineering**
**9419 Crocus Court, Fort Myers, Florida 33967, (239) 571-4619, rgeiger@hydrologic-irr.com**

## Abstract
*A case-study including a performance audit survey was conducted at Heritage Palms Golf and Country Club in Fort Myers, Florida during the summer of 2017 to determine if there might be significant increases in water-use efficiency after redesign and retro fitting from larger rotors to smaller rotors. The number 10 tee complex was evaluated, both prior to and after the retrofit, with subsequent averages calculated for Distribution Uniformity values.*

*Average values for D.U. were observed to have increased from .60 to .76 ("Before/After") as a result of the installation and upgrade to the new design using smaller heads. Ensuing worksheets recommended a comparative scheduling decrease in Adjusted Run Time of 25% (from 93 minutes per week to 70 minutes per week) based on the increase in the Distribution Uniformity values. This represents a reduction in irrigation gallons of approximately 65 percent for this particular tee complex or a savings of water equal to approximately 162,000 gallons per year based on comparative usage. Theoretically, if all 18 tee complexes were upgraded a water savings of nearly 3,000,000 gallons per year could be experienced.*

## Keywords
Performance Audit, Distribution Uniformity, Water-Use Efficiency, Irrigation Design, Block-Zone, Precipitation Rate, Irrigation Scheduling, Turf Quality, Tee Complex, Head Spacing, Adjusted Run-Time

## Introduction
Golf course irrigation systems typically include the application of large rotary sprinkler heads in their design. This approach is relatively appropriate for the majority of the golf course, however many constricted areas such as tee complexes require alternative irrigation design to achieve acceptable levels of water-use efficiency. The application of over-sized sprinkler heads installed on narrow sites result in much of the water being cast over the targeted area and effectively wasted. Moderate investments in retro-fitting such areas can go a long way toward conserving irrigation water and drastically improving turf conditions. Adding supplemental block zones that utilize smaller sprinkler heads in areas that are narrow, abruptly evaluated or severely sloped can significantly improve water-use efficiency and turf drought resistance. Performance audits can document, explain and measure this process.

An audit provides critical components of the water management plan including benchmarking how well the current system is performing compared to how well it could actually perform (Stetson and Mecham, 2011). From this starting point future improvements can be strategically made as a means of increasing the efficiency of the system to achieve the same result with less water, consequently providing costs savings on water, power, pump station operation and maintenance. Improving irrigation uniformity and efficiency also equates to fewer dry and/or wet spots, healthier turf, reduced runoff and leaching. (Barrett, Vinchesi, Dobson, Roche and Zoldoske, 2003.)

## The Case-Study

The case-study subject area is the number 10 tee complex on the Sable golf course at Heritage Palms Golf and Country Club located in Fort Myers, Florida. This particular tee complex has two separate teeing areas which are referred to as the "East" and "West" tees respectively. The corridor of turf that comprises the overall tee complex is relatively narrow and long, approximately 50 feet in width and 250 feet in length. The number 10 tee complex is very typical and representative of many other tee complexes in place on the two 18-hole golf courses at Heritage Palms.

Audits were first conducted on both East and West tees where existing sprinkler heads were of the large rotary type. The heads serving the area were full-circle in adjustment and located just off the tee surface with in-line or "single-row" spacing ranging from 65 to 80 feet apart. Nozzle configurations for all sprinkler heads yielded discharge rates of approximately 56 gallons per minute for each head at working pressure (80+ psi). Two sprinkler heads provided irrigation to each tee surface (East and West) respectively. Audit scores revealed relatively low values in the .58 to .61 range. Consequently, turf quality, vigor and density on these sites was also comparatively poor.

*Figure 1. Aerial image of site with new design overlay*

A new "block-zone" irrigation design was created for the same tee complex. Each of the two teeing surfaces were looped with 1.5 inch PVC pipe that tapped into the existing 2.5 inch PVC lateral pipe. The looped configuration supports more consistent pressure throughout the zone and permits the location of the new smaller rotors to be on both sides of the tee complex (opposed to previous "single-row") which significantly increases coverage and uniformity. The original larger rotors were replaced by 16 small rotors with spacing ranging from 25 to 28 feet apart. The smaller rotor sprinklers utilized a nozzle configuration that operated each head at a discharge rate of approximately 6.5 gallons per minute at a working pressure of 80 psi. Zone valves provided control for the new sprinklers and utilized the existing station wires that had previously controlled the larger original rotors with electric solenoid valve-in-head configuration.

*Figure2. Installation of new block-zone design, note the narrow width of the irrigated site*

A new performance audit was conducted for the same two teeing surfaces upon completion of the redesign and installation of the block zones. The "After" distribution uniformity scores reflected notable increases relative to the "Before" values observed in the original system design with the larger rotors. The average distribution uniformity increased from .60 to .76 or approximately 27 percent.

## Irrigation Scheduling Worksheets and Total Irrigation Gallons

Actual records indicated that prior run-times for this particular tee complex over twelve months preceding the case-study were averaging around 90 minutes per week. The ensuing irrigation schedule worksheets (based on the performance audit data) for both "Before" and "After" yielded average recommended run-times of 93 and 70 minutes per week respectively. This reduction in recommended run-time is equal to a decrease of approximately 25 percent.

Based on the respective recommended run-times and the corresponding sprinkler-nozzle configurations utilized, the total volumes were calculated on both East and West tees for scenarios representing both before and after the retrofit. Total irrigated gallons per week for the tee complex were calculated at 20,720 gallons per week for the "Before" scenario as compared to only 7,228 gallons per week (13,492 gallons per week less) for the "After" scenario that is now in place. This represents a reduction in irrigation gallons of approximately 65 percent or a savings of water equal to approximately 162,000 gallons per year based on comparative usage. Theoretically, if all 18 tee complexes were upgraded a water savings of nearly 3,000,000 gallons per year could be experienced.

*Table 1. Summary of audit data and corresponding totals "Before"*

| BEFORE | D.U. | HEADS | DISCHARGE (gal/min) | RUN TIME (min/wk) | TOT/WK (gal/wk) |
|---|---|---|---|---|---|
| EAST | .58 | 2 | 56 | 102 | 11,424 |
| WEST | .61 | 2 | 56 | 83 | 9,296 |
| AVERAGE | **.60** | | | **93** | |
| TOTAL | | | | | **20,720** |

*Table 2. Summary of audit data and corresponding totals "After"*

| AFTER | D.U. | HEADS | DISCHARGE (gal/min | RUN TIME (min/wk) | TOT/WK (gal/wk) |
|---|---|---|---|---|---|
| EAST | .72 | 8 | 6.5 | 66 | 3,432 |
| WEST | .79 | 8 | 6.5 | 73 | 3,796 |
| AVERAGE | **.76** | | | **70** | **7,228** |
| TOTAL | | | | | **7,228** |

## Impact on Turf Quality and Environment

While the main focus of the case-study has concentrated on water-use efficiency, emphasis certainly should be noted concerning the relative increased quality of turf at the subject site. Many turf managers have come to realize the benefit of saving water as a byproduct result of the effort to correct turf quality issues very similar to the actual situation described within this particular case-study. Areas of turf that suffer from inadequate hydration will persist to struggle and require additional inputs including extra fertilizer, hand-watering and pesticides. Typically, when the irrigation is adjusted or corrected the problematic issues cease and the turf performance increases.

Many natural areas that are adjacent to irrigated turf can be negatively impacted by the irrigation water. Wetlands, streams, ponds and other sites with sensitive vegetation or wildlife should not have irrigation introduced into their environment. This is especially the case if the irrigation water contains materials that are injected into the irrigation system such as fertilizer or other chemicals potentially containing caustic substances. Effluent or reclaimed irrigation water sources typically contain significant levels of chlorine as a byproduct of the treatment process that can certainly harm flora and fauna if constantly wetted with such tainted water. More precision in the control and placement of irrigation water through the use of smaller rotors will undoubtedly reduce these potential problems and issues related to unintentional irrigation impact.

## Conclusion

Adding supplemental block-zones with smaller heads in irrigated areas that are narrow, abruptly evaluated or severely sloped (such as tee complexes) will significantly improve water-use efficiency, turf health and drought resistance. This concept was clearly observed upon completion of the "Before" and "After" case-study at Heritage Palms number 10 tee complex where redesign and retrofit efforts resulted in water savings of 65% for the subject site.

Performance auditing provided the critical information necessary to thoroughly document and illustrate the significance of this process. More work and research is needed to further understand and emphasize the importance of the technique of utilizing smaller heads in block-zone irrigation design for constricted areas on golf courses such as tee complexes.


## Acknowledgments

I would like to recognize and thank Mr. Greg Kriesch (Director of Golf and Grounds), Mr. Jason Small (Golf Course Superintendent) and the Heritage Palms Golf and Country Club, Fort Myers, Florida for allowing and participating in this case study.

I also thank Mr. Trevor Brinkmeyer, Rain Bird Corporation, for assistance he provided in this study.


## References


Barrett, J., Vinchesi, B., Dobson, R., Roche P. and Zoldoske, D., 2003. Design. Chapter 5 in Golf Course Irrigation. Pp. 185 – 192,

Mecham, Brent Q. and Stetson, LaVerne E. 2011. Performance Audits. Chapter 14 in Irrigation – 6th Edition. Irrigation Association. pp. 565 – 610, Falls Church, VA.

UF | UNIVERSITY of FLORIDA | IFAS

Center for LANDSCAPE Conservation & Ecology

# Evaluation of EPA's Test for Pressure Regulating Sprinklers

**Irrigation Show & Education Conference**
**Nov. 6-10, 2017**


Bernardo Cardenas

Agricultural & Biological Engineering

University of Florida/IFAS


abe.ifas.ufl.edu

# Background

- ASABE/ICC Sprinkler & Emitter Standard

UF UNIVERSITY of FLORIDA IFAS

Center for LANDSCAPE Conservation & Ecology

## Committee Composition

- Irrigation manufacturers
- Utilities
- Irrigation Association
- Irrigation contractors
- Researchers

# Background

- ASABE/ICC Sprinkler & Emitter Standard
- Potential savings → flowrate reduction at elevated operating pressures

UF UNIVERSITY of FLORIDA IFAS
Center for LANDSCAPE Conservation & Ecology

# Theoretical Pressure Regulation Flowrate Reduction

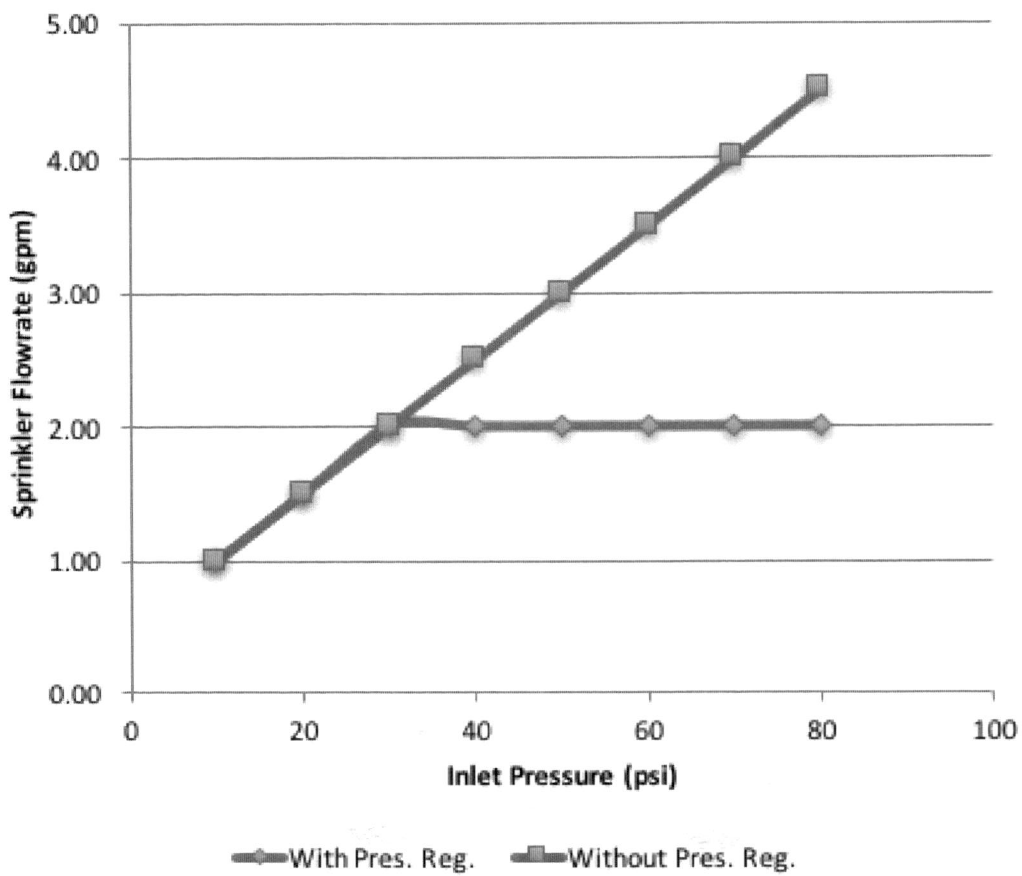


UF UNIVERSITY of FLORIDA IFAS

Center for LANDSCAPE Conservation & Ecology

# Pressure and Flowrate

**12' SERIES WITH 24° TRAJECTORY (BROWN)**

| Arc | Desc. | psi | gpm | Radius | Prec. Rate ▲ | Prec. Rate ■ |
|---|---|---|---|---|---|---|
| 90° | 12-Q | 20 | 0.40 | 11 | 1.48 | 1.28 |
| | | 30 | 0.50 | 12 | 1.55 | 1.35 |
| | | 40 | 0.60 | 13 | 1.64 | 1.42 |
| | | 50 | 0.63 | 13 | 1.67 | 1.44 |
| | 12-Q-PC | 30-40 | 0.48 | 12 | 1.49 | 1.29 |
| | | 40-75 | 0.53 | 12 | 1.65 | 1.43 |
| 120° | 12-T | 20 | 0.57 | 11 | 1.58 | 1.37 |
| | | 30 | 0.72 | 12 | 1.68 | 1.45 |
| | | 40 | 0.87 | 13 | 1.87 | 1.62 |
| | | 50 | 0.97 | 13 | 1.93 | 1.67 |
| | 12-T-PC | 30-40 | 0.64 | 12 | 1.49 | 1.29 |
| | | 40-75 | 0.70 | 12 | 1.63 | 1.41 |
| 180° | 12-H | 20 | 0.95 | 11 | 1.76 | 1.52 |
| | | 30 | 1.09 | 12 | 1.69 | 1.47 |
| | | 40 | 1.30 | 13 | 1.72 | 1.49 |
| | | 50 | 1.55 | 14 | 1.77 | 1.53 |
| | 12-H-PC | 30-40 | 0.96 | 12 | 1.49 | 1.29 |
| | | 40-75 | 1.05 | 12 | 1.63 | 1.41 |

**12** ● Green
12' radius
Fixed: ¼, ⅓, ½, ⅔, ¾, Full
Trajectory: 28°

| Arc | Position | Pressure PSI | Radius ft. | Flow GPM | Precip in/hr ■ | Precip in/hr ▲ |
|---|---|---|---|---|---|---|
| 90° | Q | 20 | 11 | 0.54 | 1.71 | 1.98 |
| | | 25 | 12 | 0.61 | 1.62 | 1.87 |
| | | **30** | **12** | **0.67** | **1.78** | **2.06** |
| | | 35 | 13 | 0.72 | 1.65 | 1.90 |
| | | 40 | 13 | 0.78 | 1.77 | 2.04 |
| 120° | T | 20 | 11 | 0.72 | 1.71 | 1.98 |
| | | 25 | 12 | 0.81 | 1.62 | 1.87 |
| | | **30** | **12** | **0.89** | **1.78** | **2.06** |
| | | 35 | 13 | 0.97 | 1.65 | 1.90 |
| | | 40 | 13 | 1.04 | 1.77 | 2.04 |
| 180° | H | 20 | 11 | 1.05 | 1.67 | 1.93 |
| | | 25 | 12 | 1.18 | 1.58 | 1.83 |
| | | **30** | **12** | **1.30** | **1.74** | **2.01** |
| | | 35 | 13 | 1.42 | 1.61 | 1.86 |
| | | 40 | 13 | 1.52 | 1.73 | 2.00 |

**12 Series MPR**

*30° Trajectory*

| Nozzle | Pressure psi | Radius ft. | Flow gpm | ■ Precip In/h | ▲ Precip In/h |
|---|---|---|---|---|---|
| 12F | 15 | 9 | 1.80 | 2.14 | 2.47 |
| | 20 | 10 | 2.10 | 2.02 | 2.34 |
| | 25 | 11 | 2.40 | 1.91 | 2.21 |
| | 30 | 12 | 2.60 | 1.74 | 2.01 |
| 12H | 15 | 9 | 0.90 | 2.14 | 2.47 |
| | 20 | 10 | 1.05 | 2.02 | 2.34 |
| | 25 | 11 | 1.20 | 1.91 | 2.21 |
| | 30 | 12 | 1.30 | 1.74 | 2.01 |

UF UNIVERSITY of FLORIDA IFAS
Center for LANDSCAPE Conservation & Ecology

# Pressure and Flowrate

**12' SERIES WITH 24° TRAJECTORY (BROWN)**

| Arc | Desc. | psi | gpm | Radius | Prec. Rate ▲ | Prec. Rate ■ |
|---|---|---|---|---|---|---|
| 90° | 12-Q | 20 | 0.40 | 11 | 1.48 | 1.28 |
| | | 30 | 0.50 | 12 | 1.55 | 1.35 |
| | | 40 | 0.60 | 13 | 1.64 | 1.42 |
| | | 50 | 0.63 | 13 | 1.67 | 1.44 |
| | 12-Q-PC | 30-40 | 0.48 | 12 | 1.49 | 1.29 |
| | | 40-75 | 0.53 | 12 | 1.65 | 1.43 |
| 120° | 12-T | 20 | 0.57 | 11 | 1.58 | 1.37 |
| | | 30 | 0.72 | 12 | 1.68 | 1.45 |
| | | 40 | 0.87 | 13 | 1.87 | 1.62 |
| | | 50 | 0.97 | 13 | 1.93 | 1.67 |
| | 12-T-PC | 30-40 | 0.64 | 12 | 1.49 | 1.29 |
| | | 40-75 | 0.70 | 12 | 1.63 | 1.41 |
| 180° | 12-H | 20 | 0.95 | 11 | 1.76 | 1.52 |
| | | 30 | 1.09 | 12 | 1.69 | 1.47 |
| | | 40 | 1.30 | 13 | 1.72 | 1.49 |
| | | 50 | 1.55 | 14 | 1.77 | 1.53 |
| | 12-H-PC | 30-40 | 0.96 | 12 | 1.49 | 1.29 |
| | | 40-75 | 1.05 | 12 | 1.63 | 1.41 |

**12** — 12' radius
● Green — Fixed: ¼, ⅓, ½, ⅔, ¾, Full
Trajectory: 28°

| Arc | Position | Pressure PSI | Radius ft. | Flow GPM | Precip in/hr ■ | Precip in/hr ▲ |
|---|---|---|---|---|---|---|
| 90° | Q | 20 | 11 | 0.54 | 1.71 | 1.98 |
| | | 25 | 12 | 0.61 | 1.62 | 1.87 |
| | | **30** | **12** | **0.67** | **1.78** | **2.06** |
| | | 35 | 13 | 0.72 | 1.65 | 1.90 |
| | | 40 | 13 | 0.78 | 1.77 | 2.04 |
| 120° | T | 20 | 11 | 0.72 | 1.71 | 1.98 |
| | | 25 | 12 | 0.81 | 1.62 | 1.87 |
| | | **30** | **12** | **0.89** | **1.78** | **2.06** |
| | | 35 | 13 | 0.97 | 1.65 | 1.90 |
| | | 40 | 13 | 1.04 | 1.77 | 2.04 |
| 180° | H | 20 | 11 | 1.05 | 1.67 | 1.93 |
| | | 25 | 12 | 1.18 | 1.58 | 1.83 |
| | | **30** | **12** | **1.30** | **1.74** | **2.01** |
| | | 35 | 13 | 1.42 | 1.61 | 1.86 |
| | | 40 | 13 | 1.52 | 1.73 | 2.00 |

**12 Series MPR**

*30° Trajectory*

| Nozzle | Pressure psi | Radius ft. | Flow gpm | ■ Precip In/h | ▲ Precip In/h |
|---|---|---|---|---|---|
| 12F | 15 | 9 | 1.80 | 2.14 | 2.47 |
| | 20 | 10 | 2.10 | 2.02 | 2.34 |
| | 25 | 11 | 2.40 | 1.91 | 2.21 |
| | 30 | 12 | 2.60 | 1.74 | 2.01 |
| 12H | 15 | 9 | 0.90 | 2.14 | 2.47 |
| | 20 | 10 | 1.05 | 2.02 | 2.34 |
| | 25 | 11 | 1.20 | 1.91 | 2.21 |
| | 30 | 12 | 1.30 | 1.74 | 2.01 |

UF UNIVERSITY of FLORIDA IFAS
Center for LANDSCAPE Conservation & Ecology

# EPA Estimated Savings

- Avg. house using 50,500 gal/yr saves 5,600 gal/yr
- 2.3 yr ROI retrofit
- 1.5 yr ROI new install

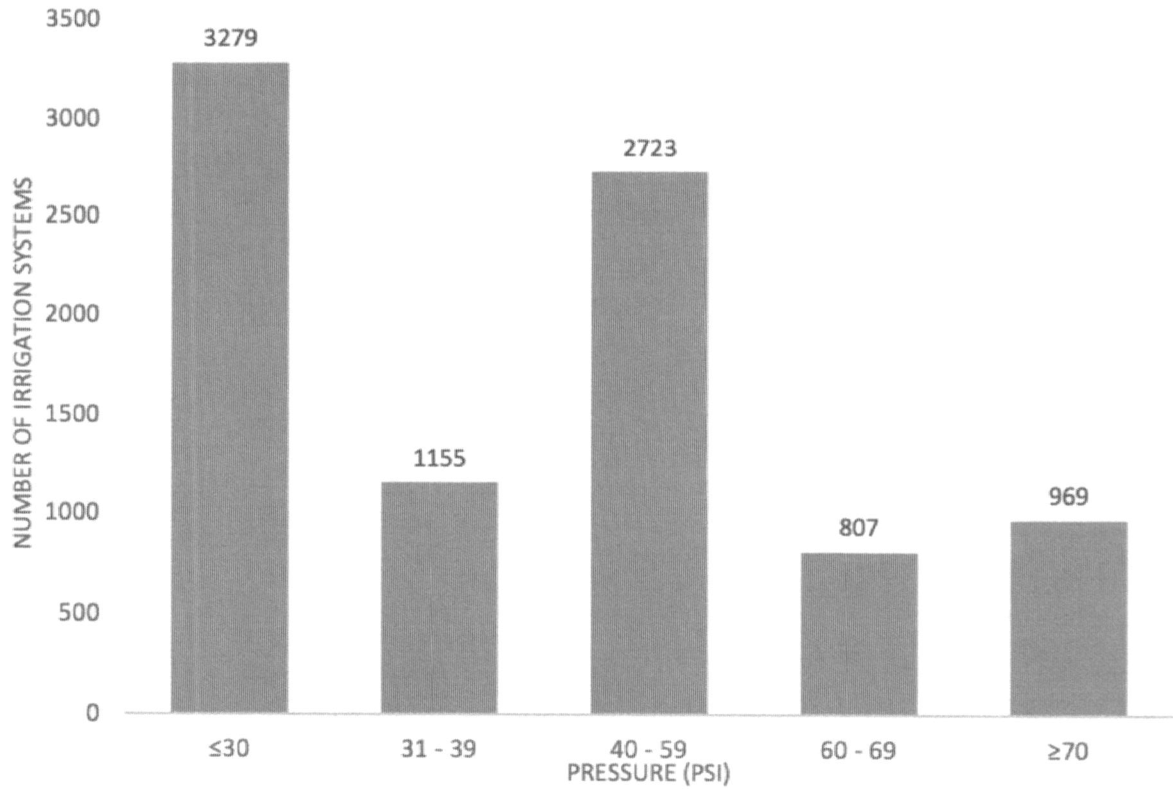


Irrigation System Pressure Data, Utah State University and Center for Resource Conservation

# Misting and Drift

# Pressure Regulation

# No Pressure Regulation

UF UNIVERSITY of FLORIDA IFAS

Center for LANDSCAPE Conservation & Ecology

# How Do They Work?

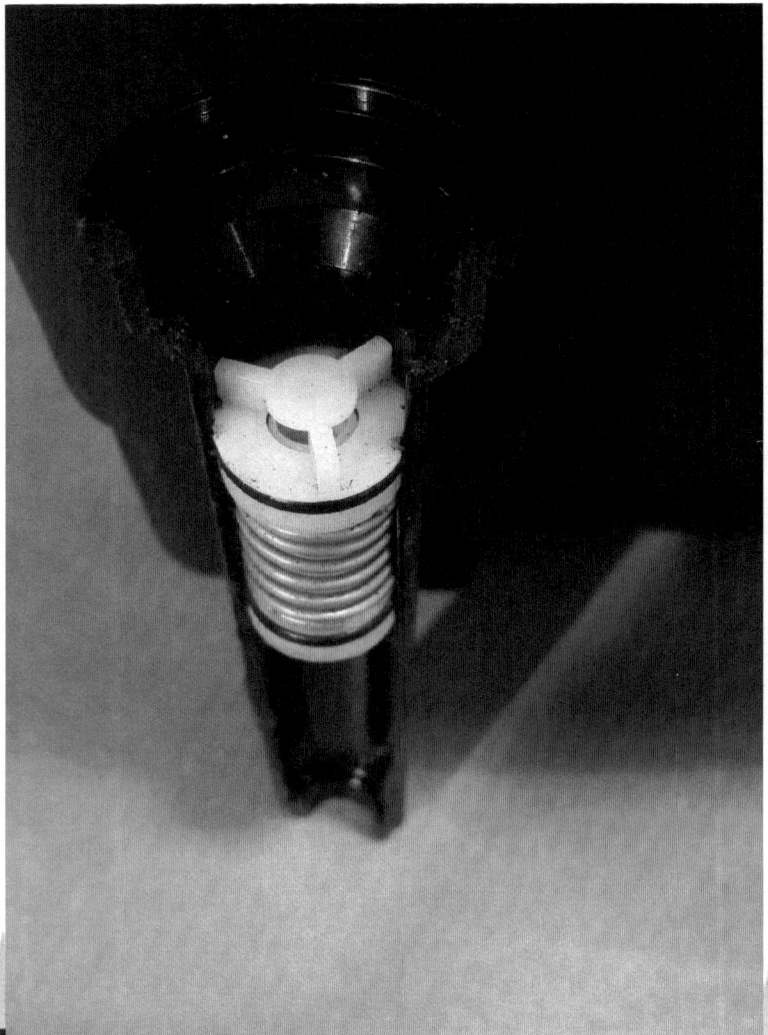

UF | FLORIDA
IFAS

LANDSCAPE
Conservation & Ecology

# EPA WaterSense Initial Testing

- Three labs
- Outlet device
  - Standardized orifice in 802
  - Ball valve/gate valve
  - Variable arc nozzle
  - Needle valve
- Increasing pressure/decreasing pressure
  → hysteresis

UF UNIVERSITY of FLORIDA IFAS
Center for LANDSCAPE Conservation & Ecology

# Initial Testing Observed Hysteresis

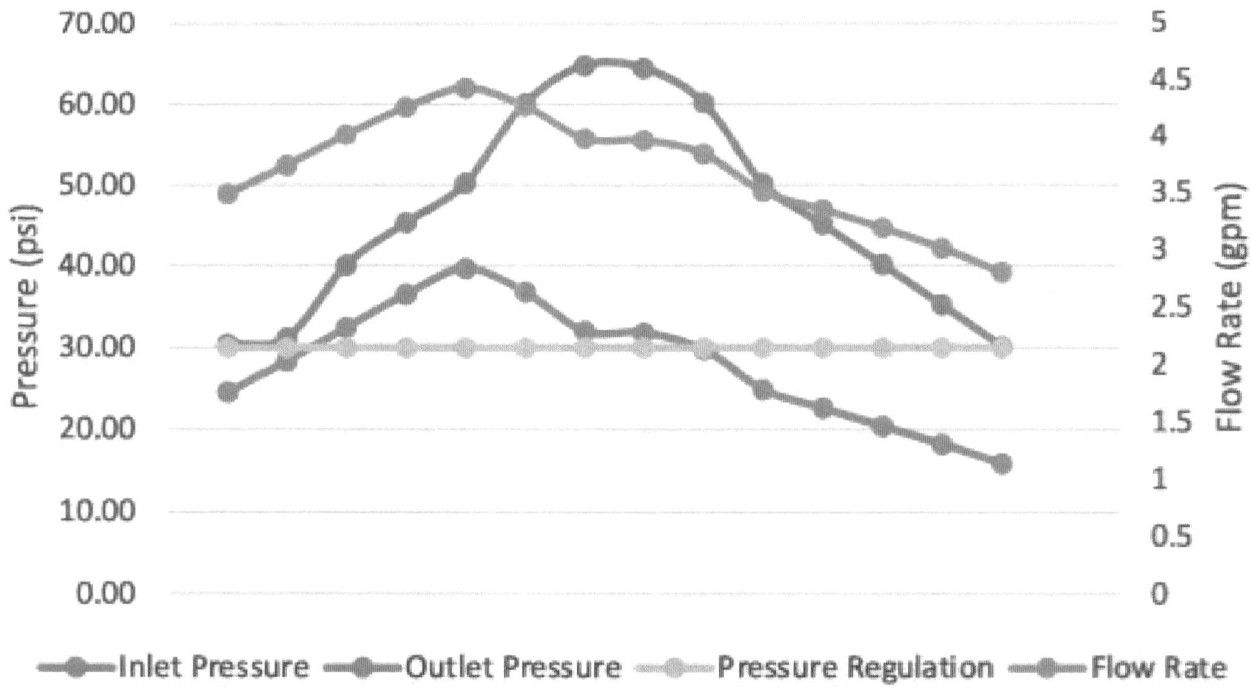

# Outline

- Test equipment
- Test process
- Modifications
- Results
- Recommendations

UF UNIVERSITY of FLORIDA IFAS

Center for LANDSCAPE Conservation & Ecology

# Test Equipment

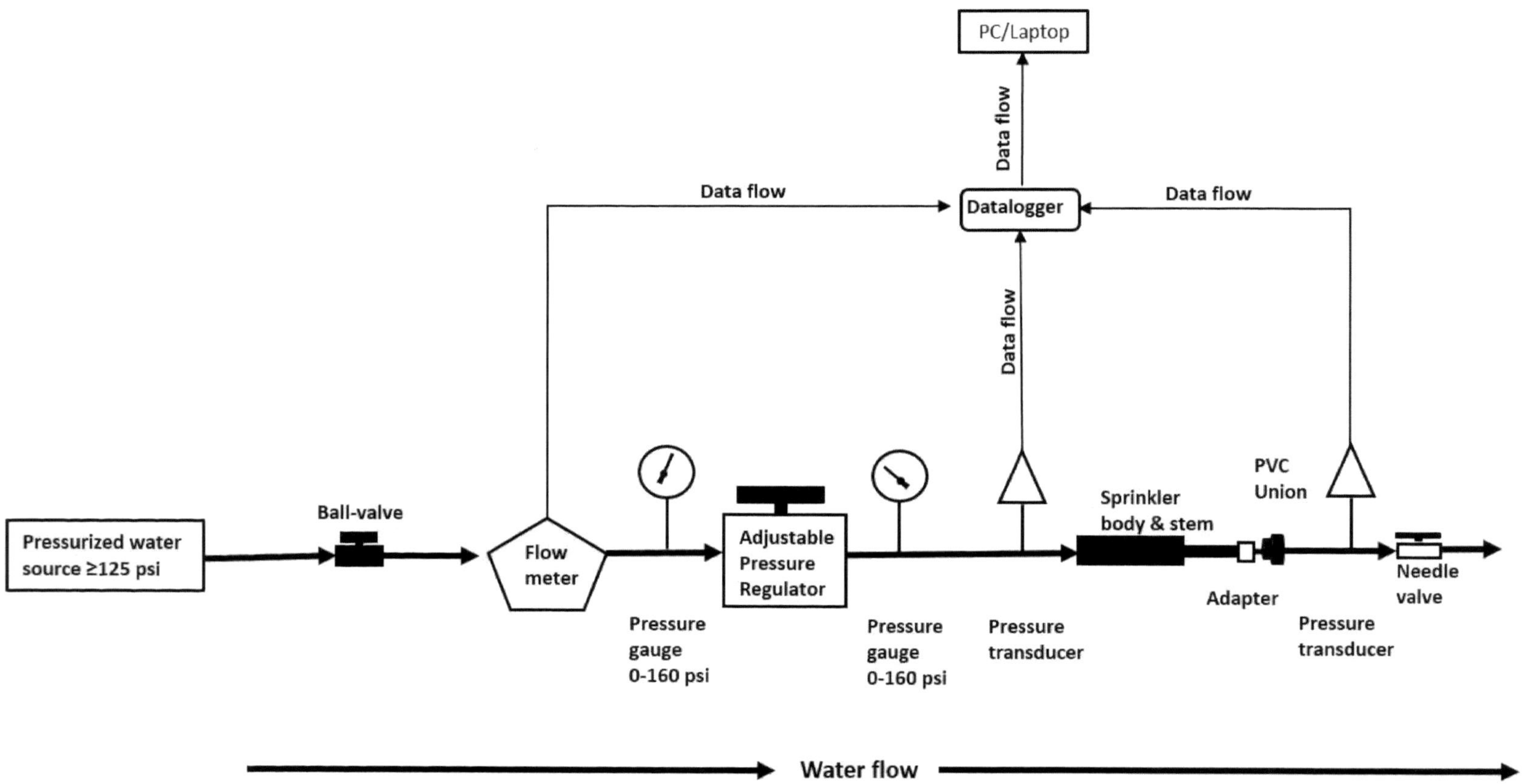


UF UNIVERSITY of FLORIDA IFAS

Center for LANDSCAPE Conservation & Ecology

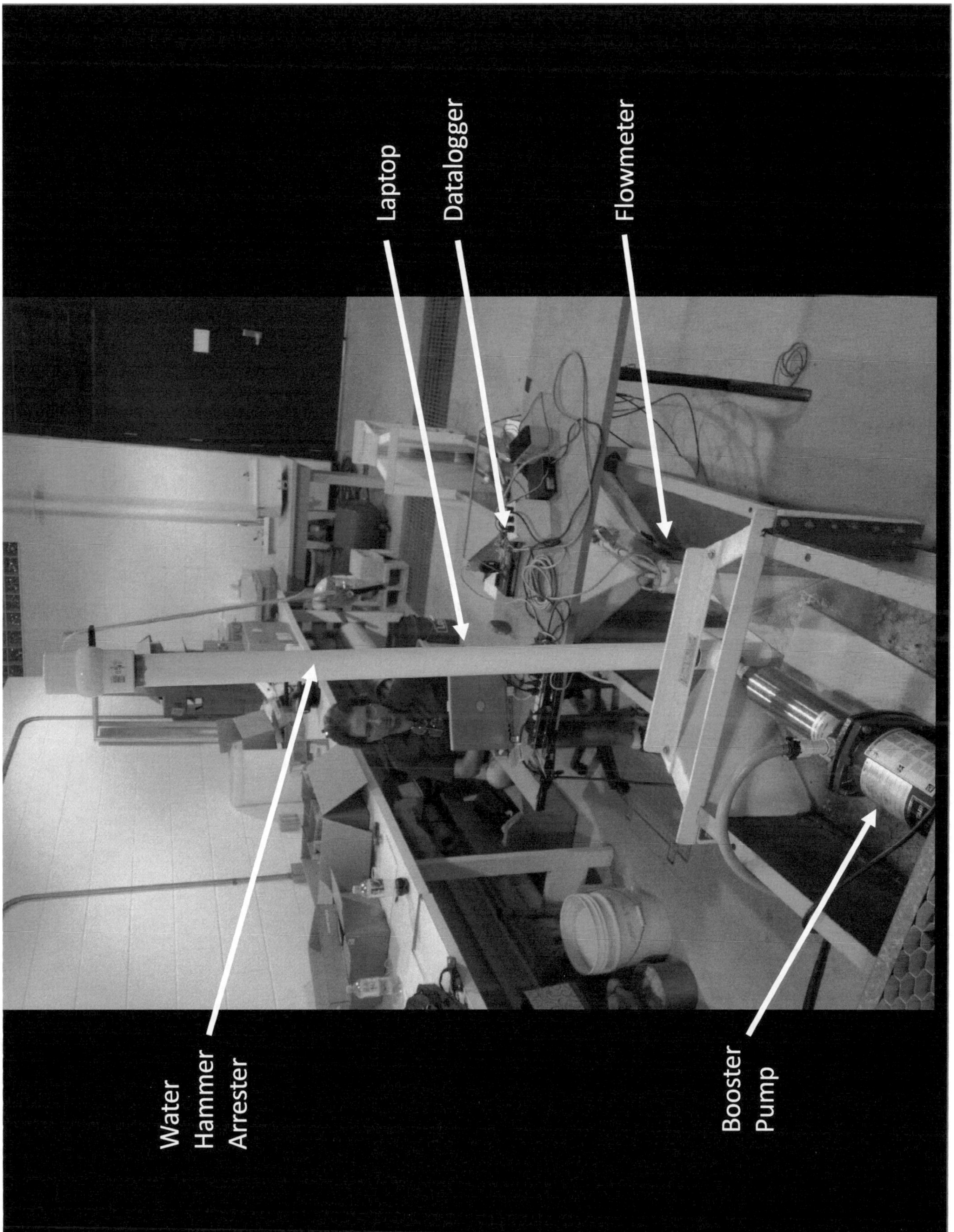
Laptop
Datalogger
Flowmeter
Water Hammer Arrester
Booster Pump

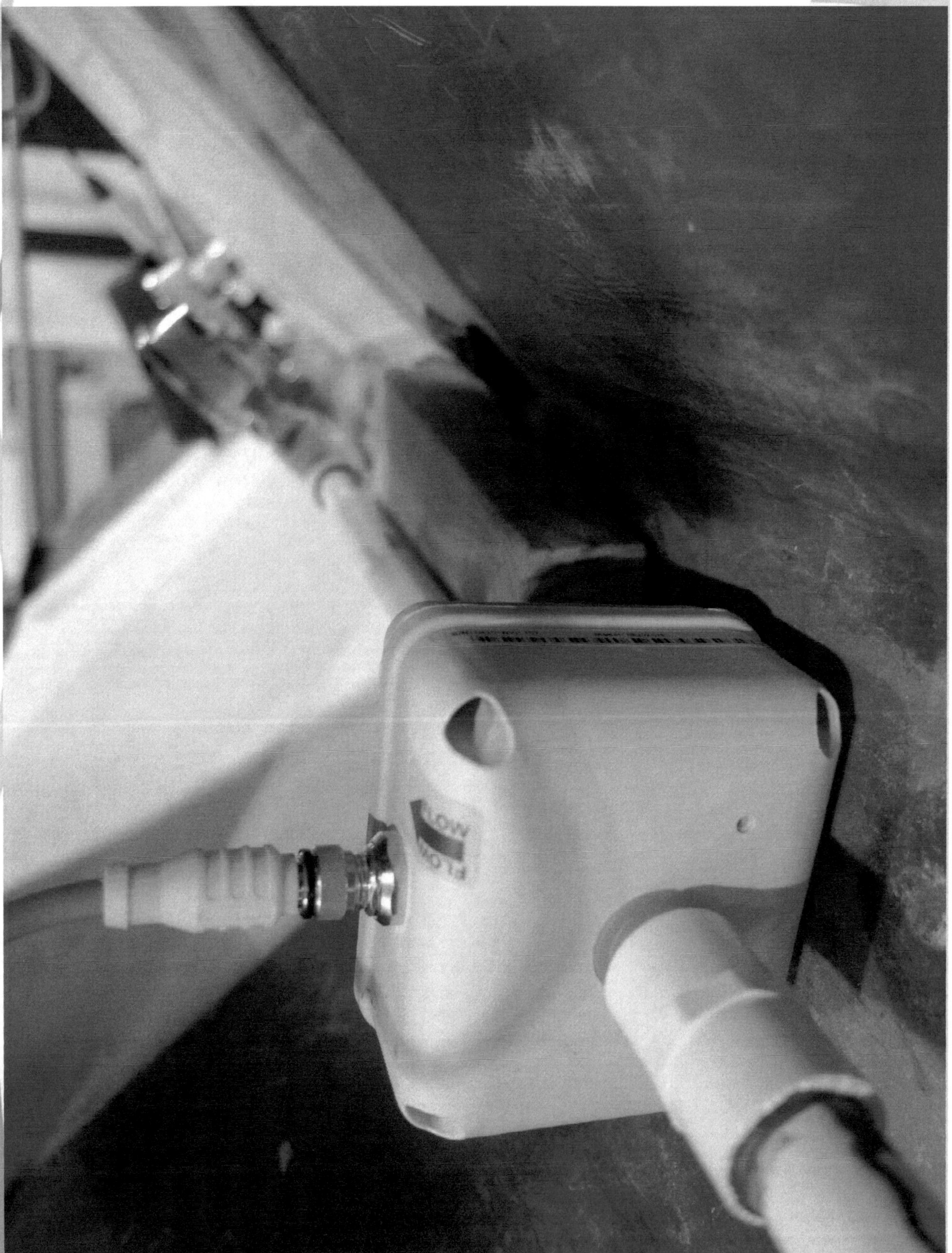
FLOW
FLOW

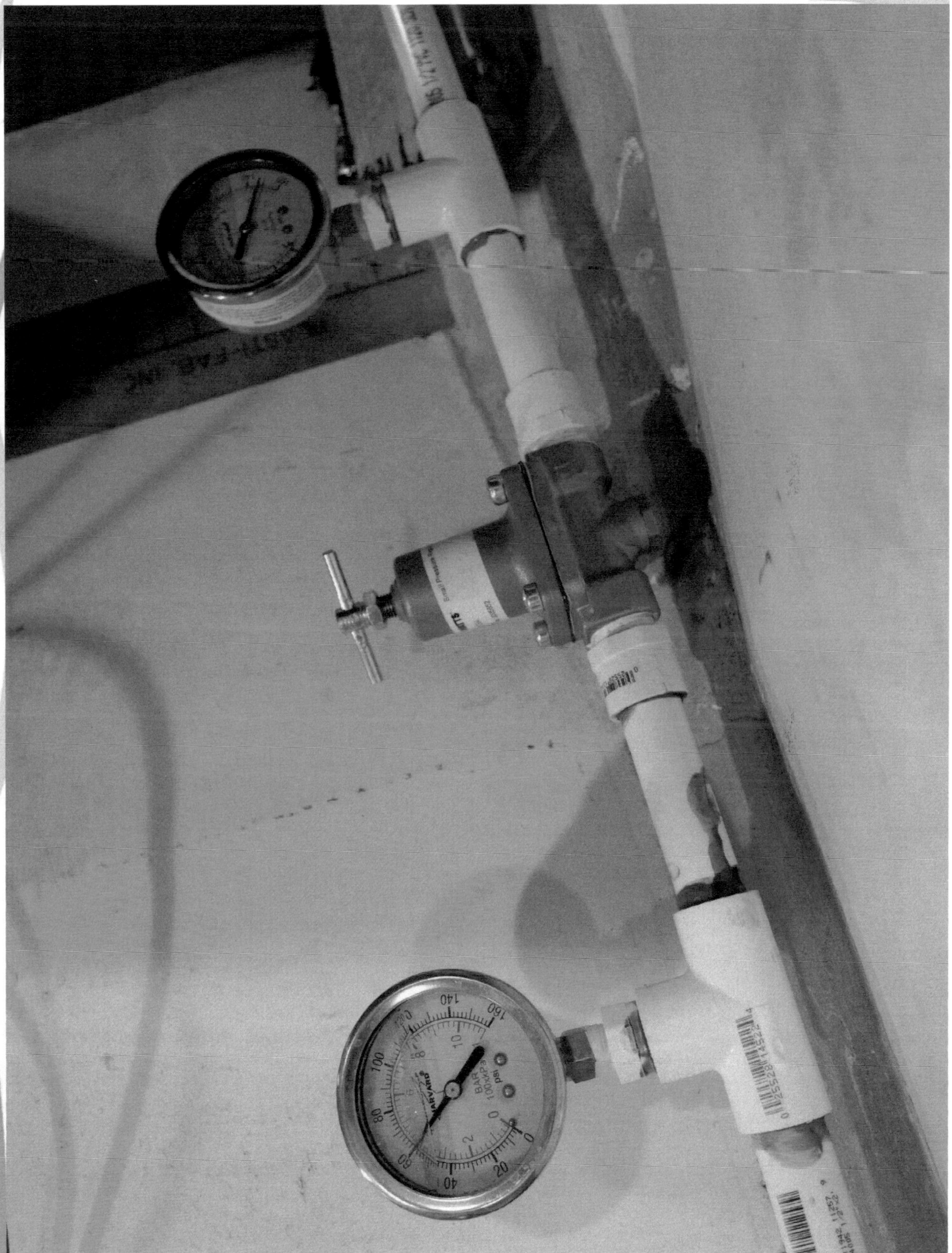

# Test Sample, pressure transducers, needle valve

Pressure Transducers

Adapter

Needle Valve

Test Specimen

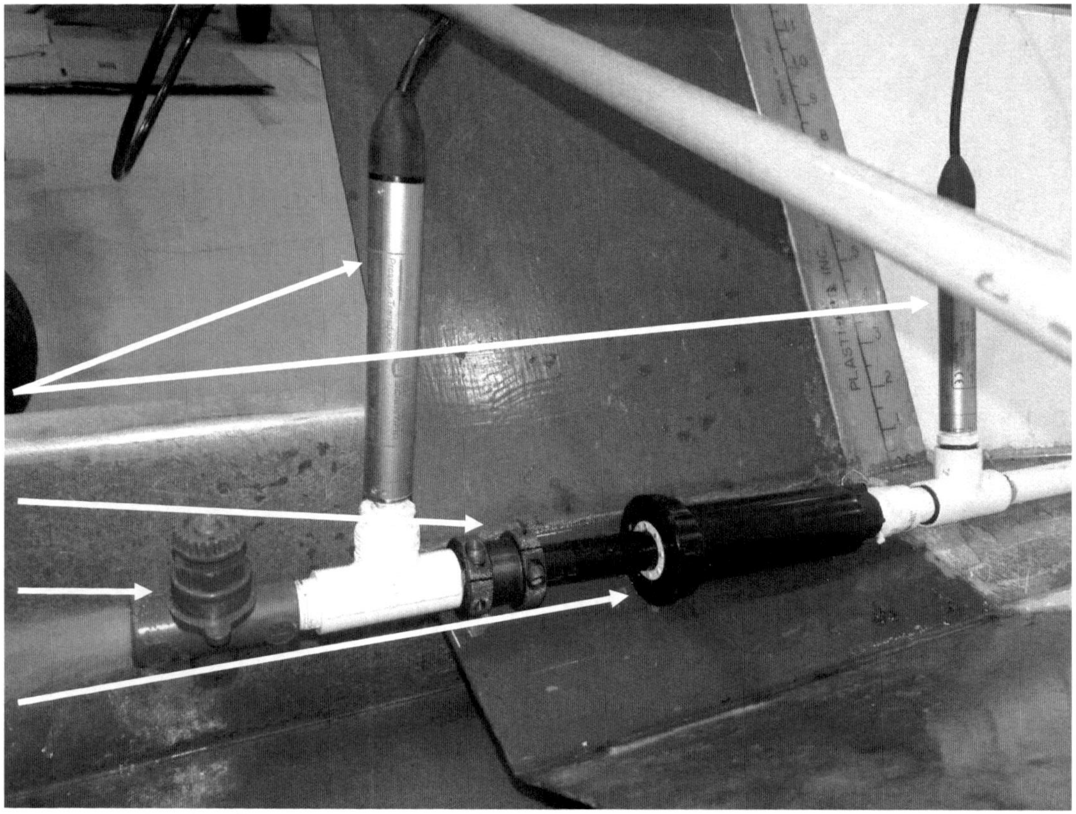

UF UNIVERSITY of FLORIDA IFAS

Center for LANDSCAPE Conservation & Ecology

# Test Process

- Verify flowrate at rated pressure (3 consecutive readings) 30 psi +/- 1 psi, 1.5 gpm +/- 0.1 gpm
- Reduce pressure to zero (for at least 1 min)
- Increase pressure to rated+10 psi (3-5 min test, 30 sec recording)
- Reduce pressure to zero
- Increase pressure to 60 psi
- Reduce pressure to zero
- Increase pressure to 70 psi
- Repeat for 60 psi, rated+10 psi

UF UNIVERSITY of FLORIDA IFAS
Center for LANDSCAPE Conservation & Ecology

# Test Modifications

- All piping ½" SCH 40 PVC, not ¾"
- First test point at regulated pressure to verify test conditions
- Accepted a 0.2 gpm deviation at 3.5 gpm test point

UF UNIVERSITY of FLORIDA IFAS
Center for LANDSCAPE Conservation & Ecology

# Models Tested

- 6 manufacturers
- 11 models tested, 3 samples each
- Brands A-C, PR and non-PR models tested
- One check valve model
- Two flow reduction models

UF UNIVERSITY of FLORIDA IFAS

Center for LANDSCAPE Conservation & Ecology

# Brand A Pressure Regulated vs. Non-Pressure Regulated – 1.5 gpm Test

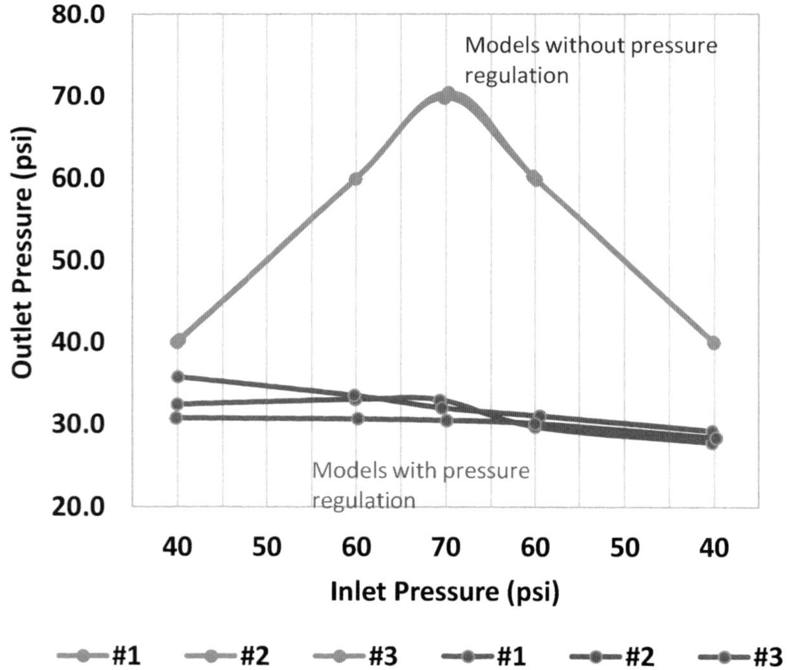


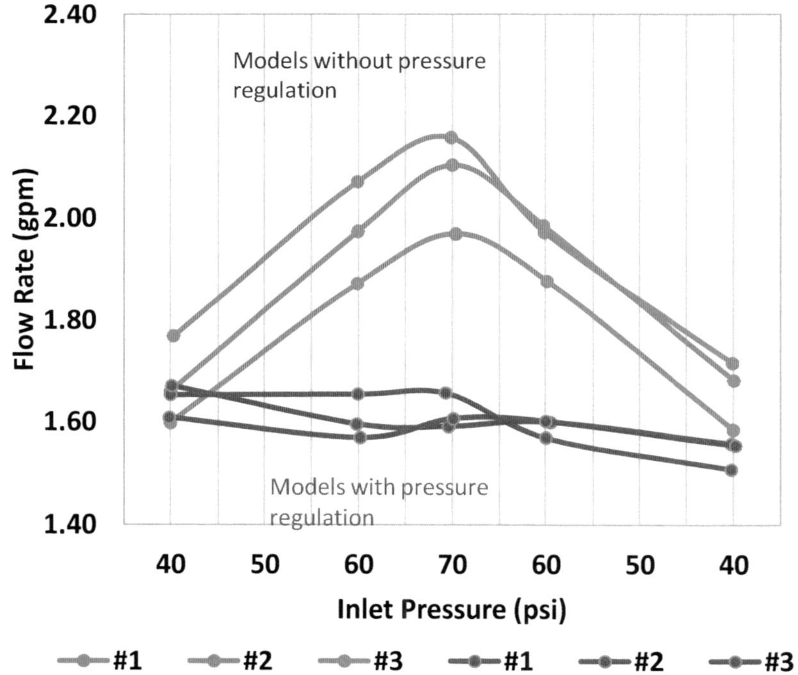


UF UNIVERSITY of FLORIDA IFAS

Center for LANDSCAPE Conservation & Ecology

# Brand A Pressure Regulated vs. Non-Pressure Regulated – 3.5 gpm Test

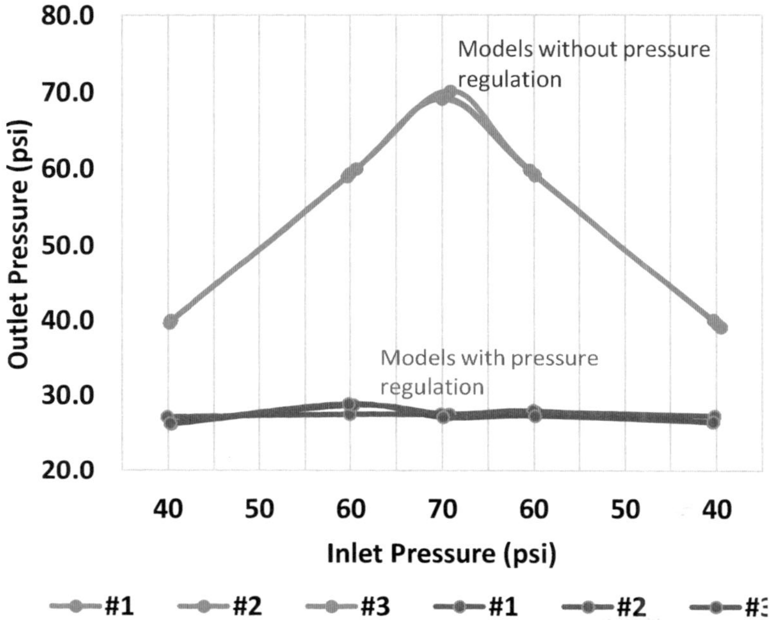


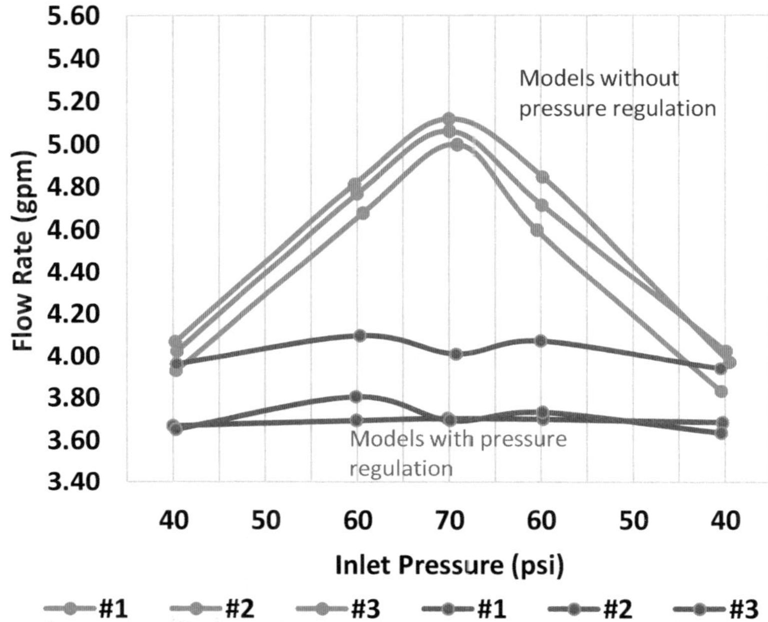


UF UNIVERSITY of FLORIDA IFAS

Center for LANDSCAPE Conservation & Ecology

# Brand B Pressure Regulated vs. Non-Pressure Regulated – 1.5 gpm Test

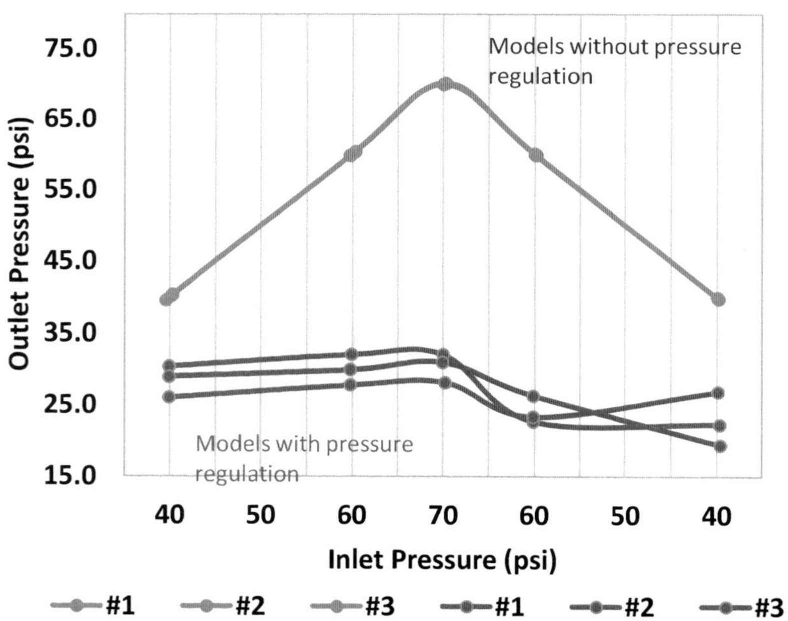


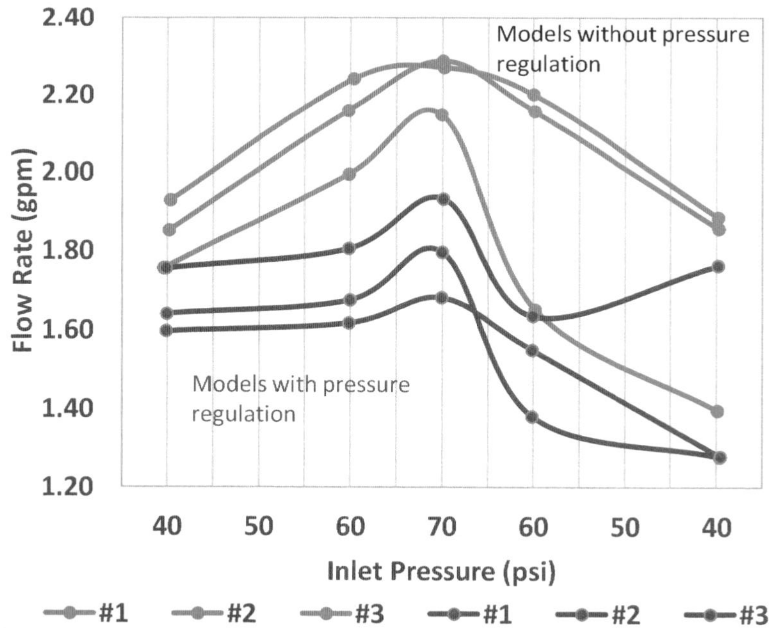


UF UNIVERSITY of FLORIDA IFAS

Center for LANDSCAPE Conservation & Ecology

# Brand B Pressure Regulated vs. Non-Pressure Regulated – 3.5 gpm Test

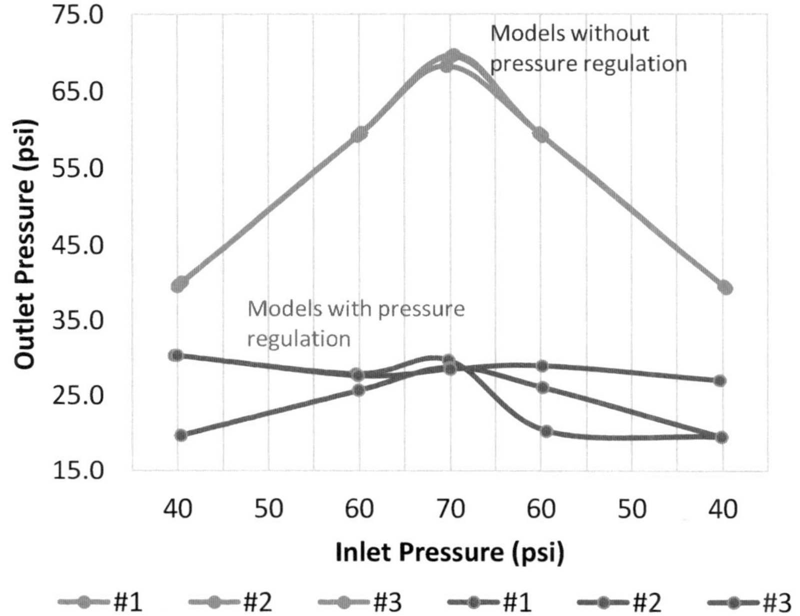


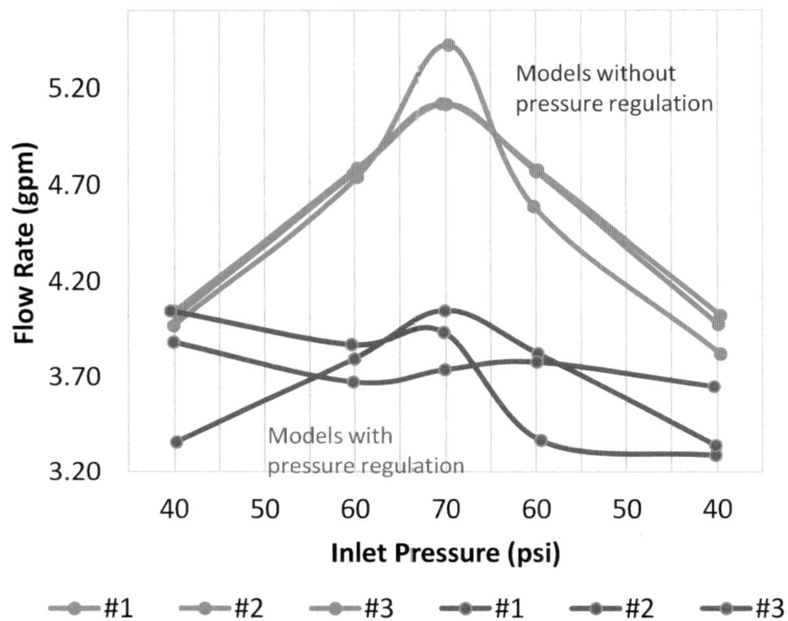


UF UNIVERSITY of FLORIDA IFAS

Center for LANDSCAPE Conservation & Ecology

# Brand E PRB & Check Valve – 1.5 gpm Test

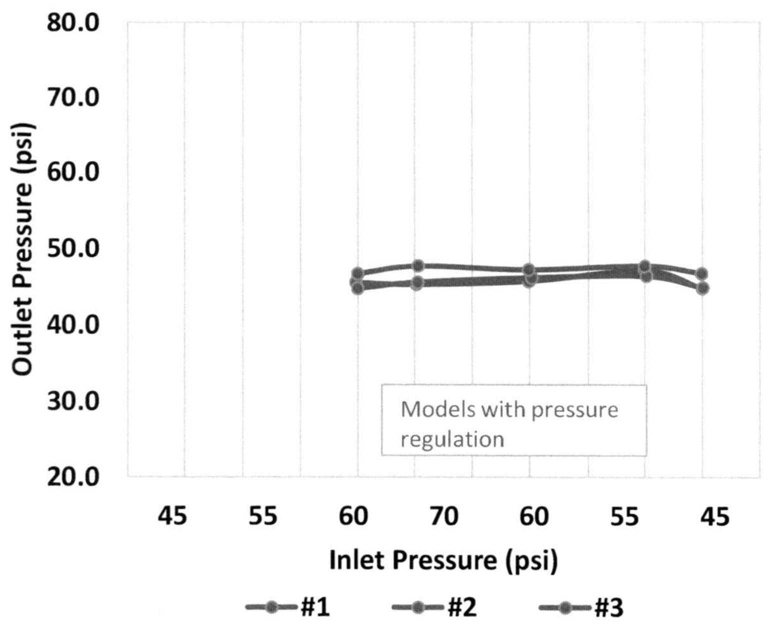


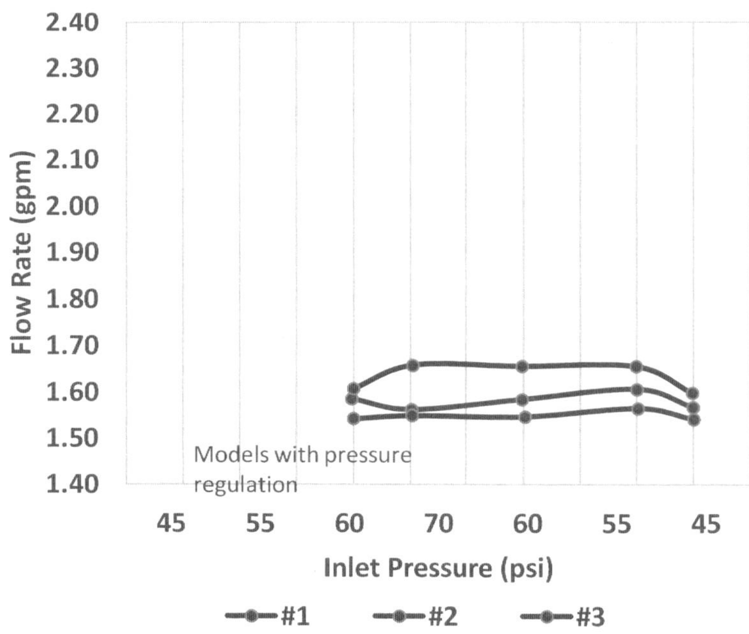


UF | UNIVERSITY of FLORIDA | IFAS

Center for LANDSCAPE Conservation & Ecology

# Brand E PRB & Check Valve – 3.5 gpm Test

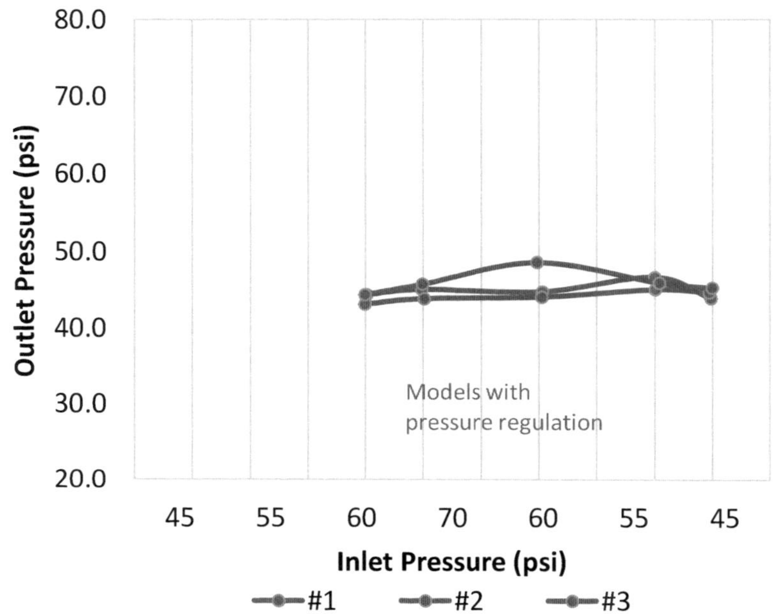


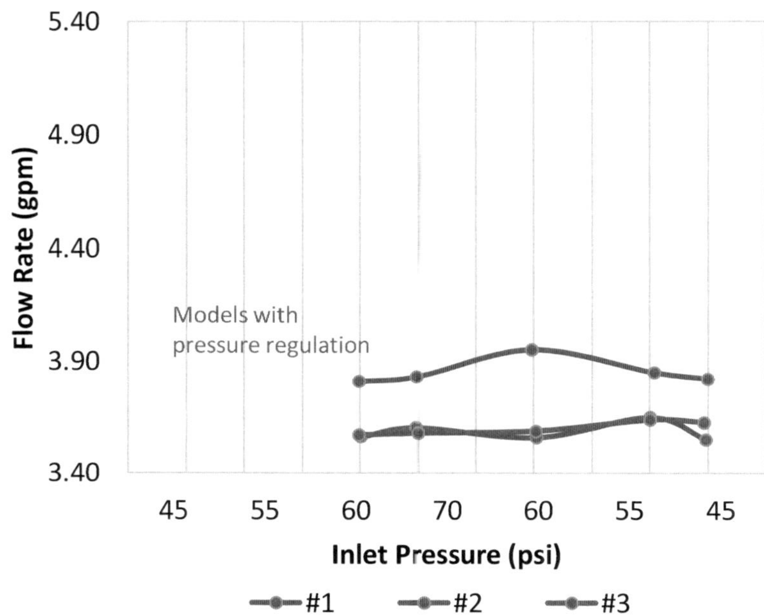


UF UNIVERSITY of FLORIDA IFAS

Center for LANDSCAPE Conservation & Ecology

# PRB Replicate Tests– Brand A Sample #1

# 1.5 gpm

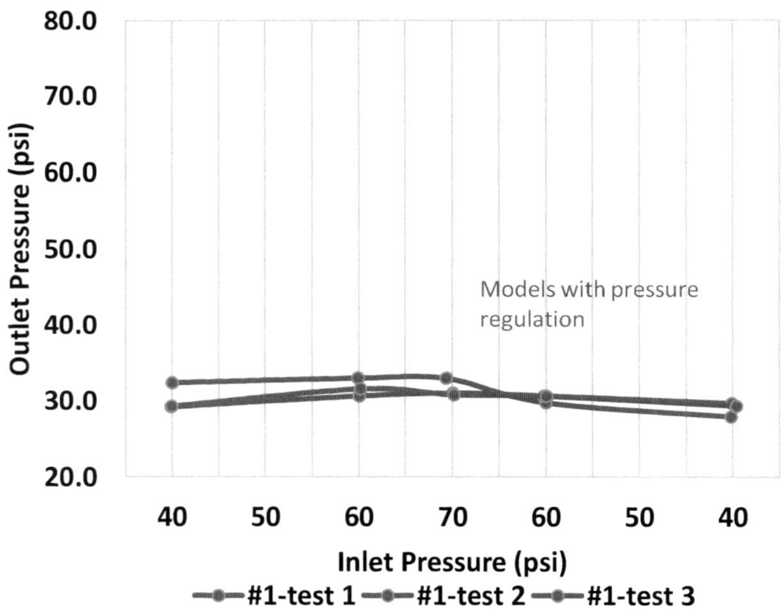


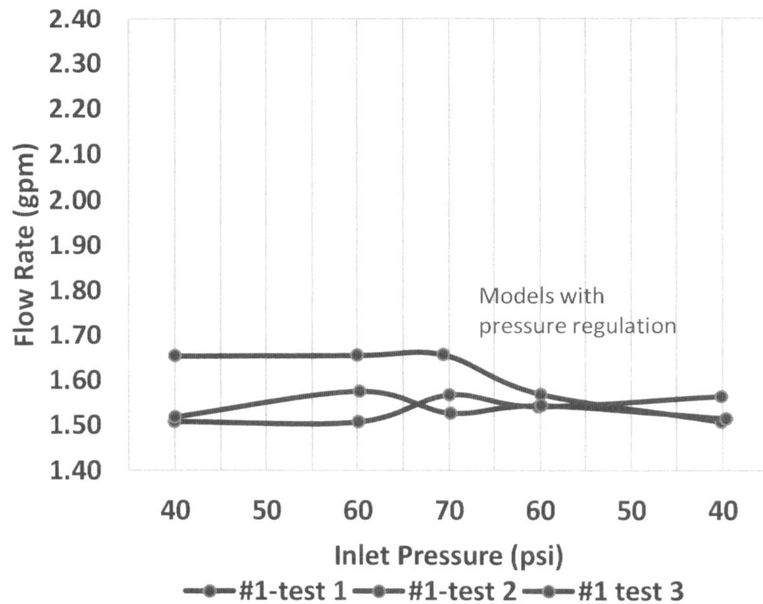


UF UNIVERSITY of FLORIDA IFAS

Center for LANDSCAPE Conservation & Ecology

# PRB Replicate Tests– Brand A Sample #1

# 3.5 gpm

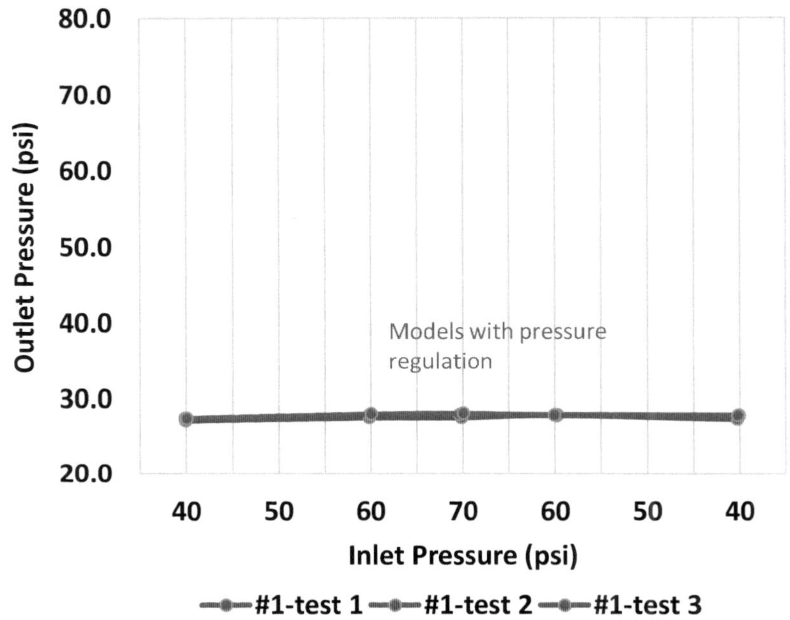


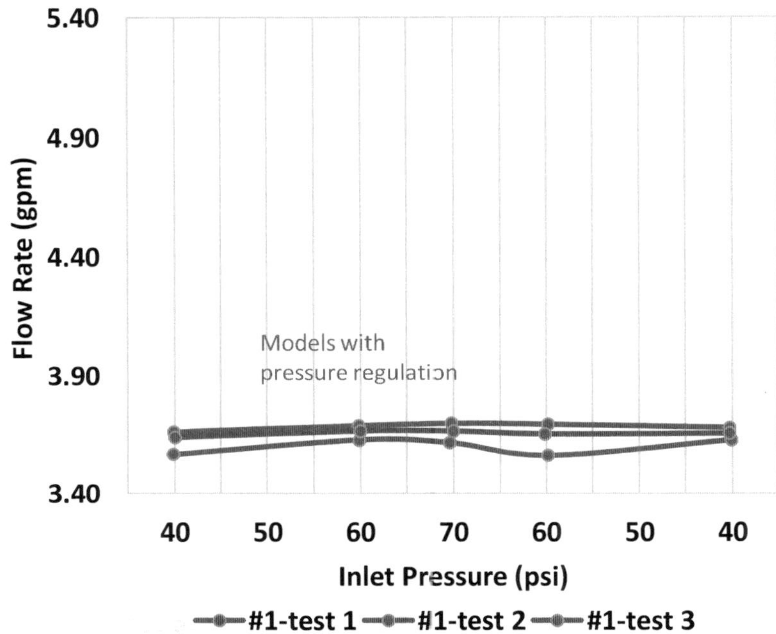


UF UNIVERSITY of FLORIDA IFAS

Center for LANDSCAPE Conservation & Ecology

# Flowrate Reduction – PRB vs. Non-PRB @ 1.5 gpm

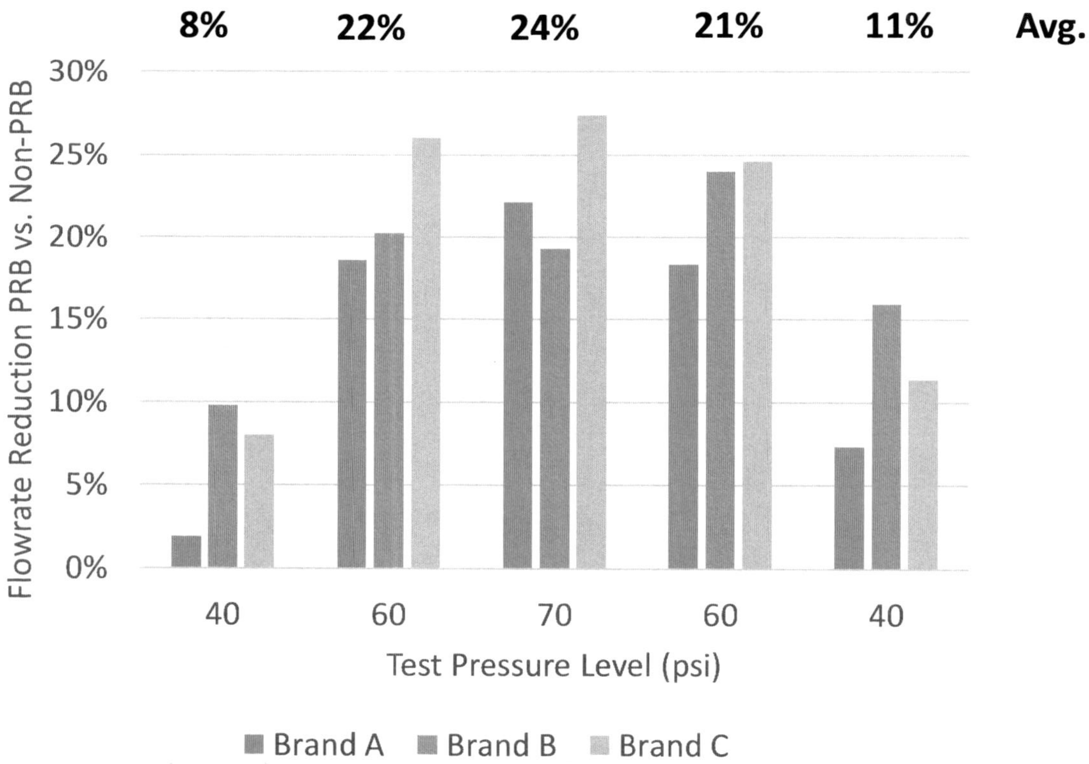

# Flowrate Reduction – PRB vs. Non-PRB @ 3.5 gpm

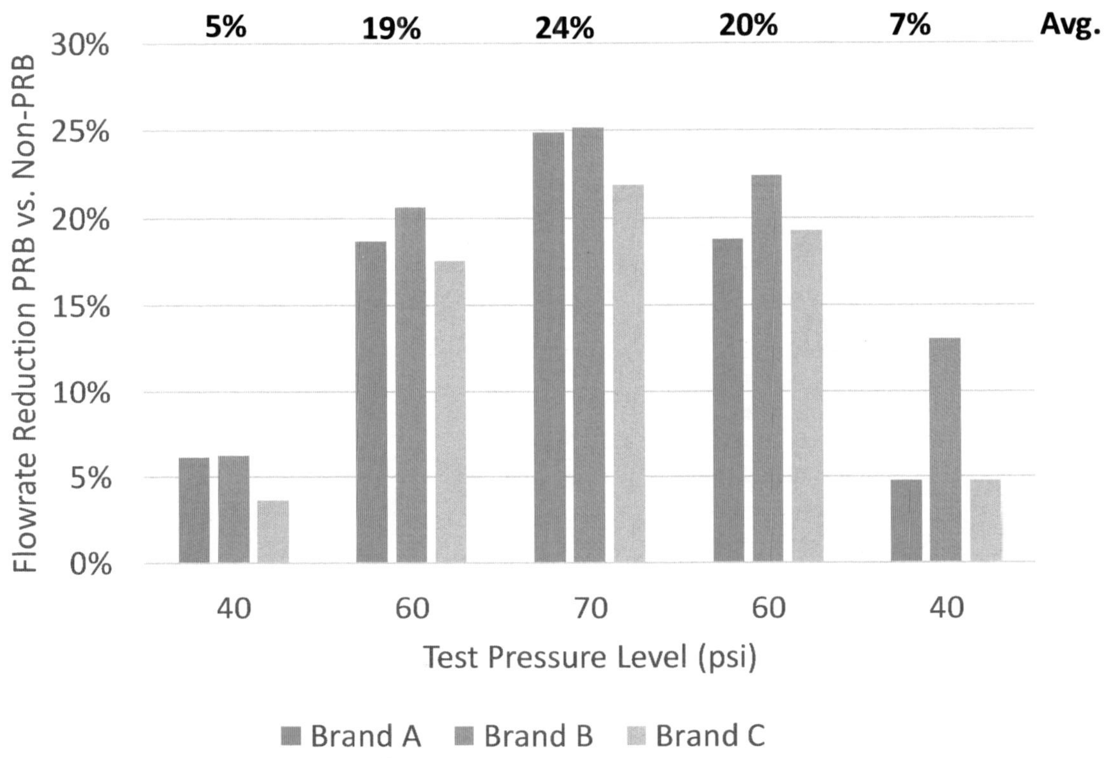

# Average Flowrate Reduction – PRB vs. Non-PRB

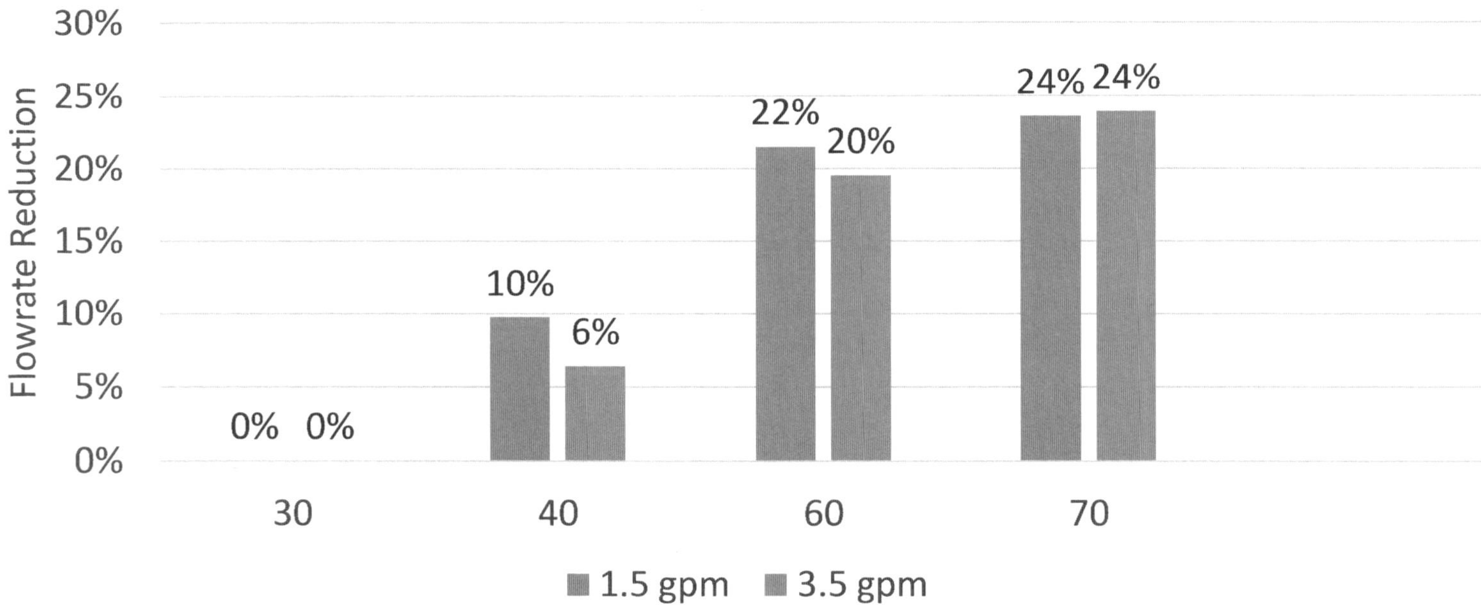


UF UNIVERSITY of FLORIDA IFAS

Center for LANDSCAPE Conservation & Ecology

# PRB Outlet Pressure

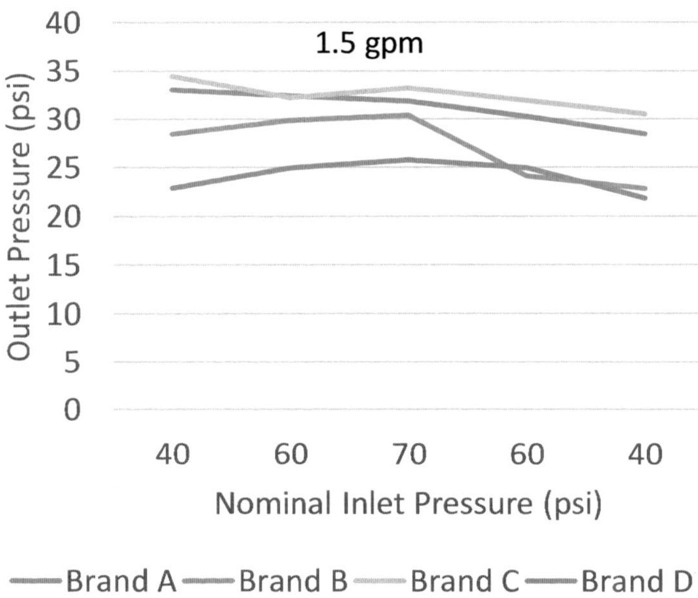


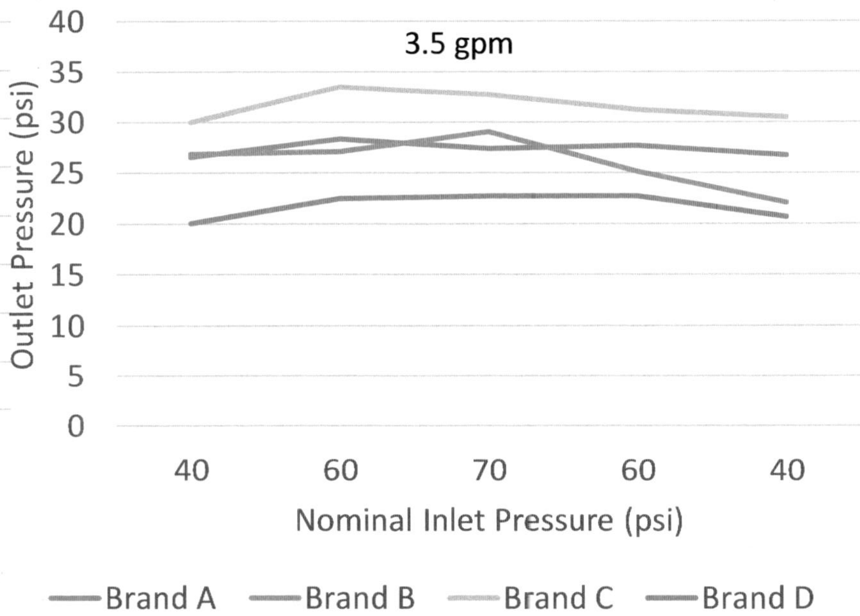


UF UNIVERSITY of FLORIDA IFAS

Center for LANDSCAPE Conservation & Ecology

# Average Pressure & Flowrate Error – 1.5 gpm

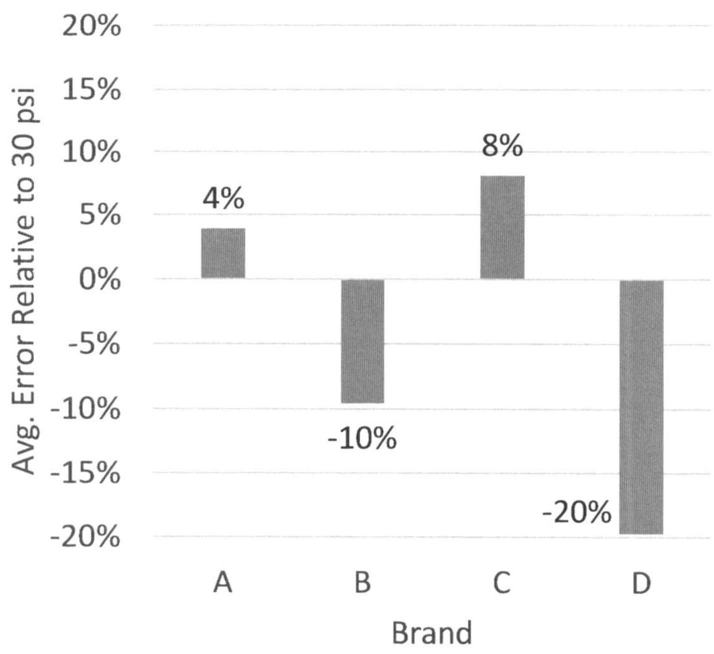


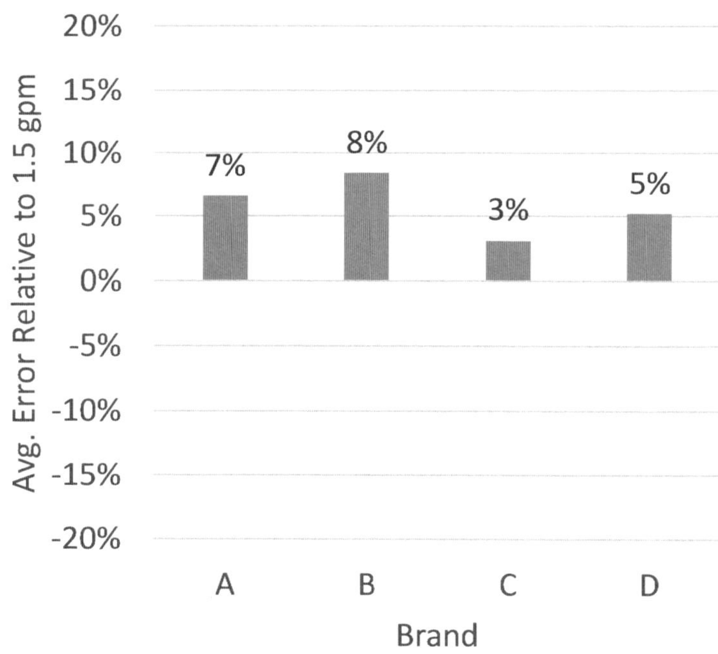


UF UNIVERSITY of FLORIDA IFAS

Center for LANDSCAPE Conservation & Ecology

# Average Pressure & Flowrate Error Rising Limb – 1.5 gpm

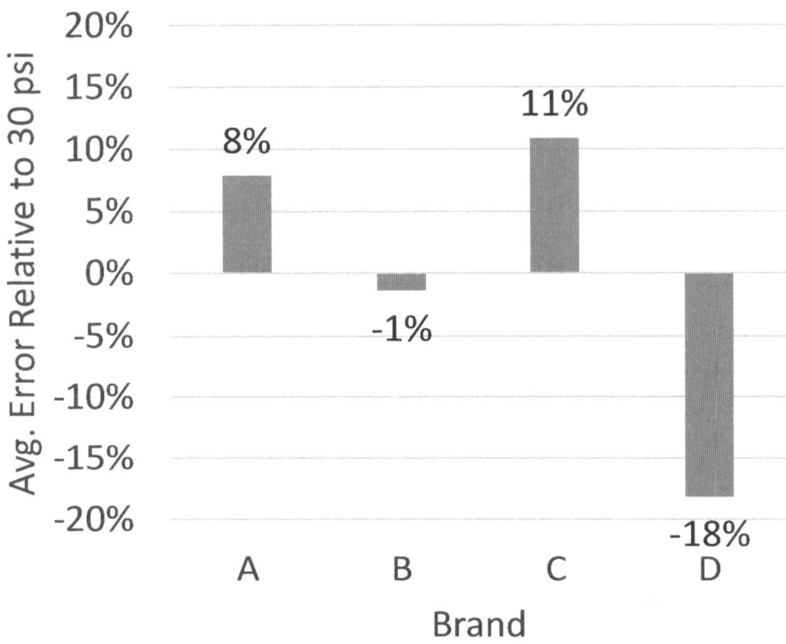


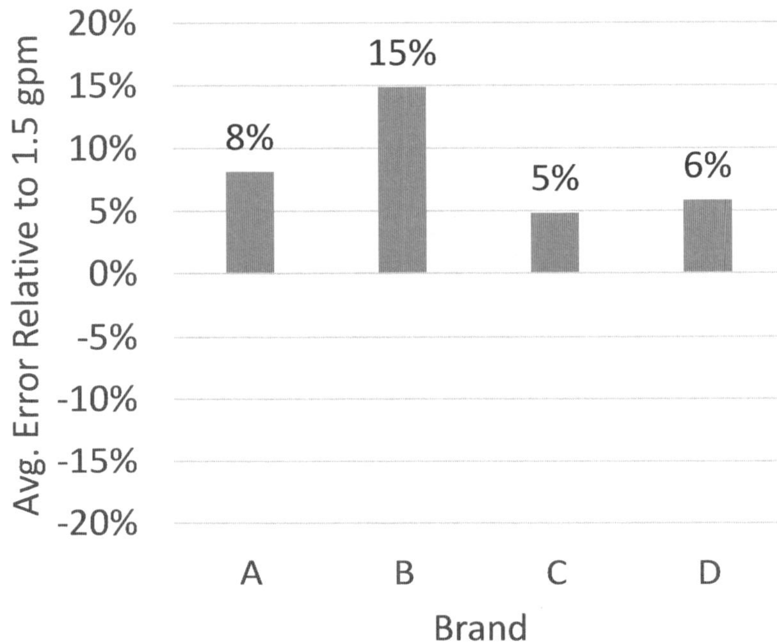


UF UNIVERSITY of FLORIDA IFAS

Center for LANDSCAPE Conservation & Ecology

# Average Pressure & Flowrate Error – 3.5 gpm

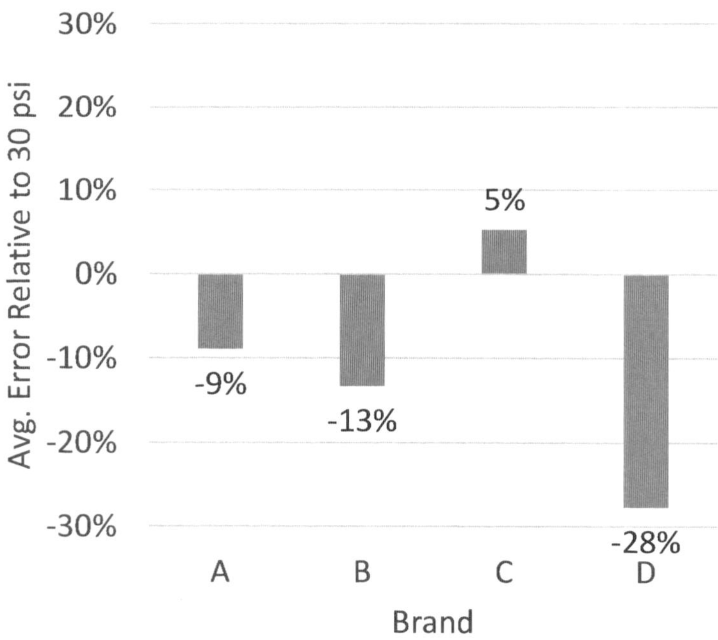


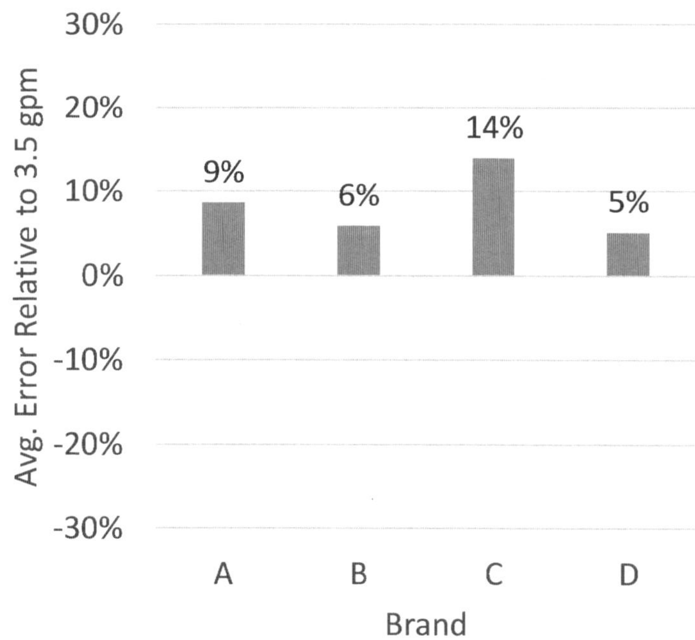


UF UNIVERSITY of FLORIDA IFAS

Center for LANDSCAPE Conservation & Ecology

# Average Pressure & Flowrate Error Rising Limb – 3.5 gpm

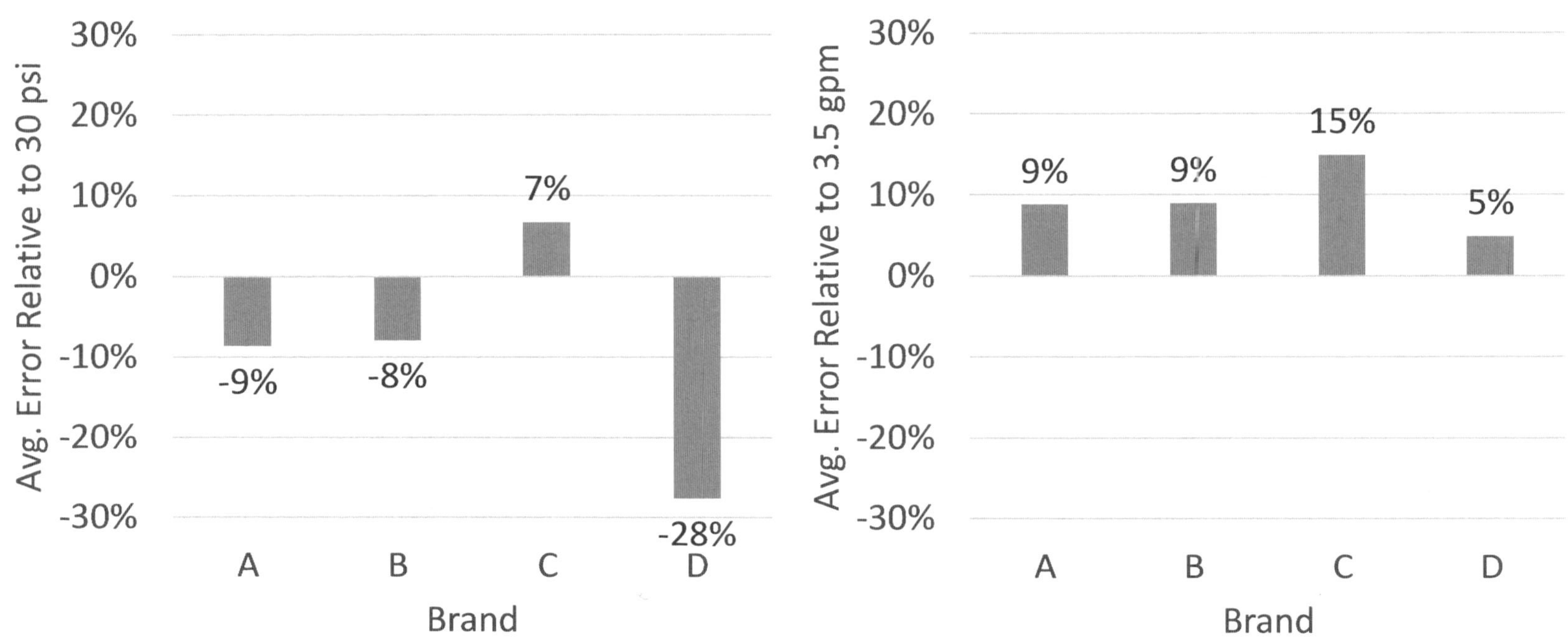


UF UNIVERSITY of FLORIDA IFAS

Center for LANDSCAPE Conservation & Ecology

# Recommendations

- Consider testing only the rising limb of pressure, e.g. for a 30 psi PRB, 40, 60, 70 psi test
- No compelling difference between 1.5 gpm & 3.5 gpm results
- Consider testing only 1.5 gpm since this flowrate is similar to the majority of sprinklers in the field
- Consider a maximum of 10-15% plus/minus deviation in peak flowrate at 1.5 gpm
- Consider average flowrate deviation maximum of 10-15% plus/minus at 1.5 gpm

UF | UNIVERSITY of FLORIDA | IFAS    Center for LANDSCAPE Conservation & Ecology

# Error Analysis on Individual Samples Criteria: 1.5 gpm actual flowrate rising limb

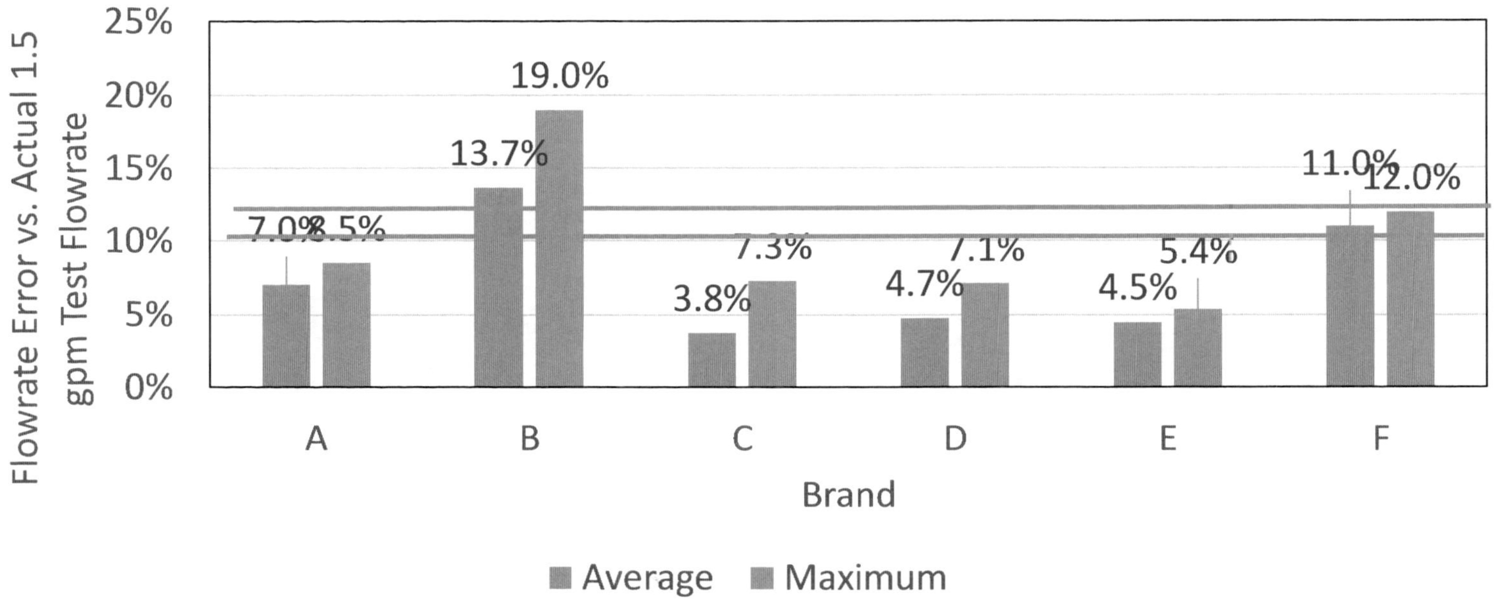

# EPA Spec Criteria

- Flowrate at max operating pressure compared to calibration flowrate shall be within +/- 12.0%
- Average of all test flowrates compared to calibration flowrate shall be within +/- 10.0%
- Average outlet pressure at initial calibration point shall not be less than 2/3 regulation pressure

UF UNIVERSITY of FLORIDA IFAS Center for LANDSCAPE Conservation & Ecology

Acknowledgements: EPA WaterSense Program

UF UNIVERSITY of FLORIDA IFAS

Center for LANDSCAPE Conservation & Ecology

# A New Generation of Smart Controllers

Parry Webb, CLIA
Weathermatic
3301 W Kingsley Rd, Garland, TX 75041

*Abstract: In the last three decades since the introduction of irrigation "smart" controller technology, the most dramatic innovation has been the recent ability to connect irrigation control to the internet yet few landscape professionals are taking advantage of this technology. With adaption on this new affordable cloud-based smart irrigation controller technology, a landscape contractor will see a dramatic improvement in crew efficiency, repair revenue, and contract accountability. The benefits of an internet connected controller include reduced trips, proactive system alerts, visibility to program changes and system functions, automated communication to a client about their irrigation systems and remote system access. All these benefits contribute to a healthier bottom line for a landscape maintenance professional.*

## Saving more than water

All the best irrigation technology in the world means nothing unless we know what's going on in the landscapes we manage. The challenge is how to keep our eyes on our landscapes without having to hang out at the site everyday all day.

Fortunately, there are a few irrigation systems which allow the landscape professional to affordably monitor his irrigation systems without having to gas up the F-350 and drive out to the site. In the old days, these were called "central control" systems and meant that the contractor had to load software on a dedicated computer and install highly technical and expensive communication devices to provide monitoring capability to any landscape. Today those systems are pretty much obsolete and have been replaced with simple, more affordable cloud based systems enabling contractors to remotely access irrigation systems from anywhere in the world via laptop, tablet, and even smartphone.

Today's systems have evolved into less "central" control and more "remote" control. The end result is still the same...in that it allows the end user complete oversight of the system and management of all aspects of an irrigation system from a remote location. The key differences with today's solutions are their ability to enable the end user to manage their system from anywhere and the "data" for their system to reside in the cloud vs. on a lone computer which needs to be backed up and whose hardware needs to be upgraded every so often.

Connecting an irrigation controller to the internet has a number of inherent benefits for both property manager and the landscape professional. The key is understanding how to communicate the value of these types of systems to customers and to the landscape maintenance organization.

For those in the landscape maintenance business, even if an irrigation technician is charging the client for driving to the site to make these irrigation adjustments, most are likely not charging enough for the service call. The average billing rate for many is less than half of their total cost burden which makes having profitable irrigation services a real challenge. One of the best ways to improve this is to raise the labor efficiency of the irrigation tech. Taking advantage of affordable remote access irrigation technology is the easiest and most efficient way to do this.

Having the ability to remotely manage an irrigation system saves fuel and hours of travel time to and from locations to adjust irrigation schedules, change time and date for daylight savings time, shut down for rainfall, and other programming issues. Remote control enables these changes to be made quickly. And by installing flow sensing, having remote access gives the contractor the ability to "see" what's happening with irrigation breaks and proactively schedule repairs and order parts quickly. This eliminates the need for the irrigation tech to have to seek out irrigation breaks and thus reduces his or her time spent looking for problems.

Additionally, using remote access to eliminate trips to the site to adjust irrigation doesn't necessarily mean maintenance professionals have to forego billing the client for a service call. As we have seen in other industries, customers are willing to pay for valued services even if those services are made much more efficient through the adoption of new technology (think computer hardware and software support). Taking advantage of technology still allows the irrigation tech to provide a service and therefore bill for the service; but he can do so at a much more efficient cost. The idea of a lower cost, higher profit "virtual service calls" is something a maintenance provider could begin offering clients. Most clients would also be supportive of an effort to reduce the amount of dollars they spend on irrigation service calls. As an example: if you originally charged $80/hr for a service call but your labor burden was $75/hr (actual cost burden is likely much higher than $75/hr), you made $5. But even if you charge a flat $50 for a "virtual service call" but your labor burden was reduced to $25/hr because of remote access, your net gain would likely be $25. And reducing your irrigation service calls from $80/hr to a flat $50 for irrigation adjustments would likely please your client as well.

Having the ability to remotely make adjustments to a client's irrigation controller wouldn't entirely eliminate the need to drive the property to look at irrigation but it can eliminate a good number of trips per year per controller.

Overall, eliminating any trips to the controller reduces labor costs, fuel use, maintenance on vehicles and a company's carbon footprint therefore contributing to a sustainable company image.

## I've got my eye on you

This brings us to the point of this discussion. The success of any smart control system is dependent on our ability to keep the system operating as intended. To do so we must insist that the system is monitored by multiple users to verify the system is doing what is supposed to do. A connected controller allows multiple users to monitor and change the programming. By doing so we introduce a level of accountability that is not only beneficial for the end user but also for the maintenance contractor. Essentially it insures both end user and landscape professional are on the same page regarding what is happening in the irrigation system. Visibility to how the connected controller is programmed and whether the controller is running in "smart" mode or the rain sensor is working can be monitored and remedied should any settings be changed. Too many "smart" systems have failed to accomplish any significant water savings because it is too easy to intentionally, or unintentionally, turn the smart features off without anyone's knowledge. This is often a reaction to the system doing something the user doesn't understand.

Therefore, the more eyes we have on the site, the more efficiently we can manage what we see. Visibility by multiple users insures our landscape and irrigation system is monitored regularly and alarms are attended too quickly. After all, what good is a flow sensor which reacts to a sprinkler break or a pump failure if it has no way of telling us what is going on? Without remote communication, our irrigation system has no way to talk to us to tell us something is wrong. In such a case, we wouldn't know we had a sprinkler break until someone notices the plant material slowly dying because it isn't getting any water (due to the flow sensor repeatedly shutting down the zone during each irrigation cycle).

The challenge we face in enabling multiple users to have access and more importantly to know when a system is doing what it was designed to do, is our ability to interact with the system without the need to seek out information about system status. Essentially, we need a system which proactively "talks to us" when there is a problem. A cloud-connected control system provides the benefit to a maintenance provider of proactively alerting a user to system operation and system failures. Knowing about system problems early also has the benefit of increasing the number of repairs needed on a given site because issues on an unmonitored system can go on for weeks if not months before someone notices.

Key to efficient monitoring is automated, system generated, notification of alerts and alarms requiring user attention. These types of site alerts can include flow sensor alarms for high and low flow, irrigation program changes, changes from smart mode to standard mode, rain shut down, system off mode, irrigation component failures, and other issues which require user attention for the system to operate efficiently.

Irrigation conditions such as high and low flow alarms, electrical issues, communication issues, valve failures, and notification for automated rain shutdown, change to non-ET mode, and controller off mode, are important circumstances that site managers and owners need to know about quickly.

Even if most stake holders don't monitor every detail of a system each day, having the ability to proactively notify them of important changes to the system can eliminate missed opportunities to manage the system as expected.

## We have something to prove

Complementing the ability to remotely monitor our irrigation system is our ability to document our progress toward water conservation. Ideally, we need to prove we are accomplishing the intended goals for the site regarding water savings and restriction compliance. To do this our system needs the ability to generate basic water use data and reports. Too many times irrigation systems are installed or upgraded to smart control with the promise of water savings only to fail to deliver because no one was monitoring day to day water use information. A key component of a successful water management strategy has to include benchmark objectives that can be verified throughout the year to insure goals are met at the end of the year and beyond.

Using a cloud-connected controller's software environment allows for the implementation of various valuable tools to enhance the systems reporting capability. Among these are the ability to document water use data and compare it to benchmark data to verify water savings and make adjustments throughout a growing season. Water use by zone, by controller, by site, and global for custom time periods is typical of the reporting tool. Without this information, the user must rely on the monthly water bill to determine his water conservation process is saving water.

An additional tool of a cloud based system allows the user to map the different assets of a site using GPS coordinates. This "Asset Mapping" enables the user to document the different elements (or assets) within a landscape adding pictures and notes. These assets can be backflow devices, water meters, pumps, irrigation zones, and any other landscape related items.

A complete cloud-based site profile enables the landscape professional to save a lot of time and frustration. Mapping an asset's location on a site can save a landscape professional a lot of time if he is unfamiliar with the property. With pictures and other important asset related documents saved in the cloud, a technician can also retrieve and review information related to these assets without having to bring the information with him or wait until he returns to the office to look something up.

As many in the irrigation industry can attest, most new irrigation systems are designed and installed with the best intentions, only to fail to meet expectations long term because no one verified that the system maintained operational efficiency through the next few years.

Unfortunately, many of us still don't trust the "smartness" of these systems which, ironically, are only reacting based on the preferences we have set up. As the well-known quote from the 1950's comic strip Pogo immortalized; "we have met the enemy and he is us".

This is the biggest issue with the water saving goals of LEED, Sustainable Sites Initiative, California's MWELO (AB 1881) and others. Without verification or enforcement to maintain the certification status of these irrigation systems, most will fail to achieve what the system was designed to achieve.

I recently had the opportunity to audit the LEED Silver Certified sustainable flagship location for a large hotel chain. Although the design for the project was well intended, there was a significant issue which would have disqualified the property from being considered landscape water efficient.

Most of the property was watered using efficient drip irrigation however a few zones near the front entrance utilized rotating nozzles on 4-inch pop-up sprayheads. These front zones were irrigating a 21-degree slope with cool season turf grass planted in clay soil. I was curious to see how these zones were programmed. What I discovered was similar to what most of us see every day in our business. Although the run time was segmented into 4 cycle times per day, the run time for each was programmed for 16 minutes. After reviewing my soil/slope chart I determined the run time was 13 minutes too long for each cycle and thus the system was wasting 80% of the water it was applying each time it ran. After some quick calculations, I determined these few zones were wasting $4,200 worth of water per year. In addition, the 1-year old parking lot which the slope drained to, was already showing signs of water damage.

Adding insult to injury, while checking the run times at the controller location, I noticed a couple of wires disconnected from the controller. Tracing these wires back to their source I discovered the rain sensor had also been disconnected. Needless to say, the general manager of the hotel was anxious to have us remedy these issues as quickly as possible and have monitoring and reporting established to verify the irrigation system is maintained efficiently.

The most important point of this scenario is how these issues could have been prevented had there been a process in place to verify system efficiency after the initial installation.

By remotely monitoring a client's site and providing water use reports to the client, most irrigation professionals can easily insure most of their landscapes stay at a water efficient peak performance.

## Site Unseen

Even with all the technology and efficiencies of a smart irrigation system, each property being managed still needs to be visited regularly. No amount of technology will reduce the need for observation of what is happening on the site. Sprinklers can still be misaligned so they are watering the parking lot or the high efficient nozzle can be clogged or obstructed by a tree or a sign. Plant material as it grows may start to interrupt the sprinkler's nozzle stream leading to brown spots or flooding of an area. The maintenance crew may have replaced a broken sprinkler with a new one with the wrong nozzle.

A sprinkler and landscape walk-through can quickly diagnose many of these types of problems and eliminate the embarrassment of promising a client a smart system that wastes water, floods a street, or wipes out plant material.

Without a reliable, regular and thorough sprinkler check, a site's irrigation system cannot be maintained at its peak performance. Many maintenance contractors don't realize that by taking advantage of technology to do regular sprinkler checks, they can improve not only their sprinkler techs efficiency, but also dramatically improve their bottom line.

For example, with some cloud-based irrigation control systems, the sprinkler check process is an integral part of the programming software. With a cloud-based sprinkler check process, a sprinkler tech can quickly and efficiently run through all zones and document problem areas (including pictures) for his follow up repair and even send a quick email work order proposal (using the integrated pricing component) from the field to the property manager for approval.

In addition, having a calibrated flow sensor eliminates the need for a tech to have to look for breaks. This strategy alone has a huge impact on a contractor's ability to be profitable in his maintenance activities. Most sprinkler techs who are regularly checking each of their system's sprinklers, can waste inordinate amounts of time just finding problems. This "search and fix" strategy is not only inefficient but typically limits the number of sites a sprinkler tech can check in the average day. This is why many sprinkler breaks are first spotted by a tenant rather than the sprinkler tech.

On a side note, I've often heard many landscape maintenance company owners tell me that their mow crews are an integral part of the process to spot broken sprinklers. While some breaks (or the damage left behind) can be spotted by driving through the property on a mower, the reality is most mow crews are moving so fast that they miss a lot not to mention the fact that most sprinkler breaks can only be spotted when the system is running.

The bottom line is; there are more problems with the typical irrigation system than a very busy sprinkler tech or mow crew can identify and fix in a given day. Taking advantage of technology greatly reduces the amount of time needed to check an irrigation system and may dramatically improve a sprinkler technician's ability to do more in less time thereby increasing repair revenue.

As landscape industry professionals look for ways to add value to their service offering, technology can go a long way toward helping differentiate from competitors. Offering the value-added services to an existing clients which a cloud-based irrigation system can provide gives a landscape professional a "sticky factor" with his clients. This makes them less willing to bid out their projects for fear that they won't get the same quality of service. This is especially key when a contractor is already saving a client water and reducing the headache of on-going irrigation problems.

Another important distinction of today's cloud based systems is that they can be updated remotely when changes are needed such as adding a new feature or fixing a software bug. Typically done through an overnight update and usually is unseen by the end user. Having the ability to update a system has a significant value in keeping the system current and eliminating the need to upgrade in the years to come.

Of significant benefit is also the ability for manufacturers to assist in troubleshooting issues in the Irrigation system. Remote access allows a manufacturer's technical support representatives look at what a contractor is seeing and help make adjustments or correct issues over the phone thus eliminating costly trips to a supplier to assist in remedying a system.

## Get smart

Today's central control systems have truly evolved into "smart" control systems. Smart has been redefined in recent years to mean more than ET. For a system to be smart today, it must be able to manage the many aspects of an ever-changing landscape by automating the appropriate reaction to a given situation. It is all part of what is becoming known as the Internet of Things (IoT). In other words, smart devices do the thinking and reacting for you based on a user's preferences.

Lastly, because many of these cloud-based systems are modular in their architecture, the cost for the average size system is much more affordable than even a few years ago. This enables landscapes of any size to be upgraded to smart technology which can greatly improve the system overall efficiency as well as lead to improvement in the maintenance crews' ability to manage the site more profitably.

## The Future of Cloud-Connected Irrigation Controllers

The future of cloud connected controllers will also enable us to connect other parts of the outdoor environment. We are already seeing this with landscape lighting. In addition to irrigation and landscape lighting, this will likely expand into connecting pumps, water features, holiday lighting, sound systems, and outdoor kitchens gadgets. Providing the end user and the landscape professional with a "dashboard" of all the connected devices enables better management of systems and sensors and encourages regular communication between customers and property owners

## Understanding cloud-based technology

Over the past 24 months, the irrigation industry has seen a dramatic rise in the number of these options based on Wi-Fi technology. With this "new to the industry" technology comes different challenges for the landscape professional. These challenges are created by the need for the anyone installing or maintaining these controllers to understand the technology behind them being connected to the internet.

The smart home savvy consumer is now driving the growth we are now seeing in connecting irrigation controllers to their Wi-Fi networks. For most contractors, this means venturing into the strange new world of internet connected devices. While most of us understand the basics of Wi-Fi connected devices, the real challenge for us is understanding that its less about Wi-Fi and more about connecting the irrigation system to the cloud.

While most of the industry is focused on "Wi-Fi" as a solution of connecting the controller to the cloud, there are other alternatives. The type of connection device is largely dependent on the scope of the project and what type of internet connection is available at the location.

**WIFI:** Wi-Fi works great where a client will allow access to their Wi-Fi network. Set up is typically simple by joining a Wi-Fi network and entering the network password. Most end-users have a basic understanding of how Wi-Fi works so they should be able to reset the network should a connection issue arise. Some Wi-Fi controllers also allow set up to be configured as an ad hoc connection which basically means the controller and smart phone application are connected directly to each other through the wireless network without an internet connection thereby creating a secure closed network.

Deciding whether or not to use Wi-Fi to connect an irrigation controller is largely determined by the distance between the controller and the Wi-Fi router and the stability of the Wi-Fi network.

**900 MHz:** Another option being offered by some manufacturers is 900MHz. This type of communication also requires use of the client's internet connection but bypasses the need to connect to the Wi-Fi router. The 900 MHz device is connected directly to the internet modem using a standard Ethernet cable. The device then broadcasts its own 900MHz wireless network thus enabling any irrigation controller with a 900MHz radio to connect to the internet.

While a 900 MHz connected controller is connected to the same internet connection as a Wi-Fi connected controller, it broadcasts its own 900MHz wireless network so interference from other devices is limited to old cordless telephones and baby monitors which used the same frequency. Also, the range of a 900MHz signal is much greater than the 2.4GHz of a Wi-Fi signal. Typically, 900 MHz can travel up to 1,500-2,000 feet and has very good diffraction abilities which allows the signal to bend around obstructions. 900MHz is a good choice for locations where the irrigation controller needs to be located beyond the range of a Wi-Fi router.

**CELLULAR:** A third communication option for connecting an irrigation controller is cellular. The advantage of cellular is that it does not rely on connecting the controller to the client's internet connection making it an ideal choice for locations where the client does not have internet available or does not want to allow a third-party connection to their wireless networks. The range of cellular is measured in miles but is dependent on the type of cellular technology used (CDMA and GSM) and the location of the local cellular towers. The benefit of a cellular connection is that the network equipment is maintained and serviced by the companies offering cellular service making the cellular platform the most reliable and up-to-date option for connecting irrigation controllers.

**Other Options:** Other communication options are just over the horizon. These new methods to connect and irrigation controller to the internet will only add to the industry's ability to provide affordable and reliable communication options to the client. Technology such as Z-Wave and Zigbee will allow multiple communication options to be used on a single site or allow one internet connection to be shared with multiple devices.

Future irrigation connectivity solutions must be able to support either network configuration and enable users to access their systems as preferred by each individual project.

For each project, a contractor or end user will have to determine the best options for their system. Deciding on whether to use Wi-Fi, 900MHz, or cellular should be based on whether the site is a single user or multiple users; a single site or multiple sites; how secure the network needs to be; whether to use on-site or off-site weather and whether to incorporate a flow sensor for logging water use and shut down. While Wi-Fi-based irrigation controllers provide easy access to remotely monitor and adjust the irrigation controller's performance, Wi-Fi won't work in every application. Alternatively, other methods of connecting the irrigation controller to the internet are available and should be considered before a connected controller option is selected.

## Summary

As the momentum for connecting irrigation controllers continues to accelerate, the irrigation industry will be challenged with "getting up to speed" on all this technology. The tech savvy customer will expect the irrigation professional to be more knowledgeable and proficient in implementing smart connected solutions and the opportunity to provide expertise to our customers has never been greater.

All this being said, water is becoming much more valuable and deserves to be managed appropriately. We can no longer waste water as if it is an abundant resource. With hundreds of thousands of people in communities around the country each year in a severe drought and 750 million people around the globe living without clean and safe drinking water, we as a landscape industry can no longer justify or afford to NOT be the leading stewards of this resource. Good water conservation practices represent a golden opportunity to improve the profitability of most irrigation professionals simply by managing water appropriately.

Many of the water conserving strategies we all know have been around for more than two decades. Yet with all this water conservation knowledge, most irrigation systems continue to waste absurd amounts of water. Thousands of "smart" controllers and water efficient sprinklers are installed each year, some resulting in a reasonable amount of water savings. The vast majority, however, fail to provide a long-term water saving solution and further damage the reputation of the landscape professional to solve the problem. It is with this in mind that we as an industry must do better.

Installing cloud-based smart technology is an easy way to fix what's broken in our irrigation systems and greatly improve any landscape professional's business opportunities and profitability.

The moral to this story is simple...fresh water is finite and if we are to use our drinking water to water our plants, we must do so judiciously and with the next generations in mind.

*Why do we think water matters more than any other resource?*

*Because water is life. And it is the one resource we cannot reproduce.*

# Evaluating Sprinkler Operational Efficiency using SMS


## Authors

### Gary Klinefelter

CEO and Founder, IrriGreen Inc.
gary.klinefelter@irrigreen.com

### Michael D. Dukes, PhD., P.E.

University of Florida, Agricultural and Biological Engineering Dept.
mddukes@ufl.edu



**Abstract.** *Evaluating irrigation sprinkler operational efficiency with catch cups is a labor-intensive process that provides data regarding the output of the irrigation system using mechanical spray and rotary sprinklers. Catch cups are a proxy for the available moisture in the soil due to many factors such as: soil saturation, soil type, run-off, etc. While soil moisture sensing (SMS) can automate the measurement process, much needs to be learned about the limits and benefits of using soil moisture to measure irrigation sprinkler operational efficiency. This presentation will cover examples of using soil moisture to measure sprinkler operational efficiency; limits in making before and after soil moisture measurements; advantages in using soil moisture sensing and what we still need to learn.*



**Keywords.** SMS, sprinkler, irrigation, efficiency, soil, moisture, measure.


## Introduction

For many years, the irrigation industry has used catch-cups to sample water output from irrigation sprays and rotors to evaluate water application efficiency. The most common method is to compute distribution uniformity lower quartile (DULQ). DULQ is computed from a set of data by taking the average of the lower 25% of the data set compared to the average of the entire data set. More recently, the industry has defined the SWAT protocol which defines methods of placing catch-cups along with calculations for operational efficiency. SWAT operational efficiency is computed by subtracting percolation losses and overspray losses. The percolation loss is computed by taking sum of the differences of the upper 75% of the data points minus the 75% data point divided by the sum of the entire data set. For this paper, the term *efficiency* will simply mean how efficiently a sprinkler is applying water in the field. The term *digital* refers to controlling a sprinkler rotation and throw distance using digital electronics.

### Smart Water Application Technologies (SWAT)

According to the Irrigation Association website: "*Smart Water Application Technologies is a partnership of water providers and irrigation companies, working to promote landscape water-use efficiency through innovative technology.*" The SWAT *Spray Head Sprinkler Nozzles Performance Characteristics* testing protocol defines test methods, test shapes and calculations such as: precipitation rate, distribution uniformity, overspray losses, percolation losses and operational efficiency.

## New Technology Influence on Sprinkler Measurement

In 2013, IrriGreen began offering a digital irrigation sprinkler that used multiple streams of water to uniformly water the soil by setting stream volumes that match the area at each stream distance. This new technology created challenges using catch-cup measurements to measure water application efficiency because some streams miss the catch-cups while others are deflected by the catch-cups. Like previous digital irrigation sprinklers (Figure 1) which used digitally adjusted spray or streams, catch-cups did not tell us what is happening in soil when the sprinkler output is streaming and moving. Do we need an alternative measurement of digital sprinkler efficiency that uses soil moisture measurements?

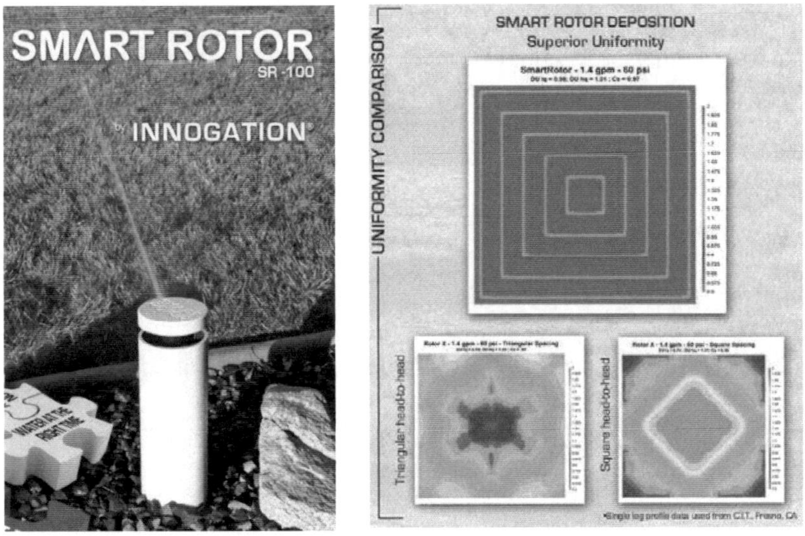


"Based on the dryness distribution plot in Figure 3, an intelligent system can reduce water use by 40% to 50% in most cases."
- Robert Walters

**Figure 1 – Adjustable Spray Sprinkler by Innogation**

### Multi-Volume, Multi-Stream Nozzle Catch-Cup Measurements

In 2012 and 2013, IrriGreen did extensive catch-cup testing to refine the performance of a multi-volume, multi-stream nozzle. Figure 2 shows an example of how early testing was performed using 6-inch catch-cups. The catch-cups close to the nozzle lost water due to stream deflection while the catch-cups farther away missed water falling between the catch-cups. To improve the result, measurements were done with catch-cups adjacent to one another. While this works fine for research, it is not very practical in the field.

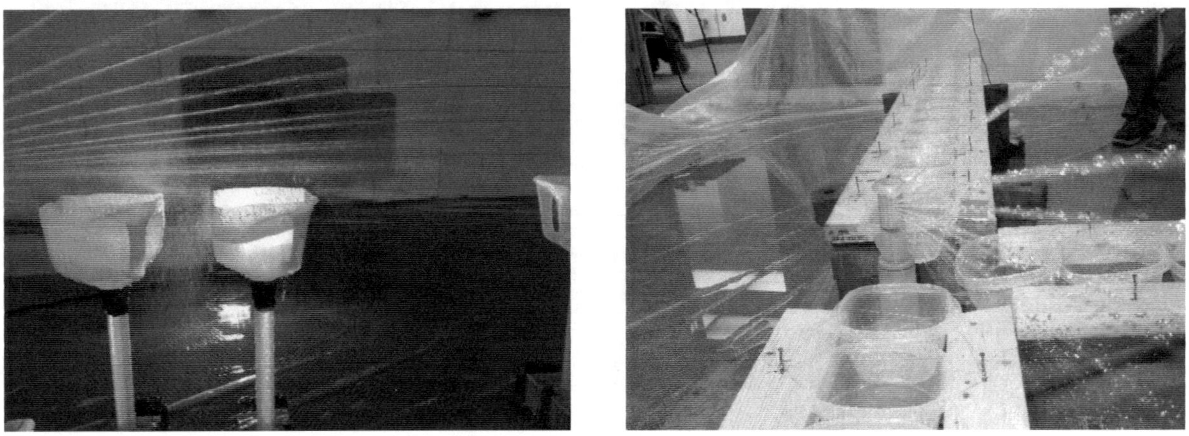

**Figure 2 – Multi-Stream, Multi-Volume Nozzle**

The resulting data from this early development was used to refine the IrriGreen nozzle to design the distance between the streams so the soil would be able to fill in the gaps like drip irrigation. Figure 3 shows how the water volume changes with distance for the multi-stream nozzle.

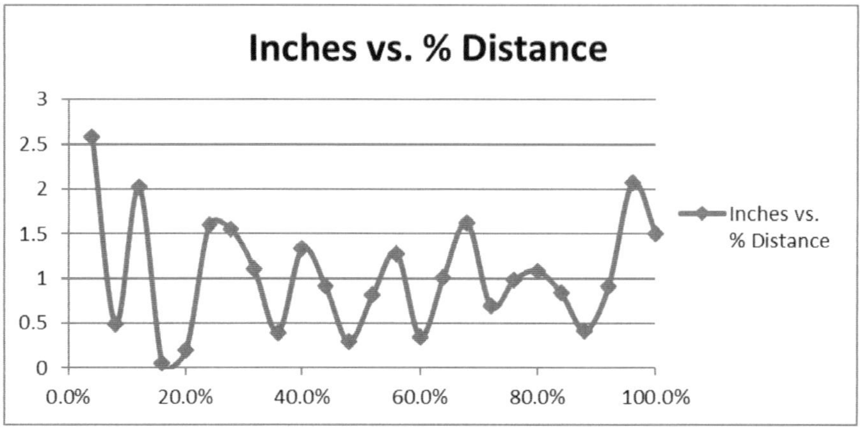


**Figure 3 – Multi-Volume, Multi-Stream Digital Sprinkler**

## Uniformity Testing with Brian Horgan, PhD. University of Minnesota

In 2014, IrriGreen worked with Brian Horgan, PhD. from the University of Minnesota to perform catch-cup and soil moisture testing on turf. He used both catch-cups and a Spectrum TDR-300 soil moisture probe to make measurements. The goal was to sample test areas of turf grass where 3 mechanical rotors overlapped versus one IrriGreen digital sprinkler. He tested three 10 x 10 plots of turf grass. In his report, the coefficient of uniformity for catch-cups was 0.91 for a mechanical rotor and 0.68 for IrriGreen. When using TDR soil moisture measurements both mechanical and IrriGreen had 0.85 coefficient of uniformity. (Full report available at www.irrigreen.com)

## Mark A. Crookston, P.E., D.WRE, Northern Water, Berthoud, CO 80513

Mark Crookston, Irrigation Management Department Manager at Northern Water heard about the IrriGreen system in 2016 and took the approach of using very large catch-cups (5-gallon pails) to overcome the difficulty of measuring streams. This yielded a DULQ of 0.58 and a SWAT sprinkler operational efficiency of 67%, not a very practical method to use in the field.

**Figure 4 – Digital Sprinkler Test, Northern Water**

## Digital Sprinkler Testing at the Center for Irrigation Technology (CIT)

In 2016, IrriGreen worked with the Center for Irrigation Technology at Fresno State to test the IrriGreen digital sprinkler. We ran 3 tests: 30 x 60 rectangle, 30-foot square and 30-foot circle with the latter 2 tests defined by the SWAT protocol. CIT selected best-in-class mechanical irrigation sprinklers to be tested on the same plots as digital sprinklers. In each case there was one digital sprinkler in the center compared to 6-9 mechanical sprinklers around the edge. We tested on turf so we could perform catch-cup testing and soil moisture testing side by side.

**Figure 5 – CIT Testing Digital Sprinkler**

### CIT Testing Considerations

The team at CIT used mechanical rotor sprinklers in the 30 x 60 rectangle to water the turf for 58 minutes using 2 GPM nozzles. Using mechanical rotors sprinklers as a reference, the IrriGreen team calculated the mechanical sprinklers low quarter and set the IrriGreen system to match at 5 revolutions, about .275 inches. At the time, we asked several experts including Dr. Michael Dukes about when to make SMS measurements and decided to make them immediately after making catch-cup measurements and at 3 hours. We wanted to make sure we did not over saturate the soil for either mechanical or digital sprinklers. For instance, the center areas of Figure 6 represent areas of high precipitation for mechanical sprinklers.

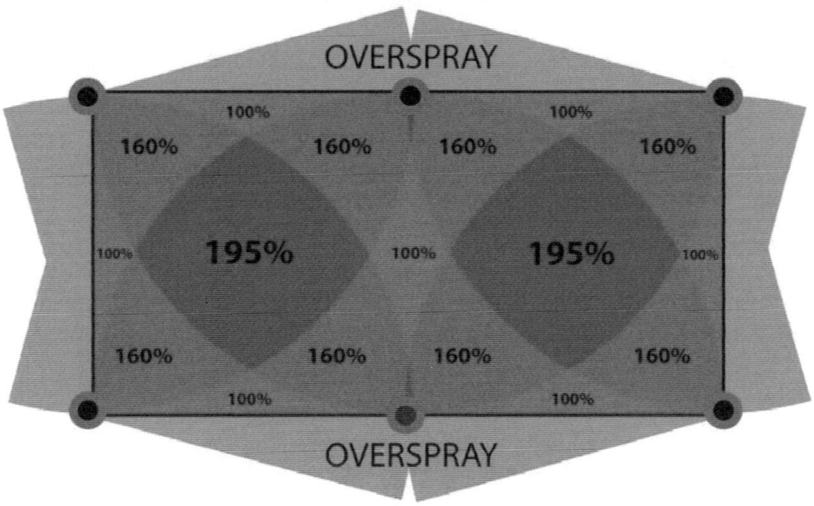


**Figure 6 – Overlap and Overspray from Mechanical Sprinklers (0.65 DULQ)**

Unlike a mechanical sprinkler that overlaps and is positioned head-to-head, a digital sprinkler is placed in the center of the landscape and waters uniformly from the inside out. For the digital sprinkler, areas of high and low soil moisture measurement occur between the streams as shown in Figure 3.

**Figure 7 – Example Mechanical and Digital Sprinklers**

## CIT Test Results

As expected due to the challenge of using catch-cups with streams, the distribution uniformity and operational efficiency numbers for the digital sprinkler catch-cup measurements are lower than mechanical measurements in Table 1. The same uniformity and efficiency measurements for the digital sprinkler using SMS data are closer to mechanical measurements. Assuming SMS readings are valid measurements for operational efficiency, then the digital sprinkler performed nearly as well as the mechanical sprinkler. (Full CIT report is available at www.irrigreen.com)

| **CIT Test** | SWAT Operational Efficiency | DULQ |
|---|---|---|
| Mechanical Sprinkler CC | 80% | 0.71 |
| Digital Sprinkler CC | 54% | 0.43 |
| Mechanical Sprinkler SMS | 70% | 0.37 |
| Digital Sprinkler SMS | 65% | 0.55 |

**Table 1 – CIT 30 x 60 Test Result Summary Mechanical and Digital Sprinklers**

## Differences in gallons collected

What is even more dramatic than the differences in SMS versus catch-cups measurements, is the difference in the volume of water collected given a similar change in soil moisture, about 40% less volume (gallons) for digital versus mechanical sprinklers. (With catch-cups it is unnecessary to collect volume because the cup is a volume measurement). Volume and SMS measurements may tell us more accurately what is happening in the soil. These results led IrriGreen to plan additional research into using SMS as a measurement for sprinkler efficiency.

## Working with Dr. Dukes, University of Florida on SMS measurements

It became clear from the work done at CIT that in the future the irrigation industry may need an alternative way to measure sprinkler performance using soil moisture measurements. Given Dr. Michael Dukes past research in soil moisture measurements in the field, IrriGreen chose to work with Dr. Dukes and the University of Florida to perform further research into using soil moisture as a performance measurement and use the resulting protocol to test mechanical and digital sprinklers with more replication.

### What SMS Measurements Are Needed

Since there are physical boundaries for using SMS as a measurement tool, we decided to define what those are. Like a catch-cup, soil can only hold so much water so we needed to define those measurement boundaries. Here is the process we chose for doing the research:

1. Installation and set-up of 1 IrriGreen Genius® Sprinkler (30 x 60 ft. turf plot).
2. Installation and set-up of a conventional 6 rotor system (same 30 x 60 ft. turf plot).
3. Estimate Volumetric Water Content (VWC) range based on starting an irrigation event at 8% VWC (maximum allowable depletion) and ending it at 12% VWC (field capacity).
4. Calculate and run 1 IrriGreen Sprinkler from 8% VWC (maximum allowable depletion) and ending it at 12% (field capacity).
5. Calculate and run 6 mechanical rotors from 8% VWC (maximum allowable depletion) and ending it at 12% (field capacity).
6. Compare results from 4 & 5 above, and re-calculate and re-run if necessary.
7. **INTERIM RESULT:** Have a good idea of which is the maximum allowable depletion point and which is the average field capacity of the area.
8. Run 1$^{st}$ set of comparative tests for delta VWC percentages based on the designed irrigation event determined in Step 7 above (includes tests for catch-cup results – immediately after irrigation, SMS probe – immediately, SMS – 3 hours after and SMS – 24 hours after).
9. Run 2$^{nd}$ set of comparative tests for delta VWC percentages based on insufficient water than designed irrigation event determined in Step 7 above (includes tests for catch-cup results – immediately after irrigation, SMS – immediately, SMS – 3 hours after and SMS – 24 hours after).
10. Run 3$^{rd}$ set of comparative tests for delta VWC percentages based on excess water than the designed irrigation event determined in Step 7 above (includes tests for catch-cup results – immediately after irrigation, SMS – immediately, SMS – 3 hours after and SMS – 24 hours after).
11. **FINAL RESULT:** Write a report that can be distributed by UF and IrriGreen showing study results.

## University of Florida Field Measurements Steps 1 – 6

**Figure 8 Test Areas of Turf Grass at the University of Florida**

The first half of the research process has been performed by the University of Florida yielding the results shown in Figures 9 and Table 2. Figure 9 shows the average VWC at 0, 3 and 24 hours. Figure 9 shows that making immediate measurements will yield better results because the water is rapidly draining and redistributing in the soil. Table 2 shows mechanical and digital sprinkler measurements at 0.5-inch application yielding 8%-10% change in VWC and 1-inch application yielding 15%-16% rise in VWC.

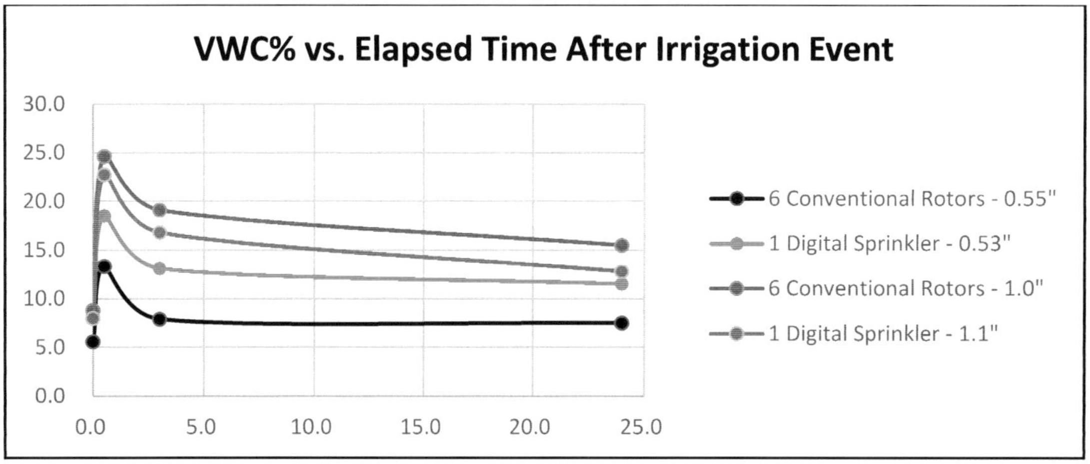


**Figure 9 – Soil Volumetric Water Content Over Time**

| **UF SMS Test** | Delta VWC | SWAT Operational Efficiency | DULQ |
|---|---|---|---|
| Mechanical Sprinkler 0.5 in. | 8% | 53% | 0.45 |
| Digital Sprinkler 0.5 in. | 10% | 67% | 0.38 |
| Mechanical Sprinkler 1 in. | 16% | 83% | 0.74 |
| Digital Sprinkler 1.1 in. | 15% | 77% | 0.66 |

**Table 2 – UF 30 x 60 Test Result Summary Mechanical and Digital Sprinklers**

Figure 10 graphs the data for the results in Table 2. While the curves are similar between digital and mechanical sprinklers, the digital sprinkler has more low and high data points, possibly due to using streams instead of sprays. Both mechanical and digital sprinkler curves become flatter at high soil saturation.

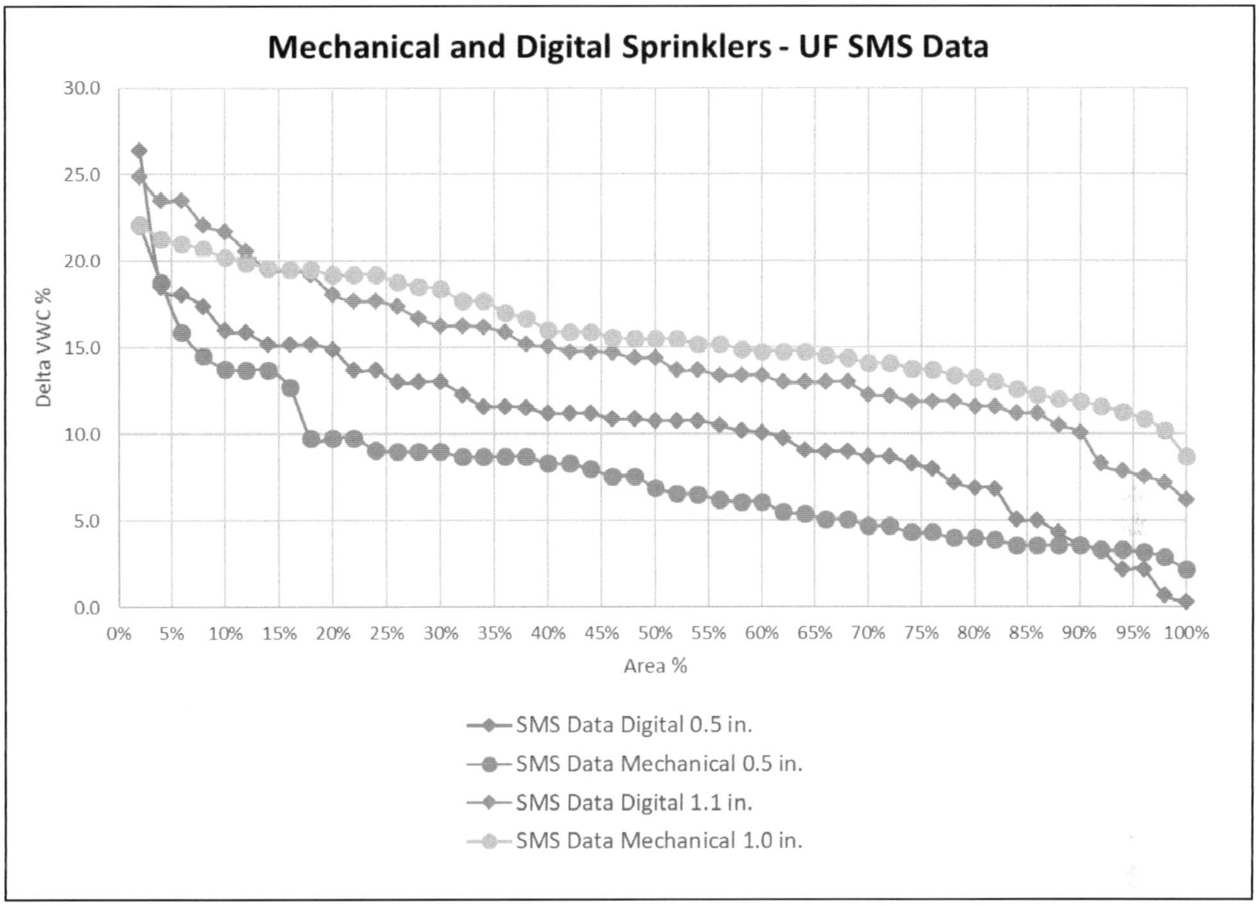


**Figure 10 – UF 30 x 60 Test Results Mechanical and Digital Sprinklers**

### University of Florida Field Measurements Steps 7 - 11

Based on the interim results in steps 1-6, the final testing will be run at 0.25-inch, 0.5-inch and 1-inch of water application to give us a range in which to determine the optimal test result. Catch-cup tests will be taken alongside of soil moisture tests for reference. These results will be reported when available.

## Conclusion Using SMS Data to Measure Sprinkler Efficiency

Volume and soil moisture measurements show promise as an alternative way to measure sprinkler efficiency. There is more to be learned about how the application rate and volume affect SMS measurement results. After we complete our research and based on our findings, IrriGreen and the University of Florida will run additional turf grass tests with both digital and mechanical rotors adding more replication to better understand the use of SMS data to measure sprinkler efficiency. We will publish these results in 2018.

## Acknowledgements

www.innogation.com

Dr. Brian Horgan, PhD., Turf Grass Consultant, University of Minnesota

Mark A. Crookston, P.E., D.WRE, Irrigation Management Department Manager, Northern Water

David Zoldoske Ed.D, Tim Jacobsen, Edward Norum, CID, Fresno State, Center for Irrigation Technology

Dr. Michael D. Dukes, PhD., P.E., University of Florida, Agricultural and Biological Engineering Dept.

Jeffrey Woodson, Director of Operations, IrriGreen, Inc, Edina, MN.

# Rainfall Effects on Seasonal Weekly Irrigation Schedules- A Case Study of the Water My Yard Program

**Charles Swanson, Extension Program Specialist-Landscape Irrigation**
Texas A&M Agrilife Extension Service
2117 TAMU, College Station, Texas 77843-2117

**Dr. Guy Fipps, Professor & Extension Agricultural Engineer**
Texas A&M Agrilife Extension Service
2117 TAMU, College Station, Texas 77843-2117

Abstract

*Proper accounting of rainfall is necessary when calculating weekly irrigation requirements. Rainfall can either reduce the amount of irrigation required or eliminate the need completely. The Water My Yard program was started in 2015 as a simple tool to provide guidance to homeowners on whether irrigation is required each week and if so how many minutes to run their irrigation systems. A review of weekly water recommendations of service areas in the Water My Yard Program in Texas has shown that as often as 50% of the seasonal weekly recommendations issued no watering required as a direct result of sufficient rainfall having been received. This paper will evaluate and compare the weekly watering recommendations of service areas within the water my yard program as a direct result of using localized measured rainfall.*

## Background

The Water My Yard program and website (http://WaterMyYard.org) was developed using simple, intuitive images and information prompts for homeowners to receive recommendations on how long (in minutes) to run their irrigation systems. The program was launched in May 2013 as a joint effort of the Irrigation Technology Program of the Texas A&M Agrilife Extension Service (Extension) and the North Texas Municipal Water District (NTMWD). To support the program, NTMWD purchased and installed 8 scientific (ET Type) weather stations within their member cities service area. Density of weather stations and accuracy of weekly watering recommendations has been a concern to promote homeowner confidence in following the program.

## Watering Recommendations

To help address questions on variability between weather station density and weekly watering recommendation, an analysis of weekly watering recommendations from the WaterMyYard Program for the North Texas Municipal District service area for the typical irrigation seasons in 2016 and 2017, respectively. The number of weeks that irrigation was recommended and not recommended was quantified from each weather station in the service area. Additionally the total amount of irrigation required and the total rainfall received was calculated. A summary of each year's analysis is shown in Table 1.

Table 1. Summary of Watering Recommendations for the North Texas Municipal Water District from 2016 and 2017.

| | 2016 | | | | 2017 | | | |
|---|---|---|---|---|---|---|---|---|
| | Irrigation Recommended (Weeks) | No Watering Needed (Weeks) | Total Irrigation (Inches) | Total Rainfall (Inches) | Irrigation Recommended (Weeks) | No Watering Needed (Weeks) | Total Irrigation (Inches) | Total Rainfall (Inches) |
| Farmersville | 18 | 21 | 6.69 | 35.24 | 18 | 12 | 5.27 | 20.09 |
| Garland | 21 | 18 | 7.21 | 29.92 | 18 | 12 | 8.21 | 20.98 |
| Mckinney | 17 | 22 | 7.19 | 31.04 | 19 | 11 | 7.2 | 21.01 |
| Mesquite | 19 | 20 | 7.61 | 32.22 | 14 | 16 | 6.57 | 50.57 |
| Plano | 18 | 21 | 6.29 | 35.47 | 17 | 13 | 6.22 | 38.74 |
| Richardson | 16 | 23 | 6.4 | 31.1 | 18 | 12 | 6.96 | 21.67 |
| Rockwall | 16 | 23 | 5.64 | 39.49 | 15 | 15 | 5.21 | 30.48 |
| Wylie | 24 | 15 | 8.09 | 30.07 | 19 | 11 | 6.86 | 22.19 |

Figure 1. 2016 North Texas WaterMyYard Summary

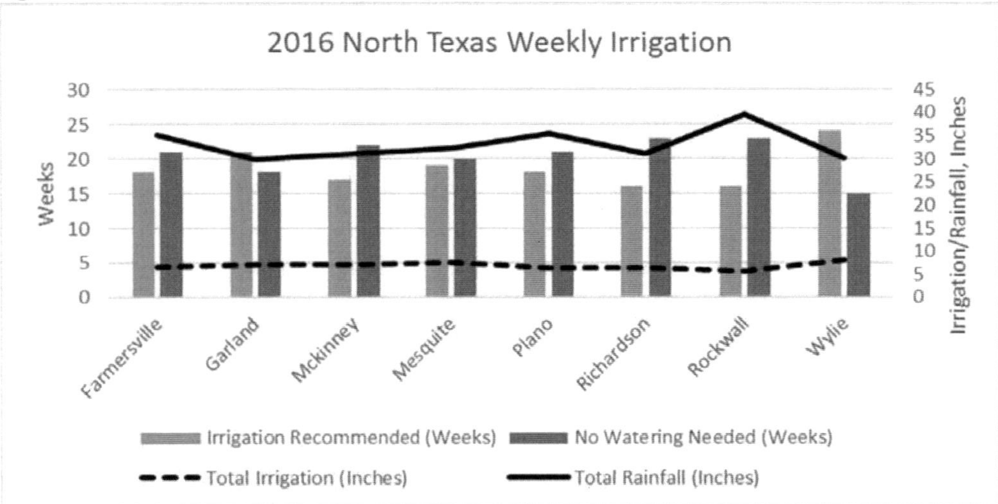

Figure 2. 2017 North Texas WaterMyYard Summary

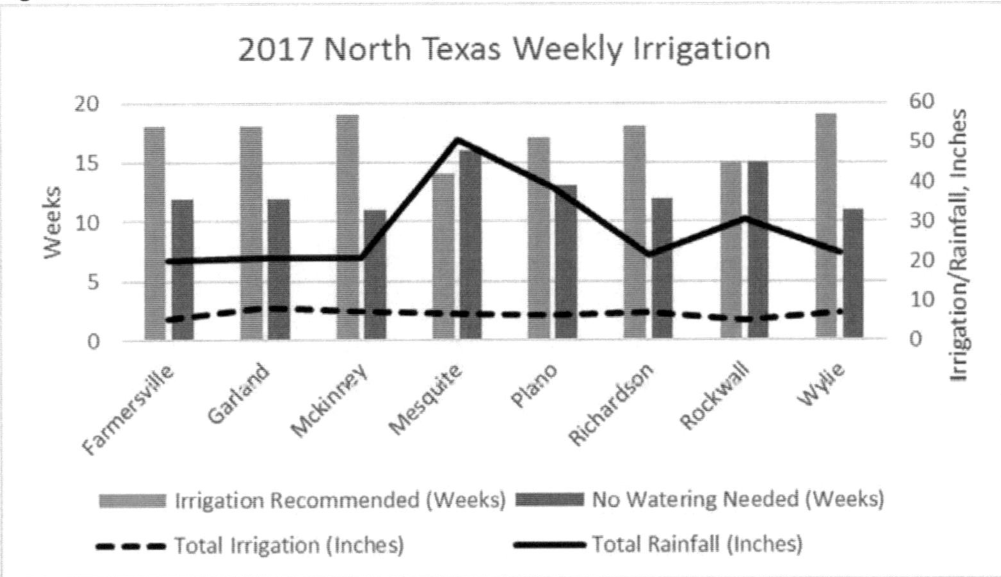


Figure 3

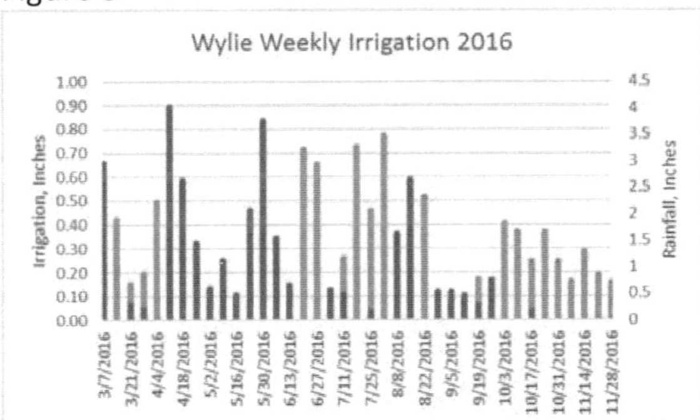


Figure 4

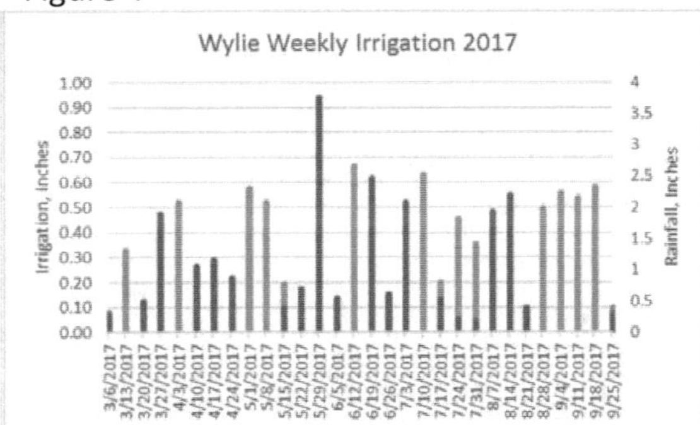


Figure 5

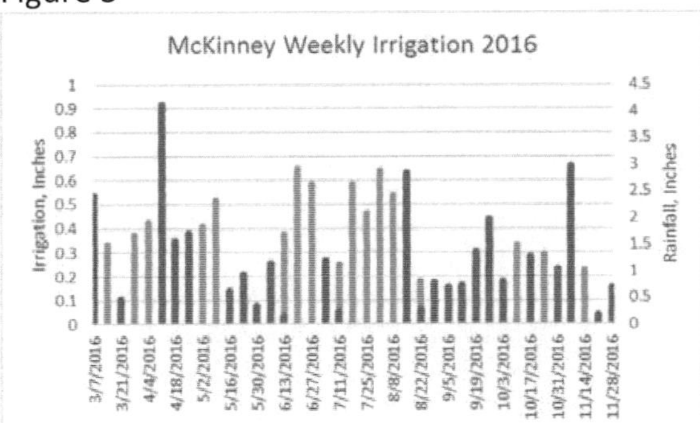


Figure 6

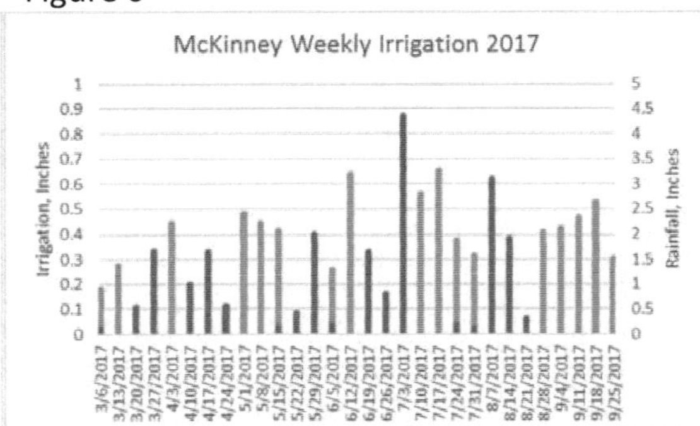

Figure 7

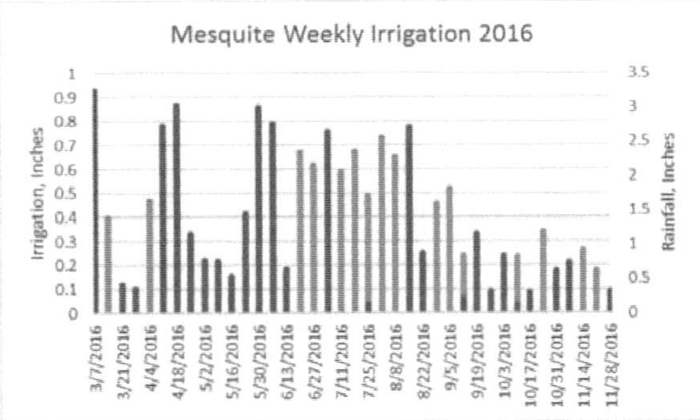


Figure 8

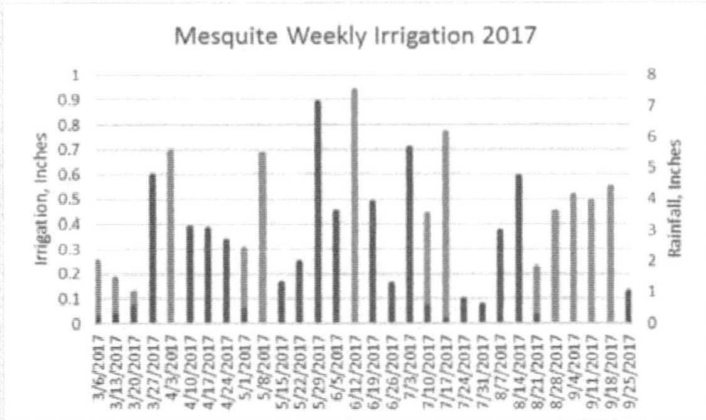


Figure 9

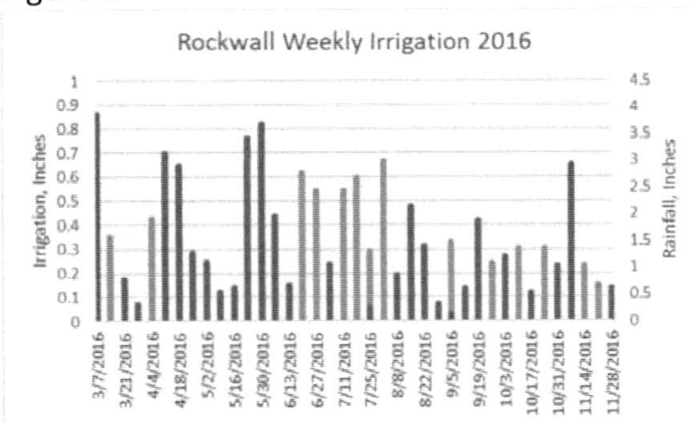


Figure 10

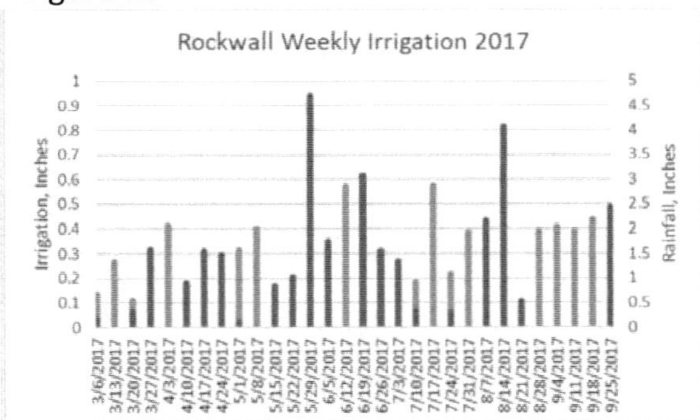


Figure 11

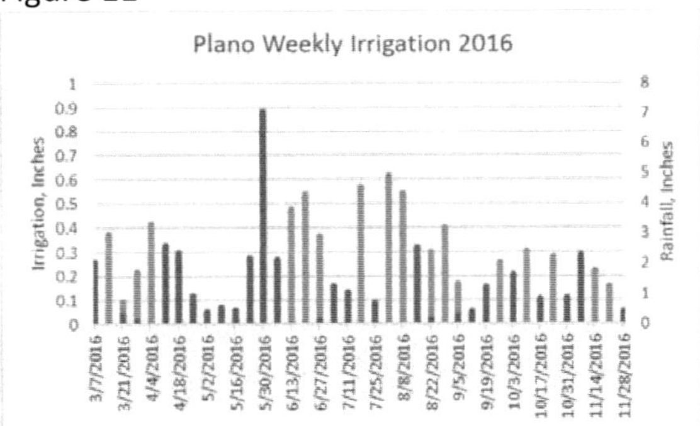


Figure 12

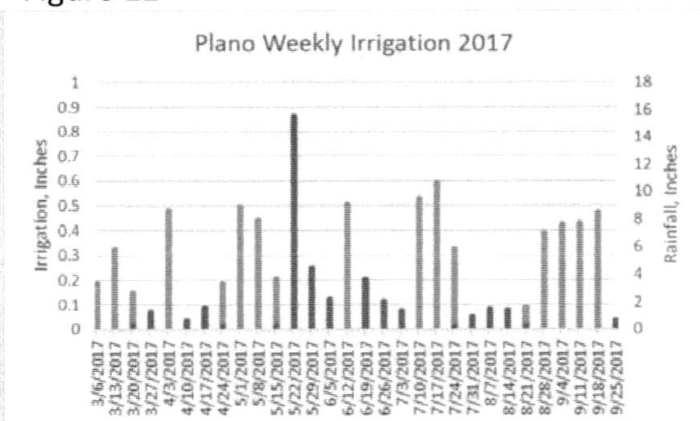


Figure 13

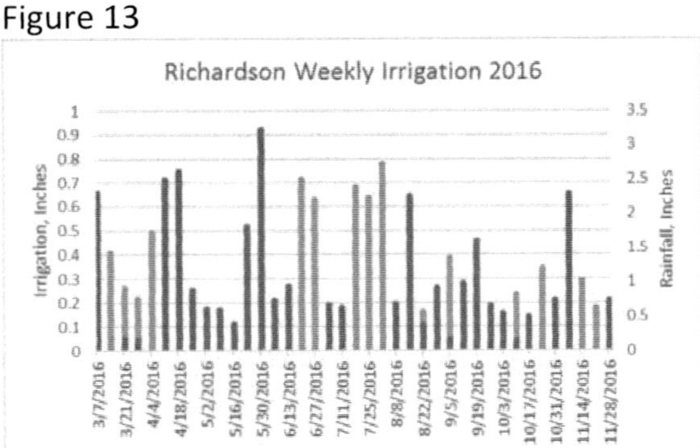


Figure 14

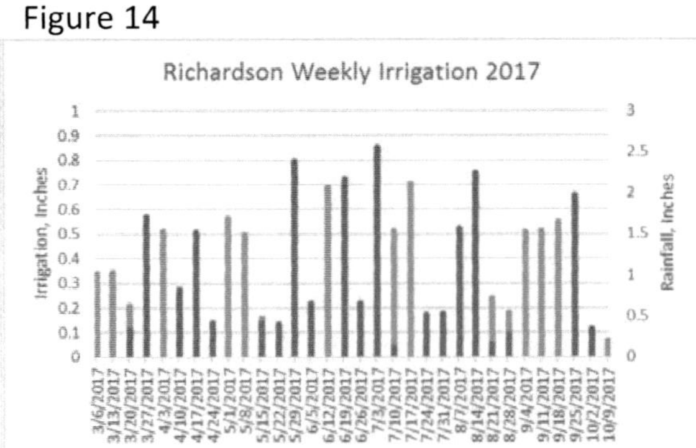

Figure 15

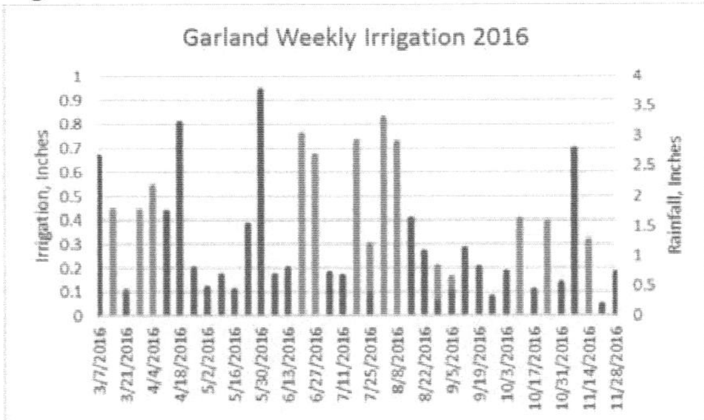


Figure 16

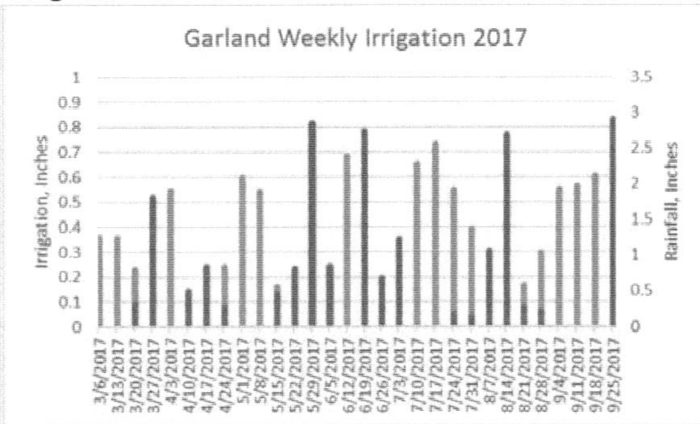


Figure 17

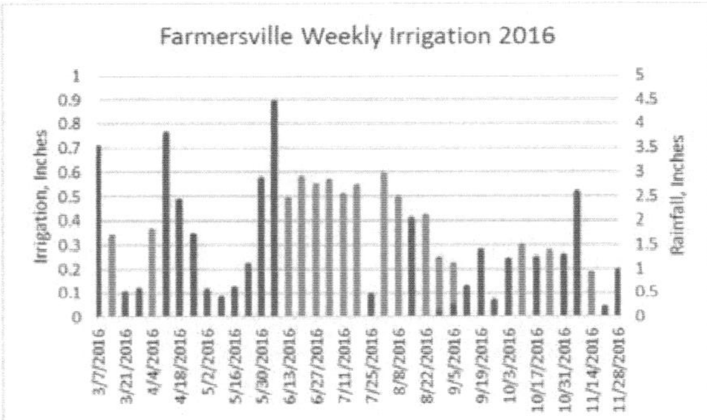


Figure 18

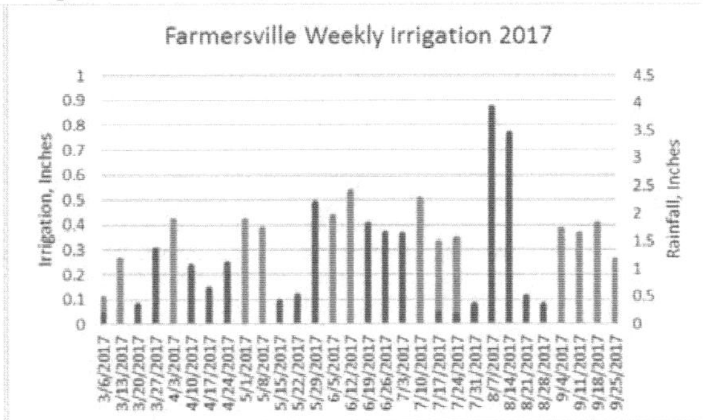


# Summary

A summary of 39 weeks of watering recommendations from 2016 and 30 weeks in 2017 showed that no watering was required on average 52% of the period in 2016 and 42% of the period in 2017 due to rainfall. In 2016, the irrigation recommendations varied from 5.64-8.09 inches (Difference of 2.45 inches) however the total rainfall received varied from 29.92-39.49 inches (Difference of 9.57 inches). In the 2017 period, irrigation recommendations varied from 5.21-8.21 inches (Difference of 3 inches) while the total rainfall varied from 26.09-50.57 inches (Difference of 24.18 inches). Analysis shows that rainfall is more variable over the service area than the amount of irrigation recommended. Based on this analysis, it can be concluded that over time, the importance in adequate accounting for rainfall is just as significant as the amount of irrigation recommended within a service area.

# Evaluating Forecast Reference Evapotranspiration For Florida

**Kati Migliaccio, PhD, PE**
University of Florida, Agricultural and Biological Engineering, PO Box 110570, Gainesville, Florida

**Xiaozhen Shen**
University of Florida, Agricultural and Biological Engineering, PO Box 110570, Gainesville, Florida

**Rick Lusher**
University of Florida, IFAS Information Technology, PO Box 110350, Gainesville, Florida

**Sharmin Siddiqui**
University of Florida, Agricultural and Biological Engineering, PO Box 110570, Gainesville, Florida

**Abstract.**
The National Weather Service has released a product that estimates forecast reference evapotranspiration (FRET) at a 2.5 km resolution. The objective of this study was to compare FRET data to calculated Florida reference evapotranspiration (ETo) data and evaluate their differences. Daily ETo values were calculated with data from 42 UF/IFAS Florida Automated Weather Network (FAWN) locations using two methods, FAO56 and ASCE-EWRI. These values were then compared to FRET data for one week during September 2017. Results from this limited data analysis showed FRET data were similar to measured data, with differences between values decreasing as the number of days between forecasted and calculated values decreased. In addition, FRET data were similar to both calculated data sets with ASCE-EWRI values being more similar than FAO56 values. Similarity between FRET predictions and calculated data suggests that FRET data could be used to help water managers anticipate future water demand and improve irrigation management. Additional work is needed with a larger data set to confirm these results over time and space in Florida.

**Keywords.** Reference evapotranspiration, web-based tools, forecast, FRET (forecasted reference evapotranspiration)

## Introduction

Irrigation decisions can be made using a variety of information. Example information used in irrigation scheduling decisions includes near real-time data collected at the irrigation site, historical trends, and/or weather forecasts. With the advancement in data acquisition methods, the Internet, and the proliferation of handheld Internet-enabled devices; greater opportunities exist to connect data to irrigation decision makers in a useful way.

The National Weather Service (NWS) has developed datasets of several forecasted weather parameters that can be used in irrigation decision making. For example, a gridded product that provides forecasted reference evapotranspiration (FRET) was released in 2014 on an experimental basis. This product provides daily 7-day FRET forecasts on a 2.5-km resolution. The values are calculated using the

Penman-Monteith Reference Evapotranspiration Equations from ACSE-EWRI for a short canopy (12 cm grass; ASCE-EWRI, 2004) and NWS gridded forecasts of temperature, relative humidity, wind, and cloud or sky cover. Solar radiation is derived using day length, sun angle and eccentricity and is based on the method established by Doorenbos and Pruitt (1977). The methodology used by the NWS to create FRET originated with Richard Snyder form University of California – Davis.

Evapotranspiration (ET) is one method that can be used to estimate the water consumptive use of a system. Irrigation schedules based on ET methods have been shown to provide greater water use efficiency than scheduling methods that do not use real-time data (Singh et al., 2009; Kisekka et al., 2010; Migliaccio et al., 2010). Thus, using real-time ET with FRET has the potential to further improve irrigation scheduling and water use efficiency.

Little has been published in literature regarding the accuracy of FRET and its potential to be used in irrigation decision making. Our objective was to compare measured reference ET (ETo) to predicted ETo at the 42 weather stations operated by the UF/IFAS (FAWN). Results of this comparison will help guide potential integration of FRET data into irrigation scheduling decision-making tools.

## Methods

Reference ET (ETo) can be calculated using different methods. The two methods used in this study were the FAO56 Penman Monteith ETo (Allen et al., 1998) and the ASCE-EWRI method (ASCE-EWRI, 2004). Daily temperature, relative humidity, solar radiation, and wind speed from the 42 FAWN stations were used with both FAO56 and ASCE-EWRI methods to generate daily ETo values. The R package "Evapotranspiration" was used to calculate daily ETo for these two methods. R is publicly available and has wide application for data analysis, model simulation, and statistical comparisons. A relatively short data set, which covered the one-week period from September 20 through September 26, 2017, was used for this analysis.

During the one-week time-period, FRET data were collected daily via the public open API at preview.weather.gov using a corresponding list of FAWN station latitudes and longitudes. Grid cells corresponding to the FAWN station locations were then identified and used for comparison purposes. An R script was scheduled using Crontab to run automatically daily to update information from the dev endpoint XML (eXtensible Markup Language). Each day, FRET data for the following 6 days were used in the analysis. Data were evaluated visually using box plot comparisons.

## Results and Discussion

ETo was calculated daily at the 42 FAWN station locations using the FAO56 and ASCE-EWRI methods from September 20 to September 26, 2017, and then compared to FRET data for the same locations (Figure 1). Average values for FAO56, ASCE-EWRI, and FRET calculations for all sites were 3.9, 4.7, and 4.0 mm over the time-period. FRET data showed less variability compared to the calculated FAO PM and ASCE-EWRI values using FAWN data as input. Interestingly, FRET uses the ASCE-EWRI method, but values were closer to those estimated using the FAO PM method. Overall, values calculated using the FAO PM method were lower than those calculated using the ASCE EWRI method.

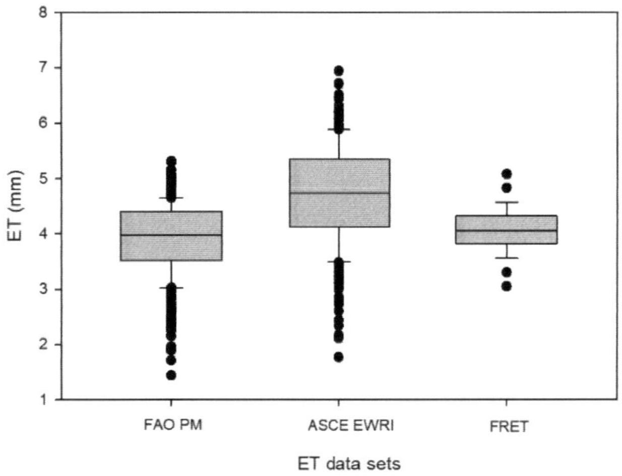


Figure 1. Box plots showing the distribution of ETo data for the three datasets.

FRET data were also evaluated at all FAWN station locations using the daily forecast starting six days prior, with the day 0 being September 26. Day -1 was the value forecasted on September 25$^{th}$ to occur September 26$^{th}$ (Figure 2). The ASCE-EWRI method was compared to the prediction to identify if predictions improved closer to event occurrence. Average absolute difference between FRET and the ASCE EWRI method decreased closer to the actual event occurrence. For example, average absolute difference six days in advance between the two ETo values was 0.67 mm, and then decreased to 0.52 mm one day in advance. Differences were less than 1 mm for 85% of the data for the 6 days prior to day 0 considering all weather stations.

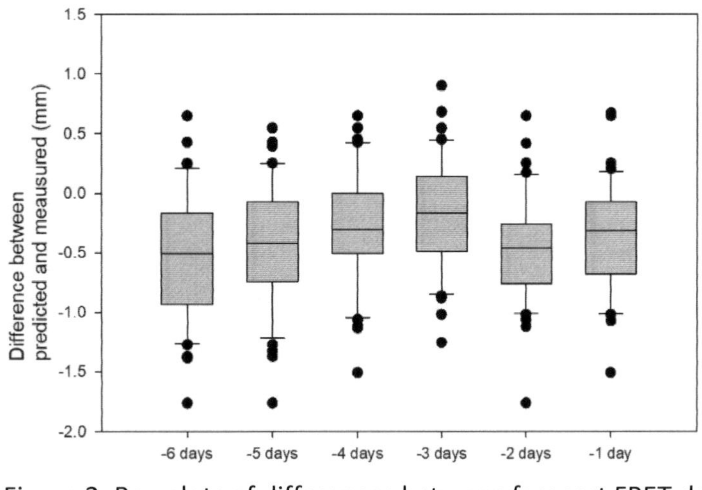


Figure 2. Box plots of differences between forecast FRET data and measured reference ETo for FAWN stations considering September 26, 2017 data.

Results suggest that the FRET data product could be used to predict ETo in Florida. However, additional evaluation is needed for a longer time series. Some quality control of the data may also be needed, as FRET data are considered experimental at this point. FRET provides a forecast data source for ETo that would be viable for FAWN and other irrigation decision making tools.

## Conclusions

FRET data were obtained and compared to ETo calculations for 42 locations in Florida. Results showed that predictions were similar to measured data. Thus, FRET may provide additional information that can be used in water management tools in Florida. Further evaluation of a longer time series is needed to validate findings.

## References


Allen, R.G., L.S. Pereira, D. Raes, M. Smith. 1998. Crop evapotranspiration – Guidelines for computing crop water requirements – FAO irrigation and drainage paper 56. Food and Agriculture Organization of the United Nations.

ASCE-EWRI. 2004. The standardized reference evaporation equation. Technical report to the Environmental and Water Resources Inst. of the American Civil Engineers. from the task committee on standardization of reference evapotranspiration. 173 p.

Doorenbos, J. and W.O. Pruitt. 1977. Crop Water Requirements. FAO Irrigation and Drainage Paper 24. http://www.fao.org/3/a-f2430e.pdf

Kisekka, I., K.W. Migliaccio, M.D. Dukes, B. Schaffer, J.H. Crane. 2010. Evapotranspiration-Based Irrigation scheduling and Physiological Response in a Carambola (Averrhoa Carambola L.) Orchard. Applied Engineering in Agriculture 26(3):373-380.

Migliaccio, K.W., B. Schaffer, J.H. Crane, F. Davies. 2010. Plant response to evapotranspiration and soil water sensor irrigation scheduling methods for papaya production in south Florida. Agricultural Water Management 97(10), 1452-1460.

Singh, R., K. Satyendra, D.D. Nangare, M.S. Meena. 2009. Drip irrigation and black polyethylene mulch influence on growth, yield and water-use efficiency of tomato. African Journal of Agricultural Research 4 (12):1427-1430.

# The Benefits of Consumable Water Management Technologies

**James Spindler, Director of Agronomy, CPAg, CCA, CPSS**

Ecologel Solutions, LLC, 4060 SE 45th Court, Ocala, FL 34480, jim@ecologel.com

**Abstract.** *Water conservation has become a major concern over the last decade or more. Influenced by rising costs, recurrent drought, use restrictions, politics and social pressures, turf and landscape managers are expected to do more with less. Beyond the advancements in irrigation hardware and software, there are a variety of consumable technologies and products designed to manage water in the soil and further maximize water use efficiencies. These products and technologies include hygroscopic humectants, surfactants and superabsorbent polymers as well as mulches, compost, antitranspirants, plant growth regulators, and hormones/biostimulants. Understanding each technology and soil/water interactions will help landscape and irrigation managers decided which strategy is best suited for their specific situation. The content will include how and why to use the various technologies, how to maintain healthy turfgrass and landscapes during periods of drought or watering restrictions and the economic benefits of using water management solutions as part of a regular program.*

**Keywords.** Hygroscopic humectants, wetting agents, surfactants, superabsorbent polymers, drought, watering efficiency, water conservation, consumable water management technologies, mulches, compost, ground covers, anti-transpirants, plant growth regulators, plant hormone, biostimulants, watering restrictions.

Water conservation has become a major factor affecting landscape and irrigation design over the last decade or more. Influenced by rising costs, recurrent drought, use restrictions, politics and social pressures, industry professionals are expected to do more with less. In many regions, landscape have become a political or social target for water conservation and movements to reduce irrigated acreage are growing in popularity nation-wide.

The cost of water in many markets, including "water-rich" regions, has increased dramatically since the turn of the century. Surveys have revealed that the cost of water has risen by 25 to 30 percent in many municipalities, with increases reaching as high as 300 percent or more in some regions. These cost increases coupled with the threat of drought shaming have caused many property owners to reconsider landscape watering practices. Is money better spent on other budgetary items than on irrigation?

Beyond the advancements in irrigation hardware and software, there are a variety of consumable technologies and products designed to manage water in the soil and further maximize watering efficiencies. Combining the use of consumable technologies with modern irrigation hardware provides a greater opportunity to sustain the industry without succumbing to sacrifices in landscape quality or quantities. Consumable products and technologies include hygroscopic humectants, surfactants and superabsorbent polymers as well as mulches, compost, antitranspirants, plant growth regulators, and hormones/biostimulants. Understanding each technology and soil/water interactions will help landscape and irrigation managers decided which strategy is best suited for their specific situation.

# Technologies for Optimizing Soil Moisture Management

## Hygroscopic Humectants

Though they are not new to the industry, hygroscopic humectants are continuing to gain favor with landscape professionals as products that are very effective at reducing overall water requirements. With a history in golf and sports turf, these products are becoming more popular for municipalities as well as residential and commericial properties, particularly those in areas with recurrant drought, watering restrictions or high water costs.

Hygroscopic humectants manage and conserve water through two modes. As the name suggests, there is a hygroscopic compontent and a humectant component. Each has a critical function in the performance of the technology. The mode of action of the hygroscopic component is to condense soil water vapor or soil humidity back into liquid droplets of water. The hygroscopic ability of these materials can be compared to condensation or "sweat" that occurs on the side of a cold drink. Rootzone humidity cannot naturally be absorbed by plant roots. Hygroscopics convert this unavilable humidity into plant usable micro-droplets of water.

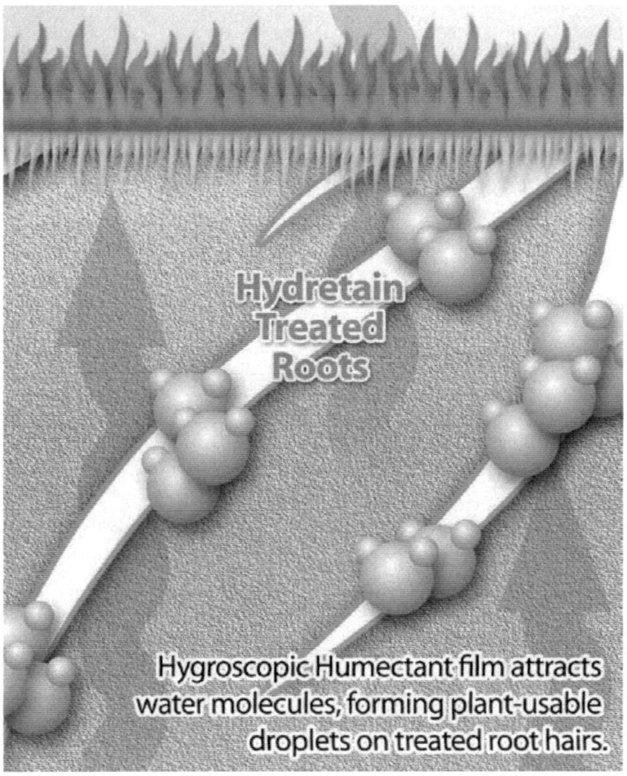


Figure 1. Diagram of the hygroscopic humectants on a soil particle. *(Courtesy of Ecologel Solutions, LLC)*

The humectant components hold the water droplets condensed by the hygroscopic components. Do not confuse a humectant with a humate. They are completely different substances with different molecular structures. The humectant component holds the droplets tightly enough to prevent it from leaving the proximity of the root, but lightly enough to allow the root to absorb the water through osmosis. The humectants in hygroscopic humectants are also utilized in cosmetics, shampoos, and other body care products where they help hold moisture in the skin and hair.

Available in both liquid and granular options, hygroscopic humectant technologies must be watered-in, at which point the active ingredients will coat plant roots, soil particles and organic particles in the root zone. The hygroscopic humectant molecules are too large to be absorbed by the roots. Once these components attach to the roots and soil particles, they remain attached and are resistant to further movement in the soil. The ingredients are primarily derived from plant byproducts (some brands are USDA BioPreferred Certified for biobased content*). Therefore, they are eventually broken down by soil microbial activity. Research and users have demonstrated that the most effective hygroscopic humectants products have been able to reduce water use by up to 50 percent or more and will typically perform for up to 90 days. In addition to providing general conservation of water, hygroscopic humectants aid in seed germination, transplant establishment and in establishing sod and sprigs. Hygroscopic humectants have also been used to suppress dust on baseball infields, horse arenas, dirt race tracks, dirt roads, etc.

Figure 2. Sports field under water restrictions. Plot treated with a hygroscopec humectant compared to adjacent untreated plot in the foreground. (Courtesy of Hydretain)

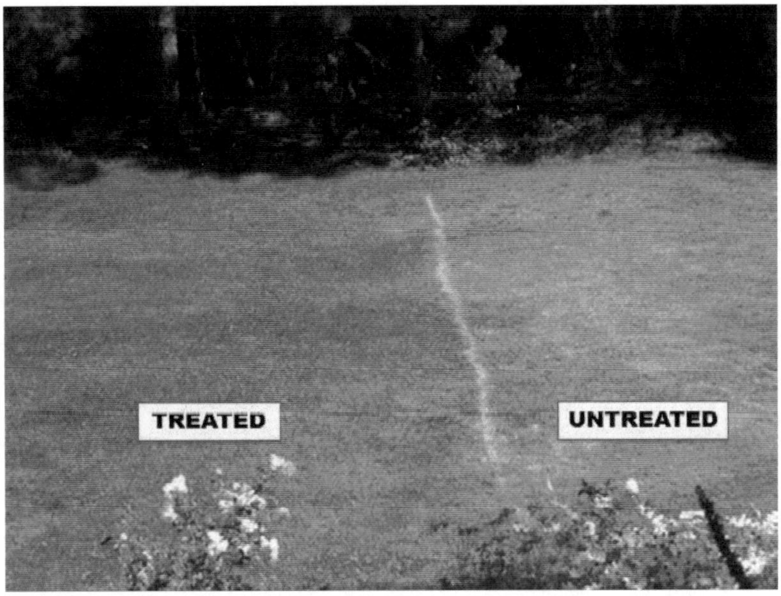


Figure 3. Golf course slope treated with hygroscopic humectant under water restrictions. (Courtesy of Lesco Moisture Manager)

## Superabsorbant Polymers

Superabsorbant polymers are a technology that track their orgin to a patent by Monsanto in 1963. They described polymers as "strings of large molecules that chemists use like Tinker Toys, adding, subtracting or linking them together to create diverse uses ranging from filling for disposable diapers to dental products" (Messina, 1991).

Polymers have been adapted for use in soil to improve water availability to plants. They are utilized to "increase a soil's water holding capacity, increase pore sizes and numbers in the soil, increase germination rates, and decrease or mitigate the effect of soil compaction on plant growth" (Orzolek, 1993). The five main types of soil polymers available commercially include:

- Cross-linked polyacrylamides (gel forming)
- Non-cross-linked polyacrylamides (water soluble)
- Polyacrylates
- Polyacrylontrile
- Starch-grafted copolymers

The most commonly used polymer is the cross-linked polyacrylamide. Soil polymers occur in a crystalline form. When exposed to water, they expand into a gelatin-like block. When used in soils, they function as mini-reservoirs of water. They absorb water and hold it until the plant removes the water. The literature indicates that cross-linked polyacrylamide polymers used in the field will absorb and hold 80 to 200 times their weight in water or more. Their ability to hold soil water is influenced by the amount of polymer in the soil, the type of polymer utilized and soil characteristics, such as salt content (Polhemus, 1992). The lifespan of polymers is thought to range from 2 to 10 years, depending on the type of polymer and soil conditions (Polhemus, 1992).

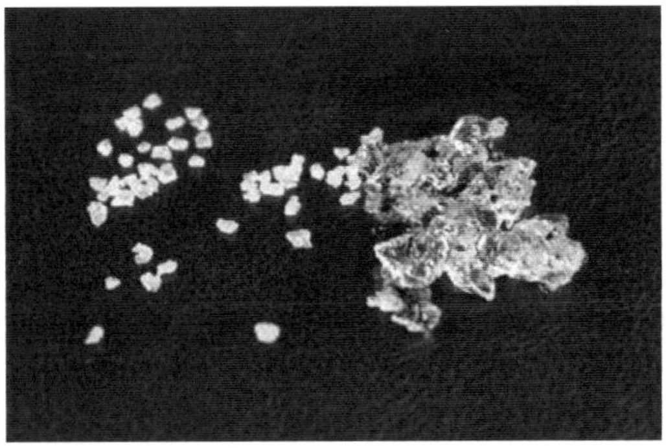

Figure 4. Comparison of dry vs. hydrated superabsorbent polymer crystals

The literature reports that the time between irrigation events can be extended with the use of polymers, but the actual water savings with use of these products is dependent on application rates and soil conditions. Cost of these products may be a limiting factor for effective application rates.

Initially, polymers were used to help reduce water use in potted plants, ornamental beds and in planting trees and shrubs. Over the years, soil-applied polymer use has expanded to turf applications. They are utilized in the establishment of sod and sprigs, improving seed germination and in general turf use. The challenge in utilizing polymers on established turfgrass is delivering the polymer crystal to the root zone. Some turf manager will aerate the turf and drag the crystal into the holes. In addition to this practice, there are now machines that will inject the polymer crystals into the soil.

## Surfactants/Wetting Agents

Surfactants or wetting agents are probably the most commonly known products used to manage soil moisture. These materials are utilized for a number of applications in turf and plant management, including relief from localized dry spots, improved drainage, assisting the efficiency of various pesticides, reduced dew and frost accumulation, improved seed germination, reduced fairy ring damage, alliviation of soil compaction, improved irrigation efficiency, and more (Karnock, Xia, & Tucker, 2004).

Surfactants stand for **SURF**ace **ACT**ive **A**ge**NTS (SURFACTANTS)**. These are agents that affect the surface of a liquid or solid. Understanding the nature of water is critical to understanding the function of surfactants. Water molecules are naturally polarized: the two hydrogen atoms attached to one side of the oxygen atom causing each end to have a slight charge. Just as opposite charges of magnets attract to each other, the positive end of the water molecule is attracted to the negative end of another water molecule. This is called cohesion and is why we see a drop of water bead up when placed on wax paper. The waxy surface of the paper has no charge or is non-polar; therefore, the only place for the water molecules to be attracted is to each other. In soils, waxy coatings are the result of the decay of organic materials and certain species of fungi that exude waxy substances. The formation of waxy, non-polar coatings on soil particles is the primary cause of hydrophobic conditions. The non-polar soil particle surface will not attract, and may actually repel, the polar water molecule, which prevents irrigation water or rainfall from infiltrating soils to hydrate plants.

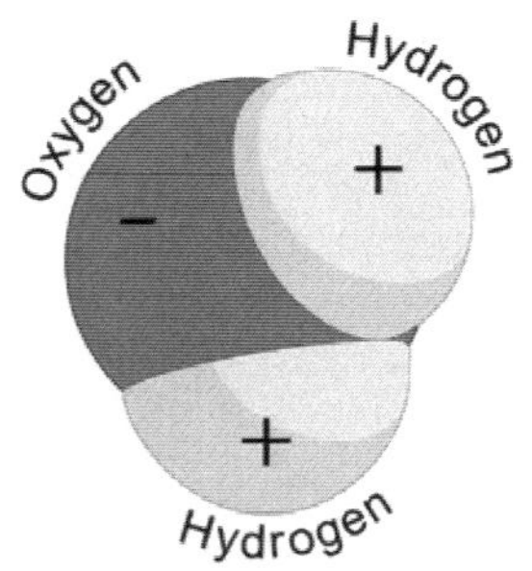


Figure 5. Graphic rendering depicts the polar nature of water molecules.

Creating a polar surface allows water molecules to enter and fill the soil. The surfactant has a non-polar and a polar end on the molecule. The non-polar end of the surfactant molecule aligns with the non-polar surface of the organic soil coating, leaving the polar end exposed outward from the soil particle. This allows the polar water molecules to be attracted to the polar surfactant molecules therefore overcoming the hydrophobic condition (Karnock, Xia, & Tucker, 2004).

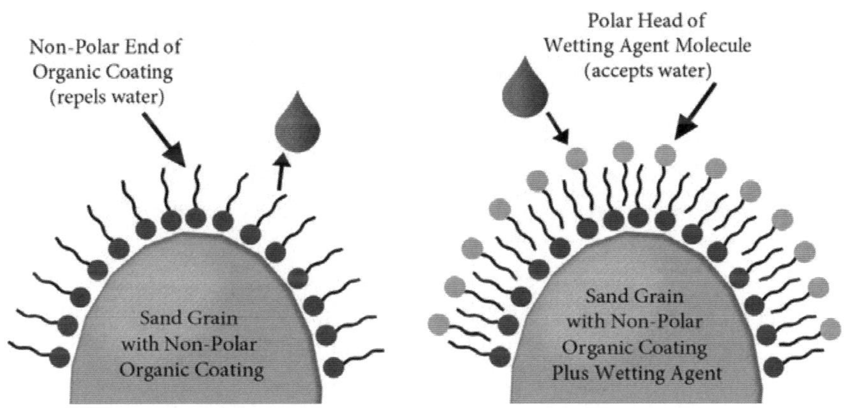


Figure 6. Diagrams illustrating wetting agent activity

There are many different kinds of surfactants, most of which fall into these four basic categories:

- Anionic—Form negatively-charged ions in water
- Cationic—Form positively-charged ions in water
- Nonionic—Does not ionize in water
- Amphoteric—Take on the ionization of the water

Non-ionic surfactants are the most common products used in the turf industry due to their safety, compatibility with other products and ease of use. As technology has improved, a number of catetories of non-ionic surfactants have been developed. These include:

- Polyoxyethylene (POE)—This is older technology originally developed to treat localized dry spots. They can be phytotoxic.
- Block Co-Polymer Surfactants—These are the most commonly used turfgrass surfactants. They are safer and are effective in treating soil water repellency, improving soil water content and plant available water. This category has two sub categories: Straight Block Co-Polymers and Reverse Co-Polymers.
- Alkyl Polyglucoside Surfactants—These are made from sugar molecules reacted with a fatty acid and are considered naturally derived. When blended with a block co-polymer, the performance appears to be better than either technology alone. These blended technologies appear to increase water infiltration, improve water availability and enhance irrigation efficiencies.
- Modified Methyl Capped Block Co-Polymer—This is a class of surfactant that is a modification of the co-polymer class. This technology forms a thinner, more continuous film around the soil particle.
- Humic Substance Redistribution Molecules—"These molecules allow water penetration through the soil profile by disrupting the hydrophobic supramolecular humic association, most prevalent in the top one to two centimeters on the soil, which lead to localized dry spots."
- Multi-branched Regenerating Wetting Agents—Most surfactants have linear molecules. These products have a much higher molecular weight and multiple branched molecules. Each branch essentially functions as wetting agent itself. (Zonteck & Kostka, 2012)

Surfactants/wetting agents have been demonstrated to possess many functions in the management of water in and around turfgrass and other plant systems. When discussing the maximization of water use efficiencies, these products tackle the barriers (non-polar coatings in the soil) that prevent water from moving into and distributing throughout the soil. Research has shown that surfactants/wetting agents can significantly improve soil moisture content and reduce variability in soil water content, improving soil moisture uniformity. In addition, they have been shown to "reduce localized dry spot incidence, allow for longer periods between irrigation events, and reduce hand watering in isolated areas" (Karchner & Richardson, 2014).

Surfactants/wetting agents are available in liquid and granular forms. The amount of water conserved, longevity of the product and cost may vary based on product type and local conditions.

### Soil Barriers and Composts

Soil barriers such as mulches, pine straw, plastics, landscape fabrics, and gravel have long been used to conserve water in landscape beds. These products retain soil mositure by acting as a physical barrier to moisture loss through evaporation. They also help reduce soil erosion and compaction, which can negatively affect water use efficiency. Furthermore, they reduce weeds, which steal water resources from desired plant materials.

Compost incorporation is also used to increase water use efficiency. Similar to the use of superabsorbent polymers, composts are known to increase soil water holding capacity. Amended soils are estimated to require up to 60% less water depending on the quality of the compost and quantity incorporated into the soil.

### Antitranspirants

Antitranspirants are substances sprayed on the leaves of plants to reduce the rate of transpiration. Antitranspirants function using there a three known modes of action. First, they reflect radiant energy way from the plant resulting in lower temperatures and transpiration rates. Second, emulsions of wax latex or other film are used to prevent the escape of moisture from the plant. Third, they prevent the stomata from opening fully and decrease the loss of water vapor from leaves. While not all experts are in agreement on the benefits of anti-transpirants, studies have shown that they can conserve water and minimize wilt and drought stress.

**Growth Regulators and Biostimulants**

Growth regulators and hormone biostimulants represent another class of chemistries that can be used to reduce plant water requirements. Plant growth regulators (PGRs) are designed to reduce excess vertical plant growth. While the intended benefit of these products is lower trimming and mowing maintance, slow growth has also proven to reduce plant water requirements.

Hormone biostimulants containing cytokinins have proven to increase root mass and depth. By encouraging root development, biostimulants improve water efficiency by helping plants source more water from soil. They have also provent to increase plant tolerance to a number of stresses, including drought.

**Conclusions**

There are a wide variety of consumable products and technologies available to help manage and conserve water. Using these technologies in conjunction with modern advancements in irrigation hardware offers greater opportunities to maximize water use efficiency. The key to success is to recognize the value of consumables and learn which product(s) are the best fit for each specific situation. It is adviseable to remember to not think linearly. Often, there is not one single issue with one single solution. The best solution for the management and conservation of water may be to combine consumable technologies. A very common example of this is the combination of hygroscopic humectants with surfactants technologies. In this situation, the surfactant will allow water with the hygroscopic humectant to enter and disperse throughout the soil where hydrophobic non-polar organic coatings exist. Water can uniformly disperse throughout the rootzone. Then, the hygroscopic humectant can reduce evaporative loss for maximum plant water use.

Thinking outside the box and using all tools available gives landscape and irrigation managers the ability to maximize water use efficiency and optimize turf and plant performance.

*Products brands Hydretain® and LESCO Moisture Manager™ have been certified to contain 93% biobased contents by the USDA BioPreferred® Program.

## References

Karchner, D., & Richardson, M. (2014, December). Wetting Agents Improve Root-Zone Moisture Distribution. *Golf Course Management*, pp. 80-83.

Karnock, K. J., Xia, K., & Tucker, K. A. (2004, June). Wetting agents: What are they, and how do they work? *Golf Course Management*, pp. 84-86.

Messina, F. (1991, March 18). Absorbing Issue: Polymers are Saving Grace for Large Water Unsers, but Some Question Long-Term Effects. *Los Angeles Times*, p. Internet.

Orzolek, M. (1993, January/March). Use of Hydrophylic Polymers in Horticulture. *HortTechnology*, pp. 41-44.

Polhemus, D. (1992, June). Soil Polymers for Turf Areas: A Technical Review. *Sports Turf*, pp. 14-17.

Zonteck, S. J., & Kostka, S. J. (2012, July). Understanding the Different Wetting Agent Chemistries. *USGA Green Section Record, Volume 50 (15)*, pp. 1-6.